ALAN PATON

ALAN PATON

A BIOGRAPHY

Peter F. Alexander

Oxford New York
OXFORD UNIVERSITY PRESS
1994

Oxford University Press, Walton Street, Oxford OX2 6DP
Oxford New York Toronto
Delhi Bombay Calcutta Madras Karachi
Kuala Lumpur Singapore Hong Kong Tokyo
Nairobi Dar es Salaam Cape Town
Melbourne Auckland Madrid
and associated companies in
Berlin Ibadan

Oxford is a trade mark of Oxford University Press

First published 1994

British Library Cataloguing in Publication Data
Data available

Library of Congress Cataloging in Publication Data
Alexander, Peter (Peter F.)
Alan Paton : a biography / Peter F. Alexander.
p. cm.
Includes bibliographical references and index.
1. Paton, Alan—Biography. 2. Authors, South African—20th century—Biography. 3. Paton, Alan. I. Title.
PR9369.3.P37Z58 1994 823—dc20 93–45420
ISBN 0–19–811237–8

1 3 5 7 9 10 8 6 4 2

Typeset by Best-set Typesetter Ltd, Hong Kong
Printed in Great Britain
on acid-free paper by
Biddles Ltd
Guildford and King's Lynn

For Roland

Acknowledgements

Boswell, in his *Life of Johnson*, describes the zeal with which Johnson's friends gave him information about his subject, 'resembling in this the grateful tribes of ancient nations, of which every individual was eager to throw a stone upon the grave of a departed Hero, and thus to share in the pious office of erecting an honourable monument to his memory'. This comparison accurately describes the motivation of the many friends and Paton family members who have contributed to the making of this book.

My greatest debt is to Alan Paton's widow, Mrs Anne Paton, who not only gave permission for me to see and quote all the Paton material of which she is the copyright-holder, but repeatedly welcomed me to her home, submitted to days of questioning, allowed me to read the typescript of her then-unpublished book *Some Sort of a Job: My Life with Alan Paton*, supplied me with hundreds of addresses of Paton friends and contacts, organized interviews for me, and encouraged me steadily during the years this book took to write. She read the completed first draft, and proved to be an eagle-eyed editor, not only spotting errors but suggesting many helpful additions of material I had missed. It is no exaggeration to say that this book could not have been published without her help and support.

Paton's son Jonathan, from whose inspired teaching I had benefited years before at the University of the Witwatersrand, generously shared his memories with me, supplied me with photocopies of the dozens of letters in his possession, and gave me access to many photographs. His wife Margaret and son Anthony also gave me many insights into Alan Paton as a family man.

Dr David Paton and Maureen invited me to their home to meet not just themselves, but other friends who could shed light on Paton's life, and made available to me hundreds of Paton photographs.

Mrs Athene Hall, Paton's stepdaughter, answered my questions and supplied me with important photographs.

Among those who undertook the onerous task of reading my manuscript at various stages of production I should mention five in particular. Professor Colin Gardner, Professor Douglas Irvine, and Mr Peter Brown took the time to read this book in manuscript at a time when each of them was under great pressure of other work, and their expertise saved me from many blunders. Miss Mary Benson read sections of the book, not once but several times as it went through draft after draft. And my wife, Professor Christine Alexander, took time off her own work on the Brontës to read the final version. To each of these hard-pressed and diligent readers I owe a great debt.

Mr Reg Pearse talked at length into my tape-recorder, and subsequently answered my many written queries while suffering a series of personal crises. He also supplied me with invaluable early Paton letters, diaries, and photographs. Without this material my account of Paton's university years would have been much poorer.

Mr Victor Harrison, another of Paton's oldest friends, displayed astonishing powers of recall in bringing alive for me Paton's schooldays of seventy years before, as well as filling in the events surrounding the murder of Paton's father.

Like all researchers on Paton, I owe a particular debt to the doyen of Paton scholars, Professor Edward Callan. In addition to this debt I have a personal one, for he and his wife Claire welcomed me as their guest for two days and introduced me to Michigan. He shared with me his memories of Paton, gave me access to his very large Paton collection of letters and other documents, and made me aware of a good deal of Paton material I might otherwise have missed. My debts to his writings are acknowledged in the notes to this volume. In addition he undertook to read the first draft of the second half of my typescript, and his detailed editorial corrections saved me from many errors.

Mrs Constance Stuart Larrabee welcomed me as a guest in her beautiful home in Maryland, shared her memories of Paton, and gave me access to her matchless collection of Paton photographs, some of which are reproduced in this book.

Mr David Philip and his wife Marie supplied many details of the period when Paton began publishing (and staying) with them, and I am grateful for their help, support, and hospitality.

Others who kindly agreed to interviews, or who supplied me with information, written memoirs, or copies of letters or photographs, include Dr Ray Adie, Mrs Jenny Akal, Mr Anthony Akerman, Mrs Dorrie Arbuthnot, Dr Anthony Barker, Mrs Maggie Barker, Miss Mary Benson, Mrs Frankie Braden, Mrs Shirley Broad, Mr Peter Brown, Mr Andrew Campbell, Professor Michael Chapman, Mrs Dot Cleminshaw, Mr Gerhard and Mrs Gertrude Cohn, Mr David Craighead, Mrs Mabel Dent, Mrs Bunny Duggan, Mrs Dorothy Durose, Mr Leif Egeland, The Very Reverend Gonville ffrench-Beytagh, Mr R. T. Foster, Miss Victoria

Francis, Mr Deryck Franklin, Mrs Nancy Fraser, Mr Edmund Fuller, Professor Colin Gardner, Dr Audrey Glauert, Mrs Eileen Goldberg, Mrs Ida Grant, Mr James W. Hargrove, Mr Mark Henning, Mr Neil Herman, Mr Edward W. Holmes, Bishop Trevor Huddleston, Mr David Johanson, Mrs Margaret J. (Peg) Kinnison, Mr Joe Kirk, Mr Peter Kohler, Mr C. J. W. Kriel, Dr Adrian Leftwich, Mrs Margaret Lenta, Mr Bernard Levin, Mrs A. S. McDowell, Mrs Rachel R. Mather, Mrs Barbara Moran, Ms Pauline (Pondi) Morel, Dr Tony Morphet, Professor Ned Munger, Mr Lionel Ngakane, Mrs Ursula Niebuhr, Mrs Jessie Northcott, Mr Harry Oppenheimer, Mrs Elizabeth A. Patterson, Reverend John B. Pesce, Father Geoffrey Plant OSF, Mr Pat Poovalingam, Mrs Holly Reck, Mr Richard F. and Mrs Beatrice Robinow, whose grateful house-guest I was in Toronto, Mr Peter Rodda, Mrs Janet D. Rood, Mrs Holly Rood Reck, Mr Leslie Rubin, Mr Roy Rudden, Mr David L. Stitt, Mrs Vi Swann, The Right Reverend Cabell Tennis and Mrs Hyde Tennis, Ms Jo Thorpe, Mr William D. Toomey and Mrs Odette S. Toomey, Mr R. H. ('Harry') Usher, Sir Laurens van der Post, Mrs Sally Vasse, Mr Bob Ventress and Mrs Guinevere Ventress, Mr René de Villiers, Mr Randolph Vigne, Mr Roy C. Votaw, Mrs Sandra Weiner, Professor David Welsh, Mr Ashley Wills, Mrs Margaret Worthington, and Mr Ian Wyllie.

Among the librarians who have helped me with my research I pay particular tribute to Mrs Joicelyn Leslie-Smith, Manuscript Librarian of the Alan Paton Centre at the University of Natal, Pietermaritzburg, who stopped at nothing to make my repeated visits to that institution both productive and pleasurable. She supplied me with thousands of photocopies from this, much the most extensive and rich collection of Paton papers, she answered endless questions, put me in the way of Paton publications I should otherwise have missed, conducted interviews on my behalf, did research on the Paton homes in Pietermaritzburg, gave advice on where it was safe to travel during this troubled time, and she and her husband Pat put me up in their own home. She also proof-read my Bibliography, and pointed me to several items I had omitted. In summary, she did more than any librarian should be called upon to do for a scholar, and my warmest thanks are inadequate.

Mrs Billie Farina of the Alan Paton Centre painstakingly made thousands of photocopies for me and, when a large packet went astray, made them again without complaint.

I am also grateful to the staffs of the following institutions for extensive help in the course of my research: The University of Cape Town Library;

The Library of Congress, Washington; The Harry Ransom Humanities Research Center, University of Texas at Austin; Kent School, Connecticut; Maritzburg College, especially Adam Rogers, Assistant Archivist of the College; The Oppenheimer collection, Brenthurst Library, Johannesburg; The Hofmeyr Collection, University of the Witwatersrand Library; and The Weill-Lenya Foundation, New York.

For financial help with the research that went into this book, I gratefully acknowledge generous grants from the Australian Research Grants Scheme and the Arts Faculty of the University of New South Wales. The task would have been impossible without such assistance.

I am indebted to the President and Fellows of Clare Hall, Cambridge, who awarded me a Visiting Fellowship for the academic year 1990–1, who allotted me the quietest of studies, and who provided an atmosphere in which congenial, inquiring company and proximity to the University Library proved ideal for my writing.

I owe a further debt to the Fellows of Christ's College, Cambridge, and particularly to Dr John Rathmell, for electing me to membership of their high table, and for the hospitality of their senior common room.

To others who gave me help, but whose names I may inadvertently have omitted, I offer sincere apologies.

P.F.A.

Sydney
September 1993

Contents

Illustrations

Introduction

Alan Paton was an almost mythical figure during my childhood in South Africa, like the unicorn or the prophet Isaiah, of whom I heard often, but whom I never expected to see. He was commonly spoken of as South Africa's most famous novelist by those who admired his political stance, and as a dangerous radical by those who did not. He wrote frequently for the newspapers, particularly the liberal English press, and the message of his vigorously expressed articles was unvarying: South Africa was heading for a precipice, black nationalism and Afrikaner nationalism were on a collision course, the country must do a U-turn away from apartheid before time ran out. These articles were received by the bulk of the white population with the sceptical irritation of British pacifists reading Churchill's warnings of war in the 1930s. Even if not dangerous, Paton was as annoying as an alarm clock to a sleeper who wishes to lie in.

As a schoolboy I read *Cry, the Beloved Country*, and was profoundly moved by it, but its author did not become more real to me as a result; rather the reverse: I could hardly believe that anyone could be so good, and loving, and forgiving, as the voice of the narrator of this extraordinary book seemed to be. It was like listening to a story of deep sorrow and love told by the Recording Angel. For all that, it had a deep impact on me, if only because it was the first novel set in my own country that I had ever read.

Then, at the end of 1974, I found myself quite unexpectedly driving to meet Alan Paton in the flesh, and I was filled with anticipation and tension. By this time I was a student at Cambridge, writing a thesis on the life and work of South Africa's best-known poet, Roy Campbell. I had had great difficulty in finding biographical information on Campbell, whose autobiographies were a farrago of hilarious lies, interspersed with grains of truth almost impossible to sift out. Through a friend who had an interest in South African literature, I heard that Alan Paton was working on a biography of Campbell, and would have all the facts at his fingertips. 'Why don't you go and see Paton?' said my friend. 'He's said not to bite the heads off students who call on him.'

Thus encouraged, I wrote to him in February 1974, very tentatively, and was bowled over when I at once received a brief but generous letter, containing an invitation to visit him when I next came to South Africa. So in December that year I phoned his home in Natal, and was told by a female voice, obviously English, that I was expected in a couple of days' time. 'Be here at twelve sharp, I don't like people wasting his time.' I promised to be punctual. Alan Paton clearly had a most efficient secretary.

I followed her meticulous instructions on the trip down in a rickety borrowed car, taking the highway from Johannesburg to Durban, turning off after Pietermaritzburg, following the signs to Hillcrest. I knew I was close when the road began winding up into the hills, rolling rounded hills green with thick grass in which the wide-horned red-and-white cattle wandered knee-deep, the valleys between the hills dark with native bush. When I stopped for a moment to look at a particularly striking view, the silence beat on my ear like a drum, except that below I could hear a stream tinkling over its black stones as it dropped into the Valley of a Thousand Hills. On the hillsides and surrounded by patches of bare earth, red as wounds, were the beehive huts of the Zulus, and occasionally a euphonious voice was upraised in a long, high-pitched cry designed, like a yodel, to carry from one mountain to another. This was Alan Paton's Natal, the area in which he had spent most of his life, and which he had transformed, in *Cry, the Beloved Country*, into one of the sacred places of the mind.

To my relief I was on time when I reached Botha's Hill, a dormitory town of luxurious houses and broad shaded streets. The Paton house was easy to find. 'Lintrose' said the name on the gate. At the bottom of a looping drive was a long, low house, painted white. From a garage near the front door came the sound of a carpenter at work. As I approached, a pleasant-faced woman emerged from the shed, carrying a plane, and with curls of wood-shavings clinging to her hair and clothing. 'You must be Peter Alexander,' she said, giving me a strong handshake. 'I'm Anne. Go straight in, he's waiting for you.' And with that the door opened and Alan Paton stood on the step to welcome me.

My first impression was of a truculent gnome. He was short, very white-haired, his skin a raw pink, his eyes very blue. But those eyes were glaring alarmingly at me over his glasses, and his face was puckered in an apparently angry scowl. 'Peter Alexander?' he said, shaking hands. 'Alan Paton.' The accent was rough, backveld Natal with a hint of something else I couldn't place; American? I was to listen to that accent many times, but there was nothing I could compare it to: it was unique to him.

He led me down a long passage towards his study. Suddenly he stopped and faced me. 'That was Anne, my second wife,' he said gruffly, though I had not asked. A pause, then: 'I couldn't live without a woman, you see. I found I couldn't live alone. And I certainly couldn't do without Anne.' And without waiting for an answer he led off down the passage again. It was a curious and touching moment. He hardly knew more about me than my name, and the fact that I was interested in Roy Campbell, but he was sharing with me a matter that went to the heart of his personal life.

In the study, surrounded by books and pictures of a little black boy praying, Chief Albert Lutuli, J. H. Hofmeyr, the originals of cartoons of Paton sweeping apartheid legislation into the sea or trying to divert the locomotives of white and black nationalism from a collision course, we talked about Campbell and, increasingly, about Paton. I was in awe of him and he, as I discovered when the second volume of his autobiography appeared, thought me constrained and over-serious. In spite of this, he spoke with apparent freedom of his first marriage, of his sons, particularly Jonathan, who had taught me at Witwatersrand University and of whom he was obviously especially fond. He did not need more than an occasional question from me to keep the conversation going; he spoke with wit and fluency about everything from his political articles ('Too many, many too many', shaking his head over a drawer of a filing cabinet stuffed with the carbon copies), to his earliest unpublished novels ('Three of them I have still, in that cupboard') and to the state of the country ('Even now the Nats can't see they've brought us to ruin'). And as we talked, I realized even at the time, he was weighing me up.

Towards the end of the afternoon he invited me to walk around the garden, and I admired the stone-walled terraces that led down to a splendid croquet lawn. As we went he said to me, 'Now Peter. This Campbell book I'm supposed to be writing. I'm having trouble with it as I told you. I don't like Campbell. And I don't like his family. How would you like to do it?' If he had hit me on the head with a spade he could not have achieved more of an impact. My reaction brought to his face a grin that transformed him. 'Come back again tomorrow morning, and we'll talk about the details,' he said.

In the event, 'the details' took many months to see to, for Paton took his responsibilities to the Campbell family very seriously, but they made no great objection to his handing over the biography to me. 'You don't realize what you're taking on,' he warned me, and he was right. But during that and many subsequent visits, he changed the course of my career and set me on the biographer's road, a road that was to lead me from Roy Campbell to William Plomer, from him to his friends Leonard and Virginia Woolf, and from them back to Alan Paton himself.

For as my friendship with him deepened, I found him more and more mysterious, and though I was no longer awed by him (he was too consistently friendly and amusing for that) I came to think him the most interesting person I knew. The opening revelation in the passage was not repeated. Instead, like many of his friends, I found him ultimately a man of deep recesses. He weighed every word he wrote, and his omissions spoke. When he published the first volume of his autobiography, *Towards the Mountain*, I told him that it gave the impression of telling the truth, the whole truth, and nothing but the truth. 'Does it?' he said, and gave a chuckle of pure, impish pleasure at my simplicity. Then, very satirically, 'That's good.' And I realized that I had been far from the mark. Quite how far, I did not realize until I began to write this book.

1

BEGINNING
1903

In my beginning is my end.
T. S. ELIOT

Alan Stewart Paton was born in Pietermaritzburg, in the British colony of Natal, on 11 January 1903. Natal was one of four British-administered territories which would in 1910 be united as the Union of South Africa, but at Alan Paton's birth his 'beloved country' did not yet exist as a national or political entity. His early allegiance was to a region, not a country: more precisely, to a city. Even after he had long since ceased to live there, he continued to regard the city of his birth as his home town. 'Pietermaritzburg, the lovely city', begins his autobiography,[1] and he grew testy when one critic suggested that he was only proving that beauty is in the eye of the beholder. 'My beauty was already created', he insisted, 'in that place where I was born.'[2]

It is true that Pietermaritzburg has the best collection of Victorian buildings in South Africa; it has broad generous streets lined with a multitude of flowering trees, large and lovely parks and not one but two rivers. It is the capital of South Africa's greenest and most fertile province, and the farms round it produce a wide range of crops. But a town, like a text and any other human artefact, can be read if the reader knows the language; and Pietermaritzburg when deciphered tells a dark tale of conquest, racial hatred, and displacement, a tale of which Alan Paton as a child knew nothing.

The town was founded in 1838 by Boer Trekkers led by Piet Retief, seeking somewhere they could escape British rule and organize their affairs, particularly affairs between white master and black servant, in what they considered to be the proper manner. They displaced the traditional Zulu owners of the site on which the town was to rise, and after the Battle of

Blood River, in which Zulu power sustained a defeat, settlement of the town proceeded rapidly.

The site was and is magnificent. The city lies about 50 miles inland from the port of Durban, in a bowl of hills in lush, well-favoured countryside, with two rivers, the Mzinduzi and the Dorp Spruit, flowing through the town itself. A spacious grid of streets was laid out and named by the Trekkers: Pietermaritzstraat, Bergstraat, Kerkstraat, Boomstraat, Loopstraat, and so on. This central grid, about eight blocks by eight, remains the heart of Pietermaritzburg today. But around it now run streets with names that tell of a second displacement, this time of Boer by Briton: Victoria Road, Prince Alfred Street, Pine Street, Bulwer Street; and beyond them again are whole new suburbs with names redolent of Britain: Scottsville, Wembley, Athlone, Montrose. For in 1843 the Boers, who had thrust aside the Zulus only a few years before, were themselves displaced by the British annexation of Natal, and most of them set off again bitterly, northwards into the Transvaal. The very map of the town is a palimpsest of dispossession.

It was in British Pietermaritzburg that Alan Paton grew up, then, not hearing a word of Dutch or its offspring Afrikaans, mispronouncing the Afrikaans names of the streets he walked every day, and knowing hardly more Zulu than he did Afrikaans. The British had made the town a place of importance: 'It was the seat of the British Governor, of the Natal Legislative Council, of the Supreme Court of Natal, of the Bishop of Natal, and it was the garrison town of a British regiment,' Paton was to note proudly years later.[3] It was to lose the British Governor when Natal united with the other three South African provinces in 1910, and the British regiment went at the same time; but to this day it retains an air of grace and dignity beyond its size.[4] The Supreme Court was of particular importance to Alan Paton, for his father was employed there as a lowly shorthand writer. Had the Supreme Court not been in Pietermaritzburg, Alan Paton would not have grown up there.

And in that case he would have missed the real beauty of Pietermaritzburg, which is the countryside in which it lies. It is a beauty of lush green hills, sometimes rolling so gently that as you crest one you see the top of the next many miles in front of you, sometimes rising so steeply and ruggedly, as they do when one begins to climb the escarpment immediately to the north of the city, that they appear formidable mountains. And there are unquestionable mountains within reach of the city, notably Natal's Table Mountain, a massive, flat-topped peak startlingly like Cape Town's landmark from a distance, and from the top of which the climber is rewarded by one of the greatest views in Africa, down to the Indian Ocean in one direction, and up to the Drakensberg range in the other, while below lie the glories of the Valley of a Thousand Hills.

This was the landscape that formed Alan Paton. His prose, when he writes of this natural world, has the drive of poetry and is filled with Wordsworthian echoes:

> I cannot describe my early response to the beauty of hill and stream and tree as anything less than an ecstasy. A tree on the horizon, a line of trees, the green blades of the first grass of spring, showing up against the black ashes of the burnt hills, the scarlet of the fire-lilies among the black and the green, the grass birds that whirred up at one's feet, all these things filled me with an emotion beyond describing . . . A rare joy was to find the magnificent orange clivia, now protected, then rapidly disappearing. A glade of clivias in flower, in one of the larger stretches of bush that might be called a forest, is a sight not to be forgotten. The stream would run over black stones, descending rapidly, sometimes with little waterfalls. And one might, though perhaps only once in a lifetime, catch a glimpse of the small mpithi antelope, shy and delicate.[5]

This idyllic area, so utterly different from the aching stony plains of the Karoo, or the highveld's sea of grass, or that curious, self-contained island that is the Cape peninsula, was Alan Paton's South Africa, the only part of his radiant country he could respond to until he was in his twenties. And he was awakened to its beauties by his father, who loved long walks and introduced his children to the delights of the natural world. 'My world was the world of hills and grass and rain and mist, and of birds seen and unseen, and of the crying of the trains on their way to Johannesburg,' Paton was to write years later. But if he first learned this love of the natural world from his father, he also learned hate, a hatred which, he gradually came to recognize, was to form his character as powerfully as love had. And to examine the nature of that hate, we must enter the Paton home.

In 1903 Alan Paton's parents were living at 19 Pine Street, Pietermaritzburg, and all his life Paton believed that he was born under its roof; he says as much in his autobiography, and if visitors asked him to take them to his birthplace, it was outside 19 Pine Street that he would pose for their cameras.[6] But in this he was mistaken. According to his birth certificate, the details of which would have been provided by his father, he was born at 9 Greyling Street, just around the corner from his own home: the back gardens of the two houses adjoin. 9 Greyling Street was the home of a Scottish couple, Mr and Mrs Ridley, with whom James and Eunice Paton had lived when they first married, and Eunice Paton returned to Mrs Ridley's for the birth of her first child.[7]

According to an unpublished memoir written by his sister Dorrie,[8] who drew on her mother's memories, the baby was frail, and at first there was doubt that he would survive. Breast-milk was thought inadequate for him, and he was also given an infant formula popular at the time, Mellin's Food.[9] Whether or not the outcome can be attributed to Mellin's Food, he rallied, thrived, and survived for 85 years.

19 Pine Street, the house in which he places his earliest memories, still stands, though it has suffered alterations since Paton's parents sold it in 1914. It was then and is now a small, narrow house, with rooms both cramped and dark, and the Paton family, as it grew, filled it beyond capacity. Alan was the eldest child, but his brother Atholl was just a year younger, born on 31 January 1904; and the two boys were followed by two girls, Eunice (always known as Dorrie), born on 4 December 1907, and Ailsa, born on 6 July 1910. The house had a small open verandah or *stoep* in front and another at the back; there was a sitting room, a dining-room, and three cramped bedrooms. By 1910 one of these was occupied by Paton's parents, one by his two sisters, and the third by his paternal grandmother Elizabeth Paton. There was no bedroom for Alan Paton and his brother Atholl, and accordingly they slept on the enclosed back verandah. Fortunately Pietermaritzburg winters are generally mild.

Almost wholly dominating the children's world as they grew was their father, a small, intense man with a walrus moustache and a tormented personality. Young Alan in particular was deeply influenced by him, for both good and ill, and everything he wrote about his father is deeply inhibited by his desire to do justice to a man for whom he had felt passionate dislike. Some time in the late 1930s, after the terrible death of his father, Alan Paton set out to sketch some aspects of the relationship between his parents in an unfinished, untitled play about 'Mr and Mrs Kingsley'.

> Scene: Kensington, Johannesburg—A lounge, but not a formal one—Photographs of King & Queen, Gen. Smuts, Deer in Scotland, Cattle in Scotland, Red-coated Huntsmen, etc—
>
> MRS KINGSLEY [*sitting with sewing*]. You fuss too much, father.
> MR KINGSLEY [*standing, huffily*]. Haven't I a duty to fuss?
> MRS K. No one has a duty to fuss.
> MR K. Haven't I a duty to be anxious?
> MRS K. No one has a duty to be anxious.
> MR K [*sarcastically*]. Then what's my duty?

MRS K [*wounding him*]. You've done your duty. You've brought up your children as best you could. Now they're grown up they must run their own affairs. [*She turns her finger in the wound*] Now you can only watch them; & if they go wrong, you can't stop them. You just have to say I must've gone wrong somewhere myself.

MR K. The girl's not yet twenty-one.

MRS K. Really, James, even I sometimes think you're old-fashioned. She's got her degree. You can't treat her like a child.

MR K. I didn't treat her like a child. But I never wanted to see her marrying a Dutchman.

MRS K. He is *not* a Dutchman. Your children have been trying to get you to drop that word as long as I can remember.

MR K. And why is he not a Dutchman? Didn't his own father's father come from Amsterdam?

MRS K. He is not a Dutchman for the same reason that your children are not Scotchmen.

MR K [*in pain*]. Ada, after all these years, must I tell you again there are no Scotchmen?

MRS K. Scotsmen, then.

MR K. And why are my children not Scotsmen?

MRS K [*impatiently*]. You know quite well they're South Africans, by law & by choice.

MR K [*gloomily*]. Aye, you're right.

MRS K. You *are* depressed. And what about, I wouldn't know. You've got splendid children. [*He refuses to be comforted*] Do you know what Andrew said to me?

MR K. No.

MRS K. He said 'Mum, I reckon Dad's all right.'

MR K [*brightening*]. He did, did he?

MRS K. Yes, he did.

MR K. He's a good lad. What else did he say?

[*Mrs K suddenly giggles. Mr K looks at her suspiciously.*]

MR K [*sternly*]. What else did he say?

MRS K [*defensively*]. I didn't say he said anything.

MR K [*with authority*]. Ada, what else did he say?

[*Mrs K returns to her seat & takes up her sewing. She is uncomfortable under her husband's stare.*]

MR K. Ada . . .

MRS K. Well, James, he said it was a pity that your I.Q. was sometimes so low.

MR K. He did, did he?

MRS K [*embarrassed*]. You forced it out of me, James.

MR K. He's an impudent young scoundrel, that's what he is. When did he say it?

MRS K. It was the night you all got so angry, about the man who was looking at the photograph, & said 'this man's father is my father's son'.

MR K. And I said he was looking at a photograph of himself, didn't I?
MRS K. Yes, you did.
MR K. And was that right? Or was it wrong?
MRS K. Really, James, must we start it all again?
MR K. I'm asking you, was I right, or was I wrong?
MRS K. Well if you must know, I think you were wrong.
MR K. Then whose photograph was he looking at?
MRS K. He was looking at his son's photograph.
MR K. Now listen here. 'Sisters & brothers have I none, but this man's father . . .'
MRS K. James, I refuse, absolutely & finally . . .
MR K. Of course you do. Because Andrew says I'm wrong, well I must be wrong. Because Andrew has a degree . . .[10]

For all its fictionalizing alteration of such details as names and places, this is the most intimate account Alan Paton produced of relations between his parents. In particular it is the clearest picture of one side of the person who was by far the most important influence on his early life, his father.

James Paton, as his son's dramatic fragment suggests, was a Scot who was immensely proud of his heritage, and who lamented the fact that his children were growing away from it. His daughter Dorrie remembered his stories of Bonnie Prince Charlie and Rob Roy all her life. She also remembered his 'vibrant readings' of Scott and Burns, delivered in the broad Glaswegian accent which he never lost; all the children knew his favourite passages from these two writers well even before they could read. Among his particular favourites were four novels which were to lodge permanently in his eldest son's mind: Stevenson's *Treasure Island* and *Kidnapped*, and Scott's *Ivanhoe* and *The Talisman*.[11]

Alan Paton knew little about his father's background, and seems never to have troubled to find out; in later years he would reply to enquiries about his ancestry with frank admissions of ignorance,[12] and it is a sign of his disaffection from his father, or of his lack of interest in this side of his heritage, that he never tried to establish even such a basic detail as his father's date or place of birth.

James Paton was born in Glasgow, in the densely crowded slum suburb of Anderston, on 4 April 1872, and subsequently lived in another Glasgow suburb, Dennistoun.[13] He was the son of Elizabeth Meicklejohn and David Smith Patton.[14] It is not known when the spelling and pronunciation of the surname were changed, but in South Africa James Paton seems always to have pronounced it with a long 'a', as in Baker, and his children naturally followed suit. David Smith Patton had died when James was 14, and the boy, finding himself the sole breadwinner for his widowed mother and his sisters Grace and Elizabeth,[15] had abandoned his education and gone out to work.

It is plain that his early manhood was a period of great difficulty for him. He did not speak much about his life in Scotland nor encourage his children to enquire about it, and he remains a mysterious figure in many respects. Neither Alan Paton nor his sisters ever knew what their father had worked at as a young man between 1887, when his own father died, and 1895, when he came out to South Africa, but at some stage he trained as a shorthand writer, the only occupation his children knew him to have.

He arrived in South Africa alone, and went to Johannesburg just in time to be caught up in the unrest of the Jameson Raid and its aftermath in 1896.[16] His daughter Dorrie believed that he had worked at the Magistrates' Courts in Johannesburg, but there are no records of this.[17] Hearing that there was a job for a shorthand writer at the Supreme Court in Pietermaritzburg, he made his way there on the outbreak of the Boer War in 1899; in later years he told his children that he had come to Pietermaritzburg 'as a refugee'.[18] He was not too proud to sleep at first on the floor of a cobbler's shop in Church Street, but once he got permanent employment at the Supreme Court he saved enough money to marry a local girl, Eunice Warder James, on 31 March 1902, and not long after that he brought his mother, Elizabeth Paton, out from Scotland to live with himself and his wife in the house in Pine Street. The salary of a shorthand writer was small, and for many years the family had a hard struggle to make ends meet; in any case, the privations James Paton had suffered after the death of his father made him frugal to the point of meanness, and Alan Paton was born into a house in which there were no luxuries.

James Paton was a man who loved literature; he not only read it with passionate engagement, but wrote and published verse, enjoying a local reputation as a poet. Many examples of his writing survive to show that he was a competent versifier. One poem, which he published in the annual Natal *Eisteddfod Book*, is enough to show his style:

South Africa

I would not boast of thee, O Land of Ours,
Thy snow-tipped mountains, tow'ring to the sky,
Thy thousand valleys, leaping to the eye,
The veld aflame with summer's radiant flowers.
Nor would I boast of conquests, warlike powers,
Of thousands slain, and thousands yet to die,

Of nations forced beneath thy heel to lie:
Mercy and Truth should flood thy fragrant bowers.

And, oh! I love thy landscapes and thy streams,
God made them, and His work stands as of old;
Another greatness see I in my dreams,
Then, Land of Mine, would'st thou have heart of gold:
That Thou, though tempted sore, deal fair and just
With those whom heaven hath portioned to thy trust.[19]

This sonnet, written in 1921, is not much more than a formal exercise; but for all its archaic language and dull phrasing, it does suggest James Paton's passionate response to some aspects of the Natal landscape, and also that piety which was such a marked aspect of his character. The last lines also indicate a desire to deal fairly with South Africa's subject peoples, though it has to be said that in 1921 they are more likely to have been thought of as the Afrikaners (conquered in the Boer War) or the Germans of South West Africa, than the Zulus, or the Indians whom the British had brought to South Africa as indentured labour in the sugar-cane fields.

James Paton had other skills and other interests. He loved music, and would organize musical evenings to which many university students came. At these gatherings his wife Eunice would play the piano, and everyone would sing parts. His children in later years thought he was making a fool of himself, but they may have been mistaken, for the university students continued to come as often as they were invited.[20] He was also a keen debater, and would take his children to hear his speeches at the Pietermaritzburg Debating Society, an outing his daughters found almost intolerably dull; he encouraged them to join the debating societies at their schools, and at first was proud of his oldest son's talents in this direction.

But if James Paton was a man of intellectual ability and cultural pretensions, he was also a man who felt his own lack of education bitterly, and the more bitterly as his clever son Alan ('Andrew' in the unfinished play) climbed the academic ladder. This jealousy of his own children contributed to the less attractive side of James Paton's personality, and grew more marked as he aged. This side is downplayed in the portrait of 'Mr Kingsley', and it was the aspect which was the source of the fear and the loathing he inspired in his three older children. In Alan Paton's play, he seems a harmless old buffer, slightly pathetic in his inability to understand a simple riddle, and his ability to cow his wife is faintly puzzling. Filial piety has censored this sketch of him.

James Paton was in fact a domineering man who enforced his will on his wife and his children. Outside the family home he was a rather humble figure, noticeably short, and a figure of fun to the Zulus he met on his long walks over the hills wearing shabby clothes and the large flat cap of the Scottish working class.[21] Inside the home, however, he was formidable. As his son put it, 'At home he was an autocrat, but outside he was diffident. Whether he was ambitious I cannot say. But he must have felt a deep sense of frustration at never having been able to progress beyond the machinelike work of a court reporter.'[22] This frustration showed itself in what Alan Paton described as 'authoritarianism maintained by the use of physical force'.[23] And looking back from the vantage point of old age, he remarked of his father, 'His use of physical force never achieved anything but a useless obedience. But it had two important consequences. One was that my feelings towards him were almost those of hate. The other was that I grew up with an abhorrence of authoritarianism . . .'.[24]

Although his writing about his father appears completely frank and largely objective, there is plenty of evidence that Alan Paton played down the brutal aspects of his father's behaviour. Perhaps he preferred not to remember the details, though his memory was generally extraordinarily tenacious well into old age. Many children exaggerate their parents' faults for a time, but Alan Paton seems to have done the opposite. He carefully and consciously downplayed his father's less attractive characteristics, for he took the Fifth Commandment very much to heart.

In his autobiographies he gave only one example of what he suffered at the hands of his father, a brief description of being beaten with an umbrella, the catch of which wounded his leg so that it bled. But an umbrella is an innocuous weapon compared to a cane in the hands of a practised flogger, and it was to the cane that James Paton routinely turned. According to Alan Paton's sister Dorrie, the boys were beaten often, and even the girls were far from immune. Nor was James Paton's wife safe from physical violence.

His daughter Dorrie's earliest memory was of seeing her father chasing Atholl, then still a small and chubby boy, into the large orange tree that grew at the back steps of the house in Pine Street, and lashing with a cane at the plump legs as the child tried to climb out of reach.[25] She herself as she grew older was severely beaten at times, notably on the one occasion when she tried to save her submissive mother from James Paton's abuse:

As I got older I began to stand up to him. One day I came and he was bullying my mother, so I got behind him and I gave him such a whack. And he turned round and nearly murdered me. And from that day I was no longer his favourite, my sister became the favourite.

Oh of course he would hit me; he would hit anyone. He used to bully my mother, he really used to bully her, and she was a—well we all loved her dearly.[26]

James Paton's harshness was to contribute to the portrait of Jakob Van Vlaanderen in *Too Late the Phalarope*.

Alan Paton's mother Eunice was a schoolteacher, South African born though of English stock, and she seems to have been a gentle and unassertive person. Her father, Bristol born, was a clerk in a firm of Pietermaritzburg waggon makers, Merryweather & Son; her mother was Edith Frances Mason, who had been born in Pietermaritzburg of English parents. There exist photographs of Eunice Paton in later life; they show a stout, large-featured woman in a pork-pie hat, smiling rather humbly and uncertainly at the camera. Her son's friends remember her as a sweet, gentle person, rather plump, and with a winning smile; she did not make a deep impression on them.[27] In the home she was powerless to resist her husband's will, although at times she fought a sly guerrilla action of the kind Paton was to picture 'Mrs Kingsley' using. She was no fool; it was from her that her eldest son learned to read and write well before he went to school; but he broke away from her influence early on, and she gets very little attention in his autobiographies.

She inspired great love in her daughters, however, who came to see her as patiently enduring great strains. Her housework seemed to them endless, particularly the elaborate laundering of the clothes her husband wore to court. A characteristic scene was of her boiling his shirts in a great copper in midsummer heat, and then washing and glazing and trimming all round the collars. She must have been unusually patient too in enduring the presence of her mother-in-law in the house all through her marriage, and her children remembered no cross word between them. In fairness it has to be added that Grandmother Elizabeth Paton was herself a woman of culture and tact, and when it was necessary to remove herself from a tense or violent situation between the parents, she would read, or pretend to read, a large volume of Herodotus for hours on end.[28] In later life one of her granddaughters read that volume to find out why it had so interested her over so many years, but in vain.

James Paton did not just beat his wife and children; he had other methods more insidious and perhaps even crueller. He submitted his children to verbal humiliation, hurtful jokes to which they could find no response and which he continued at great length, and psychological bullying of the type by which Mr Kingsley is seen forcing an admission out of his wife. And he seemed to try to provoke his children into revolt by imposing on them trivial but subtly degrading rules. 'My father had rules about everything,' his daughter was to recall, and she did not exaggerate.[29]

These rules became more restrictive when his children started school and began to go out into the world; they grew worse when Alan reached high school, for James Paton himself had never attended one. In the home there were many rules, such as that no expletive, no matter how innocuous, might be used in conversation. Swearing was a crime bringing instant retribution, but not even 'gosh' or 'gee' were allowed to pass. These rules applied also to the children's friends, who were chosen by James Paton, and who could fall from favour for a small misdemeanour. On one occasion a friend of Alan's sister Dorrie called out, 'Oh golly!' James Paton happened to overhear, and he tore out of the house in a rage: 'Doris, go home at once. I won't have people swearing here.' Doris went home and was never allowed to come again. 'That's how all my friendships ended,' Dorrie was to recall many years later.[30]

When his sons grew old enough to have bicycles, he made one of his humiliating rules about how these should be mounted:

> We were not to put the left leg on the left pedal, and swing the right leg over the saddle. We were to mount by means of the 'step', which was a small metal rod which projected from the hub of the rear wheel. We would then be the only two boys at College who mounted by the use of this grotesque method, probably the only two in South Africa. If I remember rightly my father could relate gruesome instances of boys in Scotland who had done themselves severe testicular injury by the pedal method of mounting.[31]

His sons naturally disobeyed this rule when they were out of their father's sight, just as they subverted his rule that they were to wear undervests, which they considered garments worn only by 'sissies', by stuffing these into a bush on their way to school, and donning them again in a plantation on the way home.

One of his most humiliating and resented rules concerned birthday parties of the sort every child in their small society got invited to. His rule was that they might go to the party if it were that of a child he approved of, but that they might not take a birthday present. 'I don't believe in people having parties for presents,'[32] he would say, and he either did not realize what embarrassment this exposed his children to, or did not care. His wife both realized and cared, and she would sneak money out of the housekeeping funds to buy a gift, which would be smuggled at the last minute into the hands of the child going to the party, so that James Paton did not know.

His own children's birthdays passed without parties, with the exception of Dorrie's ninth birthday, which for some reason he decided to mark with splendid festivities. But he took care not to tell the children and adults he invited what the occasion for the party was, and as a result his daughter received no gifts.[33] 'I was not very pleased, you know,' she said mildly in old age. 'But that was typical of him.'[34]

There were rules too as to who might eat what in the Paton home. James Paton's salary was small, £40 a month, his daughter Dorrie thought, and it probably reached this level only towards the end of his career. Accordingly food in the Paton home was plain and not over-plentiful. 'On a Sunday night we would each have a piece of cheese,' his son was to record, 'and not a big one. I could have finished it in one or two bites, but I would cut it into twenty pieces. If we had pickles we would get one each, and I would deal with mine in the same way.'[35] If they chanced to have a friend to a meal, the Paton children would watch in silent fascination as the guest, blissfully unaware of the household rationing, took several pickles, or put butter, jam, and cheese all onto the same slice of bread. But within the family this frugality was relaxed for only two members of the household, James Paton and his mother.

Breakfast for the rest of the family consisted of porridge and bread, but grandmother Paton had bacon and eggs every morning. Atholl, who sat next to her at table, would be tormented by the smell of her meal, and she would take pity on him and offer him half of it, pretending she did not want it. As he grew older he came to find this humiliating, and would reply loftily, 'No thank you, I don't take leavings.'[36]

James Paton always got the choicest parts of any meal, but his eldest son was to record charitably that 'he certainly did not indulge himself where we could not'.[37] His daughter Dorrie contradicted this flatly. According to her, what particularly rankled with the children was the way he dealt with the cakes his wife sometimes baked. 'Now this is terrible. If my mother made a Victoria sandwich cake it was put in front of my father, and he cut slices off and he ate and ate and ate, we never got any.'[38] It is no wonder that resentment in his children increased and hardened into hatred, mingled, as they grew towards adulthood and understood more of the world, with a kind of contempt for a man who could behave in this way. Nor could they understand him, try though they might. This was an era of stern fathers, but James Paton was more than stern. No doubt his humble position in society, his lack of education, his frustrated ambitions, even his physical shortness, had something to do with his behaviour, but there seemed more to it than that. 'It was as though an evil humour came on him at times, as it came on King Saul, causing him to try to kill David,' Alan Paton wrote long after his father's death.[39]

This demon-haunted man had one other aspect worth noticing, for it also had a profound effect on his brilliant son: he was deeply but narrowly religious. His religion was that of a small, puritanical, close-knit sect. James Paton had been brought up a Christadelphian, and as soon as he came to Pietermaritzburg he made contact with the small but active Christadelphian community there. Christadelphians had their origins in the United States, where they were founded in about 1848 by an immigrant English doctor, John Thomas. A Christadelphian community, known as an 'ecclesia', admits no distinction between clergy and laity, rejects orthodox Trinitarian doctrine, and regards the Bible as the only source of authority. In particular, the theology of Christadelphians is strongly fundamentalist and millenarian in character; they keenly await the Second Coming of Christ, and expect him to rule for a thousand years in Jerusalem. The sect admits new members through a profession of faith and baptism by immersion, and does not encourage close contact between outsiders and members of the ecclesia.

When James Paton fell in love with Eunice James, he fell out with the Christadelphians, for his fiancée was a Methodist and therefore an infidel. The marriage accordingly was performed by a Methodist minister, Mr Pendlebury, in the house of one of the bride's aunts, and relations between James Paton and the Pietermaritzburg ecclesia were broken off for nearly thirty years. Even when his wife, and subsequently all his children, became faithful Christadelphians and the community was willing to receive James Paton once more, he spurned them.

By way of compensation, he constructed a form of religion of his own, a religion in which he was the high priest, and his family the congregation. He decreed that Saturday was a holy day, and on Saturday no profane activity was allowed. Even walking, and James Paton loved walking in the hills around the town, was forbidden.[40] Yet religious activity took up most of Sunday as well. In the morning the family would walk to the Christadelphian service, leaving behind James Paton and one child whose task it was to keep him company. When they returned from the long ceremony, there would be a family meal, the best of the week, and then James Paton would gather the children together in his bedroom for the service over which he presided.

Each of the children in turn, starting with Alan, would preach to him a sermon on a text assigned the previous week. The younger children gave only short commentaries, and would be praised or rebuked according to their performance. But Alan's talks were thoroughly prepared and carefully researched with the aid of concordances and commentaries, and even his siblings came to enjoy them. In later life he was to write a biography of the politician J. H. Hofmeyr, and remark on Hofmeyr's having preached a sermon to his mother at the age of 6; he did not mention that he himself had preached before his father at an even earlier age, and not once but many times.

As a result his own use of English, almost from the time he could read, was deeply imbued with the superb rhythms and cadences of the Authorized Version of the Bible, and his view of life was based on biblical morality. His father's trivial and humiliating rules and codes of conduct he very gradually shrugged off as the lesser moralities, but to the greater moralities, which he learned not from his father but from the Bible, he held to the end of his life. As he put it,

> One did not lie or cheat. If the tram conductor did not come for the fare, you took it to him. One was not contemptuous of people because they were black or poor or illiterate. Justice was something that had to be done, no matter what the consequences . . . Murder, theft, adultery, were terrible offences, but to be cold and indifferent to the needs of others was the greatest offence of all.[41]

James Paton's bedroom service was followed by Sunday afternoon tea, and then would come a long walk over the hills, mandatory for the children old enough to manage it. On these walks a different side of their father would emerge, and they would see a man who loved nature and knew a great deal about it, a man who could name many of the veld flowers and identify most of the birds by their calls alone, a man who knew the secret places in the hills and who conversed in atrocious Zulu with every black passerby they met. This side of him his children could almost love, and all of them came to share his passion for nature. But at the end of these idyllic rambles the little party would march back into Wemmick's Castle in Pine Street, and have to deal again with the father who could make the whole household tremble.[42] Such was the man who laid the foundation of Alan

Paton's character. The hatred was one result of his influence, but in his son's own view there was another much more important.

> The other was that I grew up with an abhorrence of authoritarianism, especially the authoritarianism of the State, and a love of liberty, especially liberty within the State.
>
> Another thing happened too. . . . I grew up to have—eventually, say thirty years later—a will which, though not inflexible or implacable, was in matters of principle unshakeable.[43]

2

CHILDHOOD AND YOUTH 1903–1918

Fear was my father, Father Fear.
His look drained the stones.

THEODORE ROETHKE

After his sickly infancy Alan grew into a rather timid child, small for his age; he would remain relatively small all his life, his height in maturity less than 5 feet 7 inches.[1] His eyes were pale blue, and his gaze had a peculiar directness; his light brown hair was straight and stiff so that it stuck out in spikes, and from childhood on he tended to grease it down in a vain attempt to make it behave. His complexion was very fair, and the Natal climate was the wrong one for him; he burned on even a short exposure to the fierce African sun, and having peeled he would burn again, until, scarcely out of his childhood, he had damaged his skin beyond repair so that it would always look pink and raw.

He was not a bold or rough little boy. He was thoughtful, gentle, and strongly imaginative, and would tell himself elaborate stories and act out parts, for which his brother Atholl would mock him mercilessly, so that he learned to play in secret when he could.[2] Atholl was the one who liked vigorous games, rowdy gangs, and fighting, and who thought his brother a sissy. Whereas Atholl rebelled against their father secretly early on, and openly as soon as he was physically big enough to resist a threatened beating, Alan continued to show him respect all his life. His reaction to his father was one of fear and docility, at least until he went to university. Alan was his mother's favourite, for she found him quiet, responsible, and tractable. She taught him to read, write, and calculate well before he was of school-going age, for from the first he seemed to find learning a delight. In the world of books he felt safe.

Shortly after his sixth birthday he was sent to Berg Street Girls' School, at which his mother had taught before her marriage,[3] and which was an easy walk from his home. The school, despite its name, took many small boys; his oldest friend, Victor Harrison,[4] was a pupil at Berg Street Girls' School when Paton arrived in February 1909. 'They put him into sub A of the kindergarten, where he stayed for two weeks, then they moved him into sub B where he stayed three or four weeks, and then he was moved to standard 1,' Harrison remembered years later. 'At the end of the year he topped the class.'[5]

There exists in Paton's handwriting an unfinished novel almost certainly written in 1934, and in which Paton involves his hero, John Henry Dane, in many of his childhood experiences. The novel is untitled, but we may call it *John Henry Dane*. In it, Paton tells of Dane's miserable first day at school, in which he starts in the second kindergarten class because of his mother's teaching, but is disgraced when he wets himself from sheer nerves, and is sent down to the lowest class in the school. He tells, too, of his fear of the other children who torment him at playtime.

> Next morning I found myself down in Class (i). But after some days there the classroom became a haven of peace to me, a refuge from the devilish imps who spent their leisure torturing me in the playground. At least I found respite within four walls, & the work became a delight for me. A week, & I was back in Class (ii). Another week & I was in Standard I: I would have gone one step further, but my mother, already gratified & secretly proud, both of herself & me, decided otherwise. And at the end of the year I took home the Standard prize; my mother kissed me proudly, but my father said never a word.[6]

One effect of these rapid moves was that young Paton skipped infant education completely because of his precocious abilities; another was that he found himself, from the start of his school career, two years younger than his classmates. Later in his school career he was again advanced a year, skipping Form III, so that thereafter he was three years younger than the others. In these circumstances his small size and his remarkable cleverness stood out even more clearly than they otherwise might have, and did not always endear him to bigger, slower boys.

Perhaps for this reason school terrified him, and for a long time he remained a solitary. To the terrors of his home he had become accustomed; the fears of the streets and of the playground were new. His particular fear was of the bigger boys at the school. He felt himself to be small, timid, and bullyable, and he did his best to escape attention. Since he could not fight back, camouflage was his best defence. His strongest memories were of the times when he failed to avoid being noticed. The first occurred when one of the bigger boys spotted his shoes:

I wore a pair of shoes that did not lace up, but had a strap that went over the instep and was fastened by a button. These shoes were pronounced by one of the older boys to be girls' shoes, but I denied it, I am sure not hotly and angrily, but no doubt quietly and gently. I had no idea that such a small matter could attract such great attention, but soon there was a crowd of boys around me, and there was general agreement that I was wearing girls' shoes.[7]

This humiliation was soon followed by another, when a big boy told him to push a little girl off the pavement and he obeyed, being accustomed to obedience. The next day he was reported to the headmistress, who sentenced him to eat his lunch on the girls' verandah. 'I have now lived for seventy years since that punishment,' he wrote in 1975 with the amused detachment of old age, 'and never again have I pushed a girl off the pavement.'[8] At the time, though, he found nothing amusing about this humiliation.

The third incident was also one of being brought to the attention of bigger boys, but the emotion it evoked was not humiliation but shame, an emotion not imposed from without but coming from within. In its essentials it was a simple enough event, and many another boy would have put it from his mind almost at once. Paton was never to forget it, and instead of fading as he aged, the memory grew in significance for him until it became in retrospect a life-changing event almost akin to St Francis's sharing his cloak with the beggar. What happened was, simply, this: on a cold day his mother sent to school for him a basket containing some scones and a hot drink. When the black servant boy with the basket asked for him in the playground, Paton, horrified to find himself singled out in this way, denied that the food was for him, and allowed some of his older schoolmates to eat it. And when the servant saluted him on leaving, he ignored the greeting.

He gave an account of this event many years later, in his strangely moving story 'The Gift', and he told it again in his autobiography *Towards the Mountain*.[9] But there is an earlier, unpublished account, in *John Henry Dane*, which, since it was written closer to the time, has an interest of its own:

One cold biting day I had an experience of which even the philosopher of twenty-eight [Dane's age at the time he tells the story] is a little ashamed. My mother was in the village, staying with her bachelor brother, Mr Thompson. And at the short-break, which we country bumpkins called 'little play-time', a native boy approached with a basket of warm buttered toast & a jug of hot tea. It was the very day for such a gift, & I looked longingly at it, not knowing it was mine.

'Dane, you lucky devil, here's some grub for you,' one of the bigger boys told me.

'It's not mine,' I said.

'But the nigger says it's for you.'

'It's not mine,' I said vehemently.

Some obscure motive—fear of eating something that would single me out from the shouting personless crowd where I was content to lose myself—perhaps fear of owning to a mother & the fact of being loved—who knows?

'You can't waste it, you fool.'

'You can have it. It's not mine,' I said doggedly.

'Here goes,' said the lucky one, & I tried to watch carelessly the sharing of my mother's gift. But the feeling of cowardice, the knowledge of my own strangeness, set me drifting to a place where I was hidden from the scene of this incomprehensible treachery.

I told my mother the story years afterwards, & felt even then the shame & the need for forgiveness. Of all my queer actions it still remains the most incomprehensible.[10]

The action did not remain incomprehensible for Paton. The later accounts of this seminal event are altered in inessentials—the toast and tea becoming scones and cocoa, for instance—but the vital element of the story remained the same. It is a story of the betrayal of a dearly-held principle in the face of social disapproval.

Here it is a mother's love, the only love that bound Paton's family together, which is betrayed, and the fear is fear of being singled out by the bigger boys; but it is the embryo of that situation Paton would face over much of the rest of his life: the temptation to betray his belief in the unity of all human beings in the face of a menacing and powerful state preaching a doctrine of division. And his brooding on the shame of this betrayal would make him ever more determined not to slip into treachery of this sort again, on a bigger scale. He gained strength from the view that he had something to expiate. He clearly believed that the child is father to the man, but that the man should learn from the child's mistakes. It is significant that in the later versions of this story, particularly the one in his autobiography, the betrayal focuses itself not on the mother's gift denied, but on the black boy who brings it:

I went further along the wall so as not to be seen by the boy. I had not been standing there long when some of the schoolboys came to me and said, your boy's here with a basket. But I, though inexperienced in lying, denied that such a boy could be there. So they brought the boy, and of course it was our boy, and he smiled at me uncertainly because of the strangeness of the place, and I denied all knowledge of him. But he told them he was certainly our boy, and that I was the son of the house, and that my mother had sent me something warm to eat and drink. I denied him the second time.[11]

Childish embarrassment at being picked out in public has, in the mind of the mature writer looking back, assumed the significance of Peter's denial of Christ. Such moments as these were to form his character, and extreme moral scrupulousness and honesty were to be two of his most prominent traits.

After this first year he moved, at the beginning of 1910,[12] to a new school, Havelock Road Boys' (later known as Harward High School), where he had to rub shoulders with boys as old as twelve, and as a result was more fearful in the playground than before. Many of these boys, he wrote years later, 'were decidedly rough'.[13] It may have been in an attempt to impress them that he one day wrote on the lavatory wall the word 'fuck', of whose meaning he had not the least idea. A big boy took him to the headmaster and reported his offence, and the headmaster told him to tell his parents. The result, on his return home, was a terrified confession of guilt to his mother, a tense wait for his father to come home from work, a further wait for his mother to tell his father, and then the summons to his father's room, the shutting of the door, and the beating, the severity of which he erased from his mind in later years.[14]

Life was not all severe moral lessons, however, nor is the picture of little Paton as a shrinking violet the whole picture. Towards the end of his first year at Havelock Road Boys' School the school put on a concert, and to his mixed terror and pleasure he was chosen to take part. He was to dress as a pixie in a bright red cap with a long tassel and to sing a song beginning 'Up the airy mountains and down the rushy glen'. As this, the first of his stage performances, grew closer, he looked forward to it more and more, and according to his sister that red cap was the joy of his life.[15] But at the last minute came a hitch. James Paton was deeply doubtful of the moral value of any kind of stage performance. He asked to hear the words of the song, and having heard he strongly disapproved of it. Its references to pixie power and occult occurrences he considered blasphemous, and no son of his was going to sing them in public. Little Alan was devastated, and his mother intervened on his behalf. For once she was successful, and the performance not only went ahead, but with the whole Paton family in the audience.[16] Alan Paton seems to have enjoyed this stage appearance, for it was to be the first of many, both at school and later at university. It was one situation in which his imaginative power and his love of role-playing could be given free rein without provoking the mockery of his brother and his school-mates.

In later years he would often write and stage elaborate performances for his siblings, and if his father approved of them, he would also put them on at the musical evenings in front of an audience of university students. These performances were often more dramatic than literary; his sister remembered one as 'a rather brilliant item, depicting a man suffering agonies after a heavy meal of cucumbers and other indigestible salads. He lay on the platform with his pillow and blanket and rolled in agony from side to side as the pains increased in violence.'[17]

At the beginning of 1914, having just turned 11, he made yet another move, this time to Pietermaritzburg's most famous school. He had sat a competitive examination for a Natal Provincial Bursary, and won a scholarship that paid his fees and books at Maritzburg College.

The College traces its foundation to 1863, which makes it one of South Africa's oldest schools, and in Paton's view the most venerable. Its site, on a rise near the river, is spacious and leafy; its buildings, chiefly of red brick with terracotta tiles, and centring around the very fine Victoria Hall, are a happy colonial adaptation of nineteenth-century British architectural styles, not grand but gracious, and perfectly attuned to their setting. Like other South African government schools, it was in organization the result of a fusion of two models, the Victorian public school and the Scottish day school. From the Scottish institutions it took such habits as tasselled caps and the 'dux', or boy achieving highest marks.[18]

Paton was daunted by the place: not surprisingly, since he was considerably the youngest boy there, a child who had only just turned 11 and was underdeveloped for that age, still wearing short trousers. After his usual frightened reaction to the bigger boys, the older of them already hairy young men splendid in the school uniform of red, black, and white, and to the masters, who taught in mortar boards and academic gowns and seemed to wield the cane with great freedom, Paton settled down to his usual easy mastery of the work. In time he came to love Maritzburg College, and it was a love he retained until the end of his life.

In class he continued to be eager, hardworking, and clever, as good at languages as he was at mathematics and science. And at once he found a friend here, his cousin Noel Griffin, son of his mother's sister May. He had another friend too, Victor Harrison, who had been with him at Berg Street Girls' School. Although he was fifteen months older than Paton, Harrison was now in his class because of the younger boy's rapid progress, and proved a great support to him. Harrison remembers him at this time as small, pink, bright-eyed, and very untidy. His hair, bleached fair on top but darker beneath, was irrepressible, and his quiff was continually falling over his face and continually being brushed back.[19]

A second important event in 1914 was the family's move from Pine Street to a new and much larger home, at 551 Bulwer Street, on the other side of the town. The Pine Street house had long been too small for them, and now James Paton's frugality had made it possible for him to buy a bigger home, for £850. The Bulwer Street house, which still stands,[20] though its garden has been subdivided and built on, is a large and low red-brick house, with French windows opening onto a wide verandah running round three sides. When the Patons bought it, it was set in a most spacious garden. In later years Alan Paton described the new house like this:

> It was for us a big house. A wide passage ran through it, with three bedrooms on one side and a sitting room and a dining room on the other. At the top of the wide passage was a small bedroom on the left for my brother and me and a pantry. The small passage [sic] itself led into the kitchen. The bathroom had been converted into our bedroom and the bath had been transferred to the kitchen, and remained there until my father built a bathroom on the very spacious verandah that ran round three sides of the house.
>
> We had no water sewage at first. The earth-pit lavatory was a little way from the house, and in the rainy weather it was an unpleasant trip to go there. Our servants, a man and a woman, had no lavatory at all in their quarters, and had to walk several blocks to a communal and unspeakably filthy convenience supplied by the Pietermaritzburg Corporation. This was a common arrangement in those days.[21]

The house, which James Paton named 'Dennistoun' after the Glasgow suburb in which he had grown up, stood in a large garden of nearly an acre, full of mature trees, including a most productive orchard of about thirty peach, plum, pear, quince, and apple trees. There was also a persimmon and, against the walls of the house, many grapevines. What was more, Bulwer Street offered immediate access to open country. 'One opened the back gate, crossed over East Street, and there a few feet below was the Dorp Spruit. Between that and Mountain Rise, a distance of between one and two miles, was unoccupied country.'[22]

The garden had a drawback, however, for although the Patons were now financially comfortable enough to have two Zulu servants instead of the single small *umfaan* (boy) they had previously employed, James Paton made new rules, one of which was that his sons were to work in the garden every Saturday and every holiday. They bitterly resented having to labour in this way, particularly in the blazing heat of a Natal summer, on the few days free from school or church, while their friends passed by on their way to swim or loaf about the hills.[23] On the occasions when they were allowed out, the Paton boys had always to be supervised. The girls were on an even shorter leash, and were seldom allowed out at all.

Exceptions were the long walks which they continued to take in company with their father; on occasions these would be extended to occupy a full day, on several occasions as far afield as Natal's Table Mountain or Swartkop, a beautiful landmark to the west of Pietermaritzburg. The family would take a picnic lunch on the top, and these picnics were a particular joy to Alan Paton, who planned them elaborately in advance.[24] They would collect wild flowers among the rocks as they climbed Table Mountain, and when they reached the grassy top would eat their sandwiches looking across the rich folds of Natal to the city of Durban and the Indian Ocean.

Watching the wealth of native birds became a particular passion with young Alan, who in later life acquired a copy of the standard bird-watcher's guide, Austin Roberts's *Birds of Southern Africa*. This was the start of a lifelong hobby for him, and he became an authoritative amateur ornithologist. *Birds of Southern Africa* was to play an important part in his second published novel, *Too Late the Phalarope*, and a battered copy of the Roberts book, dog-eared and crammed with Paton's detailed notes of birds he had seen, was among his treasured possessions at his death.[25]

Perhaps because of James Paton's love of debating, the Patons were an argumentative family, and when the topic of conversation became controversial, centring usually on some aspect of religion, Alan would get up and wander off on his own with his sandwiches. 'Watch Alan,' Dorrie was to recall her mother saying on one such occasion. 'He doesn't like to face up to anything unpleasant.'[26] His later career suggests that his mother did not understand his motives very well.

His character was now developing, and he showed an increasing love of horseplay, particularly in the absence of his father. James Paton was sometimes away from home for days or even weeks at a time, because he had to accompany the Judge-President on Circuit Court. On one of these trips he took his wife with him to give her a much-needed holiday in Durban, and he got his sister, who had followed him and her mother out from Scotland

and was teaching in Durban, to help grandmother Paton take care of his children. Aunt Elizabeth, as the children called her, proved unable to keep control of her nephews, and her mother was little better.

On one occasion they terrified their aunt, but frightened themselves even more, by a feat of daring that went wrong. In summer Pietermaritzburg, like most of the summer rainfall areas of South Africa, is subject to dramatic thunderstorms, often accompanied by strong winds and sometimes hail. During one of these, when a tall pine tree which stood at the bottom of the Bulwer Street garden was swaying wildly, the two boys decided to climb it to enjoy the motion. Each daring the other to climb higher, they were soon near the top, and then the wind rose further and the lightning began. When their aunt came out on the verandah to look for them, she found them clinging in terror near the top of the tree, being thrown around as it whipped in the gale, and much too frightened to climb down. They had to stay there, soaked by the storm and terrified by the stabs of lightning, until the wind abated enough for them to creep down, penitent and exhausted. They escaped punishment on this occasion because their aunt felt that for once they had learned a lesson.[27] The incident was to find a place in *Too Late the Phalarope* when Pieter van Vlaanderen terrifies his aunt in this way.[28] Aunt Elizabeth, who was much loved, contributed a good deal to the portrait of Tant Sophie in that novel.

It was when he was about 10 that he made his first garden. His brother Atholl and he, through being forced to garden, had slowly gained an interest in growing things; Atholl proved to have green fingers, and produced beautiful vegetables from his own small plot. Alan, stung to emulation, decided on a formal garden with neat footpaths and carefully trimmed borders, with a great variety of flowers in small beds carefully laid out in symmetrical squares, triangles, and diamonds. In the four corners of the plot were laid out four square rock gardens, each ornamented with river stones laboriously carried up from the Mzinduzi, and bright with small and interesting native plants the boys found on family walks.

This was the first of his many gardens; when he came to have a house of his own, he always paid more attention to the garden than to the house itself, and this, like his hatred of tyranny and his love of the Bible and of nature, was one of his father's legacies. The garden shows another of his marked characteristics, his tendency to do whatever he did with immense thoroughness. He already had apparently endless energy, and he combined this with a great appetite for taking pains in all he did. It is a recipe for success, as he had already shown in his schoolwork.

He says very little, in his autobiographies, about his relations with his brother, but it is plain that they were dominated by sibling rivalry. Atholl, although a year younger, was already taller and stronger than Alan, and able to beat his brother in a fight. According to their sister Dorrie they fought often, these contests reaching a climax in a ferocious battle early in 1914 soon after they had moved into the Bulwer Street house. Their father was away at work, and their mother was powerless to part them. They fought with hatred, drawing blood from each other's noses, wrestling, punching, gouging, gasping, while their mother and grandmother vainly begged them to stop, and 6-year-old Dorrie, terrified, prayed fervently that they would not kill each other.

At the end of it Atholl had won, and he remained physically dominant thereafter, to the deep and lasting humiliation of his older brother. When Atholl joined him at Maritzburg College in 1917, Alan's classmates were soon asking him, 'Is it true that your brother can give you a hiding?'[29] Adults tend to regard children's emotions as trivial and transient, but for the child the matter can be one of deadly seriousness. In *John Henry Dane*, the narrator is tormented by a bigger brother, and some of the most powerful and effective sections tell of his passionate loathing of this less intelligent, but physically stronger sibling. 'My brother Richard I hated with a fierce fearful hatred. There were times when I would have killed him had I dared.'[30]

But 1914 was a year of more than domestic quarrels: on 4 August 1914 Britain declared war on Germany, and South Africa accordingly was at war too, though a great part of her Afrikaans-speaking population, with memories of the Boer War still fresh, had no desire to fight for their former enemy. Within a few weeks of the outbreak of hostilities Britain asked the government of Louis Botha, himself a former Boer general, to attack German radio stations in South West Africa, a request which was at once met, although at the cost of a short-lived rebellion by the most anti-British Afrikaners.

The Paton home, with its British background, was passionately involved in the conflict, and the family read every newspaper with keen interest. James Paton had always called Britain 'Home'; so did his wife, South-African-born though she was, and all his children, including his older son, at least until the late 1920s.[31] The fate and prestige of the Empire were matters of the greatest importance to them; patriotic lumps rose in their throats at the very mention of the Royal Family or the sight of the ships of the Royal Navy in the newsreels they watched at the new local cinema,

The Rinko. The movements of troops, the probable objectives of the armies, the chances of a swift victory, were the topics of their dinner table for many months. And they shared in the wild excitement towards the end of the year when the local regiment and its new volunteers were formed up in the centre of town and marched away with their band playing bravely, and Paton recorded the scandal of the population when it was noticed that white soldiers were winking at black girls.

Maritzburg College, too, was swept by enthusiasm for the fight; at least six of the masters enlisted, and three[32] were to be killed. Even before the war the school had been fiercely patriotic. The school badge, the crest of which bears crossed rifle and assegai, has the words 'Pro Aris Et Focis' meaning '[Fighting] For Hearth And Home'.[33] Both badge and motto were consciously derived from the memorial to the Old Boys who had given their lives in the wars against the Zulus, and their names were to be seen proudly recorded in the school hall.[34] Patriotic fervour was such that many of the senior boys cut short their school careers to volunteer, and at least three of the younger boys ran away in an attempt to do the same.[35] The College Magazine reflects the early enthusiasm, and its grim aftermath: by 1916 its pages were beginning to fill with little black commemorative boxes, each containing the name and details of an Old Collegian who had died in action.

A sign of the excitement in the Paton home is the fact that the children at this time gave each other names drawn from accounts of the war. Three of these names stuck for life. When an early and minor naval action took place near Cuxhaven (a German port at the mouth of the Elbe) Alan began calling his younger sister by that name, later shortened to 'Cooks'; he went on calling her Cooks all her life, though she disliked it.[36] He called Atholl by an insulting German name. His siblings reciprocated by calling him 'Kaiser Bill', and his sisters continued to call him 'Bill' until his death, much to the later puzzlement of his children.[37]

Yet even on such a matter as the war the household was divided, for Christadelphians are pacifists, the sect having taken its name and organization in the 1860s in order to formalize its members' opposition to serving in the American Civil War. When, in 1916, Alan was required to join the school cadet corps, there was a debate as to whether his father should allow him to do so. In the end, inexplicably, James Paton gave his permission.[38] Alan Paton himself was divided on the issue, for while he shared in the excitement of the first year of the war, he gradually decided that if he were called upon to fight, he would be a conscientious objector. In later years he attributed this decision to a combination of Christadelphian influence and his own natural timidity.[39] However, the chief influence on him, he came to believe, was his own reading of the Scriptures:

I do not suppose that I reasoned it all out, but I do suppose that the way that I reason it all out now has its roots in my earlier years. I am sure that the forgiving by Jesus of those who reviled Him, the rejection of the third temptation in the wilderness, the Sermon on the Mount, made it difficult for me to contemplate taking up arms.[40]

Fortunately for him, the war was to end before he turned sixteen.

The faraway conflict had little effect on day-to-day life in Natal. There was no distant muttering of guns, no threat of invasion, no food shortages. The only arenas of conflict Paton had to face were playgrounds and sports fields. He was obliged to play cricket in summer and rugby in winter, and he enjoyed neither. But his character was gradually asserting itself. One sign of this came in 1914 when a drunken soldier snatched off his hat as he passed in the street, and flung it into one of the wide *sluits* (gutters) that carried water round the town. Paton never forgot what followed.

I rescued my hat and was filled with a burning sense of affront and injustice. Could I let this insult go unpunished? I decided that I could not. Clutching my hat firmly in my hand I went after the soldier. I now exhibited the courage that is born of anger. I launched myself at the soldier and delivered a kick at his back. Whether I reached his bottom is doubtful. But I certainly kicked him in the back of his thigh. With angry curses he turned to grapple with his assailant, but the boy was winging his way home, with un-Christian joy in his heart.[41]

This instinct to strike back at injustice was to grow into one of the dominating features of his character, but it seems never to have led him to revolt openly against the tyranny he was exposed to every day, that of his father. Instead, he began turning his back on the religion he associated with his father, Christadelphianism, and though he could not yet openly announce that he intended to break with the ecclesia, he was already planning to do so.[42]

He was undergoing the usual small moral tests of the schoolboy, but for him they loomed large. One afternoon, left alone in a shop, he stole a single toffee from a large glass jar and put it in his pocket. Overcome with guilt, he then slipped into a lane, Collier Street, and thrust the stolen toffee into a hedge. 'I never stole again,' he was to write soberly.[43] About telling lies he was much less sensitive, and he records that in order to escape one of his father's humiliating rules he would lie unhesitatingly.[44]

He tended to take an unsympathetic attitude to the failings of others at this time, too. His friend Victor Harrison remembered that Eunice Paton was generous to the tramps who used often to call on them asking for charity, and one day, completely convinced by a moving story about a starving family, she gave a tramp a ten-shilling note, a large sum for that frugal household. Paton, quite sure the man was going to spend it on liquor, followed him, and without surprise saw him turn into a pub on Boshof Street. Filled with moral indignation, the boy entered the pub too, and when the tramp put the note on the counter he sprang forward and seized it, crying 'That's my mother's!' and then ran back home with it.[45]

He was developing rapidly, and in more ways than one. For the first time he encountered sexual temptations, some filling him with guilt, others not. Once he reached puberty he discovered masturbation, the pleasure of which filled him with anxiety. He fought to resist it, and records that he kept 'a secret chart of defeats and victories'.[46] He was very nervous of girls, perhaps because some of his father's most rigidly-enforced rules were designed to protect his daughters against all boys, including their brothers. One of these rules specified that on no account were his sons to enter their sisters' room or vice versa; infringement of this rule meant a severe beating. 'In our house, boys don't go into girls' rooms, girls don't go into boys' rooms,' he would repeat, a sentence that worked its way deep into Paton's memory, and was to appear most unexpectedly many years later.

He learned little about sex at home but plenty at school. One Christmas Eve he, Harrison, and Noel Griffin walked to the railway station with three girls; the six of them sat in an empty coach, and there was what Paton describes as 'smoking, kissing, and fondling within fairly strict limits'; but he and the girl the others had brought along for him, May Wyatt, got nowhere. 'My time had not yet come,' he concluded wryly.[47] In fact it would be a long time coming, for his father's teaching combined with the puritanism of the Christadelphians put a most effective mental chastity belt on him for years, where women were concerned.

His sisters he treated with great respect and affection. The earliest-known example of his writing is in the form of a cartoon strip, written and painstakingly illustrated with skilful pen-and-ink drawings for his sister Dorrie on 8 December 1919. In this cartoon, entitled 'The Cheeky Goblin', a male goblin finds a very feminine tube of prussian blue paint (of all things), complete with beribboned hat, sunning herself on a rock. 'You are so sweet, I must squeeze you', he says, but as he does so the lady, with the words, 'You take too many liberties, sir', covers him in paint and makes off, daintily swinging her parasol and bidding him 'Good morning!'.[48] The 16-year-old Paton regarded girls as untouchable.

He did, however, have homosexual experiences with one of his two friends, whom he tactfully calls 'C.', though the details he gives make this a disguise not hard to see through:

It is probable that boys who are abnormally shy of girls enter more easily into homosexual relationships. I entered one such relationship with C., who was in the same class as I but was some two years older. He and I were eating our lunch on the terraces. Just how our conversation proceeded I cannot remember, but it ended up by my feeling his penis, the size and rigidity of which astonished me.

After that I spent the day with him, perhaps two or three times, at his parents' small farm outside the rectangular confines of the city, on which occasions we fondled and masturbated each other. Some two years later, when we were together at the Natal University College, I found a note from him on the notice board. In this letter he expressed his sorrow for what had happened and begged my forgiveness, which I gave freely and briefly by word of mouth.[49]

It is noticeable that Paton did not feel it necessary to ask the friend's forgiveness; neither then nor later was he to give any indication of thinking homosexuality particularly reprehensible, and in old age he was to deny that he had ever condemned it.[50]

On the other hand, when the family dentist, who had for years given him small gifts of money, tried to fondle his penis during the regular annual checkup, the 13-year-old boy repulsed him firmly with the words, 'I don't like that.' The man gave him half a crown, saying, 'That is the reward for virtue.' Subsequently he continued going to the dentist each year, but the advance was never repeated and, Paton added, 'To my regret there were no more rewards for virtue.'[51]

Only well after he had been thoroughly instructed by his schoolmates did his parents make any attempt to tell him about sex. His mother, following the fashion of the time, offered him a booklet entitled *What a Young Boy Ought to Know*, saying rather wistfully, 'I suppose you know all about it already.' His father made no approach to him on the matter at all, and this probably spared both of them a great deal of embarrassment, for they had little to say to one another.

James Paton's alienation from his children, or rather his alienation of his children, was by now almost complete, and he began to reap what he had sown. Atholl's rapid physical development soon made open rebellion against his father possible, and he made full use of this freedom, disobeying openly and answering back without fear. Presently, towards the end of his school career, the time came when he would curse his father to his face. James Paton's only recourse was to wait until his son was out and then lock the doors against him, forbidding his wife to let the boy in; but she would wait until he had gone to bed, and then disobey him.[52]

James Paton's biting humour was now one of the few ways he had of asserting himself against his sons. 'We did not always appreciate my father's jokes,' wrote Alan with restraint.[53] Instead the older man asserted his authority with increased rigour over his daughters, who were not allowed to play cards, to dance, to wear party frocks, to play tennis, or even to learn music. Dorrie, who longed to play, had to teach herself without his knowledge out of an old hymn book, with the aid of her mother.

When Mr Kingsley in the play tries to keep control of his daughter even when she is on the point of marriage, he is accurately reflecting James Paton's attitudes to his daughter Dorrie. When she was in her twenties she had to be smuggled out of the house to attend her first ball, chaperoned by an uncle sworn to secrecy. She wore a dress secretly made by her mother and silver shoes given her by Alan, and on her return undressed at the side door and ran naked down the passage to her room, preferring to meet her father in that state than risk his seeing her ball-gown.[54]

Later still, at a social gathering of the Ramblers, her father's walking companions, this earlier secret ball came back to haunt her when a young man asked her to dance. 'I daren't,' she said, for her father was in the room. 'Oh come on, I'll say I'm teaching you,' said the young man, and she gave in. Her father watched narrowly for a few minutes and then came up to them. 'Here. Come here. Go get your things. We're going home,' he said, and when they got home he flew at his wife. 'Don't tell me that girl doesn't know all about dancing,' he raged.[55]

During the war years Alan Paton advanced through the school, at this time under the headmastership of Mr E. W. ('Pixie') Barnes, with apparent effortlessness. He was in Form II in 1914, under a tyrannical master, Mr F. C. Sutcliffe, 'Sucky' to his suffering class, whom he used to urge to greater efforts with vicious thrusts of a pointer. When Paton was unable to remember 'portare', the Latin for 'to carry', Sutcliffe made him get down on all fours: 'He piled books on my back and ordered me to take them from one side of the classroom to the other. The Latin sentence for this was "portat asinus libros", which was "the ass carried the books"'.[56]

He was not a man to trifle with, and Paton records that one day when he came in to find 'Sucky is Bucky' written on the board, a sentence the boys regarded as the quintessence of wit, 'Mr Sutcliffe was extremely angry, but he was unable to discover the author, who preferred to remain anonymous.'[57] One of Sutcliffe's little quirks of humour was to beat two boys simultaneously by making them bend down with their bottoms in close proximity, and he would then oscillate his pointer violently and painfully between them.[58] Maritzburg College depended for discipline on severe thrashings, a tradition which Paton would carry away from the place with him.

Sutcliffe was a repellent man, a racist who tried to drum into his pupils his beliefs about the danger of blacks and Asians to white civilization. One of his many unpleasant oddities was to draw a chalk line on his door and get each of the boys to kick it, explaining that this was 'the level of a coolie's backside'.[59] Fortunately his political views had little perceptible effect on his pupils, perhaps because he, English-born, also despised 'colonials', a group that included most of them.

Paton did not top Form II as he had each of his classes in previous years, but he did well enough to be promoted straight to Form IV at the beginning of 1915, skipping another year. His friend Victor Harrison was similarly advanced at the same time. Paton was now beginning to play the fool in class, as smaller boys often do to win the approval of their classmates, but at first he did so with extreme caution, choosing teachers who were safe. One of these was the only female teacher they had, Miss Norman, grey-haired and kind, who taught the boys shorthand. Paton scarcely needed to be taught; his father, a professional, had taught him years before, and according to Harrison Paton could write shorthand at a speed of 120 words a minute at the age of 13.

As a result he was bored by Miss Norman's lessons and decided to do something to break the tedium. In the first class, while laboriously copying the circle S from his Pitman book, Harrison heard a groan and a thump and turned to see Paton lying slumped on the floor, apparently in a faint. Miss Pitman ran to him with a cry of alarm: 'Paton, what's the matter?' As the delighted class gathered round, feigning concern, Paton's eyelids fluttered open: 'Oh, Miss Norman, I can't get round these circle S's!' Miss Norman, completely taken in, had him carried out by his eager friends to be revived in the washrooms.[60]

On other occasions he would make sexually ambiguous remarks, along the lines of urging her to 'bang' or 'poke' a boy for some misdemeanour, and he never knew whether Miss Norman understood them or not. In later years he was not proud of these jokes.[61] He made them with other teachers too, notably with a new master who taught them Dutch. This man was

Afrikaans-speaking, and his grasp of English idiom was imperfect; once he realized this, Paton began exploiting the weakness with devastating effects. On one occasion the class was reading a text about agriculture, and the master chanced to mention that he had been brought up on a dairy farm, and that when a cow was dehorned he and his brothers would compete for the horns and treasure them. Paton's hand was up at once: 'Sir, did you ever get the horn?' And to the man's great puzzlement the class nearly burst.[62]

Paton was only 13 in the penultimate school year, Form V, which he finished at the end of 1916, and it was plain that unless something was done he would finish school at 14. His headmaster, Mr Barnes, thought this a mistake, and with the agreement of his father he was made to repeat Form V in 1917. This may have been an error, for he found the work so unchallenging during this repeated year that he clowned more than ever and began to lose the habits of application which had been natural to him.

His form master during his final year, Form VI, Mr William Abbit, known to the boys as 'Fluff', was an alcoholic. This weakness did not prevent him from being a fine teacher, in Paton's opinion at least. Abbit was certainly no fool, though he was not as clever as Paton thought he was. Abbit claimed to have been senior wrangler, or best student of his year, in mathematics at Cambridge, a distinction which he advertised widely to the boys, though it was fantasy: in 1892 he was in fact classed in the maths tripos at Downing College, Cambridge, as a Junior Optime, meaning that he attained only a third class,[63] a very long way from being senior wrangler. All the same, when Paton pointed out a mistake he had made on the blackboard, Abbit spluttered in rage, 'Do you argue with the senior wrangler of Cambridge?'[64]

His eccentricity and vanity gave his class many opportunities for ragging him, opportunities which Paton enjoyed to the full. Certainly to all appearances Fluff was a figure of fun, for he smoked huge, foul-smelling cigars, and, in spite of the warmth of the Natal climate, in winter wore a very long scarf wound many times round his neck to ward off the colds from which he often suffered, and which he treated with endless small tins of chicken soup. This soup he would drink from the tin with a spoon as the boys got on with their maths exercises.[65]

Paton became and remained one of 'Fluff' Abbit's favourite pupils. Abbit particularly admired Paton's writing, and on several occasions did him the honour of reading his essays out to the class as a model for them to follow. Harrison remembered Abbit, during one of these readings, shaking with laughter at Paton's description of an election campaign.[66] Afterwards one of his classmates said to him with candid admiration, 'If I could write like that, Paton, I wouldn't care two hoots about passing matric. I'd be a writer.'[67]

In 1917, the year in which he repeated Form V, Paton, who was not going to sit the exams at the end of the year, took to needling three bigger and duller boys with subtly sarcastic jokes of the sort with which his father had so often tormented him. A good deal of the time his shafts went over their heads, but others in the class chortled. Presently he went too far, and both he and Harrison, who shared a front-row desk with him, saw that Paton was in for trouble when the lunch break came. When the midday bell went, Paton got out of his seat, opened the door for the master and dashed out after him. He was well away by the time the three bigger boys emerged, and they had to chase him all over the College's extensive grounds before they caught him and brought him back to the classroom, where they had decided to make an example of him. The room had large fanlight windows high up, and to the catch of one of these the three slung Paton up by the ankles with his head on a desk, so that he appeared to be doing a handstand. They then stood guard over him until the master came in, no doubt in the hope that he would be caned for fooling.

When Abbit appeared he found the class very innocent and apparently unaware of Paton's predicament, but they were watching with keen anticipation for Abbit's reaction. When he saw Paton he made no sound, but his eyes closed and his whole body shook for some time. And when he had finished laughing he said, 'Well Paton, I think someone has caught you out at last.' A long pause. 'Take him down.' And with that he resumed his teaching.[68]

These schoolboy japes were evidence of high spirits, wit, and humour. But they were also evidence of a growing tendency in Paton to challenge authority wherever he saw it, whether it was the authority of masters, or that of other boys who thought they could enforce their will on him through superior strength. And it is noticeable that from the start his challenges were posed, not in the realm of crude force or direct opposition, but in the form of subtle intellectual undermining of his opponents. To this form of challenge a violent reply was meaningless. The boys who upended him had responded to his taunts, but they had neither replied to them nor silenced him. In later years the Nationalist government, when it menaced him with imprisonment, searched his house, destroyed his car, and took away his passport, would find its response similarly ineffective.

He grew, if anything, more confident in class because of his intelligence. The Maritzburg College system stressed continual assessment. Every three weeks the class sat what were aptly named 'trials', and every three weeks

Paton finished first or second. In sport, however, he never rose above the second team in rugby or cricket. His brother, by contrast, achieved first-team status and swaggered about, magnificent in the black, red, and white 'colours' blazer and cap which were the prerogative of members of the First Eleven and First Fifteen. Alan envied him miserably, according to their sister.[69] 'At the age of ten a boy would willingly be struck a moron', reflects John Henry Dane amusingly, 'in exchange for a team photo to hang upon his walls.'[70]

Their father, himself a keen soccer player, refused ever to come to watch either of his sons playing rugby, a form of rejection which hurt Atholl as much as their father's failure to remark on his academic achievements hurt Alan. 'My father could have won Atholl over if he'd ever come to watch him,' Dorrie was to say years later, 'but he never would.'[71]

Alan's one athletic distinction was his ability to run tirelessly for hours. He had a peculiar long, loping stride, and Harrison believed he would easily have beaten all comers at a marathon. Unfortunately the school at that time had no such race.[72]

He finished his school career at the end of 1918, when he was 15, matriculating with great distinction. He took the school prizes for English, Mathematics, and History, and in Science came second to a boy named Julius Goldman, who also just edged him out as Dux of the school. Paton was two years Goldman's junior, but he attributed this failure to carry all before him not to his youth, but to his beginning to enjoy school and 'playing the fool'.[73] Perhaps it was as well that he did not strain himself too much at such an early age; a warning was provided by the victorious Goldman who, his mind overtaxed by the effort of his final school year, was certified insane the next year and died not long after.

In spite of this, Paton's parents, instead of showing pride in their brilliant son, were disappointed in him for not being made Dux, a reaction which he recorded without comment many years later, but which must have hurt him bitterly at the time.[74] It seemed that, in his father's eyes at least, he could do nothing right. Success caused jealousy while failure brought rebukes. But he was increasingly freeing himself from his father's direct influence—not rebelling with the anger of his brother Atholl, but more and more inclined quietly to ignore or subvert his father's rulings. He was getting ready to go his own way, and nothing would hasten this process more than the next few years. For at the beginning of 1919, having just turned 16, he was off to university.

3

NATAL UNIVERSITY COLLEGE
1919–1922

> Aye, 'tis well enough for a servant to be bred at an University. But the education is a little too pedantic for a Gentleman.
>
> WILLIAM CONGREVE

Alan Paton intended to serve his fellows, and all through his childhood he had wanted to be a doctor. He seems to have felt that medicine would give him the opportunity to help other people in a practical way, and his outstanding abilities in mathematics and science encouraged him to plan in this direction. This ambition lasted at least until his last year at school, when it suffered two setbacks. The first, oddly enough, came when his brother's pet owl escaped from its cage and flew into a high tree some distance from the house in Bulwer street. Atholl, after long searching, spotted it and climbed up to catch it, but he lost his grip and fell a considerable distance to the ground. When he recovered consciousness he staggered home, very concussed and bleeding profusely. The alarm had been raised by his long absence, and his brother was searching for him; but when Atholl appeared at the gate, his face a mask of blood, Alan turned up his eyes and fell in a dead faint. 'Alan,' said his mother firmly when he came round, 'there'll be no doctor for you.'[1] It was true that he could not bear the sight of blood, though no doubt he could have overcome this in time.

The second and more serious setback came when he discovered how much it would cost, and how long it would take, to qualify as a doctor. James Paton was no longer a poor man; a lifetime of the most rigorous frugality had paid off, and when he was murdered in 1930 he would leave his wife a large house and £2,000, a considerable sum at the beginning of the Great Depression. He could probably not have afforded to send his brilliant son to Oxford or Cambridge, which was the natural destination of clever Natal boys with rich parents; but in 1919 he could almost certainly have paid for medical training in South Africa had he wanted to. He did not want to. 'My father explained that he just could not do it,' wrote Alan uncomplainingly.[2] His academic success was already difficult for his father to swallow, and James Paton had no intention of paying for him to study any further. Alan had put himself through Maritzburg College on scholarships, and he was going to have to do the same at university. Had he not been able to support himself, his fate is likely to have been that of his brother, who, though he matriculated, won no bursaries, was refused all financial help by his father, and in due course was obliged to become a clerk with the electricity department of the Pietermaritzburg City Council.

In 1919 the most readily available bursaries, and the most generous, were provided by the Natal Education Department for training teachers. These covered the university fees and books, and paid £80 a year in a direct grant. To Paton, just 16 and the product of a most frugal home, this was wealth. His parents intended to share in it, for he would go on living at home, and pay them £5 a month rent, so that £60 of the £80 would go to them.[3] Even the mere £20 remaining to him each year for clothing and all other expenses seemed to him munificence. In each of his novels there are strangely affecting descriptions of the care, almost reverence, with which poor people fold up banknotes and store them away. It was a gesture he knew from deep experience, and he never lost that attitude to money even when he grew rich.

He later said he had given up the dream of becoming a doctor without much regret, and came to consider that he had been born to be a teacher.[4] His mother was a teacher, so were several of his aunts, and both his sisters were eventually to become teachers; it was a natural course for him to follow. In South Africa at this time the social standing of teachers was very high, and it remained so into the 1960s; in a country town the headmaster was the social equal of the doctor and the ministers of religion. In addition Paton seems already to have seen teaching as a way of influencing young minds towards ethical behaviour, and he was determined to make the world a better place by his passage through it; certainly by his second year at university he would be talking and writing in these terms. In February 1919, then, he entered Natal University College on an Education Department bursary, to read for a Science degree.

His friends, knowing his writing ability, were astonished that he chose Science, and he was later to say that since the Education Department needed science and mathematics teachers above all, he had decided to give up the arts.[5] But this was not the whole truth. His other reason was a curious one: the temporary weakening of his faith. According to Victor Harrison, Paton was alienated from his father's religious beliefs by the last year of his school career, and at this time he often said that it was not religion which offered the best way of improving the human condition, but science. This view was not hard to understand in the context of the time. The years immediately following the Great War were, in many ways, a great age of scientific achievement, or, more exactly, a period when a great many scientific achievements obtruded themselves rapidly on the popular consciousness.

Chief among these achievements was the triumph of the internal combustion engine. While Paton was at school, the number of cars running through the wide but dusty streets of Pietermaritzburg could have been counted on the fingers of one hand, or at most two. Only one master at Maritzburg College, for instance, owned a car during the years Paton spent there. After the war, though, many masters had cars, and for those who could not afford one, motorcycles became commonplace. The roads improved too: in the early 1920s the first streets in Pietermaritzburg were tarred, a transformation which the entire population of the town came to watch in their spare time, and soon the glutinous red summer mud of the streets, churned by many hoofs, and the thick winter dust of Paton's childhood, were becoming memories.

Other inventions became commonplace too, though they altered everyday life less radically. The first aeroplanes were seen over Pietermaritzburg; radio ceased to be a mysterious means of communication largely confined to shipping; and the cinema, previously a marvel only heard of, or perhaps glimpsed on holiday trips to Durban, became a reality when The Rinko opened its doors just before the war, followed by the Excelsior Bio.[6] The very names tell of the naïve excitement generated by this new entertainment. The twentieth century had arrived: the age of the machine, the age of applied science.

Paton was tremendously impressed by it. When, in 1920, Victor Harrison asked him to address a group of young Methodists in Pietermaritzburg, he preached to them an extraordinary sermon in which he affirmed that science offered the best hope for mankind.[7] Quite where God fitted into the picture was not clear to his hearers, nor, probably, to himself. Science, then, would be his course of study, and in February 1919 he began reading for a B.Sc. degree at the Natal University College.

It was not at this time a particularly august institution. In many ways it resembled a high school rather than a university, and it was in fact an offshoot of Paton's school, Maritzburg College, in whose grounds it had opened its doors in a two-roomed corrugated iron shack as recently as February 1910, some of the masters having been elevated to the dignity of lecturers. Its development from these most modest beginnings, however, was rapid. In a few years it had moved to a spacious campus in the Pietermaritzburg suburb of Scottsville, where by the time Paton began attending it, the College was comfortably housed in a single handsome building, a domed block with a high central hall surrounded by lecture rooms and offices for the university staff. The youthful and eager Paton, just turned 16 and very proud of his first pair of long trousers and a pair of hideous socks with bright green clocks (the first of his bursary payments had arrived), was older than his university.

There were only 115 students in the University College in 1919, so that it was a considerably smaller institution than Maritzburg College.[8] The result was that it was quite possible for even a new student to get to know all his fellows, and for one remarkable character to influence a great many others. And despite the shortness of its history, Natal University College had already had some remarkable students, among them the poet Roy Campbell, who in a few years would be famous for writing *The Flaming Terrapin*, and who during 1918 had sat in lectures making detailed, upside-down drawings of buck and birds and not listening to a word the lecturer said. Campbell had sailed for Oxford at the end of the war, but had left behind a reputation for such eccentricities as staging a public face-slapping competition with a half-crazy student named Van Rooyen.[9]

In 1919 the University College had two quite different sets of students in the first year. There were those who, like Paton, were straight from school, boyish, fresh, and eager to learn about life; and there were the returned servicemen, some of them five or six years older than the ex-schoolboys, and many of them only gradually recovering from their hideous experiences in the trenches of Flanders. These two groups, predictably enough, had little to do with each other because they had so little in common. And yet it was from an older man that Paton in the next three years was to learn some of the most important lessons of his life.

Railton Dent, whom Paton and his other friends always called Joe, had been born in 1897 and was therefore six and a half years older than Paton,[10] but his maturity was such that he seemed a full generation older. He provided Paton with a father-figure by comparison with which James Paton was largely nullified, for Dent lived by a larger and much more generous ethical system than James Paton's tyrannical network of petty rules. Dent had not been on active service, but had been the headmaster of an African school, Edendale High School, during the war years, and had now come to the University College, on a bursary like Paton's, to take his degree.

He was a man of very strong character, and held pronounced religious and ethical views which he managed to convey to those around him without appearing to preach. He was big, well-built, and as good-looking as any film-star. Not surprisingly, he was very popular with women, but men felt his magnetism too. Paton very quickly came to hero-worship him; he actually used the word 'adoration' to express his feelings for Dent.[11]

Even towards the end of his life, when Dent's friendship with him had cooled somewhat, Paton affirmed that of all those who had influenced his life, Dent was far and away the most important.

> I could see no fault in him. He was I think the most upright person I ever was to know, and his influence on me was profound. He did not make me into a good man, that would have been too much. But he taught me one thing . . . that life must be used in the service of a cause greater than oneself. This can be done by a Christian for two reasons: one is obedience to his Lord, the other is purely pragmatic, namely that one is going to miss the meaning of life if one doesn't.
>
> How Railton Dent *taught* me this, I don't quite know. I suppose that my reverence and affection for him was so great that I caught it from him. And I must have caught it thoroughly, because in the course of a life which I have not considered conspicuously good, I have never given up *trying to be obedient* . . .[12]

Dent was a man of the type that has rather sneeringly been labelled 'muscular Christian'; he exemplified to Paton a group of virtues whose very meaning seems as vague and questionable towards the end of the twentieth century as it was clear and admirable at its start: manliness, decency, honour. And he exemplified something else, whose lasting value is clearer: he lived by a conviction that each human life has a purpose, directed towards helping others, and the primary duty of each person is to seek out that purpose and fulfil it.

He taught Paton a further important truth too: that the communication of moral purpose is best not done through words. This was a lesson Paton had already begun to learn for himself. He had sat through enough sermons, and delivered them too, in his father's bedroom, to know how slight is their worth as a means of changing human lives. Dent taught him that man does not live by words alone, and that religion that has no immediate practical outcome tends rapidly towards religiosity. Belief in the message of Christ should result, not just in the moral betterment of oneself, but in the practical improvement of the lot of one's fellows. From a belief in this social gospel Paton was never to swerve, though towards the end of his life he came to question extreme interpretations of it.

Although they were in the same year, Paton's main contact with Dent came through their joint membership of the Student Christian Association, which had been started at Pietermaritzburg in 1918 by a student named Eric Pennington, together with George Gale. Paton had had to ask his father's permission to join this organization because of the Christadelphian disapproval of close contact with non-believers, but his father had agreed, in part because several SCA members were among the students who came along to sing at the Patons' musical evenings. And with that agreement the Christadelphians, and ultimately his own family, lost Alan Paton.

Student Christian Associations, where they still exist on campuses, are at many western universities reduced to small, more or less embattled groups. But at Natal University College immediately after the Great War, there were four chief centres of student intellectual life. They were the Students Representative Council, which was in effect a students' union; the *Natal University College Magazine*, an elaborate publication for which every student with literary pretensions tried to write; the debating society; and the Student Christian Association or SCA. The SCA in particular functioned as a centre of social as well as religious activity, and the members were not all quiet, sober, and intellectual. 'Even the rugby players belonged to it,' a woman student of this period remarked in some wonder.[13] Almost all of Paton's university friends were active members, and several of them remained his friends for life. Paton was brought into close contact with them largely because they were all under the influence of Railton Dent. Given the tiny size and relative obscurity of the College, they were a remarkable group of young people.

There was Douglas Aitken, one of those who sang around Paton's mother's piano, and who was in due course to qualify as a doctor and spend his life as a medical missionary in the Transvaal. He was older than Paton, who always had the feeling that Aitken could see his deepest thoughts, and who was wary of him for that reason.

There were the two founders of the SCA in Pietermaritzburg, Pennington and Gale. With Pennington Paton did not become particularly close, but George Gale was another matter. He was another medical student, son of a Methodist missionary in Zululand. George Gale was a brilliant man who worked as a medical missionary himself, founding the Gordon Memorial Mission at Tugela Ferry, before going on to become Secretary for Public Health in Smuts's wartime cabinet.[14]

Aitken, Gale, and Dent began a vitally important change in Paton, a change originating in religious belief, but running gradually and ineluctably into political action: the long process of awakening him to the plight of South Africa's black population. He himself, in later life, would remark with some wonder on the extreme slowness of his coming to consciousness in this regard, but in truth it was a very rare white South African at this time who had any recognition that a large part of the population was suffering terrible and increasing injustice.

It was George Orwell who was to remark, in the 1930s, that black people were invisible, and the truth of this strange observation can be demonstrated everywhere in South African literature up to this period. Although blacks outnumbered whites four or five to one, and the disproportion would grow steadily through the rest of the century, they figure as little in the writing of white South Africans as the working class does in the novels of Jane Austen. White South Africans were intent on building a European country in Africa, and even to people of goodwill and optimism it was hard to see how blacks were going to fit into the picture. Their still-widespread adherence to tribal practices, combined with their large and growing numbers, meant that it was far from clear how they could be accommodated in the developed and prosperous vision of the future, yet excluding them was plainly both immoral and dangerous. As a result they were a problem that would not go away and that could not be solved, an embarrassment from which it was simpler to turn one's eyes and one's thoughts. Whites not of good will towards Africans both despised and feared them.

Although Africans were not despised in the home in which Paton had been brought up, there was no thought of working for their upliftment either. The little black boy who brought Paton his mother's gift of food at school is a case in point: no one in the Paton home would have considered that he should have been at school himself, much less have taken steps to achieve that end. The two black servants whom the Patons employed once they grew able to afford them were provided with no toilet, though they lived on the premises; no one in the Paton home would have thought to provide them with one, much less share the household toilet with them.

In fairness to Paton's parents it has to be said that this was the common attitude of white South Africans at the time, and it need hardly be added that it was mirrored in white attitudes throughout Africa, and in the southern United States, Australia, and Asia. It was also the attitude of high-caste Indians towards what were then called untouchables, of Japanese towards all other Asians, of black settlers towards indigenous blacks in Liberia, and so on, for racism comes in all colours. But at university Paton found himself for the first time in touch with people who thought of blacks as human beings very like themselves, but with even greater needs, and whose existence was, not an insoluble problem, but an extensive field for helping one's neighbour. The seeds sown in Paton through contact with Dent and the others grew slowly but steadily through the next decade and more, and were to bear rich, varied, and unexpected fruit.

It was through Dent that Paton first met Africans who were neither servants nor labourers, on a visit to the African school at Edendale, where Dent had been a youthful headmaster. They were teachers, and they were certainly the first blacks Paton had met as equals, and almost certainly the first he had ever touched except by accident: 'I called them Mr and Mrs and Miss (there was no Ms in those days), and I shook hands with them. It was not a giant step for mankind, but it certainly was a big step for me.'[15] A great many white South Africans at this time went through their whole lives without taking such a step.

But in addition to these enlargements of his life, there were two other friendships made or strengthened in this eventful year, 1919, and kept throughout life. There was Cyril Armitage ('Army' to his friends), who had been at Maritzburg College with Paton and who would become a headmaster of a high school at Port Shepstone on the Natal coast, and then of the Parkview School in Durban, an unassuming young man with a strong sense of high moral purpose.[16] And, importantly, there was Reginald Pearse, who was to become his closest friend and confidant at University College. Pearse was the son of a lawyer in Ladysmith, and he came up to the University College intending to read law and go into his father's business.[17] He changed course at university and read for a BA instead, on an Education Department bursary like Paton's and Armitage's. Pearse, like Armitage, was to become a headmaster, and in his spare time to become the great explorer and mapper of the Drakensberg, and write the classic book on that superb mountain range, *Barrier of Spears*.

Another influential friend of Paton's at Natal University College was Neville Nuttall, whose parents lived in Durban, and who was deeply religious; in time he was to become an Anglican priest, the only one of Paton's friends to go into the formal ministry. However, Paton met him only in 1921, for he was two years behind Paton in his university career.

Pearse was one year ahead of Paton at NUC, but they shared a love of literature which drew them together, with Armitage often making a third member of their group during 1919 and 1920. They shared an interest in horseplay, practical jokes of a schoolboyish kind, and playing with words. They used to compete at producing ludicrous puns, an activity for which Armitage had something approaching a genius. On one occasion he boasted that he could produce a pun instantly on any word they cared to suggest. 'Potato', said Paton after a moment's thought, in the hope of baffling him. And without missing a beat, Armitage said, 'I woke up potato clock this morning.'[18]

Paton himself loved puns, an activity alluded to in a verse composed chiefly by Armitage in the July holidays of 1919. Entitled 'Varsity Yzz [no doubt pronounced "wise"] of 1919', it went through the alphabet naming some of the students of that year ('D demands Dent, among those who daren't stew / At debates he is Sec. & our Treasurer too'). The couplet for P ran, 'P pierce[s] Paton on a pitiful plight, / For he perpetrates puns all the day & the night.'[19]

But Paton's powers of language did not just show themselves in jokes and rather feeble juvenile poems. He could also, on occasion, show a sureness of touch, and a delicacy of feeling, which struck his friends as setting him apart. Sometimes it showed itself in unexpected ways, as when his friend Victor Harrison's mother died in 1923. Harrison never forgot that of all his university friends, Paton was the only one who wrote to him. Nor did he ever forget the simplicity and brevity of Paton's letter.

> Dear Vic,
> I'm terribly sorry.
> Alan.

'That was all he wrote', Harrison was to recall in old age, adding, 'and it was enough.'[20]

Paton took a little time to settle into his university work, as at first he and his friends tended to regard the university as just another school, and an arena for schoolboy pranks. Paton told of one of these, perpetrated by Armitage on a mathematics lecturer, when Armitage entered the lecture room, saluted the lecturer, slipped out of a window at the back, re-entered by the door and saluted again, and then repeated the procedure a third time. He would no doubt have gone on indefinitely had the lecturer not smashed a box of chalks in a sudden rage.[21]

As usual, once he settled down to working steadily, Paton found the task pleasant, and though he did not entirely enjoy mathematics he had no difficulty in coping with it. His one setback was to come in 1921, when he failed physics, the first (and last) time he failed an examination. The shock was profound: to the end of his life he had nightmares two or three times a year about this experience.[22] But he picked himself up, and by hard work not only passed physics the following year, but achieved a distinction in it. Increasingly confident in his studies, he began devoting his time to the many other activities that make up student life, and he thoroughly enjoyed his four years at the College. The variety of his activities was extraordinary. Towards the end of his university career he was to write, 'I have gone the whole hog, acting, joking, magazining, processioning, circling,[23] tennising, rugbying, running, & I feel jolly tired.'[24] But he had left a good deal out of this list.

One of his favourite pastimes was reading poetry, and he and his friends urged each other on to wider and wider reading. Like all their generation, they had been brought up chiefly on nineteenth-century English poets. When Paton looked back on his reading, it was on the work of Tennyson or Browning that he dwelt; when he finished school he would not have known any poet more recent than Kipling. Of Yeats's writing he knew only the earliest volumes. He knew virtually no American literature, and certainly he had never heard of Eliot or Pound, though *Prufrock and Other Poems* had been published in 1917, and Pound had published five small volumes of verse by this date. Not until 1922 are there references to Eliot in his writing.[25]

Now, with Pearse and Armitage, Paton began to discover the Georgian poets through the annual volumes of Georgian verse, as well as Rupert Brooke and the war poets, Owen, Sassoon, and Julian Grenfell. The Georgians he disliked, thinking them 'not worth the money',[26] while Brooke he considered a great poet, though he was to change his mind later. Masefield he considered 'above everything in modern verse'.[27] Each of his friends tried to contribute to the education of the others, drawing on what he was studying himself, but it was particularly with Nuttall that he read steadily and widely.

His father had perforce relaxed the rule on not going out unsupervised once he went to university, and at night he, Pearse, and Armitage, in characteristic student fashion, would go to the town cemetery in Commercial Road, where there is a curious little arched building set among the older graves. In this structure they would read poetry by candlelight to each other, at first mainly their new discoveries of poets one thought the others should read too, but increasingly their own verse. The tone and content of these discussions, though not their slightly macabre setting, can be reconstructed from Paton's letters to Pearse, all of which Pearse preserved. They are an amusing bundle, addressed variously to 'Dear Reg', 'My darling Reginald', and 'Idiot', and they confirm the impression that Paton was still very much a boy even at the end of his teens. The first of them is dated 3 January 1922:

The fact is this, young man; literature today tolerates Shakespeare & Milton, because, apart from all form, they were great. But it tolerates no other, in fact, recognises no other poets, until the Romanticists; even today, Reg, the trend is still further to sheer beauty than even Keats & Shelley went. That is why Dryden & Pope, except for 'Alexander's Feast', are fit for naught; the only reason that 'Alexander' lives is probably because it was the only time in Dryden's life he yielded to impulse.

I have studied modern poetry fairly carefully, & I find that the trend is more & more to write things which are purely beautiful, not things sensible or deep. I think that that verse of De la Mare's, which I consider the most beautiful thing since Tennyson, is typical.

'Who knocks!'—I, who was beautiful,
Beyond all dreams to restore,
I from the roots of the dark thorn am hither,
And knock on the door.

And Paton then goes on to quote a four-stanza poem of his own, of which the first stanza will convey the flavour:

Ah! why do you come in the hours of dark,
And hold out your arms at the window there,
And your face is so sad, & so wondrous fair,
And your voice speaks so low, & so pleadingly,
The sacred words that were all to me,
In the days that are long passed by.

It is plain that his view of 'modern' poetry was dominated at this time by de la Mare, by Yeats's world-weary early verse, and perhaps by Ernest Dowson. For the work of the Georgians and the early Modernists, he had little respect, as he showed by a parody following his quoted poem, of which he remarks:

Pardoning my cockiness, I consider that effusion typical of the modern movement, & an *extremely* conservative type at that. To-day produces

'And the roar
'Of the moth-eaten sea
'That hungers her faint passion to the wind,
'And—shloomp!
'180 kilometers.

Do you agree?

And he then goes on to talk about Browning and Tennyson as poets worthy of a modern critic's admiration. Before one writes him off as an ignoramus stranded in a colonial backwater where the work of the Modernists had scarcely been heard of, it is worth remarking that very few students at Oxford or Cambridge would have heard of the name of T. S. Eliot either before he published *The Waste Land* in 1922.

Through discussions like this one, in the Pietermaritzburg cemetery, Paton was encouraged to consider poetry as a serious activity, and he, Pearse, and Armitage composed many doggerel verses together. Since the University College was in Scottsville, they called themselves The Three Scots Villains. Then, working presumably on the principle that if a thing is worth doing it is worth overdoing, they added three further *noms de plume*: Xerxes D. O'Shantonville (Paton), A. Orthagoras Pharnabozas (Pearse), and Callisthenes J. Exercitus (Armitage). A little of their verse goes a long way. This is the opening of 'Euclid's Bed', written on 1 November 1919, its target a mathematics lecturer recently arrived from England, who had had difficulties finding lodgings:

A Practical, Praiseworthy Proposal for
Procuring Precise & Permanent
Premises in this Present
Precarious Period

'Tis the rhyme of old Euclid we're singing this day
Just stepped from the puff-puff now snorting away
He has come from far England from Dover's fair strand
An dover [sic] the sea to this heathenish land.
Hotels he has rung up the only result
He has run them down too—with intent to insult
For the answer's the same from each place of renown
'Hotels all full up & no houses in town!'
'By fair means or fowl' then this Euclid quoth he
'A mansion I'll find or a chicken I be!'[28]

And so on, for more than sixty lines of callow humour and outrageous puns. Perhaps a dozen of these very early productions survive, each worse than the last, but clearly a source of joy and pride to at least one of those who produced them, for Armitage preserved them to the end of his life.[29] What is more, The Three Scots Villains submitted several of their verses to the local paper, the *Natal Witness*, and had the thrill of seeing them published in South Africa's oldest newspaper on four occasions at the end of 1919 and early in 1920.[30]

Perhaps spurred on by the unexpected (and, it has to be said, undeserved) publication of these joint productions, Paton began writing poetry on his own, and submitted it to the student publication *Natal University College Magazine*, which at this time appeared twice a year. In the October 1919 issue there appeared a poem signed 'A. P.' which, though Paton had no memory of it in later years, was almost certainly his first publication. It is a translation of a poem printed immediately before it on the page, a rather poor Latin verse by a fellow student, L. Sormany. Paton's title, 'Idem Anglice Redditum', meaning 'The same rendered in English', makes this plain. The first verse runs,

Fair Natalia's land we hymn,
Where Phoebus holds his sway benign,
Our Alma Mater's praise declare,
And venerate the Morning Star.[31]
    Gracious Mother, live for aye,
    Thou that dost dower us bounteously!
    For ever chanting with great glee
    Vivat, vivat NUC!

Encouraged by the publication of this cliché-riddled effort, he published an original poem in the *NUC Magazine* in November 1920, and in later years rather reluctantly (and mistakenly) acknowledged this as his first appearance in print. At the time he signed it 'Ubi'.

To a Picture

He gazes on me with his long-dead eyes,
And dumbly strives to tell me how he died,
And shows the hilt-stabbed dagger in his side;
I see mad terror there; the murd'rous cries
Draw near—more near—half-tottering he tries
To reach the door—one step!—'unbar, 'tis I'.
But none unbar—I hear the broken cry,
I see the mirrored anguish in his eyes.
So conjure I the tale; the faded print
Hangs on the bedroom wall, and there I see

Those wild eyes ever gazing on my bed.
They lead me to strange wondering; what hint,
What sign, what tragic muteness will there be
In mine own eyes, when they do find me dead?

It was not just the influence of his University friends that spurred Paton into trying to write poetry. His father, as we noticed, wrote and published verse, and was particularly proud of his appearances in the *Natal Eisteddfod Book* each year. On several occasions James Paton's poems were printed first in the book, meaning that they had been judged best in their class, and this was a matter of great pride to him. Alan Paton's literary activity, like so much else in his life, was sparked off in part by his father's influence.

Paton and his father met head to head, in a sense, in the Eisteddfod competition towards the end of 1921, when James Paton, as usual, entered several poems, one in each of the four set divisions: bardic poems, lyrics, sonnets, and ballads. Alan Paton entered a lyric, without telling his father. On the morning the published volume was delivered, early in December 1921, James Paton opened it eagerly at breakfast, and the family watched as he scanned the contents in silence. His bardic poem had been rejected and so had his lyric. But among the lyrics was a poem entitled 'You and I'. The opening runs:

You and I—
Ah! what care I?
Let the whole world die,
Let the leaves fade,
And all that's made,
Wither and die.
Let the birds not fly;
Let the sky be blue,
Or the sky be grey,
Or any such hue
Adorn the sky.
For what care I,
When I have you?

And at the foot was the name of the poet: Alan S. Paton.

James Paton's family was watching him covertly, but he put the book aside and said nothing, either then or later, to his son. It was another of the rejections, the sins of omission, that so alienated his children, almost more than the beatings and the bullying. When they came to look at the volume themselves they realized why he had been silent, and exulted. What James

Paton felt can only be guessed at, but his daughter Dorrie had no doubt. 'My father was nearly sick,' she said years later. 'His own poem wasn't there at all.' And she considered that her brother had triumphed over the older man, a psychologically crucial victory in which she shared with secret rejoicing. She kept a copy of 'You and I' for many years, and when they were both elderly, was surprised to find that Alan had forgotten it completely.[32]

It was this kind of success in achieving early publication, though, that convinced him he had the capacity to become a writer. He subsequently produced two brief dramatic pieces, 'His Excellency the Governor'[33] and 'Minutes of any Meeting of the—Society',[34] which, though light-hearted, showed he had a gift for natural-sounding dialogue, a gift which he was to deploy to very striking effect in his novels.

He was spreading his literary wings, and towards the end of 1922 he wrote a substantial part of his first attempt at a novel, which he seems to have intended to call *Ship of Truth*, and which he abandoned, incomplete, in December the following year.[35] It is a Forsyte-like saga of a group of rural families, and the skill of the young author is very impressive in dealing with such a large cast. In part the novel is the rather shapeless and fragmented story of Michael, an eager, active boy, very like Paton, and his love for a proud, unattainable girl, Dorothy. But there are also vivid sections concerning the Sotheran family, who run a brothel in the countryside, full of cursing and violence, and a most perceptive (and, as it was to turn out, prophetic) sketch of an unhappily married man, Robert Jenkinson, who falls in love with another woman and struggles against sexual temptation. And there are interesting reflections on the condition of the Africans, and what can be done to improve their lives. The young author's answer seemed to be that they should be converted to Christianity, and that this seminal change would be followed by knowledge of better farming methods, a view he puts when Michael and a friend are looking at a mission station:

> 'After all, you & I are Christians, it means something to us. And if it does mean something, & if it helps us, then we should help these poor devils.'
>
> 'But some people say they're happier as they are.'
>
> 'Wait till you see what they do here. It's not all school and church, like some people think. . . . Look at the huts round here, for example. They're tidy pieces of work. These people have learnt to plough deep, & sow properly, & they've planted trees & make their own clothes, & all the rest of it. Wait till you meet some of these old native ministers. Really good fellows, they are.'[36]

In passages like this one can detect the germ of *Cry, the Beloved Country*, with its view that what blacks needed most urgently, after the gospel, was practical help. And in some of the passages of the novel he expresses views on the advancement of blacks which a Nationalist government fifty years later would have rejected as dangerous idealism:

> Education is the helping of the native to share your culture & your tradition & to understand & venerate your laws; leave him alone, with his blankets & his sunshine, & he becomes a drag on your advancement & a menace to your prosperity. I'm as sure of that as I am of anything . . .[37]

By the end of the manuscript young Michael has decided to become a missionary and devote his life to African advancement. These were Railton Dent's ideals, not Paton's own views, but he puts them across convincingly enough.

He even wrote a brief burlesque oratorio, and actually got together a choir of his fellow-students to perform it in 1922, though the attempt broke up in howls of laughter during the first rehearsal. According to Dorothy Durose, who was in the choir, even Paton, conducting, thickset and determined, from a podium, could not keep a straight face.[38] The oratorio has been lost, which is probably as well for his reputation. All the same, the young writer was developing rapidly. But he was about to spread his wings in other fields too. The world seemed all before him: the only question was which way to go.

4

DISCOVERING THE WORLD 1922–1924

I can go anywhere
I seem to have been given the freedom
of this place what am I then?

TED HUGHES

Perhaps prompted by his brother's physical prowess, Paton had begun trying to develop his body. His shortness was to distress him until he was well into his forties,[1] but he could not, by taking thought, add a cubit to his stature. However, he could add inches to his biceps, and he now adopted the habit of starting each morning with what he called 'physical jerks'. Hanging from a branch of a jacaranda in the Bulwer Street garden, he would do a series of chin-ups, followed by gymnastic twirls and swings before dismounting with a spectacular leap, a performance much admired by his sisters.[2] He also began running, several nights a week, at first alone, presently in company with Railton Dent, and they would run 2 or 3 miles a night. He was soon sporting impressive muscles, and it became clear to his sisters that other girls than themselves had noticed and admired him. He seemed quite unaware of this interest, though he longed for the attention of the opposite sex, and even when two young women waylaid him on the way to a sports meeting and asked him a series of naïve questions in the hope of starting a conversation, he merely brushed them off without apparently becoming aware of their intentions, much to his sisters' amusement.[3]

He joined the University College rugby, cricket, and tennis clubs, but never rose into the first team in any of these sports. Leif Egeland, the tall and muscular student who was captain of the tennis team and a fellow-member of the SCA, remembers him as playing with dedication, 'but he never got anywhere near the top'.[4] He played doggedly rather than skilfully, but with such evident enjoyment that according to Harrison everyone liked playing with him.[5] It is not really surprising that he was not much good at tennis; it was one of the games his father considered might lead his children into socializing and sin, and he forbade them ever to play it. They had to wait until he was out on one of his long walks in the hills before they could get their old racquets out of hiding and play on one of the local courts, and his daughter never forgot the troubled pleasure of this illicit activity.[6]

But his father's grip on Paton was now weakening rapidly, and he was even allowed to join the Dramatic Society, although James Paton regarded any theatre as tantamount to a house of ill repute. Perhaps he felt that, having given way over the pixie performance at Havelock Road Boys' School so many years before, it was too late to put the brakes on now. At any rate Paton took a minor part in a play in 1920, and did so well that in September 1922 he played the title role in a popular play, *His Excellency the Governor*, by a now-forgotten Victorian playwright, Frank A. Marshall. The *University Magazine* that term gave him brief but honourable mention, saying that the best actor, a student named Ingle, was 'ably supported by Mr Paton, as the Governor'.[7] The play was such a success with its student audience that the cast took it to Durban, where their performance was very well received.[8]

Some of the most lively dialogue of the play involves the attempted seduction of the Governor by a beautiful and flirtatious girl, Stella de Gey, whom he eventually resists with an effort. Stella, in the NUC production, was played by Dorothy Turner, an extremely pretty girl who had known Paton for some years. 'Special mention', said the *University Magazine*, 'must be made of Miss Turner, who, as the flirtatious Stella, made the most of her part.'[9] Dorothy Turner had reason to put her heart into the stage seduction, for she was a keen admirer of Paton's at this time.[10]

Indeed by 1922 there were several young women who thought him a man worth attention. His sister Dorrie was brought to see the rehearsals, and remembered that 'crowds of admiring females always undertook to look after me while he was on the stage.'[11] Enboldened by his triumph in *His Excellency the Governor*, he went on to take the lead in the Dramatic Society's main production in 1923, *If I Were King*, and he seems to have been a great success, certainly with the female members of the audience.

By 1922 he had outgrown the socks with clocks and acquired a smart cream-coloured suit and a collection of bright ties, and was picked out in the College Magazine as one of the better-dressed men on campus. On the stage, in a white dinner-jacket and bow-tie, his hair plastered down in the fashion of the time, and his gaze as intense as a hawk's, he seems to have made several female hearts skip a beat. Dorothy Turner was attracted to him by his ambition and intensity as much as anything, and by the fact that he was 'down-to-earth and no-nonsense, which suited me'.[12]

She had begun paying attention to him a year or two before, and they would sometimes walk from the tram stop to university together. One morning he said in his abrupt, intense fashion, 'One day I am going to do something worthwhile'. Dorothy Turner looked at him quizzically and asked what he meant, but he turned the subject.[13] She, however, never forgot the remark, and in later years, when his name was known the world over, she wrote to remind him of it. 'Did I really say that?' he replied.[14]

In her attempts to get closer to him, however, she failed in life as Stella de Gey failed in the play. This was in spite of the fact that Paton was not unaware of her interest, and found her attractive in turn. He described her in his autobiography as 'a vivacious young lady named Dorothy who had blue eyes, and a tipped-up nose, a beautiful figure, and I should think a well-to-do father, if one judged by the way she dressed.'[15] On one occasion he escorted her home in the dark, but to her great disappointment he made no move; he did not even take her arm. She subsequently told a friend that he was 'still a child'.[16] Paton rather agreed with this judgement, telling Pearse in 1923, with winning candour, that Dent had been urging him to get engaged, but adding, 'Take it from me, Reg, that A. S. P. realises he is still a mere babe.'[17]

The fact seems to have been that the contrary impulses of his training and his instincts combined to paralyse Paton at this time. He continued to be painfully shy of women; they were for him wonderful and unattainable objects of worship. The love poem he quoted in the letter to Pearse makes it plain that for him they were creatures of dream, not of flesh and blood like himself. As late as 1923 he produced another poem, 'Gemellia', showing the same attitude:

> Once the long dark hours of sleeping
> I woke, and the dawn-wind spoke to me,
> And told me there was a woman weeping
> Down in the pines by the sea.
>
> So I went to the pines in the dawn-wind's blowing
> Coldly and keenly over the sea,
> Pale in the East, and long hair flowing,
> Gemellia passed by me.

And I, I turned and followed after
With throbbing heart and soul in me,
But there was naught but mocking laughter
Over the misty sea.

And now each morn in the dawn-wind's blowing
Cold and keen in the pines by the sea,
Pale in the East, and long hair flowing,
Gemellia passes by me.

And now each morn I follow after
With that same mad heart and soul in me,
But there is never aught but that mocking laughter
Over the misty sea.[18]

His father's influence and the puritanism of the Christadelphians had something to do with this sense of the inviolability of women, no doubt; but according to Pearse the general atmosphere at the University College was remarkably chaste and innocent. Pearse himself first kissed a woman (other than female relatives) when he was 26, and then only after he had become engaged to her.[19]

But Paton was gradually throwing off the teachings of the Christadelphians too, partly under the influence of Dent and the other members of the SCA. Dent was a Methodist, Nuttall an Anglican, and Paton began to realize that religious faith does not have to be a matter of prohibition and self-denial all the time. One by one he abandoned the lesser moralities of his father, beginning with smoking, and then moving on to drink, swearing, and sex. Nuttall smoked Westminster Magnums, and Paton, impressed by the very sound of these, took them up too. When he saw that Dent disapproved he did not smoke in front of him again,[20] but he went on smoking. For a time he affected a pipe, but then went back to cigarettes until the end of his life, smoking five or six a day.

Learning to drink took a little longer, for his father's taboos on alcohol were particularly strong. There was no alcohol of any description in the Paton home, not even whisky, despite the Scottish heritage. On visits to the house of a judge or some other superior in the Supreme Court, James Paton might drink a glass of wine, but he would never have allowed his children to do so. As a result Alan Paton grew up with a strong sense of the dangers of drink: who knew what internal restraints might fall if a glass of alcohol passed one's lips?

During 1921 he was elected Secretary of the Students Representative Council, and helped to organize a student ball at the end of the year. The question arose as to whether alcohol should be permitted at the ball, and Paton made a passionate speech against it, winning the day. Probably a majority of the other students were as ignorantly teetotal as himself. But his speech angered those few students for whom alcohol was not a demon but an everyday drink.

One of them was Douglas Saunders, the son of one of Natal's rich sugar farmers, and heir to the Tongaat sugar fortune. He was the only student to own a motor-car, and was known to his friends as 'the Duke'.[21] It was whispered among his fellows, as a matter both of envy and scandal, that the Saunders family had alcohol in their house 'at all hours'.[22] Saunders started an argument with Paton after the meeting over the alcohol issue, and before Victor Harrison, who was present, quite knew how it had happened, Saunders had pinned Paton in a corner and started a fist-fight.[23] Paton's many violent encounters with his brother Atholl now stood him in good stead; he fought Saunders off and challenged him to have it out properly outside. An excited crowd accompanied them to a secluded spot behind the Physics laboratory, and formed a ring round them. The future author and future millionaire then slugged it out with bare fists, and the result, as Leif Egeland later recalled, was a draw and gentlemanly handshakes. But Paton had won the larger battle: there was no alcohol at the students' ball that year.[24]

Saunders would have been filled with righteous indignation if he could have seen Paton a few months later, during the mid-year break in 1922. Neville Nuttall had by this time arrived at the University College, and been befriended by Paton; and in the holiday, Nuttall's parents being away from home, he invited Paton to stay with him in the family's house in Durban for some days. Nuttall's family was as teetotal as Paton's; neither of them had ever touched alcohol, and they agreed to experiment with it. They bought a bottle of port, and that evening divided it between them and drank the lot. They then, with some trepidation and a good deal of excitement, awaited the results. 'The experiment was a failure', wrote Paton in disappointment. 'We did not get drunk at all.'[25] He may have been wrong in thinking himself unaffected, though, for by way of showing their feelings the two went out into the garden and bellowed all the swear words they could think of into the velvety Natal night.[26] After this minor rite of passage Paton remained teetotal again until his wedding in 1928.

But this was not his only voyage of discovery while at university. His was a passionate nature, and his shyness of girls must have been a torment to him, for it kept him from making the amorous advances every teenager longs to make. What was he to do about the raging fever in his blood? 'Nice' girls, like his sisters, were unapproachable and certainly untouchable; his father's teaching had been beaten well into him. But he was well aware, from an early age, that there were other girls, not 'nice', and easily approachable. For Pietermaritzburg, despite its small population, had a brothel, and it was in Paton's own street, Bulwer Street, and within a very short walk from his house. He must often have watched the place covertly, for his description of it is not lacking in detail:

> The brothel was indeed a strange one; it was in the house owned by an ineffectual man, Mr B., who had a job at the market, a large blousy wife, two lusty daughters, and a third daughter of about twelve who was clearly being trained for the business. I have no doubt that the mother had been in the trade herself. There was a steady stream of men callers to the house, and it was the scandal of the neighbourhood. . . . It was the second daughter of Mr B. who initiated one of our neighbour's sons into the ways of the world. He was a friend of the family and a College boy, and my parents would have been astonished had they known of it. He told me all about it, with a strange mixed look of pride and shame. He had been afraid to go to the house itself, and had met the girl on the banks of the Umsindusi. I think it cost him a pound.[27]

As a schoolboy, Paton had neither nerve nor inclination to follow his schoolmate's example. But his student days were a different matter. Once he began breaking away, no matter how tentatively, from his father's teachings on swearing and drinking, it was probably only a matter of time before he experimented with sex.

The brothel, as a symbol of easy but forbidden sex, had an undoubted attraction for him, and he showed it in the many references to it in his poems and prose pieces over the next decades; but he was too virtuous and (no less important) too fastidious to enter one. He could have followed his schoolmate's example and got a girl to meet him out in the countryside, but he chose a different course.

Natal University College, as we have seen, had only 115 students in 1919, Paton's first year there, and only a minority of these were women. The great majority of these young women were as puritanical as Paton and his friends, but there were one or two who flew brave little flags of nonconformity. Not surprisingly, these more adventurous spirits were both popular and despised, and to be seen with them caused raised eyebrows and knowing winks. Paton seems to have taken care to avoid them. In a town as small as Pietermaritzburg, even to be seen entering a house in which one of these women lived would cause speculation. This helps to explain why one of Paton's closest friends should remember, all his life, an encounter late at night:

During my period at the University I had an uncle who was running a small private school . . . in Longmarket Street. I used to spend every Saturday evening with him, playing chess. On my way home at about midnight (I was boarding with an aunt in Berg Street at the time) I had to pass a boarding house in Pietermaritz Street [Walmsley House]. At this boarding house one of the girls at the University was boarding. She had the reputation of being rather free with her favours. One evening, round about eleven o'clock, I was emerging from a side street linking Church Street with Pietermaritz Street when I saw Alan coming out of the front door. He looked furtively first up and then down Pietermaritz Street, but not across to the side street from which I was emerging, so he did not see me. Then he turned and headed for home. My first reaction was not so much shock as surprise. I knew, of course, that such things did happen, but I did not think it of Alan. I am quite sure it was Alan. Over the front door of the house was a bright light, and I could see him clearly. The house where this occurred was towards the upper end of Pietermaritz Street, a long way away from Alan's home at the lower end of Bulwer Street, and under normal circumstances he would never have been so far from home at that hour.[28]

It is of course possible that Paton had been visiting someone else at the boarding house, or that there was some other innocent explanation for his being there. But the boarding house, for women only, had a rule that men were never to cross the threshold,[29] and the friend, who knew him as well as anyone could at that time, was convinced at the time of his guilt and remained so in old age. He never mentioned this episode to Paton, nor to any other of their friends: 'We just never talked about such things,' he said years later.[30]

Sex was in fact a forbidden subject, and remained so to Paton in spite of this episode. He unquestionably would have felt deeply ashamed if his friend had remonstrated with him, or, worse, joked with him about it. 'I have on occasions been possessed by lust,' he was to write many years later, 'and I did not like it and I felt polluted by it. When I was a boy I felt polluted because I had committed and wanted to commit a sexual sin . . .'[31] Sex, like the bottle of port, seems to have been a disappointment to him. But its fascination did not slacken.

Both the Dramatic Society, as we have seen, and the SCA offered opportunities for meeting young women with propriety. Among the most active members of the SCA were at least two women, May Graham-Gerrie, who was to marry Armitage, and Eva Grundy, who later became headmistress of the Pietermaritzburg Girls' High School. Another organization which drew men and women together was the Debating Society, in which Paton took an active part, and where he often shone. His father's lessons on sermonizing had given him a good deal of confidence in public speaking, and much practice in preparing what he was going to say. It was in the University College's debates that he honed the skills which in later life would serve him so well as a politician.

Although he was good at thinking on his feet, he developed the habit of preparing his speeches well in advance, and speaking from notes wherever possible. His pace of delivery was rather slow and this, combined with the often unexpected content and expression of his speeches, gave weight to what he said and compelled close attention from his listeners. He was capable of most vivid turns of phrase, and of the most devastating witticisms, carefully prepared in advance, and depending on timing for their full effect, so that it was usually a pleasure to listen to him. His sense of humour was highly developed.

The Debating Society was run with strict attention to rules of procedure, and some of Paton's best jokes involved ignoring procedure in the knowledge that he would be called to order. It was not permitted, for instance, to insult an opponent, and even the mildest abuse would draw a reprimand. In 1921, the year Reg Pearse was President of the Debating Society, Paton sprang a carefully prepared mine. 'Pay no attention to my opponent,' he announced contemptuously during a debate, 'he hasn't got the brains of a grasshopper.' Pearse at once intervened, demanding a retraction. 'Certainly, Mr President,' said Paton smoothly, 'I meant to say, my opponent *has* got the brains of a grasshopper.' Memories of this joke were still vivid among his friends nearly half a century later.[32]

On occasions, for variety, the Debating Society would constitute itself a mock parliament, complete with a Speaker, who had of course to be referred to as 'Mr Speaker'. On one evening the Speaker was Douglas Aitken, the young thought-reader, and Paton in the course of his speech referred to him as 'Mr Aitken'. Aitken banged his gavel vigorously: 'Mr Aitken?' he said, 'Mr Aitken? I do not know the gentleman.' 'Well, Mr Speaker,' said Paton gravely, 'you haven't missed much.' The house roared, and the Speaker himself was incapacitated by laughter for some time. 'This kind of triumph filled me with joy,' Paton wrote later, 'and I made no attempt to hide it.'[33]

His own intellectual dominance, and that of his friends, was now very marked in the College, and they were in a position to fill most of the official positions in the student body, other than sporting ones: one or other of them headed the SCA, the Students Representative Council, the debating society, and so on. Paton's letters to Pearse make this plain. 'Am looking forward to NUC next year', he wrote at the beginning of 1922, 'I shall take some old secretaryship or other. SRC! by Jingo!'[34] That year he was elected secretary of the SRC. At the beginning of 1923 he wrote to Pearse, 'I have decided, in spite of our talks, not to take the SCA . . . it is up to Joe [Railton Dent] to take a final hold of the reins before he leaves. After that we shall see.' But there was an even more influential position at the College, the Presidency of the Students Representative Council, for which Dent thought another of his protegés, a student named Frank Bush, should stand. Paton disliked Bush, whom he thought tactless and blundering, and he wrote to Pearse, 'on this I am determined, without embroidery—that if Joe doesn't intend to rule the NUC himself, he is not going to let Bush do it. The job will naturally fall on Adolph [Adolph Bayer, one of Paton's friends], Army [Armitage], & myself.'[35] Nor was his confidence misplaced: having served as secretary of the student body in 1922, he was elected President of the Students Representative Council in 1923, and helped edit the College Magazine at the same time.

His popularity was such that in 1924, when he served a second year as President of the Students Representative Council, his fellows paid him the great compliment of collecting £50 to send him to Britain as their representative at the Imperial Conference of Students, held in London and Cambridge in July 1924. He found on his arrival that he was the only student to be specially sent in this way, other institutions making use of their graduates already in England for other reasons.[36]

This was his first trip abroad, indeed his first trip further afield than Durban. He was tremendously excited by the voyage on the SS *Beltana*, and astonished by the size and gloom of London, 'wonderful London', as he called it in a letter to Reg Pearse.[37] In England his official host was a very wealthy Liberal MP and his wife, Mr and Mrs Allen, who had a large house in Queen's Gate Gardens, South Kensington, staffed by a butler, a housekeeper, footmen, maids, and a chauffeur; there was also a large country home, a grouse-moor in Scotland, a Rolls-Royce, and similar appurtenances of wealth. Paton was overawed by this style of living, and felt very much a crude colonial, born and bred under corrugated iron. This feeling sharpened when he knocked over his morning tea-tray and smashed what seemed to him priceless china, which the Allens would not let him replace. That was just as well, for it would have ruined both his budget and his holiday.[38]

His painful feelings of inferiority continued, for the Conference delegates were prodigiously fêted:

Lunches and dinners were given by the Government, the Lord Mayor, and the Worshipful Company of Carpenters, teas at Ranelagh and on the terrace of the House of Commons. Then we adjourned to Trinity College, Cambridge, where we lived in the Great Court, but not before my exuberant spirits had again been brought low. After the Government dinner at Lancaster House I retrieved my raincoat and was putting it on as I walked out. But at the door a very imposing person stopped me and said, 'One does not leave Lancaster House improperly dressed'.[39]

So he finished putting on his coat indoors in humiliation, and then crept away. Such events might impress the colonial with the splendours and superiority of the British ruling class, but they did not make him love it.

This was his first experience not only of visiting the Imperial centre, but of other people's views of South Africa's racial policies. There were no black African representatives at the Conference, but there were Indians. These, politely but firmly, took the opportunity to make it plain to Paton and the other South Africans that they deeply resented the treatment meted out to Indians in Natal. Paton, who had probably never thought about this question any more than he had about the position of Africans in his country, fought back, though he found it hard going. As he put it in a report to his fellow students in Pietermaritzburg in October 1924,

The Indians I have met here are fine, dignified, clever, and in many cases handsome men, but it was necessary to point out to them that their compatriots in Natal are not of the same class. They agree that it is just to exclude Indians from Natal on economic ground, and also to withhold the franchise until they are fit to receive it.

But there was a further line of argument from the Indians which Paton could not resist:

But what they do complain of, and as far as I can see, rightly too, is that chances of improving themselves are denied to them; they are given no chance of higher education, and as became alarmingly apparent to the Conference, India will secede from the Empire unless her wrongs are redressed . . . After all Natal opinion against the Indian is in many cases unjust; when one sees the big issues at stake one realises that European opinion must be educated, and it is the Universities that must lead . . . it rather embarrassed me to be embraced heartily by the Indian delegates on saying goodbye. Of course they realise that we labour under tremendous difficulties, but they look to us to do something.

It is as well for him that there were no black African delegates, so that he was able to conclude his report, rather airily, 'The native [black] question also came up, but of course it is not so urgent.'[40] In just a few years it would be impossible for a South African to make such a remark, and that Paton could in 1924 raise and dismiss the 'black question' in a single sentence shows the extent to which he still remained unconscious of his country's central dilemma.

He must have been quite relieved to finish the conference and get away to see a little of the country his parents still called 'home', as indeed he did himself until after this visit. He bought a two-stroke Douglas motorcycle and rejoiced in the freedom it gave him. How and where he learned to ride it is not clear. In company with another Natalian, Herbert Maunsell, mounted on a similar motorcycle, he toured Britain, combining Paton's two chief interests in the choice of itinerary. Literature demanded visits to Stratford for Shakespeare, Tintern Abbey for Wordsworth, Ludlow for Housman, Abbotsford for Walter Scott. Piety required pilgrimages to the great cathedrals, Salisbury, Exeter, Wells, Durham, York. He spent a few days in Birmingham, and was shown over Cadbury's chocolate factory.[41] His London hosts, the Allens, had invited him to their grouse-moor in Perthshire, and he stumped about it in plus-fours, refusing to shoot.[42]

The countryside which is one of the chief glories of Britain left him curiously unimpressed, as his comments to Pearse suggest: 'the incomparable English countryside, beyond waking faint admiration, does not call to me.'[43] The truth was that there is an element of defensiveness in this refusal to be impressed, and in addition he was already longing for the beauties of Natal. 'I get mighty homesick for South Africa at times, and am mighty thankful that I am not over here for two or three years,' he reported to his fellow students.[44] When he sailed back to South Africa on the SS *Ballarat* in October 1924 he found he had learned as much about South Africa as about Britain:

> The sight of Table Mountain rising from the sea overwhelmed me. I doubt if I put my thoughts into words, but it was clear that at the age of twenty-one I had, for better for worse, for richer for poorer, given myself to this strange country, to love and to cherish till death us did part.[45]

Though his was to be a life full of overseas travel, and though he was occasionally to flirt with the thought of living in other countries, he never lost his sense of profound loyalty to South Africa. Exile was not for him.

He had come back from this first brief excursion with an expanded view of the world and of himself. He made a verbal report to his fellow students, 'which while serious in parts kept his audience in gales of laughter', according to an account one of them wrote many years later.[46] He was beginning to sense that he had great gifts, but their variety made it difficult to know in which direction he should move. Was teaching really to be the life for him, or should he consider going into politics in the hope of influencing others on a larger scale as he so manifestly could in debates? He needed a sense of purpose, all the more because in shrugging off his father's yoke at this time he was in danger of shrugging off his religious teachings too, and losing spiritual momentum altogether as a result. From this he was saved by Reg Pearse and his newly won freedom to explore the countryside as he chose.

For three years, in the winter holidays of 1921, 1922, and 1923, Paton and Pearse undertook marathon walks as a cheap way of having a holiday. The first of these, in July 1921, was from Pietermaritzburg to Ladysmith, in company with Cyril Armitage; the second was a climb to the top of Natal's Table Mountain, followed by a ten-day camp at the bottom of Griffin's Hill between Pietermaritzburg and Ladysmith, in July 1922; and the third was a twelve-day walk from Pietermaritzburg to Ladysmith, via Greytown and Dundee. On these immensely energetic expeditions, in which Paton and Pearse would often cover as much as forty miles a day, they trundled their tent, cooking equipment and bedding in a handcart made out of a packing-case and two bicycle wheels, starting before dawn and pitching camp when it began to grow dark. According to Pearse, they were inspired by the exploits of George Borrow and Robert Louis Stevenson.[47] It was not practical to take a donkey with them, but they could and did push the cart themselves, though in fact they paid a Zulu four shillings to help them start up the Town Hill stretch out of town.[48]

Pearse subsequently wrote a detailed account of their walks, made up from notes by all three of them and illustrated with photographs of the walkers and their various campsites, through which many days of their activities during these walks can be traced in detail. Their energy was prodigious. They commonly started in the darkness of a chilly July morning, sometimes soon after 3 a.m. and walked 9 or 10 miles before pausing for breakfast; then they walked steadily till midday, taking turns to pull the cart on the level, but harnessing all three of them to the task with ropes when they reached steep slopes. The ropes proved effective as brakes when descending. Brakes were needed: the contents of the cart weighed 200 pounds.[49] By nightfall they would have covered 25 or even 30 miles, and this pace they kept up day after day. Pearse's account is detailed enough for one to calculate that they and their cart averaged 5 mph even over very hilly country. On the last day of this first trip, having sent their cart on by rail when one wheel broke, they did 52 miles, arriving at their destination in the small hours of the morning, and after a good sleep they played a game of tennis the next day.[50]

Nor was this the record; on a subsequent occasion, in July 1922, Pearse managed 60 miles, walking steadily for 21 hours,[51] though Paton did not complete the distance because of a boil on his chest which forced him to stop after a mere 46 miles.[52] On yet another occasion Paton cycled from Pietermaritzburg to Durban, 54 miles, to play tennis. When he reached Gillitts station, about half way, he decided to stop for a sleep, but found the station locked. It was a very cold night, and the only covering he could find was the enormous and unutterably filthy station doormat. This he pulled over himself, and proceeded to sleep very soundly. In later years, when they lived at Botha's Hill near Gillitts, his second wife Anne could never pass that station without thinking of the future world-famous novelist peacefully lying under the doormat in the litter of dust, dead insects and cigarette-butts.[53]

The distance from Pietermaritzburg to Ladysmith by the main road in the early 1920s was 110 miles, and in addition they made many side-trips to find good campsites. What is now a major highway carried virtually no traffic in those quiet days; they lay down in the road when they felt like a rest, and they greeted the rare appearance of a car with a cheer, most drivers returning the salutation.

But these walks were much more for Paton than marathon rambles. During the course of them he and Pearse talked through the problems of belief Paton was having, and the two of them reinforced each other's theological views. They were really remarkably intense, and, for all their joking, serious young men, and even when they were joking they seem not to have touched on ribaldry. Many years later Pearse wrote a brief account of these walks, no doubt drawing on the detailed contemporary account already mentioned, in which he touches on their chief concerns at this time. It is interesting that he feels a need to explain why these concerns did not apparently include sex:

> Their main topics of conversation round the campfires were, apart from blisters, God and literature. This is not to say they were not interested in young women. But Puritans think rather than talk about young women.[54]

They were Puritans, and there was little licence on these walks and camps of theirs. Swimming in the streams near their campsites was a favourite occupation in warm weather, but they did not swim naked: a bathing suit was among the items Paton would remind Pearse to bring along.

Though they were two of a kind, at times they severely tested one another's patience, as in the account Paton gave of his trip with Pearse to the top of Natal's Table Mountain:

> Pearse had catered very parsimoniously and we finished all our food at the evening meal, and were still hungry. It was the month of May, and we did not expect rain, but rain it did. We spent a cold and miserable night.
>
> I was wakened in the hour before the dawn by the crackling of the fire. I opened my eyes to see that Pearse was toasting a crust of bread, our last scrap of food. In my starved and cold condition I jumped to a terrible conclusion, namely that he intended to eat it all. This dastardly act inflamed me, and when Pearse told me that he had intended to wake me and give me half the crust I treated his story with contempt. I was in a thoroughly bad temper, and I put him in one too.
>
> I ate my half-crust full of hatred. We packed, and started on our walk to Umlaas Road. Our first task was to descend from the mountain down a steep and narrow defile. Pearse went first, and I realized that one push from me would send him to grievous injury or death. I played with the idea but did not proceed to the act. One thing however was clear to me, that it would be a proper end for a mean and ignoble life. Only the thought of the death penalty deterred me.
>
> Then the sun came out and we were warm again. What is more, Pearse had two shillings, and bought a calabash of sour milk, half of which he offered to me. When I had finished my share, I realized the contemptibility of my behaviour. I mumbled an apology which Pearse graciously accepted.[55]

The barrage of jokes in their conversation and letters partly concealed a deep earnestness which bursts out occasionally in their correspondence. 'May the year be prosperous and happy for you, my friend,' wrote Paton to Pearse at the beginning of 1922; 'may we meet all again next term in health & happiness, the same old friends, the same old jokes, above all, the same ideals, the same aims, the same God.'[56] In particular, it seems, the ten days he and Pearse spent camping together at Willow Grange, between Pietermaritzburg and Ladysmith, in the last fortnight of July 1922, had a deep influence on Paton. Their talks here have not been recorded, but eight months later Paton reflected on them in a letter to Pearse:

> Reg, old man, our ten days at Willow Grange did more for me in the highest sense than any other time in my life; whatever I am & may be, I owe to our trip—it seems that in the watches of the night we penetrated deeper into the heart of the essential than we have ever done before. And I am hoping to repeat this at Easter.[57]

This high moral tone, a little cloying to the modern reader, is perhaps the result of his continuing doubts, for he went through another and more severe spiritual crisis in 1923, when he read psychology as part of his teacher's diploma studies. One of the set texts was the work of J. B. Watson, founder of behaviourism, the study of the effect of outside forces upon animal and human behaviour. Watson for a time seems to have persuaded Paton to adopt a determinist viewpoint. As Paton put it in an essay on behaviourism written some years later,

> It [Behaviourism] insists that the outside forces are all that matter, and the man himself is simply a physiological structure upon which the outside forces act; that he simply can't help behaving, and that to assert that he has any control over his behaviour is so much nonsense.[58]

This now discredited point of view seemed to free human beings from responsibility for their own actions; indeed, to deny that they can have any. There is a degree of attraction about this for a man who has been brought up in a home where he is called to account for all his actions; it sets him free to do all the previously forbidden things he has half-longed to do. In 1981, in his old age, he would tell Bernard Levin that behaviourism had shaken his faith more seriously than anything else in his life.[59] This may well have been the period in which Paton paid his visit to the generous girl in Pietermaritz Street. Certainly he felt he was in danger at this time of falling away from the high moral standards he had set himself:

> I . . . had no doubt that an acceptance of the dogmas of behaviourism would lead some people who had hitherto believed in purpose and striving to attempt them no more, and to deteriorate in conduct and character. I believed that this would happen to me. *Therefore I chose to reject the extreme dogmas of determinism.*[60]

Victor Harrison saw evidence of Paton's doubts and his spiritual struggle when he again invited Paton to preach a sermon during a student service at the Methodist church Harrison regularly attended in Maritzburg. Paton's address was about the relentlessness of God in wanting his people, and he spoke with a passion that hinted at his own struggles, quoting freely from Thompson's *Hound of Heaven*. As Harrison recalled in old age,

> That was an outstanding sermon. So much so that when it was over, I told him how it had appealed to me, and he reminded me not very long ago what I had said to him. I said What did I say, Alan? He said You said I wish I could speak like that.[61]

Paton's intense moral struggle was still going on when, towards the end of

1924, he began looking forward to his first permanent teaching post. He had taken his B.Sc. with distinction, travelling to Pretoria for his graduation in April 1923, and had then taken a higher Diploma in Education during 1923 and the first half of 1924. He would have to begin at a school to be nominated by the Education Department at the beginning of 1925, and the Department decided to send him to the small rural settlement of Ixopo, a decision which he heard with regret.[62] There is a sense of forced cheeriness in the letter he wrote to tell Pearse of his intentions as a teacher on the verge of his new career.

> Well Reg, I am on the last stage now; then for the world & work. So far my life has been soft as you like, & overmuch softness does not conduce to overmuch to be proud of. God grant that I can raise the standard of everyday manliness & cleanliness in our public schools; I would die happy.[63]

The tone of this letter, hearty and a little over-enthusiastic, suggests that his new pupils were going to find the arrival of this earnest and idealistic, but morally tormented young master something of a shock. The experience was, in fact, going to prove very unpleasant for them, and for Paton too. The move into the world of work was going to be a voyage of self-discovery, and Paton would not always like what he found.

5

IXOPO
1925–1928

O never give the heart outright.
W. B. YEATS

Ixopo lies in beautiful countryside about fifty miles south west of Pietermaritzburg. The hills surge like great Pacific rollers, in parts covered with the grass and bracken that Paton was to describe with such love, in parts now given over to vast plantations of eucalypts and wattles, so that one might be in eastern Australia. The climate, though prone to occasional devastating droughts, is usually delightful; the winters are mild, the summers not unbearably hot; and after a bright summer day the mist will flow in at evening, steeping the world in a mysterious and silvery silence.

Ixopo itself is a pretty village, its streets shaded with jacarandas and other subtropical trees, its many gardens bright with flowers. It was founded in 1877 by a settler named John Grant, and for the first few decades of its existence was officially known as Stuartstown, after a local magistrate, Martinus Stuart, before the Zulu name Ixopo gradually established itself in the mid-1920s.[1] It was a backwater when Paton first went there at the beginning of 1925, and it is a backwater now, but one with real charm. It is possible to sit on the verandah of the Plough Hotel, as Paton often did, and watch the main street (along which almost all the place's main buildings are concentrated) filling with mist through which the lights glow as night comes on, and hear hardly more traffic passing than Paton would have heard seventy years earlier.

Nowhere is the open countryside more than a block or two away, and it is country that calls to the walker. Paton wrote of the village years later, in a description printed in the official history of Ixopo School:

> In those days people made their amusements for themselves. By far the most popular sport was tennis, followed by golf, rugby, and far behind, cricket. At night

the village often lay shrouded in mist, and not a soul stirred; nothing was to be seen but the ghostly lights of the Off Saddle and the Plough in the eerie darkness. It caused a great sensation when one New Year's Eve the magistrate's clerk climbed nude on to the roof of the butcher's shop, and would not come down.

It was a quiet and pastoral life, full of simple pleasures. When one visits the buildings of the present [1964] Ixopo High School, one thinks back with nostalgia and astonishment to those far-off days, and remembers those perilous journeys through the mud from hostel to school and back again, enlivened by the raucous laughter of the boys and the shrill shrieking of the girls.[2]

For someone with Paton's Wordsworthian passion for natural beauty, it could have been an ideal place to work. He came to love the place, and would in time record the countryside round about in loving detail; but the job itself he did not enjoy, certainly not at first. He was to spend only three years in Ixopo, and with good reason.

He had had several short-term experiences of teaching before he came to Ixopo, and should not have been taken by surprise. As early as the end of 1921, while still an undergraduate, he had done some part-time teaching of mathematics at the Epworth High School for Girls in Pietermaritzburg, to make some pocket money. 'My success [as a teacher] was minimal,' he wrote candidly later.[3] In June 1923 he had been asked to act as Professor of Physics during the illness of the single Physics staff member, Paul Mesham, and had taught three undergraduate classes and five practical classes, enjoying himself hugely, occupying Mesham's large office and making himself and his friends endless cups of tea. At the end of 1923 he spent a week invigilating examinations at Centocow, a black mission school near Braecroft in Natal, a job he did not much enjoy, and he advised Pearse, who was considering working with Africans, not to work in the black school system.[4] This view ran directly contrary to the views he had put forward in his first attempt at a novel, *Ship of Truth*, finished shortly before he went to Centocow. Finally, at the end of 1924 he spent eight weeks filling in for an absent teacher at the high school in Newcastle, a centre for steel-making in Natal.[5]

The effect of these various experiences of teaching was to give him some feel for the requirements of the job before he took up his post at Ixopo. But a temporary post was quite a different proposition from the one he now found himself in, and in towns like Pietermaritzburg and Newcastle he had been able to preserve his own privacy in a way that would simply not be possible in a village, where the new master soon found that he was known to every white inhabitant.

The Ixopo school itself was very small, with particularly low numbers in the senior years which Paton concentrated on teaching: Standard IX, the penultimate school year, had only three pupils, and Standard X four. Ixopo School had been founded in 1895, but the secondary school had opened only in 1920;[6] the year Paton arrived, 1925, was the first in which a complete class of the pupils had reached the final year, and he taught the school's first matriculants. He was not only to teach mathematics and physical science to all the senior classes (Standards VII, VIII, IX, and X), and English composition to Standard II, but he also had to act as form master of Standards IX and X, and, most exacting of all, to take charge of the small boys who shared a hostel with the girls. This meant that he lived in the hostel itself, in a single room without its own bathroom, and could virtually never get away from his pupils. Big boys had a separate hostel of their own, with which he was not at first concerned. His living conditions were spartan, and he ate the meals the children got, and about which they complained a good deal, particularly on such occasions as when they found maggots in their morning porridge.[7] Ixopo School was no Dotheboys Hall, but it offered few comforts and little free time to a new young science master.

In his autobiography, written half a century after his experiences in Ixopo, he was to give no hint of his difficulties at the school. He describes himself as an easy-going teacher whose discipline was anything but strict:

> My most obstreperous pupil was Pat Kippen in Standard I, age about nine or ten. If I asked a question he would leave his desk at the back and come advancing up the aisle, with contorted face and flicking fingers, consumed by zeal and exhibitionism. I would shout at him, 'Kippen, if you leave your seat again, you will suffer the extreme penalty,' whereupon he would retire scowling to his seat. But not for long; when I asked the next question he would be up there again, snapping his fingers under my very nose . . . Eventually we agreed that he would answer only alternate questions, otherwise no one else would have had a chance.[8]

His only references to discipline were his disapproval of the discipline inflicted by others, as when he describes his first experience of making an enemy. She was one of the two matrons who looked after the girls in the hostel:

> One afternoon when I returned to the hostel from school I found that the senior matron, Miss Shimmon, a most aristocratic and needy Irish gentlewoman of about fifty, had shut two small girls in a cupboard for some offence. In my anger I told her that such a punishment was disgraceful, after which she would not speak to

me . . . In fact we became enemies, and in the end I became totally indifferent as to whether she spoke or not.[9]

The implication of this account is that Paton's own use of disciplinary measures was both humane and lenient.

Perhaps in order to get him away from Miss Shimmon, the competent and energetic headmaster, Mr Buss, moved Paton by putting him in charge of the senior boys' hostel. Here, he says, he found the older boys like brothers rather than pupils. In fact he professed himself astonished, even in retrospect, at the disciplined behaviour of his charges, and by his account he lived on terms of mutual affection and respect with them: 'I was like an elder brother to them, and on a free weekend would often walk with one or more of them to their homes, over those hills where I heard the titihoya crying.'[10] This account of brotherly affection between him and his pupils is no doubt true, for Paton was a truthful man, but it is far from the whole truth.

Not many of Paton's Ixopo pupils survive in 1992, and those who do are in their eighties. But they remember Paton well, and they agree in remembering him as a master who inflicted physical and other punishment freely and with unusual severity. Schoolboys, of course, tend to exaggerate the punishments they have suffered, and even in the mind of the grown man they may loom larger than they were in fact; but there is striking agreement about Paton as a teacher, among men who were taught by him at widely separated times. One of Paton's Ixopo pupils said that he was 'loathed' by his charges in the classroom, using about him words such as 'brutal and sadistic'.[11] He also thought Paton a Jekyll and Hyde character because of his sudden and violent rages.[12] Another, more mildly, said that 'Paton was a very tough master at school. He used to use the stick, and the boys had a pretty thin time of it.'[13]

This was a period of flogging masters, when corporal punishment was much more freely resorted to than in the later decades of the twentieth century, and many teachers 'used the stick', in spite of an Education Department ruling confining to headmasters and housemasters the right to cane.[14] But it is plain that Paton stood out in this regard in the memories of his ex-pupils. Another of these said that Paton would use sticks, cricket-stumps, 'anything that came to hand', and added that he used to subject the girls to ridicule so biting that in his view the boys got off more lightly on the whole.[15] The girls, however, did not entirely escape physical punishment, as one Ixopo old boy, Harry Usher, remembered many years later:

A little girl in the Standard V dropped a blob of ink on her page as he was passing, so he grabbed her by the hair and bumped her up & down on the bench, she was so scared of that man she would start shivering when he came into the class.[16]

Yet another pupil wrote an account of Paton's handling of the senior boys' hostel, a rather roughly-built structure known as the 'Hospital':

This building housed 15 boys and one master [Paton]. The washing facilities were very crude: one cold tap into the bath, and one piece of piping over the bath, which had a jam-tin with holes in it for a shower. The Bathroom was at the other end of the grounds and was made of wood and iron, and had a large window with no panes of glass in it. The winter mornings were not looked forward to, as quite often the ice had to broken in the jam-tin before showering!

Mr Alan Paton made a rule that all boys have a shower every morning, but he had a ten-pound jam-tin of water heated on an open fire in one of the out-buildings by one of the boys for his wash and shave.

On Sundays Mr Paton used to pin a notice on his door with the time he intended rising written on it. If anyone dared to make a noise before the time printed thereon, and he heard the person, they would be in for at least four of the best.[17]

This account was printed in the Ixopo School history; Paton read it before publication, but made no changes to it. Paton's insistence on cold showers for the boys while using hot water himself seems to have caused particular resentment; Harry Usher went so far as to allege that the obligatory cold showers had caused or hastened the death from pneumonia of a boy named 'Boet' Ferreira.[18]

Paton's own letters at the time tend to confirm his pupils' view of him as a severe flogger. They also give part of the reason for this uncharacteristic behaviour: tormented by unresolved sexual tensions himself, he came to believe that his pupils were on the verge of immorality. During 1925 he wrote to Pearse,

When I first came I was hated like poison; still am, by some. But discipline was slack; some of the girls, who have ten times the go of the boys, were pretty near immorality. You see, tho' the big boys sleep apart, they ate & did prep. together—you could go in unawares & find them spooning about the rooms, arms round necks, etc. That was a month for me—I hardly smiled, my cheeks caved in, I had the name of the demon. If there was any trickery, I spotted it, & lammed [beat] the culprit; slowly and surely ground was gained—I was the sworn pal of my thirty odd youngsters—surely gaining the friendship of the bigger girls—but hated like nothing on earth by all but a few of the bigger fellows. I was just getting pally, when I caught a dozen or so cribbing in the Easter Exams; more hidings, more unpopularity.[19]

According to an ex-pupil, boys and girls were not allowed even to talk to one another on pain of a caning.[20] Paton was right to think his pupils hated him, and a number of them retained this feeling even after his death. Even during his life, after he had achieved fame, there was a proposal to make him the President of the old boys' association of the Ixopo high school, but it was voted down, partly because of the memories of his teaching, partly because he was too politically liberal for his ex-pupils' liking.[21]

It is worth asking why Paton should have behaved in this way, when all his own writings and the memories of those who knew him in later life suggest that excessive use of corporal punishment was not natural to him. Part of the explanation no doubt lies in the fact that he was a newly-arrived master who felt the need to impress his authority on his pupils, particularly those among whom he was going to have to live for twenty-four hours a day. This need would have been exaggerated by the fact that Paton was a very immature 22, only four years older than the oldest boy in the school,[22] and that he was physically smaller than a good many of his pupils.[23] In addition he would no doubt have been given a talk by the headmaster on his arrival about the need to maintain discipline, though in fact, according to his colleague Piet Barnard, from whom he took over the older boys' hostel, discipline was very good.[24] But there can be little doubt that it was the sight of the boys and girls 'spooning' that so incensed Paton, or that his father's teaching ('Boys don't go into girls' rooms, girls don't go into boys' rooms') and his father's almost automatic reaching for the cane had taken over at this testing time.

The charge of sadism cannot be sustained, but there was a side of Paton which tended to lash out at what he regarded as moral laxity. In the unpublished novel *Ship of Truth*, written in 1922 and 1923, there is an extraordinarily powerful scene in which a young man, Dirk, is beating one of his sisters for punishing one of their siblings.

'Anna,' he crooned, in a quiet singing voice that filled the girl with terror. 'Did you hide [beat] Peggy?'

'No,' she cried, beating her teeth with her fists.

The sjambok [a whip] descended on her legs. She screamed but that dread voice quietened her.

'Did you hide Peggy?'

She writhed towards him appealingly.

'Before God, no. Dirk! No! No!'

And again that pitiless impersonal stroke. Dirk stood over her smiling.

'Did you hide Peggy?'

'Yes. Yes, I did. For God's sake, Dirk!'

The sjambok descended this time with a flavouring of passion. She screamed terribly.

'Will you again?' he asked her gently.

'No, Dirk. Before God, no. I swear.'

And again the stroke.[25]

Paton kept up his fearsome reputation on the sports field: he taught the boys cricket, and would line them up against a toilet wall 15 yards from himself, have a boy bowl balls to him, and hit them as hard as he could for the hapless fielders to catch, appearing to think it a joke if they were hurt.[26] Many of the bigger boys were disappointed that he would never play rugby, because, as one of them remembered, 'there were plenty of chaps waiting to give him some of his own medicine.'[27] At the end of his first year a group of the boys who were leaving that year planned to ambush him in a dark spot near the hostel and give him a beating, but their plan failed.[28]

Having established a reputation as a strict disciplinarian inside and out of the classroom, Paton seems to have maintained it all his teaching career, for accounts of his methods and the dislike they engendered continue through his time at Ixopo, and into his next post back at Maritzburg College. In mitigation, it should be added that his own schooling at Maritzburg College had introduced him to a culture of flogging, for the College had until recently a dreadful tradition as a flogging school.[29] A habit formed so early in a career may be very difficult to unlearn, though Paton would succeed in time.

Meanwhile his teaching kept him very fully occupied. 'I am resident master, sportsmaster, gamesmaster, & Sunday school master, & I draw the line there,' he wrote to Pearse; 'I see kids, feel kids, lam kids till 9 p.m., & at that hour I am glad to go out & have a rest, & read & a smoke or a game of bridge.'[30] Nor was he always free even at 9 p.m., when lights were put out in the 'Hospital'. Because he and another man named Burger were the only male teachers responsible for the hostel, they had to be on duty on alternate nights and alternate weekends.[31] The work must have seemed unending.

Yet Paton's energies were such that he took on extra work: for instance, he continued his interest in the theatre by organizing a school play in several consecutive years. One of the first of these was a version of *Pride and Prejudice*, which Paton dramatized himself.[32] Unfortunately this dramatization, a real literary curiosity, has been lost.[33] The following year Paton also produced *Twelfth Night*, with what he fondly believed to resemble an Elizabethan stage, bare of scenery or stage-props, but with signs announcing, for instance, 'This be a yew hedge'.[34]

In addition, he was even teaching outside the Ixopo school; as he had mentioned to Pearse, he was a Sunday school teacher now, teaching children each Sunday in the Methodist church. He had decisively abandoned Christadelphianism, and was looking for a spiritual home elsewhere: for a time he seemed to have found it in Methodism. Yet though he taught for three years in the Sunday school, he never formally joined the Methodists, a fact which suggests that he was still searching.

The Sunday school was not the only attempt he made in these days to 'raise the standard of everyday manliness & cleanliness in our public schools', as he had expressed it to Pearse. Not even in his school holidays was he free of schoolboys, for in 1924 he helped organize what would prove to be the first of many formal, annual camps arranged for schoolchildren by the members of his old University College SCA, as part of a deliberate missionary endeavour which would combine a pleasant holiday with the spreading of the gospel.

The SCA camps were begun by The Three Scots Villains, Paton, Pearse, and Armitage, who continued to maintain close contact with each other for years after they left the University College to begin teaching. Pearse played a leading role in the earliest camps. The other originator of these camps was another figure from Paton's university days, Oswin Bull. Bull was a generation older than Paton, a handsome and powerful man in the mould of Railton Dent, and like Dent he became a father figure for Paton and a deep influence on him. He was an Englishman who had a permanent position with the SCA movement, for which he had come to South Africa in the early years of the century as a travelling secretary.[35]

Under Bull's guidance the three young schoolmasters organized the first SCA camp in 1924,[36] at Umgababa on the south coast of Natal. At the last minute Paton found himself unable to attend this, since he had been selected to go to London for the Imperial Conference of Students, but he attended the second, in 1925. It was followed by a third in 1927, and by many others in succeeding years, for the camps proved immensely successful and popular. Pearse sketched the basic plan as he, Paton, and Armitage had conceived it:

> You get a group of boys from various schools together, for about a fortnight, you have the fun of camp life, and in the evening you meet in a big tent or building, waffle away, pull each other's legs and sing songs and so on, and end up with evening prayers. And we also had evening prayers in each tent. Each tent had an officer in charge of it, and it was his job to get the half dozen boys in the tent round himself before coffee was brought round in the morning.

That was the only emphasis on religion in the camp. Early morning and evening prayers.

The rest of the day was bathing and fighting and the general fun of camp life. And those camps were magnificently successful. They are talked about even today. I meet numbers of campers, even today, who say what those camps meant to them in their lives. Difficult to put it all into words, it wasn't soft religion, it was a manly sort of approach.[37]

For many years the annual camp, held over the winter, was to be a fixture in Paton's calendar; every July was given up to it, and the camps were to assume great and unexpected importance in his life.

The first of the SCA camps was held on a Scout campsite, a piece of flat ground at the foot of the railway bridge at Umgababa, then a scattered village set in dense bush along the coast, and with a tiny population. The boys were housed in borrowed bell-tents, exceedingly primitive latrines were dug, and very simple food was cooked over open fires. The routine of the camps was set very early on, and did not vary much in succeeding years. Cecil Armitage's daughter Shirley Broad retained a clear memory of it, though her memories are of the 1930s:

The daily programme at the camp was rising bell; roll up your beds; wash; make breakfast; morning prayers after that. Alan Paton was a tent leader. He would have led the bible reading and discussion. And the rest of the day was fun and games at the beach, or going for walks. Then back for lunch, which one of the tents would have taken its turn to prepare. And wash up.

And the great soul of the place was evening round the campfire. We would write jokes or skits. We used to read these out and sing. 'Widdicomb Fair', things like that from the English song book. And then choruses, Sankey and Moodey stuff. Then someone would give a talk, one of the leaders, taking it in turn.

The idea was not to be preached at, but to live brotherhood.

The camps were always in winter: the crackling of the dry grass underfoot, the quality of the air, the birds, those red berries the matingulus, always July.

You'd stuff the mattresses with straw, and stitch them up, and you'd put your head down on them, and you'd hear the pounding of the surf, the most marvellous lullaby.

The boys and girls had separate camps, first the boys and then the girls. And then men would go, and the women come. The men dug the latrines for the girls' camp which followed.

She also had vivid memories of Paton himself at the camps:

Alan was sturdy, sturdy legs; bottle legs. Hair straw coloured. Eyes blue. He had a slight stoop; the shoulders became an arena for the head. Even when he was quite young. Pursed mouth, when he was thinking, or when he was going to say something serious or very funny. He'd get this twinkle in his eye, his eyes communicated. But he could be very stern.[38]

Pearse, Armitage and Paton hoped to use the camps to give spiritual guidance to their pupils in a more concerted and direct way than could be managed at school, and each tried to influence his pupils to attend the camp. Paton, however, had made himself so unpopular at Ixopo that he was rather sheepishly obliged to admit to Pearse that none of his pupils would accompany him: 'I can only support camp with my own presence.'[39] Instead he planned to bring back a couple of the tents to Ixopo, with the idea of taking some of his pupils out during weekends to camp in the countryside.[40] This plan seems to have had gradual success, for the camps grew larger with each passing year, though how many of those attending were Ixopo pupils cannot now be determined.

It might have been thought that someone as fully occupied as Paton was during 1925 would have little time for other pursuits, but in fact he had a new obsession, and one that made his reaction to his pupils' 'spooning' even stronger than it might otherwise have been. Towards the beginning of 1925, quite soon after his arrival in the village, he fell hopelessly, headlong, in love with a married woman.

Her name was Doris Olive Lusted, and she was the wife of the part-owner of Ixopo's one garage and service station, Lusted & Johnson, a firm which is still thriving today. Doris Lusted, known to her friends as Dorrie, had been born in Ixopo in 1897, the oldest of four children of a local solicitor, George Francis, and had attended the Ixopo School before falling in love with an Englishman, Bernard Lusted. She had two sisters, Rad and Ruth ('Googie'), and a brother, Garry. Her parents had unsuccessfully opposed the match with Bernard Lusted, partly no doubt because they thought a motor mechanic not their social equal, but also, more importantly, because Lusted was evidently a dying man. He had the wasted physique, the pallor and the febrile glittering eyes of a patient in the final stages of tuberculosis, and his cough was unceasing. He had in fact come to South Africa from Somerset in the vain hope that a warmer climate might slow the course of his disease. Dorrie Francis's passion for him was as inexplicable as love always is, and her domineering mother fought hard against it. The girl was even sent to England for a time, to attend Beerbohm Tree's school of acting,[41] but when she came back she was as determined as ever. In the end she had her way and married Bernard Lusted towards the end of 1924, only months before Paton arrived in Ixopo. Such a match was naturally the talk of the tiny community.

She and Paton met through the Ixopo tennis club, which Paton joined as soon as he arrived in the village, and he describes her game in very much the terms his university friends used in describing Paton's own style of play:

> There was something in her of the urchin. She was full of mischief and zest and repartee, and played with all her heart. We played a great deal, both together and against one another. I was full of mischief and zest too, and our conversation was full of banter and teasing.[42]

She was 27 when they first met at the beginning of 1925, and Paton had just turned 22. He found her intensely attractive: she was slightly built, dark-haired and dark-eyed, with a fine creamy skin, and though she had a rather long nose which not everyone found beautiful, Paton found her bewitching. She was a fine mimic, and could tell an uproariously funny story in which she imitated Zulus and Scots settlers with equal ease; she was quick, irreverent, mercurial, and apparently uninhibited. She had a facility for satirical verse and lampoons, and a liking for slightly risqué jokes.[43] She seems to have been the first woman he had met who could match him in the japes and leg-pulling he excelled in, and he found himself powerfully attracted.

Dorrie encouraged his evident interest, ostensibly because she hoped he would take a liking to her sister Rad, 26 at this time and unmarried. She asked Paton and Rad to dinner together at her house, and subsequently Paton was often at her parents' big house, Morningview, which was in easy reach of his hostel, being just across the road from the school. Dorrie, who shared Paton's interests in literature and acting, flirted constantly with him, and he was soon deeply in love with her, hardly knowing what had happened to him. In a community the size of Ixopo this could not go unnoticed for very long. After the very first dinner at the Lusteds' house, the dying Bernard warned his wife, 'That youngster's in love with you.'[44] Other eyes noticed the same thing: according to one of Paton's pupils at that time, Dorrie Lusted's growing fascination for Paton set tongues wagging all over the village.[45]

Paton himself was quite innocent. Years later, after Dorrie had died, he wrote looking back on his feelings at this time, and using the form of a letter to her:

> It was for one thing the purest love in the world, for it did not desire to possess you. I do not remember that I ever thought of touching you; I certainly did not ever try to touch you. I, not knowing it fully, had given my heart into your hands. Although I laughed and teased you on the tennis-court, I was virginal and shy. I have a picture taken of me at that time. There was no guile in that face.[46]

He was quite right to describe himself as virginal and shy, in spite of what may have happened in the boarding house in Pietermaritz Street, and he had little idea what to expect from a woman.

The smallest sign of attention from Dorrie was treasured by him, though in fact there was plenty to treasure. On one occasion, she and her sister Rad dressed up as nuns from the Hospital of Christ the King, and came and asked Paton for a donation for their work. He, suspecting nothing though it was 1 April 1925, gave them five shillings and signed their subscription list. That evening at Morningview the two girls gleefully produced the list, and handed Paton back his money. 'It gave me much pleasure to have been thought worthy of such an elaborate deception,' he wrote later, 'especially as the two deceivers were so proud of themselves. It was sex, and it wasn't. It confirmed me in my attraction.'[47]

Paton, so inexperienced in his dealings with women, and so strongly repressing his own sexuality, must have been tormented by this developing relationship, in which he scarcely understood either his own motives or those of his beloved. He thought of her as pure and untouchable; no wonder the sight of his pupils touching or 'spooning' evoked such furies in him, particularly since he was strongly attracted to one rapidly blossoming 13-year-old girl himself, being saved from temptation partly by his purer love for Dorrie, partly by his pupil's father's fearsome reputation for shooting trespassers on sight.[48]

Then, a couple of months later, an event occurred which changed the nature of his dangerous and apparently doomed relationship with Dorrie: Bernard Lusted died on 25 May 1925. Paton returned from a weekend in Pietermaritzburg with his parents to hear the news, and to see Dorrie swathed in mourning clothes. It would have been strange if something in him had not rejoiced at the sight, for she was now free to marry him—if she were willing. To the end of his life he preserved a photograph of her taken at this time dressed in black, her large sad eyes looking out under the brim of a black velvet hat, her face solemn but still young and attractive. For all the grim attire, this was how Paton liked to remember her in later years, and the photograph of her in mourning became his favourite when he was himself in mourning, for her.

They did not immediately renew their flirtation, because Dorrie went off to Durban for some months, and when she returned Paton was away for the August holiday. But in September 1925 they began picking up the threads again, and as before it was Dorrie who initiated the moves. She asked Paton to coach her younger sister Ruth, at this time 16, in mathematics, and he was delighted to agree. The coaching was done at Morningview, where Dorrie was now living again, having sold the Lusted home, and there was ample opportunity for her meetings with Paton. They would sit on the deep verandah that ran round three sides of the big square house, or wander in the large garden, full of maturing trees.[49] But as the girls at the University College had done before her, she found Paton painfully shy and apparently unable to summon the courage to make any move toward her. It was she who invented such games as pretending to read his palm, and afterwards he could remember nothing of what had happened, except that his hands had been in her hands. And still he was puzzled by her motives: 'What were you doing, reading my hands? You never again read any person's hands as far as I remember. As for me, I did not try to touch you at all. . . . Did you know what you were doing?'[50]

The question is unanswerable. Dorrie had been deeply in love with Bernard Lusted; her feelings for Paton, as her subsequent actions made clear, were nothing like as strong. Perhaps she was just testing her power, assuring herself that she was still a desirable woman. Perhaps she did not realize how inflammable Paton's feelings were, or think through the consequences of what she was setting in train.

On the evening of 19 September 1925 Paton had dinner at Morningview, and after dinner sat alone with Dorrie. Her invitation to him had been a bold one, for her parents were away in England with her sister Rad, and she was alone in the house with a housekeeper. This woman presently went to bed, and Paton and Dorrie sat by the fire into the small hours of the next morning, alone. Paton made no move. Dorrie left her chair and sat on the floor. Paton made no move. She complained that it was uncomfortable on the floor. Paton, with what he described as 'incredible courage',[51] suggested that she should rest her back against his knees. She did so, and Paton made no move other than to put his hands on his knees. Impatiently, for was there ever a more dilatory or cowardly lover? Dorrie reached up, took his hands and pulled them down onto her breasts. It was as if lightning had struck him. When, in his second novel, *Too Late the Phalarope*, he wanted to describe complete intimacy, this was the action that would come naturally to his mind.[52] And then, as Paton described it,

> my arms were around you, and my head was pressed against yours, and your face was upturned, and we kissed, and I said '*I love you*' and you said '*I love you too*'. How long I stayed after that, I do not remember. When at last I left, you walked along the verandah with me, and when you kissed me goodnight, you pressed

yourself against me in a way that no one had ever done before. And I, knowing you to be a chaste woman, knew what it was to be loved.

The other things were left unspoken. Except for one, and that was that our love must be kept a secret for some time, out of respect for the decencies of those days.

And I walked home like a man to whom a door has opened so that he may look into heaven.[53]

But the woman who had opened the door could also shut it again, and before long Dorrie did.

Paton was painfully aware that his beloved had been married before, that all her first affections had been given to Bernard Lusted. He was naturally though absurdly jealous of the dead man, and possessive of Dorrie. Perhaps it was this that irritated her, or the sense that things were moving faster than she intended them to move, or (as Paton thought) that she suffered guilt in loving again so soon after her husband's death.

There is yet another possibility, which is that she was alarmed by the strength of the passions she had awoken in Paton. All the evidence in later years suggests that Dorrie, though flirtatious, was not a sexually responsive woman. Only very rarely was Paton to find her a passionate partner.[54] Perhaps part of the attraction of the dying Bernard Lusted was that he was not likely to be physically demanding, and Paton's second wife was to suggest that Dorrie's marriage with Lusted had never been consummated.[55] But Paton, six years younger than herself, an eager, athletic youth, was quite another matter. His own long-repressed sexuality, now suddenly offered the possibility of a legitimate means of release, must have alarmed her.

At all events she rounded on him, one night soon after their joint declaration of love, standing in front of the same fireplace, and said to him coldly, as a woman speaking to a presumptuous boy, 'There is one thing you must understand clearly, and that is that I shall never love you as I loved my husband.'[56] Paton, cruelly hurt, left Morningview without a word, and went back to his bare little room in the hostel, resolving never to go near Dorrie again. 'I was a mass of pain, like a whipped creature,' he wrote forty years later.[57]

But she now did something as dramatic and shocking as putting his hands on her breasts: she followed him into the hostel, in the dark, and found his room, and for her to have been seen there would have meant disgrace for her and dismissal for him: 'Then the door of the room was opened, and Dorrie was there at the side of the bed, weeping, and asking to be forgiven. So we were reconciled, neither of us having acknowledged or confronted the two causes of it all.'[58]

It was not the opening of the door, but the darkness, which was to be the abiding symbol of this reconciliation. Their failure to confront what had gone wrong between them, their inability to talk through the problems that lay behind what might have seemed a mere lover's tiff, was to be the chief cause of lasting sadness in their married life, and it was a failure that showed itself early on. The truth is that it was part of a power struggle between them, in which Paton longed for outpourings of physical warmth and affection from Dorrie, while remaining too constrained to ask for them except very rarely; while Dorrie controlled the physical intimacy between them so that he almost always felt she was holding something back. The germ of this future coldness lay in these early incidents, and might have been resolved had they talked them through. But they both shied away from the pain of doing so, and paid a long and bitter price in consequence. 'My advice to all young husbands and wives, and indeed to all husbands and wives, is not to be too niggardly with their words,' wrote Paton when it was too late.[59]

At the time they focused on the immediate problem, which was to get Dorrie out of the hostel unseen. The boys were all asleep and no one else was about. Paton sensibly checked to see that the coast was clear, then let her go first. 'When you had been gone a minute, and no outraged inspectors of schools had appeared', he wrote humorously years later, 'I slipped out myself, and together we walked up the dark road to Morningview.'[60] One consequence of this evening's events was that as if by consent they never again mentioned Bernard Lusted, and even when Paton finally spoke of him in 1966 he did not mention him by name. Yet the shadow of Bernard Lusted lay between them at all times, and gave Dorrie dominance in the relationship. Power, in such a situation, is in the hands of the one least in love.

When Dorrie's parents returned from England, probably in November 1925,[61] they seemed to welcome the news, but it was decided that there should be a pause of a full year before the engagement was made public, and no date for the wedding was set. In fact it did not take place for nearly three years. Even for those days that was an uncommonly long engagement, and Paton gives no explanation of it, other than to say that a mourning period of a full year was required.

It is possible that the Francises thought his income insufficient to marry on, for certainly he began taking steps to increase it. These included learning Afrikaans. In April 1924 General Smuts had resigned as Prime Minister, and the Afrikaner Nationalists under Hertzog had come to power for the first time, in a coalition with the Labour Party. One of their first acts was to make Afrikaans an official language, ranking with English. Civil servants who could show proficiency in Afrikaans received a pay increase, and Paton began studying the language with intensity and in time mastered it, speaking and writing it almost as well as he did English. His initial motive was undoubtedly the desire to increase his earning power.

The chief reason for the long engagement, though, was unquestionably Dorrie's continued doubts about the match. These were perhaps increased by a visit she paid with Paton to his home to meet his parents towards the end of 1925. It must have been a strange and strained meeting: on one side the high-spirited widow, six years older than Paton, loving pleasure, fond of dancing, acting, alcohol, and smoking to excess, and on the other side Paton's rigid father, stern to the point of cruelty, and his submissive mother. Dorrie was not the woman the older Patons would have chosen for their son, and she and James Paton disliked each other on sight.[62] Dorrie showed her feelings towards him by calling him 'Jimmy',[63] though not in his hearing, and he in his grim way called her nothing, and said not a word about her to her fiancé. Her smoking alone was enough to condemn her outright in his eyes, even though Paton had prevailed on her, with the utmost difficulty, not to smoke in his family home. Paton's gentle mother was kind to Dorrie, but her son knew well that in her opinion he should have been marrying a Christadelphian.

Dorrie was not only not a Christadelphian, she was not a believer at all when Paton met her. Bernard Lusted had been an atheist, and under his influence she had drifted away from the Anglicanism of her parents. Now under Paton's influence she gradually began worshipping in the Anglican church again, and presently began taking communion once more. 'Thank God I did not attempt to instruct you or argue with you,' Paton was to write, 'for even in the days of my young manhood I believed that this was the most barren way of communicating faith'.[64] Instead he tried to 'communicate faith' to her by loving her, and succeeded: as he put it, 'you warmed to me in gratitude.'[65]

He also made an attempt to alter his own mother's beliefs about this time, no doubt in the hope of bridging the gulf that was opening between them. His temporary adherence to Methodism had hurt his mother, whom he particularly pained by saying that he judged a person by his life and deeds, not by his beliefs.[66] He tried to convert her to Methodism, by inviting her to Ixopo to stay with Methodist friends of his;[67] she came and enjoyed her stay, but it did not have the desired effect, and the chasm between mother and son continued to widen. There is little doubt that both Patons saw Dorrie as taking their brilliant son further away from them, and her meeting with them must have made her more doubtful of the marriage she was contemplating.

Paton had to return to Pietermaritzburg alone during the Christmas holidays in 1925, and spent six weeks there getting only occasional letters from Dorrie. During this time he spent much of his accumulated pay on a new and powerful Indian motorcycle, of which he was immensely proud; the fact that he did not at once ride it the fifty miles to Ixopo to see his beloved suggests clearly that he was aware of her doubts and perhaps feared the outcome. She clearly felt she needed time away from him to think through what she was going to do.

And when at last he got back to Ixopo on the red machine for the start of the new term early in 1926, and rode at once crackling and roaring to Morningview, there was a long and embarrassing wait while she lingered in a bath and he made stilted conversation with her mother. Dorrie was plainly unwilling to come to meet him, and finally came chilly and reserved. And she told him that she had been having further doubts, had consulted a friend, Jean H., another widow, and had been advised that she had been caught on the rebound, and that her love must be suspect because of her loneliness and because there was no eligible man of her age in Ixopo. And then, Paton wrote later,

> either you told me that you yourself thought it would be wiser for me to marry some younger woman, or you asked me, *don't you think you ought to marry some younger woman*? And I said to you in spite of my pain, *I don't want to marry some other woman.*

And he added,

> I do not profess to understand why after all these years of your warmth and your comradeship, I should feel again the pain of that long-gone time . . . I am writing a story of love which began with joy and pain, and ended in steadfastness.[68]

The pain was caused partly by his sense of betrayal at her consulting a stranger about things so intimate to themselves. His distress showed itself in anger, and they had their first quarrel, followed by an icily restrained family dinner. And the pain continued, for Dorrie's doubts appear to have lasted for many months. But within a month or two, probably in April 1926, she and Paton were thrown into each other's arms by one of the accidents that so often alter human destiny.

Among the subjects Paton taught was chemistry, though he knew little of it, having read only mathematics and physics at the University College.

One of the experiments he had to perform for his classes involved bringing a small piece of sodium into contact with water: the piece had to be small because of the potential danger of the reaction. The school sodium, heavily oxidised, refused to react though Paton cut off fragment after fragment. At last he marched the class outside, got them to stand well back, and dropped water on the entire remaining piece of sodium. The result was an explosion which blinded Paton with the products of the reaction, caustic soda and burning hydrogen, and sent him rolling and screaming in unbearable pain, to the consternation of his class. He was rushed to the local doctor, who anointed and bandaged his eyes and ordered that he be taken to Pietermaritzburg at once in the hope of saving his sight. Dorrie, who had come at once on hearing the terrible news, sat with him in the car, in the sight of all the village, whispering 'I haven't any doubts any more,' and insisted on accompanying him to Pietermaritzburg. 'So the great "secret"', Paton wrote wryly, 'went up in the explosion too.'[69] And since what had been widespread private knowledge had been made public knowledge in this way, the wedding was bound to go ahead.

It is a sign of his deep unpopularity that, according to ex-pupil Harry Usher, the school celebrated the news of this accident with childish callousness: 'When school came out that day, the whole school celebrated, the boys all throwing up their straw bashers in the air, the girls jumping up & down on the verandas & playground. I have never seen a man who [was] hated by so many children.'[70]

In the event his sight was saved, though he wore glasses from this time to his death. When he was at last released from hospital Dorrie was there to meet him and take him down to Scottburgh on the Natal coast for a brief recuperation.[71] Nothing is known of this time: where did they stay, and for how long, and how did they preserve the proprieties? Paton kept no letters from this period, and he gives no information in his autobiographies. The failure to preserve Dorrie's letters is itself significant, for he was to treasure most of them all his life. The likelihood is that he destroyed her early letters after one of their quarrels during this long and troubled engagement, or after her death.[72]

One of the worst periods of tension and pain came on 25 May 1926, which was the first anniversary of the death of Bernard Lusted. On the previous day Dorrie announced to him that on the following day she did not want him to touch her, or speak to her, or see her. He was, he wrote later, 'sick at heart, and consumed with jealousy, of a man who is dead, and of the years of your life that are closed to me'.[73] He went to dinner at the hostel for the first time in many months, and drew looks of surprise and speculation from his colleagues as a result.

But the lovers got through this crisis too, and he and Dorrie became officially engaged on 20 September 1926.[74] And this official engagement, itself much delayed, was followed by a further twenty-one months of waiting before the wedding took place. Of this further delay Paton says nothing at all, and as far as his relationship with Dorrie goes, 1926 and 1927 are lost years: not a letter, note, or scrap of paper that passed between them has survived, so that the record must remain silent. But there can be little doubt that the storms in their relationship continued, and there must have been periods when they both wondered if they should go any further, for all that Paton, speaking with piety after Dorrie's death, described this period as 'halcyon days'.[75] He continued to suffer agonies of jealousy over her: jealousy not just of her dead husband, but of her habit of flirting with other young men: 'being cheeky with them' was the way Paton put it. He was particularly hurt by her 'cheekiness' with a young Scot who worked on her brother's farm, and who was probably in love with her too.[76]

Nor were his relations with Dorrie's parents always easy. Her father was almost as taciturn and mean with money as his own, and almost as oblivious of his family's real needs. He made no more mention of the planned marriage than James Paton had done. Even when Paton, in the fashion of the times, formally asked his permission to marry Dorrie, he only grunted 'Have you asked Mother?'[77]

'Mother' was not much more approachable: she was a formidable matriarch who ruled her family and her husband, deferring to him only in matters of money. Dorrie feared her; Paton treated her much as he had done his father, with deference and obedience, and in consequence got on well enough with her. But he did not love her, and she knew it. 'There is something in my nature, good or bad or neutral I do not know, that will not let me love a matriarch,' he was to say.[78]

Meanwhile his work at the school was going only a little better. At the end of his first year, 1925, he had had the sobering experience of seeing the school's matriculation candidates, four of them, sit the public examinations, and all four failed.[79] Three pupils were promptly removed from the school and sent elsewhere, one of these being Dorrie's younger sister Ruth. Paton must have felt this as a personal vote of no confidence in himself on the part of the Francis family. 'I might have thought I was not fitted for teaching,' he wrote, 'except that we did excellently in the Junior Certificate (standard VIII) and gained our first first-class pass.'[80] Fortified by this success, he continued his efforts, and in the following two years the school's results improved so much that, as he proudly recorded, parents from other districts began sending their children to the Ixopo School.[81] And he never forgot his pride at hearing that he had been praised by Mrs Francis for these improving results, with the words, 'Alan is going to make a mark in the world.'[82]

This was the period when he walked the hills with his pupils, sometimes to their parents' farms in the country, sometimes camping in one of the heavy bell tents he had borrowed from the SCA camps. His relations with his pupils must have improved for any to have been willing to accompany him, though the memories of those still alive do not substantiate this. He walked with Dorrie, too, and the two of them played as partners for the Ixopo second tennis team. They also began acquiring possessions together, the most important of these being a second hand Chevrolet car, the first Paton ever owned, and for which they paid £175.[83]

The long-awaited marriage took place in Ixopo on 2 July 1928 in the beautiful little stone church of St John the Baptist, with Reg Pearse acting as Paton's best man.[84] The church stands next door to Morningview, and the reception was held in Morningview's extensive garden. According to Pearse, Paton was so nervous the night before that he was unable to sleep, and at dawn he called on the priest for words of advice and comfort. The wedding was delayed by over an hour, for when the appointed time, 11 a.m., came round, the bride, as if reluctant to the last, was found to be still in her bath.[85]

Paton's parents had come over from Pietermaritzburg with his sisters, though they must have disapproved both of the strange ceremony and of the alcohol consumed at the reception. Dorrie Lusted, when she eventually appeared, wore mauve. She seemed to Pearse a slightly pathetic figure, obviously thinking of Lusted even while plighting her troth to Paton.[86] Paton himself, resplendent in the first tailor-made suit he had ever owned, drank a glass of champagne before his parents' very eyes, the first alcohol to pass his lips since that memorable but disappointing bottle of port in Durban. From this point on he would never be teetotal again: Dorrie would have laughed him out of it, for hers, as he justly remarked, was a morality concerned with affirmation, not prohibition.[87] For many years he would drink beer, a beverage which he admitted to not liking, and presently he discovered whisky which (once he could afford it) he drank daily until his old age.

They had decided to spend their honeymoon at the Victoria Falls in what was then Southern Rhodesia, driving there and back in their £175

Chevrolet, a bold plan. The first night was spent in a hotel near a more modest waterfall, at Howick in Natal. What should have been a time of joy was an occasion of more pain and bitterness for Paton. When they were in the hotel he noticed that though Dorrie wore his new ring on her left hand, on her right was the ring of Bernard Lusted. He said nothing until they were in their bedroom, and then he asked her, with what sadness can be imagined, 'Are you still going to wear your first wedding ring?' And she replied, 'Yes.' Even in the account Paton wrote long after the event, the sorrow of not being able to escape the dead man, even on his wedding night, comes through the attempt to put on a brave face:

What did I do? And how did I feel? I cannot remember. I do not believe that anything I might write here would be the truth. So I must relate only those things that I know to be true. I must record here that whatever the pain, it was outweighed by the joy of being alone with you, of sleeping in the same bed with you, of fully loving you for the first time, of being in the same room with you, of knowing that you were now Mrs Alan Paton.[88]

She was his in name, certainly, but her divided loyalties and his agonized jealousy and sense of exclusion were to continue for decades. Many years later their younger son Jonathan was to remark, 'The whole question of my mother's having been married first was one that lived, as a kind of guilt, through virtually my parents' whole lives.'[89] Paton made the best of a bad job, and did so with courage and in silence.

He was not a man who wrote about sexual matters with any ease, and his relations with his wife were often too painful for him to want to think about them, as when he defensively says he cannot remember what he felt on his wedding night. But scattered through his published writings and his letters are rare but unmistakable hints that he found his marriage sexually unfulfilling. This was partly because Dorrie very rarely responded to his own passion, and she sought and found ways to restrict sexual contact in the marriage. She found the most effective of these when their second child was born, as we shall see, but she seems very early on to have rationed Paton to infrequent days and particular times, an enforced routine that for him took much of the joy out of sex. There is a rare passage in one letter he wrote her when they had been apart for some months, in which he touches sadly but philosophically on the matter:

I wish I were sleeping with you. I wish indeed I could always do so, not to make love to you, but just to be near you. But that is one of the things that you have had your way in, wrongly I think, so that the making of love has become a regulated thing, & often goes astray. . . . I think you have always separated sex from love, but if you can't change, you can't. But you must never doubt my love for you.[90]

This letter was written in 1960, by which time Paton was in his late fifties. It was a position reached after long struggle with his own passions, over more than thirty years, and yet the pain and sadness are still there. For the ardent young man on his honeymoon that first wedding ring and all it symbolized must have been terrible indeed. 'It is part of the story of our life,' he was to write. 'But it is only a hundredth part of it. I hope I shall never think about it again.'[91] The fact is he went on thinking about it until the end of his life. At the age of 81 he would write a poem about the Victoria Falls, associating them with both passion and terrible loss.[92]

6

MARITZBURG COLLEGE 1928–1930

Writing is not a profession but a vocation of unhappiness.

GEORGES SIMENON

On their return from the troubled honeymoon the young couple packed up their possessions in Ixopo and moved to Pietermaritzburg, for on marrying Paton had decided to take a post at his old school, Maritzburg College. Among the advantages of this move were that a bigger school would offer more scope for his professional ambitions, for he hoped to rise rapidly. In fact he nursed the secret desire to be headmaster of Maritzburg College by the time he was forty,[1] or, failing that, to have moved into the inspectorate, using it as a stepping stone to some bigger school.[2] There is little doubt that his marriage made him a more worldly man, and the move to the greener pastures of Pietermaritzburg was part of this change. In addition it is likely that Paton saw advantages in living away from his parents-in-law, and from the wagging tongues inescapable in a small community. Lastly, he may have wanted to be near his parents, for his father was now in decline.

James Paton had been suffering headaches since 1923, and these worsened about the time of Paton's marriage. In addition he was increasingly afflicted by forgetfulness, and was reproved by the Judge-President himself. 'My Dad is anything but fit,' Paton wrote to Pearse as early as May 1923, 'in fact he is getting too old for work.'[3] At fifty-one James Paton was clearly not too old for work, but he was an embittered and disappointed man, alienated from his family, and after 1923 his decline was rapid. Early in 1925, when Alan Paton was in his first year at Ixopo, his father was obliged to retire from the court, and he never worked again, becoming increasingly withdrawn and morose. Paton, as the eldest son, may have felt it his duty to be near his mother at this difficult time.

He and Dorrie arrived in Pietermaritzburg at the end of July 1928, and he began teaching at his old school in the second half of the year. He came back to Pietermaritzburg a very different man from the carefree, jokey student who had left it for Ixopo less than three years before. His responsibilities as a teacher, and his responsibilities and disappointments in marriage, had both saddened and matured him. Instead of the constant smile one sees on his face in photographs in the 1920s, there had developed in him the habit of frowning, with a pursing of the lips like a man who has bitten into a lemon. Even his stance had altered: he developed a characteristic stoop, involving not so much a rounding of the shoulders as a thrusting forward of the head and neck. One acute observer[4] was to describe this as making him look vulturine; it also gave him an appearance of pugnacity, as if he were leading with his chin and daring life to hit him again. The appearance both of bellicosity and severity would increase as he aged: like one of the characters in *Cry, the Beloved Country*, he hid his gentleness behind a fierce and frowning mask, and it was at this time that he began to wear it. It was, in part, of Dorrie's manufacture.

The return to the town where he had been a brilliant boy and a carefree student made him vividly aware of how much he had changed. A brief poetic fragment, 'Sterility', almost certainly written at this time, expresses his emotions, and a new note of sadness sounds in his verse:

> When I was young my songs I sang
> I thought their like could never be,
> Now I am old and know the worth
> Of all the songs that are on earth
> They seem such artless melody.
>
> And now with ripe and mellow songs
> That from the womb call me and call
> I find I cannot sing at all.[5]

The young couple bought a small house in Gough Road, the conveyancing being done by Paton's school friend Vic Harrison, who was now practising as a solicitor.[6] Gough Road was then a quiet street having only three houses along one side, of which the Patons' was the middle one, and on the other side vacant corporation land. From here Paton could cycle in to work, and he was also in easy reach of the countryside, which by now he knew intimately, as only an habitual walker can get to know it. He was to buy a bigger house at 10 Howick Road early in 1932, the death of Dorrie's mother in the previous year having left the young couple better off.

He enjoyed being back at his old school, which was certainly a large step up from the Ixopo school in terms of size and organizational complexity, though perhaps not in the challenge it posed to the teachers. Paton had no task as difficult as coping with big boys for twenty-four hours a day, or taking sole responsibility for teaching three and more subjects, or having to see himself as moral guardian of blossoming girls to whom he was himself attracted, as at Ixopo. Instead he was part of a large staff who shared the load of teaching boys only, and after school he had a wife and home of his own to escape to. This should have meant that he could have relaxed his tendency towards excessive strictness, but he did not.

In his autobiography he was to assert, as he had about his teaching at Ixopo, that he wanted to be a popular master, and succeeded. His IIIA class, he relates, was always full of laughter and noise, so that the Afrikaans master, Lucien Biebuyck, called it 'the Tickey Bazaar', and reproved Paton for saying that he liked his class to like him.[7] In the same way he had asserted that at Ixopo he had been actually loved by a large proportion of the two hundred children.[8]

At Maritzburg College, as at Ixopo, he rapidly acquired and kept a reputation as a particularly fearsome master. Not many of the boys could see past his developing mask. At least one of his pupils, Deryck Franklin, was taken away from the school and sent to Hilton College because his parents did not want him to have to endure Paton's beatings.[9] Franklin, who thought Paton sadistic, remembered that he had a particular skill in coming quietly up behind a boy who was misbehaving, and suddenly seizing and twisting the short hair above his ears so as to cause exquisite discomfort. Another pupil, Neil Alcock, remembered Paton as a 'much-feared master', and when in later years his wife wrote to tell Paton of this, he seemed not displeased to have acquired this reputation.[10]

Even before Paton's death, accounts of his methods had appeared in print. The official history of Maritzburg College, *For Hearth and Home*, asserted that Paton suffered from uncontrolled rages. 'For, once aroused, Paton, like all intense men, had a dangerous temper,' the authors wrote, and they went on to record an occasion on which Paton thought a boy's answer to a question insufficiently respectful:

> Seizing the hapless and surprised youth by the scruff of the neck, the enraged Paton literally hurled him out of the classroom door with such violence that the poor fellow did not touch ground till he ended up in a crumpled heap against one of the verandah pillars. W. H. Mutton, another Old Collegian, remembers that Alan Paton had a habit of 'grabbing a tuft of hair and shaking your head that way'. On

one occasion he performed this ritual a little too enthusiastically with C. B. Smith who ended up sporting a sizeable bald patch as a result.[11]

This account was published while Paton was alive and was probably seen by him before publication, but he made no objection to the passage. Perhaps his violent tempers were part of the explanation for his behaviour, for certainly he was a passionate man in more ways than one. But in addition he was plainly continuing his father's methods of discipline, and would only gradually wean himself off them.

But this was only one side of Paton. His schoolwork and his home life alone could not absorb all his apparently inexhaustible energies. He taught mathematics to apprentices at the town's Technical Institute at night to make extra money, but was fired when he tried to apply school discipline to these resentful young men.[12] He continued his work for the SCA camps, cycling down to Anerley each winter, and was acknowledged as being, with Pearse, the main force behind them.[13] But he also began looking for other practical ways of applying Christian morality to bettering the lives of young people in Pietermaritzburg. Growing social problems in the town seemed to offer a field of opportunity.

The Wall Street crash of 1929 was slow to have an effect on South Africa, but as the 1930s advanced the Great Depression gradually came to grip even bucolic Pietermaritzburg. The economic downturn showed itself in increasing numbers of unemployed boys and youths on the streets. The 'poor white' problem, as it was known, had arrived, and drew much concerned comment in the papers. The poor black problem, even larger, was ignored.

The municipality provided relief schemes for adults, but did nothing for boys, and Paton felt called upon to act: 'It began to weigh on my conscience. Why this should have been so I am not quite sure.'[14] The fact is that despite all his own unhappiness, he had a strongly developed moral conscience that impelled him to act when he saw a need. And somehow he could not help seeing the need of poor white boys, though thousands of his fellow citizens were blind to them, as he himself was still blind to the needs of blacks.

When he thought of Africans at all, he seems to imply that lazy whites will sink to the level of blacks, as in a poem entitled 'No Responsibility Accepted', which he published in the *Natal University College Magazine* in 1929:

> Ho, Long One, what do you do?
> I build, Master, with wattle and mud,

And finished, lie me down in the sun
And watch my toes lift, one by one,
That's what I do.

Ho! ho! Long One, that's what you do.
Your fathers built with spear and blood,
And you, ho! ho! with wattle and mud,
That's what you do.

Laugh well, White One, you laugh today.
Tomorrow we will build the white man's way,
Tomorrow, Master, what'll you say?

Tomorrow, dog, will be as to-day.
Think you to walk where the white man walks?
Think you to talk as the white man talks?
Think's all you'll do.

Tomorrow, Master, will be as to-day,
Tomorrow the same—in a different way,
Your best, our best, will lie in the sun,
And watch *our* toes lift, one by one,
That's what we'll do.[15]

He deplored, as many did at this time, the growing idleness and waste of a whole generation of white youth. His willingness to serve, to do something about human need rather than merely deploring it, was an outgrowth of the moral concerns that produced such mixed results in his school work and that impelled him to help organize the SCA beach camps each year. It was the desire to make of his religion something practical, to do and not to preach. Paton took to heart Christ's prescription for distinguishing good men: 'By their fruits shall ye know them'. In consequence, his was to be a life of works first, and words second.

Among his most powerfully suggestive and ironic poems is 'The Hermit', written in 1931, in which he depicts the crippling consequences of a refusal to respond to the needs of others:

I have barred the doors
Of the place where I bide,
I am old and afraid
Of the world outside.

The poor souls cry
In the wind and the rain,
I have blocked my ears,
They shall call me in vain . . .

Who will chase them away,
Who will ease me my dread,

Who will shout to the fools
'He is dead! He is dead!'?

Do they think, do they dream
I shall open the door?
Let the world in
And know peace no more?[16]

The image of the closed door, deriving from Christ's words, 'Behold, I stand at the door and knock', was to recur many times in his writing. He firmly believed that to refuse to open your life to the needs of others and to do something practical to help them is to imprison and impoverish yourself. He who saves his own life will lose it.

He quickly realized that he could not help the street boys on his own. He looked about for a structure within which to act, and decided on Toc H, which he had first heard of in 1928.[17] Toc H was a British ex-servicemen's movement which had been started in France near the Ypres salient in November 1915 by two young British Army chaplains, Philip (Tubby) Clayton and Neville Talbot. Their aim was to spread among the soldiers the ideals of worship, fellowship, and service to one another. They started a club-house just behind the lines, and named it Talbot House: its initials, TH, or, in the signaller's alphabet, Toc H, became the name of the movement as a whole. Under Tubby Clayton's inspired leadership the ideals of Toc H spread widely after the war, first through Britain, and then through the Empire. Paton, who was to be profoundly influenced by the movement, described its goals like this:

> Its aim was nothing less than to achieve a new spirit between man and man. . . . It attracted the great and the not so great, the rich and the not so rich, the professional man and the worker in the factory. It was a religious movement without dogma, it was Christian but not closed to any man who could subscribe to its principles. Its aim was to spread the Gospel without preaching it.[18]

Since this was Paton's aim precisely, and since he had acquired in reaction to his Christadelphian upbringing a dislike of all dogma, he found himself powerfully drawn to the organization.

Toc H spread to South Africa in 1926, and by the following year had attracted so many members in Pietermaritzburg that they had to be broken into two groups. When Paton first made contact with it, he was still living in Ixopo, and it was he who organized the launching of a group in the village, in 1928.[19] When he moved to Pietermaritzburg later that year he naturally attached himself to one of the Toc H groups there. Apart from its ethos of peace and reconciliation between all human beings, Toc H involved its members in practical works of charity. The two most important officials in any group were the padre, who taught the ideals of the movement, and the jobmaster, who put them into practice, distributing among the members the 'jobs', ranging from visiting the sick to raising money for a boys' hostel to accommodate country youths drawn to the city.

Paton threw himself into working for Toc H with all his heart. His dedication was such that by the early 1930s he had become the senior administrator of Toc H for all of Natal, and for a quarter of a century he would continue to give selfless (and unpaid) service to the organization. During the war years he would rise to be the National Chairman of the YMCA-Toc H War Work Council, and in 1949 he would be appointed by the British head organization as Toc H's Honorary Commissioner for the whole of southern Africa, the most visible official in the organization.

In 1930 all this lay in the future. But Paton had already decided that if something were to be done for unemployed boys in Pietermaritzburg, Toc H was the organization to do it, and he the man to get things moving. The Pietermaritzburg Municipality had begun constructing a large wild garden in the city's biggest park, Alexandra Park, and in this work employed jobless men in an attempt to keep them off the streets. Paton came up with the scheme of collecting money among Toc H members to pay white boys two shillings a day to join the Alexandra Park work, and when the adult supervisors objected to coping with unruly boys, he moved them to the Botanical Gardens.

The scheme was a great success: the boys were kept occupied and got useful pocket money for their efforts, and the Botanical Gardens benefited from their labour. The greatest part of the work involved in this imaginative scheme lay in the collection of the money to pay the boys, and this again Paton organized, finding Toc H men willing to make up lists of subscribers and go the rounds collecting the money each month. He founded a social club for homeless boys, and by 1935 had also launched a Toc H hostel and made himself responsible for the discipline of the twenty or so boys who lived there.[20]

Between his ceaseless work for Toc H, the club, the unemployment scheme, and his extra teaching, with the SCA camp in the winter holidays, Paton must have had very little spare time, since a teacher in South Africa is expected to do a huge amount of extracurricular activity when formal classes end. Dorrie must have seen very little of him, and it is hard to resist the conviction that his outside life was a substitute for the happy home life he had hoped for during the long years of his wooing. Dorrie continued to wear Bernard Lusted's ring, and she continued to endure rather than respond to Paton's attentions. At the end of his long marriage to her he was to imply that he could remember every one of the times when he had found her sexually responsive: 'I shall not easily forget the times when you showed your affection. They were mostly times when I had been away, perhaps for many months.'[21] On another occasion he was to recall in particular just two of these happy times, and the implication is that there were not many more to remember.[22]

Paton could of course have turned to other women for solace, but he was too morally and personally fastidious to do so. All the same there is a suggestion in many of the poems he was to write over the next few years that he had been tempted to patronize prostitutes, and had put the thought from him with loathing. In 1931, for instance, he published in the *Natal University College Magazine* a poem entitled 'Poor Whites', undoubtedly inspired by his work with unemployed boys, who appear in the first stanza. But from the second verse onwards, in spite of its ostensible subject, the poem begins to focus on sex—firstly that of animals, beginning with the jackal:

> The jackal prowling thro' the tares
> Of starlight fields that once were theirs
> He knows no fret of bars,
> His limits are the stars,
> He has no cares.
>
> When body hungers with desire
> His call will waken answering fire,
> And some as-hungry mate
> Will answer him and wait
> For him to sire.

Next, even more oddly, the sexless ox:

> The ox that dumbly draws the load
> Along the sweet and lonely road,
> He has the dark for friend
> And at his journey's end
> A safe abode.
>
> Of passion is his body shorn
> As virginal as he was born,

Who knows but half-ashamed
It surges mute and maimed
And sinks forlorn?

But it is not jackal or ox that Paton chooses to end the poem on, nor even the poor white boys, but their sisters:

And in their slattern veldskoen shod
That many a sweeter path have trod,
Their women walk the street
And sell for bread and meat
The gift of God.[23]

In February 1931 he wrote an entire poem on this subject, 'The Prostitute'. In it he sees the bought woman as a deadly trap for innocent but tempted young men:

Your eyes are heavy-lidded mystery,
They mock, or drop demure, or fill with tears
To make a eunuch dream of dead delight.
What matter if some boy ere he can flee
Sees the sweet building of the dogged years
To earth go crashing in one tragic night?[24]

And in yet another poem of this period, 'Scottsville, 1931', Paton dealt with another aspect of sex outside marriage, this time rape:

Where with his heavy-swinging trunk
The elephant went up and down,
The lights are strung along his track
And trams go through the paved town.

Last night a woman ran to me
And sobbing caught me by the arm,
Gone frantic with the sudden fear
That she might come to woman's harm.

I saw no soul, I heard no sound,
But knew that in some modern man
Under the lights of town, the old
Primordial savage stirred again.[25]

There can be little doubt that easily accessible but forbidden sexual pleasure fascinated Paton, just a few years after his marriage. In these early poems it is the prostitute; in later years it would be the black woman. To be tempted is not to fall, and there is no reason to think that he resorted to prostitutes himself at this time. But equally there is no doubt that his unsatisfactory marriage exposed him to temptation which he had to fight to resist.

The one time of year when he and Dorrie could spend whole days together was the December holidays, and even these Paton on several occasions spent apart from her, as when, at the end of 1928, he travelled to the Cape to live for a month on a farm to improve his Afrikaans. This trip alone would be enough to suggest that something was badly wrong with his marriage; it is a rare young husband who chooses to spend his first extended holiday since his honeymoon living on a farm many hundreds of miles away from his new wife among a family of strangers.

This was the second effort Paton had made to learn the language, and an essay he had produced survives to show that he already wrote correct, if rather stiff, Afrikaans, and was able to draw on a large vocabulary.[26] During his stay on the Boland farm near Porterville he made rapid progress with the language, and learned to speak and write it idiomatically, and to love it. Afrikaans has all the richness, suppleness, and pungency of idiom of Chaucer's English, but it also has the sophistication one might expect a highly industrialized and modern society to produce. And with his grasp of Afrikaans, Paton's sympathies for the Afrikaners themselves grew. As he later wrote,

> I had by this time reached the judgement that the Anglo-Boer War was a scandal and that although it had been declared by President Kruger, it was really the work of Chamberlain, Milner, Rhodes, and Jameson. This sympathy for what one might call Afrikaner nationalism lasted ten years . . .[27]

His sympathy for the Afrikaners was one of the first signs in Paton's life that his theoretical Christian belief in the brotherhood of human beings was beginning to allow him to transcend racial boundaries.

South Africa is fragmented by racial, tribal, and linguistic borders that make it one of the most complex societies on earth. Among the black African population there are seven major groupings, distinguished by tradition, language, and customs from one another, and making up by far the greater part of the population. There is a much smaller group of people of mixed race, known as 'Coloureds' (of whom the 'Cape Coloureds' are a distinct subset), another of Indians introduced as indentured labour and living chiefly in Natal, and a white group. The whites are further divided into those whose home language is Afrikaans and those who speak English. And of all these many chasms, Paton had begun to bridge only the smallest, that between Afrikaners and English speakers. Even that he did not find easy, but it was a start.

His chief reason for wanting to know more about Afrikaans and the Afrikaners at this period seems to have been ambition. This urge to get on in his chosen occupation does not seem to have come from his wife. Dorrie had been quite happy, in her first marriage, to be the wife of a partner in a garage and petrol station in a tiny village, and she seems to have been equally happy to be the wife of a teacher. But Paton did not intend to remain a schoolmaster. At the least he intended to become a headmaster of his school, and as quickly as could be managed; but already his ambitions went further. He had begun to dream of a brilliant career in politics, and for this a thorough knowledge of Afrikaans, he believed, was going to be required. He dreamed that he might be capable of uniting Afrikaner and English speakers, if not single-handedly, then as part of a team; and the most likely leader of that team, he considered, was a new and powerful friend, Jan Hendrik Hofmeyr. Hofmeyr was in fact the chief inspiration of Paton's dreams.

Paton had first met J. H. Hofmeyr at the third of the SCA camps, which was the second Paton actually attended, in 1927. He was 9 years older than Paton, being 33 when they first met, and of a distinguished Cape Afrikaans family, so that he and Paton were as different from one another in background, language, and traditions as two white South Africans could be. Whereas Paton had an extraordinary knowledge of nature, Hofmeyr neither knew nor cared about it. Paton could give the Latin, English, Afrikaans, and Zulu names of the most obscure Natal veld flower. Hofmeyr, by contrast, knew only two kinds of flower, as Paton remarked in wonder; one kind was the rose, and the other wasn't.

In addition Hofmeyr was physically unprepossessing. As short as Paton himself, he was stoutly built, a gross feeder, slovenly and indeed dirty in his dress, so that a visiting dignitary, Princess Alice, is said to have asked, 'I wonder who wears Mr Hofmeyr's clean shirts?'[28] He was ugly of face, and abrupt to the point of rudeness in his manner: an unlicked bear of a man who peered out at the world through spectacles as thick as bottle glass and without which, Paton claimed, he could distinguish nothing but night from day.[29] In religion he was unctuous and puritanical, though Paton was to say that: 'he was saved from being a puritan or a prig by his sense of humour, and by his wit too. As a teller of jokes . . . he had no superior.'[30] Yet Hofmeyr attracted attention at the camps, not for any of these striking features, but because he was the Administrator of the Transvaal, and therefore one of the most powerful civil servants in South Africa. That such a man chose to spend his annual holiday at a boys' camp on the Natal south coast was curious indeed.

The more Paton learned about Hofmeyr, the more fascinated he became because of the links he seemed to perceive between his own upbringing and that of Hofmeyr. Paton, as we have seen, had had a brilliant and accelerated school career. Hofmeyr had been a child prodigy who as a schoolboy had translated the Classical languages for amusement, and who had completed his schooling at the age of 12, taken a bachelor's degree in arts at the age of 15, a science degree at 16, a master's degree in Classics at 17, and a further degree at Oxford at 21. At the age of 24 he had been chosen to be principal of the Johannesburg School of Mines, which under his leadership became the University of the Witwatersrand. Despite having made himself very unpopular there (he had mercilessly hounded out a competent academic, Professor Stibbe, for what Hofmeyr's narrowly puritanical mother considered immoral behaviour), he had attracted the attention of General Smuts, who in 1924 had appointed him to the Administratorship of the Transvaal.

Paton's rapid school career had made him feel isolated from his fellows, and with Hofmeyr this process had gone much further, so that he was almost entirely a solitary. To make matters worse, he was unmarried and completely under the thumb of his mother, who kept him in subjection and fiercely repelled any woman who showed an interest in her son. She was in her way as bad as Paton's father, perhaps worse in that Hofmeyr never managed to break away from her influence. She seldom let him out of her sight except when he was at work, and when he had gone to Oxford she had accompanied him so that the two of them might go on living together even in England. When his official position called for him to make a trip abroad, she went too, and when he became a cabinet minister in the 1930s she made the annual pilgrimage from Pretoria to Cape Town with him. Anyone who hoped to befriend him, and his position meant that there were many who did, found that they had better befriend his mother too, and that one wrong move would be the end of them. 'Her sphere was petty, but in it she moved with terrible power,' Paton was to write.

This subjection Hofmeyr seemed to submit to gladly, but it helped to explain why he so enjoyed the SCA camps, for this was one time of the year when his mother could not be with him. He had been attracted to the SCA as a schoolboy, when he fell under the influence of Oswin Bull in Cape Town, and he continued to come once Bull, Paton, and the others started the camps in Natal. He enjoyed himself at camp in the most schoolboyish way, and indeed having been a small child among big ones at school, and a boy among men at university, he scarcely knew how to relate to people in any other fashion. The boy in Paton rejoiced to see the boy in Hofmeyr liberated, and he writes of these times with unconcealed glee:

Hofmeyr came to the second camp at Umgababa [as we have seen, it was actually the third], and then year after year to the new camp site on the Idomba River at Anerley, which site he helped the Association to buy and develop. He was a camper of the first water. He wore a canvas hat of uncertain shape and great antiquity, a khaki shirt and shorts, and discoloured sandshoes, known as tackies. He subjected himself in every detail to the discipline and programme of the camp, though he put on a show of subversiveness, and through this seduced more than one problem boy into developing a liking for obedience. He played every camp game. To oppose him at rugby football was a profound physical experience; this was one of the few activities for which he discarded his spectacles . . . If he got the ball, he would charge for the goal line, two hundred pounds of concentrated material. If one knew no oriental arts, it was a clear duty to keep out of his course, for not to do so was to risk total disablement. If some extra-heavy camper should try conclusions with him, the air would be full of cries, of agony from the one, and of uncontrollable giggling from the other. It was strange that one who almost certainly in the whole of his life had never struck a living thing in anger, should revel in this gross physical combat . . . At any idle moment of the day, especially before and after meals, boys would gather round him for a game. After a day or two they all called him Hoffie.[31]

In this atmosphere Paton got to know Hofmeyr as he could never otherwise have done, and the two became good friends, in so far as Hofmeyr's mother would allow him to be friendly with anyone. Their friendship grew with each successive camp, and it was made the closer by their shared religious enthusiasm, and their determination to make the world a better place by their passage through it. But whereas Paton's sphere of activity was confined to the classroom, Hofmeyr could sway a province, and would soon sway a country.

Even in the SCA Hofmeyr could accomplish quite effortlessly what Paton, for all his energy and dedication, could not. As Paton said, it was Hofmeyr who bought the site for a new camp, which he continued to own and manage until his death, at another south-coast village named Anerley, and he quickly recognized Paton's dedication to the SCA spirit of service, as well as his intelligence. In due course Paton began to be invited to the official residence of the Administrator of the Transvaal, and to meet his formidable mother. Paton treated her with the extreme deference he reserved for such matriarchs, and after a period of judgement she approved of him.

Hofmeyr was more than a friend: he was a man capable of helping Paton in his career, with informed advice if not with active assistance, and Paton began calling on his help, telling him about his ambitions and asking his opinion. His letters are filled with expressions of appreciation for Hofmeyr's friendship, which he clearly felt to be an honour:

> I too would like to put in black & white my appreciation of the time we snatched from camp activities to talk. It is for me an honour & a privilege to have your friendship . . . with you I always feel keen & alive, & a better man for it.[32]

These expressions of warmth were perfectly genuine, for Paton not only very much admired Hofmeyr, but liked him enormously. And because he admired Hofmeyr, he hoped Hofmeyr would rise rapidly in influence and power in South Africa, and hoped to rise with him. There can be no doubt that just as Hofmeyr had risen through the patronage of Smuts, so Paton hoped to gain the patronage of Hofmeyr. So he wrote about his career, and tried to keep it before Hofmeyr's eyes:

> Confidential—It is very much on the cards that I may be put on the Inspectorate. I was in a great dilemma, but I feel unscrupulous enough to use the job for a few years, & then put in for a bigger school than I could otherwise have done. If I don't see my way clear to getting my old school before I'm 50—& there's little chance—I may make for the Superintendancy. To that end I plan to take a doctorate in London in five year's time. But this Province seems to present one closed door after another.[33]

This last remark was aimed at what he knew to be Hofmeyr's dislike of provincial rights, and his wish for strengthened powers for the central government. Paton meant to show Hofmeyr that the two of them were on the same side of the political fence, because in 1929 Hofmeyr had decided to enter politics, and Paton hoped eventually to join him. But he did not confine himself to career and political matters; he would also tell Hofmeyr about the latest novel he had been reading, for Hofmeyr was keenly interested in literature, and he would mention such family matters as the sudden death of Dorrie's mother in August 1931. To these letters Hofmeyr, who was famous for the speed and brevity of his replies, would return terse notes, but Paton was not discouraged.

Meanwhile, in 1930, Paton's family life was altered by two very different events. One was the birth on 7 December of the Patons' first child, a boy, whom they named David (for Paton's Scottish grandfather, who had died before his own birth) Francis (for Dorrie's family). The baby was a great joy to Paton, and to Dorrie, particularly when her own mother died suddenly some months later. 'Dorrie is in Ixopo with her father, and I thank God for David,' Paton wrote to Hofmeyr shortly after old Mrs Francis's death; 'his prosaic hungers & thirsts & his unintelligible chatter have proved an all-powerful diversion.'[34] David's first memories were to be of the Gough Road house, memories as vivid and discrete as old photographs, as childhood memories usually are:

> I remember living in Gough Road, my father's first house. I remember eating an icecream out of a little tub there. Sitting on the steps in Gough Road. I had a pedal car there. I remember going to see some of my father's friends. One lived up Town Hill, I don't know the name. I went on the bar of his bike. I think he cycled to work.[35]

The other event in 1930 was much more sombre: it was the mysterious murder of Paton's father. James Paton, as we saw, had been forced into retirement in 1925. Embittered and increasingly withdrawn, his one remaining solace seemed to be his marathon walks over the hills, and he would often go off into the countryside on his own for a whole day. On 1 May 1930, a Thursday, he set off for one of these walks, heading for the Town Bush Valley, and did not return. His wife did not take fright until Saturday 3 May, which suggests that she was used to his staying out all night on occasions. This in itself is a curiosity, for the missing man had taken no camping equipment with him, nor food for more than one light meal.

On Saturday 3 May his mother phoned Paton, who had gone with Dorrie to Ixopo for the weekend, and asked him to return at once. He did so, and he and his brother Atholl began searching the countryside. They hunted for two days in vain before they raised the alarm, their father having by now been missing for five days. Paton was to write that they believed James Paton might be suffering loss of memory.

On 6 May 1930, and thereafter for many days, scouts, ramblers, schoolboys, and soldiers searched the great wooded hills round the town, but there were endless places where a body might have lain undetected, and they searched in vain. It was learned that on the day of his disappearance he had spoken to a roads overseer whom he had met, and that he had also called on a friend named Sowerby Mason for a drink of water, but his conversation with them gave little guidance as to where he had intended going.[36] On 21 May, nineteen days after her husband was last seen, Paton's mother Eunice accepted that James Paton was dead, and in the columns of the *Natal Witness* thanked those who had helped look for him. 'We are just heartbroken, but seek our comfort from the Unseen Presence,' she wrote to a well-wisher at this time.[37]

Then, on 17 June 1930, an Indian boy walking in the hills noticed a large rounded object protruding from the water of a pool not far from where James Paton had spoken to the roads overseer on the day of his disappearance. Another Indian helped him draw it to the bank, and it was found to be the badly decomposed corpse of a man. He had been floating face down; the skin of his back, which had showed above the surface, had been dried hard as tortoiseshell by the sun, while the rest of his body was in a very advanced state of decomposition. Paton and his brother Atholl had the terrible task of identifying the remains, and, according to their sister, were both sick. They were, however, able to confirm that it was their father. The dead man's mother, now in her nineties, was with great difficulty prevented from seeing the body, and she herself died within a few months, never recovering from the grief of this murder.[38]

Excited rumours at once ran through Pietermaritzburg's white population. It was widely believed that the body had been mutilated and some of its organs removed in a ritual killing by blacks, or what is known as a 'muti murder'.[39] In his autobiography Paton goes to the other extreme, quoting the police as denying that any murder took place, and suggesting that his father had suffered a heart attack while bathing. The inquest findings, however, directly contradict this suggestion, and in fact his sister remembers the police telling her mother that it had been murder beyond doubt.[40]

The body had been stripped of all possessions, not just the clothes and watch that might have been found on the bank had James Paton gone swimming, but the gold ring he habitually wore, and which he would not have removed for a swim.[41] What was more, the inquest found that his lungs contained no water, as they would have done had he drowned or lapsed into unconsciousness in the water, and that they were, on the contrary, in a state of collapse. The doctor who conducted the inquest, Dr T. Albertyn, stated that death had occurred before the body was put into the water, and gave it as his opinion that the cause of death was murder by suffocation.[42] The police clearly suspected the involvement of blacks in the killing, for they searched all African huts in the vicinity but found nothing. On the face of it the most likely motive for James Paton's murder was robbery, and it is conceivable that he might have been killed for his gold ring alone. But there is another possibility, which is that his death was a crime of passion.

James Paton was, in his son's words, 'very susceptible to the charms of girls and young women'.[43] His strict Puritanism concealed and perhaps helped to control a strong libido. His eyes seem to have gone in every direction, and his fierceness in guarding his daughters even against their brothers suggests this. But his children were not deceived; when he began to invite young girls to the seaside cottage the family occasionally rented for a month, ostensibly so that his sons could be 'introduced to the world of girls', his daughters speculated that they were invited, not for his sons, but for his own pleasure.[44] And the whole family knew that their father's friendship with a family that included four attractive girls was motivated by his desire for the eldest girl, whose hand he would hold under the table when they came for tea, something Paton's mother resented deeply but endured in silence.[45] Reflecting on these things, Paton was to write,

> I do not know that he ever had a liaison, and I guess that he did not. The conventions of those days were so strict, and our own circle of friends so proper, that one would have had to defy both conventions and friends and face a first-class scandal. My father could not have done it, whether he had wanted to or not.[46]

But a liaison among the daughters of his friends was not the only option open to a white man in Pietermaritzburg. In the town propriety of a sort was observed, and the sight of a woman's knee was a rarity. In the great hills over which James Paton walked so often, though, a common sight would be a young woman, or group of young women, walking naked but for some beads and a small piece of leather strategically suspended from the waist. The tribal people who made up the great bulk of the population went about all but unclothed, and towns like Pietermaritzburg and Durban had by-laws requiring the men at least to wear trousers on passing the city limits.[47] A man susceptible as James Paton was to young female charms would find them everywhere displayed in the smiling countryside. Since 1927 marriage between black and white had been illegal, but there were many whites willing to flout the new law, as the growing mixed-race population amply demonstrated.

James Paton had plenty of contact with the Zulus on his walks; he habitually conversed with them, and his atrocious Zulu set them laughing. But a white man, even if a buffoon, was also a figure of power in this society, and a black girl would have found the determined approaches of such a person hard to resist. Because he understood the language so poorly he would not have perceived the limits of acceptable behaviour in tribal society, and it is conceivable that persistence in such behaviour would lead him to a violent end.

This possibility is given credence by the slowness of the Paton family to sound the alarm: it is hard to escape the conclusion that they feared James Paton had got himself into some scrape from which they hoped to extricate him with the minimum of publicity, and that they continued their quiet search for nearly a week before taking fright and calling in the authorities. Victor Harrison, who saw something of Paton during this terrible time, remarked that even in his autobiography Paton did not mention the way in which he had walked himself to exhaustion every afternoon after finishing his schoolwork, and during the weekend, in searching for his father.[48] The truth will probably never be known, but certainly Alan Paton shows in his later writing, notably in *Too Late the Phalarope*, a fascination with the psychology of the white man, apparently a pillar of rectitude, wrestling with his overwhelming desire for a black woman, and ultimately falling to his destruction.

A final sign is that James Paton's murder did not, as might have been expected, decrease his children's hatred of him: rather the reverse. Atholl Paton and his sister Eunice showed detestation of their father all their lives; and Alan Paton's autobiographical accounts of James Paton are marked by a determination not to speak ill of the dead, rather than by any hint of warmth. In later years his hatred for his father subsided, as he came to feel that he understood the older man better through the difficulties of his own family life, and the imaginative effort involved in being a writer and getting inside the skin of characters as alien as James Paton. But even in his own old age he could not summon up affection for the strange, cruel, tormented figure who had been his father. 'For all his jokes and jollity, his life had in some way been solitary,' he was to write, 'and he made it more so by alienating the affection of his children. Now of course I think of him with nothing but pity.'[49]

7

REVALUATION
1930–1935

A child is sleeping:
An old man gone.
JAMES JOYCE

The death of his father, and the birth of his son, seem to have made Paton re-evaluate his own life, and there are several signs that he was taking stock of himself and considering in which direction he should move. One such sign was his growing religious commitment, which showed itself partly in his renewed impulse to serve others through Toc H and the camps, partly in his decision at this time to become an Anglican. He had set about getting the padre of his Toc H branch to prepare him for confirmation in the Anglican communion, without telling Dorrie, and for the ceremony itself he chose the annual holidays which they often spent apart.

At the end of 1931 Dorrie and little David were in Ixopo, staying at Morningview, when Paton was confirmed by the Bishop of Natal, Leonard Fisher, on 10 December 1931, in the private chapel of Bishop's House in Pietermaritzburg, in a ceremony which marked his final break from Christadelphianism and the traditions of his family. Dorrie, he was to say later, was astonished and moved when he wrote to tell her what he had done. Among his motives were a desire to draw closer to her by joining what he thought of as her church: 'I did not want our children to grow up in a divided home'.[1]

His mother and sisters were deeply saddened by his final abandonment of Christadelphianism, considering him an apostate. They drew away from him, and this rift, which they felt more bitterly than he, was never to be healed. Nor is there much evidence that his relations with Dorrie, after her surprised reaction, were greatly changed by what he had done. In their relationship he remained the suppliant and the pursuer, and she continued to maintain her dominance and control. For all that, he had found his spiritual home at last, and though he was to have differences with the church hierarchy in years to come, he communicated Sunday by Sunday from this time on and remained a faithful Anglican until his life's end. Anglicanism, for him, represented freedom from the dogmatism in which he had been brought up, and in later years his faith would be a vital source of his Liberal creed.

A poem he wrote on 3 April 1932 and published in the *Natal University College Magazine* shortly after, 'Trilemma', vividly suggests his inner debate at this time as to what he should be doing with his life:

> I dreamt three students walked a road,
> Nobly degreed and capped and gowned;
> A humble labourer in a field
> Close on the roadside tilled the ground.
>
> One student wrapped in lofty thought
> Passed by with neither sight nor sign.
> I saw his face beneath the hood
> And gaped bewildered—it was mine!
>
> One student smelt the honest sweat,
> Screwed up his nose in cold disdain,
> I saw his face beneath the hood,
> Gaped more bewildered—mine again!
>
> One student leapt the roadside hedge
> And tilled the ground without a word
> Beside his mate—I saw his face,
> This dream was growing more absurd!
>
> But most absurd of all was me,
> The real me, not the other three,
> Going from hood to hood to see
> Which of the three was really me![2]

Intellectual life, worldly ambition, and the desire to serve his fellow human beings: these were the three forces which tugged at him at this time, and the struggle would be long. But already the signs of increased religious commitment were there to suggest that the third dream-ego was the likely winner.

The other sign of change in him was an upsurge in his writing. He seems to have written very little during the Ixopo years, or rather, little survived. It is probable, of course, that what he wrote then were chiefly love poems: if so, they were destroyed either by him or by Dorrie in one of their many periods of stress, just as the love letters which certainly passed between them were destroyed. Once he began teaching at Maritzburg College, however, he resumed writing, and he rapidly expanded his output to include plays, short stories, and novels. And in these, repeatedly, he made use of the people he had observed at close quarters in Ixopo, and the scenery he had come to love as he and Dorrie did their wooing. In later years he was to write to an American scholar, Professor Edward Callan,

> I was always keen on writing, wrote poems as a student, began to teach in 1924 (age 21), went in 1925 to Ixopo High School and spent three and a half years in a beautiful farming countryside in an environment which fascinated me, as indeed did the people. So when in 1928 I married, and went back to teach at Maritzburg College . . . I wrote two (or three, I cannot even remember!) novels of country life (white not black), triggered off if I remember rightly by the Rogue Herries books, set in the English Lake District.[3]

The over-coloured, formulaic novels of Hugh Walpole have today fallen into disrepute, but in the 1930s they were hugely popular, and it is not surprising that Paton should have chosen to model his first efforts on them.

The first of his attempts at a novel, *Ship of Truth*, was written during 1922 and 1923 while he was a student. The second is entitled *Brother Death*, probably dating from the middle of 1930.[4] The manuscript is in the form of five notebooks, totalling 716 pages, and the novel reflects its inspiration chiefly in its very 'English' (and therefore un-South African) feel, reflected in the place-names: Garth Place, Borrowdale, Atherton-under-Mist, Glen Asher, Windy Hill, and so on. Some of these, notably Atherton and Kaffirlands, are recycled from *Ship of Truth*; but the others are drawn from Walpole's glamorized Lake District, and made even more artificial by coming from the pen of a South African whose sole experience of the Lake District consisted of two days as a young tourist.

Concessions to the African setting of *Brother Death* included appearances by both 'Dutch' and 'natives', but the chief action involves transplanted Walpole stock characters: the gloomy old bully Cromwell, blunt and intolerant, his cheerful son and daughter Tony and Anne, the gross and lazy Fenton, the snobbish Carlton couple, and so on. One of the characters of interest is a local farmer named Jarvis, described as profane, cheerful, and of great good nature. In *Brother Death* Jarvis, who has a son named Douglas, is afflicted with cancer and feels that his life has been wasted. This character Paton would, years later, put to much better use in *Cry, the Beloved Country*.

Paton began by sketching these characters for himself, and then laid out the essence of the plot in point form, and it consisted of a series of set-pieces: court cases between Jarvis and another man named Hansen, and between Jarvis and Cromwell, with resulting enmity between them; a suicide; a marriage; a death; the news that Jarvis was dying of cancer; Cromwell, now an old man, coming to visit Jarvis on hearing of his cancer, and a reconciliation between them, with Jarvis pitying Cromwell. The novel, like those of Walpole, is formulaic and episodic, and Paton was right not to try to publish it. The world was not greatly impoverished by the loss of this second piece of apprentice work. All the same, he had begun to try his literary wings again.

The most ambitious and successful of his writings at this time was not a novel at all, but a play which he entitled *Louis Botha*. He began it in August 1932, and was sufficiently pleased with it to send it to Hofmeyr in 1933. His aim in doing so was partly to share his writing with a friend, partly a genuine interest in Hofmeyr's opinion, but partly to make Hofmeyr aware of Paton's increasing sympathy for the Afrikaners. Paton considered that Hofmeyr's fluency in both English and Afrikaans made him an ideal politician to heal the rift between the two main white groups in South Africa, and, as we have seen, he was working hard to equip himself for the same role.

He believed that the Anglo-Afrikaner gulf was rapidly widening in 1932, and in truth South Africa was approaching a political crisis. It was sparked primarily by the Depression, for unemployment and financial distress were reaching a climax in 1932 with the continuing refusal of the Nationalists to abandon the gold standard. Those who believed that the salvation of the country lay in following Britain in this regard naturally included the loyal Imperialists in Natal, and Paton believed that the Union of South Africa was in danger of breaking up again along the fault lines of this division between the whites. He wrote to Hofmeyr in August 1932,

> I have been very sore at heart these last few months, as you know, at the way things are going in South Africa. The work of Botha seems to have been for nothing after all; here, twenty-two years after Union, the country is as divided as ever it was.[5]

He was convinced that he was expressing Hofmeyr's own opinions, for he had just finished reading Hofmeyr's *South Africa*, published at the end of 1931.[6]

In this book Hofmeyr analysed South Africa's problems, and, in a chapter entitled 'South Africa's National Future', argued that the central problem facing the country was the growing split in Anglo-Afrikaner relations. He opined that the answer was a realignment of the political parties, and he warned that the greatest danger came from extreme Afrikaner republicanism. It was for just this political realignment that Paton hoped too, in order to contain the centrifugal forces he saw at work. His letters to Hofmeyr are increasingly full of political comment at this period, and he must have believed that *Louis Botha* would show Hofmeyr that Paton's heart was in the right place.

The play, as its title suggests, focuses on the career of the great Boer general who after the war worked energetically for the unification of South Africa, and served as its first Prime Minister. Paton's work was sparked by his reading of J. A. Spender's *General Botha, the Career and the Man*,[7] and, more surprisingly, Marjorie Bowen's *Brave Employments*, a romantic novel which he commended to Hofmeyr.[8] And in a letter written while he was working on the play, Paton used a discussion of Botha's achievements as a politician to raise the issue of Hofmeyr's own future. Paton was beginning to hope that Hofmeyr would break from both the established parties, the South African Party and the Nationalists, and form a third party of his own, in which Paton might find a place.

> Now that Nationalism & S.A.P.-ism have hardened (largely on racial lines), how can you achieve what you hope to achieve as an S.A.P.? Our children are now being born S.A.P. & Nationalist. Twenty years ago there was a chance of winning men's allegiance to a new thing. But it is no longer a new thing. Is it perhaps the lesson of Botha, the failure of a man to be neither one nor the other, that influenced you? On the contrary, can a *conciliator*, as I believe you are & will be, do anything if he identifies himself with one party? These are questions that disturb me; I know I need ask no forgiveness in asking them, because you know they are inspired by my affection & my confidence.
>
> I hope to send you my play very shortly . . .[9]

It is plain from this letter that Paton still considered South Africa's main problem to be enmity between Afrikaner and English speaker, and that he thought of Hofmeyr as the heir to Botha's mantle, the hope of a united white South Africa.

His play opens during the last days of the Boer War and closes in the first days of the Great War, during the rebellion of some of Botha's former friends. Its focus and sympathies are entirely with Botha and those who stand by him. They are depicted, in the opening scenes, as fighting an Anglo-Boer war which has been forced on them by British imperialism, and as suffering bravely for their hopeless cause. The play depicts Botha's attempts to persuade himself and others that surrender is preferable to complete destruction, and his efforts after the war to unite all South

Africans under one flag, in the Union that came about in 1910. He struggles against British coldness and Boer resentment, but he will not abandon his vision:

I see—I go on seeing—a new country, where English and Dutch live in peace, where the Englishman has his King and his tradition, and the Dutchman has his freedom. And I will not yield it for Englishman or Dutchman. Let the Englishman talk of my slimness [cunning], and the Dutchman of my treachery, but I will not yield it.[10]

And after he has become Prime Minister, he finds himself struggling against the growth of Afrikaner nationalism. 'They will not rest till they build their race again,' he says of his fellow Afrikaners. 'But it is not a race *I* would build, but a nation.'[11] More savagely, he says later, 'It is a crime . . . A crime that you commit. To build a race, and break a nation! You will flatter yourselves, call on your tribal God to witness your loyalties. Loyalties that split a nation asunder. In God's name, why?'[12]

This was the first of Paton's many protests against Afrikaner nationalism and its destructive power, and it is not the least eloquent. He was yet to perceive the terrible danger that would arise when Afrikaner nationalism was opposed by Black nationalism, but he clearly saw the evil and the folly involved in breaking a nation, as he put it, in an attempt to ensure Afrikaner survival in Africa.

The play was designed to carry a political message, and it is not too much to say that Paton had a single person in mind as its primary audience: J. H. Hofmeyr. The angry words spoken against Botha in the play by his former allies were daily spoken against Hofmeyr by Afrikaner nationalists who considered him a traitor to his own people and his own culture; they thought him a dangerous liberal, much too pro-British for his or his race's good. *Louis Botha*, read in this light, can be seen to be a tribute not just to its ostensible subject, but to Hofmeyr. And Paton unquestionably intended it as a defence of Hofmeyr's ideals, which he shared so fervently at this time. He often called Hofmeyr a conciliator; now he wrote to him of Botha,

As conciliator he was neither Boer nor Briton, a position I am sure he felt deeply, but which he adopted of set purpose & with a foreknowledge of the isolation in which it would leave him. It was the price he was willing to pay.[13]

This was to be his view of Hofmeyr too, a view he maintained long after Hofmeyr's death, and expressed in detail when he came to write Hofmeyr's biography.

Not surprisingly, Hofmeyr was favourably impressed by Paton's play when it reached him in April 1933. 'From the literary point of view I think it a very good piece of work indeed,' he wrote to a British friend, the actress Sybil Thorndike, when asking her to give her opinion of it.[14] She agreed to read it, and suggested changes which Paton made. Fortified with this expert advice, he subsequently entered the script in a competition for full-length plays in Johannesburg, but it was rejected on the grounds that it was too short, had too many scenes and too little action, and finished on what the adjudicating committee called 'a wrong note'.[15] He is unlikely to have been greatly distressed by this judgement, for he had won the approval of Hofmeyr, which mattered a great deal more to him than that of any committee. And Hofmeyr continued to grow in importance in Paton's eyes.

The political realignment Hofmeyr had predicted in South Africa came about in March 1933, when the financial crisis impelled Smuts, leader of the South African Party, and Hertzog, leader of the National Party, into a union which resulted in a new United Party, formed in December 1933, and controlling 136 of the 150 seats in parliament. From this new party the extreme Afrikaner nationalists under Malan broke away in 1934 to form the *Gesuiwerde* [Purified] *Nasionale Party*, and, on the other side, the extreme British loyalists under Colonel Stallard broke away to form the Dominion Party, dedicated to keeping ties with Britain. The Dominion party, in time, was to wither away and die, while the Nationalists would steadily wax in strength. But for the moment the danger of increased division appeared to have passed, and the conciliators seemed triumphant. The triumph was a personal one for Hofmeyr, who in March 1933 acquired three portfolios in the newly formed United Party cabinet, becoming Minister of the Interior, Education, and Public Health. Paton's friend, always powerful, was now a giant in the land.

Paton had not abandoned his ambition of taking a doctorate in Britain, and as a step on that road embarked on a Master's degree in Education through the Natal University College in 1930, successfully completing the first part, a written examination, early in 1934.[16] His papers were rated *summa cum laude*. The second part was a thesis which he was never to complete. This degree involved him in reading psychology (McDougall and Freud), educational theory (Comenius, Pestalozzi, Rousseau, Froebel, A. S. Neill, and Montessori), history of education, and a book which he was to say had had the most profound effect on him:[17] Cyril Burt's *The Young Delinquent*.

His degree was also to have a long-term political influence on him. One of his lecturers was Dr C. T. Loram, a distinguished educationalist and a man with a profound concern for the plight of blacks in South Africa. In 1929 Loram, with J. D. Rheinallt-Jones, Assistant Registrar of the University of Witwatersrand, founded the South African Institute of Race Relations, an organization designed to gather information on race relations, combat racial prejudice, and function as a non-political body. Around them Loram and Rheinallt-Jones gradually gathered most of South Africa's tiny minority of liberal-minded individuals: Rheinallt-Jones's wife Edith; Edgar Brookes, professor of political science at the Transvaal University College; Leo Marquard, founder of the National Union of South African Students (NUSAS); Alfred Hoernlé, professor of philosophy at the University of the Witwatersrand, and his wife Winifred; Donald Molteno, a brilliant Cape Town advocate; and a handful of others. This little group, whose names read like an honour roll of the supporters of black rights from the late 1920s on, was joined in 1930 by Alan Paton, who had been recruited to the organization by Loram.[18] For some years his membership of the Institute of Race Relations had little impact on him, but in later years he was to list his joining it as one of the seminal events of his life. In particular the constant flow of information pamphlets put out by the Institute made it impossible for any member who read them to pretend that all was well with race relations in South Africa.

Early in 1934 Paton paid a visit to Pretoria to stay with Hofmeyr and his mother in the grand ministerial mansion, a visit reported, together with his M.Ed. examination result, in the social and personal columns of the Pretoria papers. Slowly but steadily, through his contacts with Hofmeyr and through his increasingly conspicuous Toc H work, Paton was coming before the public eye. When one of the periodic Royal visits to South Africa took place early in 1934, he seems to have considered that he should have been involved in helping to make arrangements in Pietermaritzburg. Instead, he wrote to Hofmeyr, 'Gordon Watson [a civic official] has been the big chief of the whole affair, a thing which rankles preposterously in my breast!'[19]

Paton's ambition was clearly taking him beyond being a mere headmaster. He was keenly aware that some of his fellow students from the NUC were now embarking on political careers. One of these was Leif Egeland, captain of the tennis team at the University College, whom Paton had thought an inferior debater to himself. Egeland, after a brilliant university career which had taken him on to Oxford, had returned to South Africa and stood for parliament in the election of May 1933, his chief opponent being a Mrs Benson, who made a spirited appeal to the women's vote. When Egeland won comfortably, Paton sent him a witty note of congratulation:

> It's quite clear that your manners
> Have gone into retreat.
> You saw the lady standing,
> And yet you took the seat.[20]

If Egeland could embark on a political career, why should Paton not do the same?

Yet he trembled on the brink, uncertain whether he wanted to be a writer or a man of action. Perhaps as a writer he could do more good than through the slow and uncertain labour of the Toc H organizer, the teacher, or the politician. In 1931 he had expressed this doubt in a poem entitled 'The Poet', which ends,

> You with some trick of phrase
> At one leap scale the walls
> And tread the heights of truth.
> I hear your calls
> As I swim moats, climb battlements;
> I tell myself
> You know not what you do
> And all my life of days
> Wish I had gone with you.[21]

He seems to have considered that he might combine the role of man of letters and social reformer, and could become better-known at the same time, if he published a book on a political and moral subject; and to this end he began writing a volume he called *Religion and My Generation*. He seems to have started it in the school holidays at the end of 1933, and by March 1934 he was telling Hofmeyr that he was half-way through it.[22] But *Religion and My Generation* was never to be finished, and the manuscript was subsequently lost, for in April 1934 occurred another of the events which changed Paton's life. He caught typhoid fever, and nearly died.

He contracted the disease early in April, and was ill from 3 April until 16 June 1934, a total of 75 days. For a week he was on the edge of death, his temperature at one stage reaching 105 degrees Fahrenheit. Dorrie, finding him in a raving delirium, must have realized that there was a real possibility that she might be widowed for a second time, and became assiduous in visiting him in the Catholic hospital to which he was sent. The treatment for typhoid (or enteric fever as it was commonly called) in the 1930s was starvation, and for weeks he was fed on nothing but fluids. Paton's flesh, as he put it, 'fell away', and when he recovered enough to be aware of his emaciated frame and stick-like limbs, he, who had been so proud of his muscles, was filled with understandable self-pity.[23] He lost more than a third of his body-weight, which fell from 150 pounds to ninety-odd. 'I was that thing called skin and bone,' he said, and even gentle physical contact gave him pain.

He came to hate the hospital, St Anne's Sanatorium; it stood on high ground across the Umzinduzi from Maritzburg College, so that as he recovered enough to begin to shuffle about the corridors he could see the school carrying on without him. He longed to get back to real life, and looked forward keenly to his discharge. Twice the great day arrived, and twice Paton had a relapse at the last moment and was put back to bed. The second time he was so disappointed and so weakened that he broke down and cried like a child.[24]

On his seventy-third day in bed, and just before his eventual discharge, he wrote to tell Hofmeyr of his experiences, a letter filled with returning life and ambition:

> I ask no sympathy, for I wouldn't get any—you would dart a devastating glance at me with a facial expression I cannot hope to describe, enquire how I felt now, elicit that I feel thoroughly restored, & dismiss the topic forever . . . But you will be interested to hear that in my days of delirium, or to be more accurate and less sensational, of light-headedness, several things happened. (i) Smuts came to me & said he wanted a young man of brains & ability, & a good speaker, to oppose Sutton or Derbyshire or Marwick [anti-coalition members of parliament]—I forget which—at the next election. (ii) I got my wife to get folders from the shipping agents & actually planned a recuperative journey round the world—money apparently no object. (iii) I built my Dutch house—a stately and worthy dwelling. (iv) I re-wrote 'Louis Botha' & captured London. (v) I wrote a humorous novel that earned a lot of money. (vi) I wrote a memorable & moving novel that made less money but brought forth discriminating praise.
>
> It's no use your saying that these things never happened. They did happen. I stuck my head through a crack in the universe, into a new dimension, & for a golden moment saw & conquered.[25]

This account is a curious mixture of fantasy and reality, for Paton did in fact get his wife to bring him travel brochures so that he could plan an elaborate and impossibly expensive holiday for the two of them to Hawaii, Norway, and Italy,[26] and he did begin to rewrite *Louis Botha*. His plans for novels were real enough too, and (vi) is a startlingly accurate prediction.

What really interested him, though, was Hofmeyr's reaction to his vision of Smuts's offer. He wrote to Hofmeyr,

The only dream that worried me was No. 1. For somehow in my critical period of illness it seemed to me that your sanction was essential before any choice could be made. It is a hypothesis, Sir, that in my clearer moments I utterly repudiate. But then it seemed essential. I was afraid to wire to you. I feared the answer 'Be patient—the headship of Ginginhlovu will soon be vacant.' In any case, you could have hampered my political progress, & effectually barred my otherwise inevitable elevation to the Cabinet. I shirked the issue and fell back on my reserves, mainly No. 6; & have started—even on this bed of pain—my moving and memorable book.[27]

And he ended the letter, after a good deal of sincere flattery, 'I may see you in the new Parliament—if Smuts really comes to me, & you are decent about that wire!' Through half-jokes like this he tried to draw from Hofmeyr an opinion on whether he should enter politics, but he tried in vain. Hofmeyr in his rather stiff and matter-of-fact replies simply ignored the hints. But Paton was not going to give up hinting.

The 'moving and memorable book' he mentions having started at this time was almost certainly *John Henry Dane*, the most directly autobiographical of his unpublished novels, and the third to survive in fragmentary manuscript form. Perhaps Paton had learned from *Brother Death* that he needed to stay closer to his own experience. The early episodes of John Dane's life, as we saw in dealing with Paton's childhood, are drawn from events in his own schooldays. But, interestingly, he drew not just on his own experience, but on his wife's.

One of the most vivid episodes in the novel concerns Dane's birth during the Bambata Rebellion, the Zulu uprising in Natal in 1906, in which some whites were killed and the population of Natal thrown into a panic. Paton had been only 3 years old at this time, and had no memory of the uprising. But Dorrie had been 8, and vividly recalled the way the citizens of Ixopo had been both reinforced and demoralized by the panic-stricken farmers who had crowded into the village, and how they waited in terror for an attack. The danger was real enough, for the rebellion had started in nearby Richmond. All men, as well as the older schoolboys, carried rifles, and the villagers fortified the school buildings and took shelter there as darkness fell. That night the Ixopo church bells began to ring wildly; presently a noise was heard in the darkness, but after a good deal of nervous investigation it proved to be a pig. The expected attack never came, but Dorrie and a school friend slept on a table in one of the school offices for a week.[28] Paton drew on these details of his wife's experience to give an account both vivid and amusing, and one which showed his growing power as a novelist.

He wrote most of this book on the coast of Natal at Park Rynie, then a small holiday settlement, where his maternal grandfather had owned a beach house. This the Patons borrowed for three months, which was all the recuperation leave the Education Department would allow him, and he, Dorrie, and little David, now three and a half, spent what Paton later called 'three perfect months', sitting in the sun, swimming a little, and drinking a case of Stellenbosch wines they had bought with a feeling of wicked self-indulgence. When the wine was gone Paton took to beer.[29] He taught his small and clever son to read and write during these three months, and he gradually taught himself to walk again, for the hospital had provided no physiotherapy and he had to recover the use of his wasted muscles by himself. The hardest thing to cure was his flat-footed shuffle, caused by the disappearance of his calf muscles; these he rebuilt by walking on his toes for many months, an intensely painful but, in the long run, effective process.

His relations with Dorrie at this time can only be guessed at, though there is clear evidence that within a year they were to reach a particularly low ebb, during which Paton seems to have considered leaving her. Even now, during these 'perfect months' at Park Rynie, he produced a curious poem which he entitled 'Translation from the Hindustani', a language of which he knew nothing:

Life was bitter, be that said,
So I prayed my God for a small knife.
For thus I reasoned, if I be dead,
What matter how bitter be life?

God was bitter, be that said,
For He sent me never a small knife.
He sent me a woman, a woman instead,
So what matter how bitter be life?

The woman was bitter, be that said,
So I prayed my God for a small knife.
For thus I reasoned, if she be dead,
What matter how bitter be life?

God was bitter, be that said,
For He sent me a knife, a small knife.
And now I'll hang till I be dead,
So what matter how bitter be life?[30]

Whatever his marital relations were like, life was not bitter in fact; he rejoiced in simply being alive, and loved the time he could spend with his child. There is a brief reminiscence of this period, probably written as an exercise at the time, which conveys the flavour of these days at Park Rynie with the vividness of a piece of historic film:

The world was green when I fell ill; the hot days of summer slept over a steaming earth. But now I hobble gratefully into a wintry sun, & sink happily into an easy chair; and the planes shed their golden glory over me. My three year old son hovers round me; I wince when he jars the chair, but it is too small a price to grudge for his company.

'When will Dad be strong?' he asks.

'Soon,' I tell him.

'And then will we fight agen?' (This 'agen' he gets from my wife. When he grows older he will learn to say 'gel' too.)

'Yes, we'll fight, son.'

'Play-fighting, not real fighting,' he insists. He clambers on to my footstool.

'Be careful of my feet, son!'

'I'll be careful, Dad,' he says, & stands on one of them.

'David! You hurt my foot,' I say aggrieved.

'I made a mistake,' he answers gravely . . .

At eleven my wife comes out with the tea. And to have tea in the sun, with fresh buttered scones & honey, after an illness that leaves a man with the appetite of a boa-constrictor but far more speedy powers of recovery, is one of the great delights of that golden present called convalescence.[31]

As his strength grew he walked more and wrote more, and his power as a writer continued to develop rapidly. There exists an undated short story entitled 'Secret for Seven', which was probably written at Park Rynie, and which marks a further advance in his skills as a storyteller. It also reveals a growing sensitivity to the racial problems in South Africa, and is of great interest as a result.

'Secret for Seven' revolves around the marriage of Mary Massingham, daughter of a retired major and product of a genteel Natal family which farms, plays polo, and reads the *Tatler* and *Country Life*: a little outpost of Empire, in fact. She falls in love with Charles Draper, whose father is a carpenter, and whose mother is described as 'a dark-skinned woman about whose descent kindly tongues did not enquire'. They are divided, therefore, not just by social standing, but by colour. The Massinghams consider Mary's attachment to Charles Draper an unfortunate misalliance, but accept it, and allow their daughter to marry him. After that they visit the young couple 'as often as parents should', as the narrator puts it, 'but they made no attempt to bring them into their own world; for Draper had never been in it'.

There are more worlds than one in this story, for even clearer than the line between Drapers and Massinghams is that between tribal Zulus and whites. Paton draws it in terms reminiscent of *Wuthering Heights*'s distinction between children of moor and of valley:

A world apart; for when the mists came down over the mountain, they cut off white from black with their level line. Up here the swirling mist, & wattles dripping eerily in it, & gates looming suddenly out of it; down there clearness & stars, & the cries of natives from kraal to kraal, & lights here & there from hut and hut. Up here the farms of white people, houses, wireless, mist; down there the lands of the blacks, huts, singing, & stars.[32]

The distinction between whites and blacks is crucial to the theme of the tale, as quickly becomes apparent. In due time Mary bears Charles Draper a child, and with the emergence of the baby into the world the crisis of the story arrives. The midwife shuts her eyes and prays; the doctor can find nothing to say; and Draper himself is faced with the fear that must have haunted him all his life. Paton handles the scene masterfully:

In that terrible moment Draper had to make up his mind. He went out to his wife's parents.

'Mary is well,' he said. 'But it would be wiser not to see the child. It's a daughter.'

Mrs Massingham was alarmed.

'Is there something wrong, Charles?'

'Very wrong,' he said. 'She's black.' And then to their silence, 'You understand? A black child.' And then again, 'I'm sorry.'

The soldier [Major Massingham] asserted himself. 'We'll come over tomorrow, Draper, tell 'em it was a false alarm. And think out something. Love to Mary, eh?'

His wife followed bewildered, for she would have liked to have seen the daughter. Yet what could she say or do? Her husband started the car & under cover of the noise said 'Good God' several times. When he did speak on the way home, it was [to] say 'Good God' again. For he was extraordinarily moved.[33]

Draper and Mary hand the child over to Catholic nuns, and they have no more children, 'for they were afraid'.

The rest of the story concerns the nuns' puzzlement at how to bring up the child. The story comes to an abrupt end in the middle of this discussion, as if Paton had raised a problem to which he could see no solution, but the fact that he had raised it at all shows how far he had come since the days when, as a student delegate to the Imperial Conference in 1924, he could dismiss the 'native problem' in a single sentence. He was beginning to see that a system which attempted to divide a country into hermetically sealed compartments was an impossibility; and he was beginning to show in his writing that a system that made a happily married couple ashamed of their own daughter, and afraid to have another child, was a monstrous injustice.

In essence this was to be the central message in all his writing from this point on. His voice at this point is still faltering and uncertain, and it appears to fall back from the implications of the issue it has raised; but it is closer to the voice of the Alan Paton the world would come to know than anything he had written before. Even his imagery, in describing the country he loved, was approaching the vividness that would characterize *Cry, The Beloved Country*; and some passages from 'Secret for Seven' could have been early drafts of the later novel:

> She [Mrs Massingham] and her daughter would sit on the verandah of 'Emoyeni', looking down on the valley below at a different world. For there lived the natives in their reserves, a land where no mist came & no bracken grew & no titihoyas called; where the earth was red, & the thorn-tree & the aloe flourished, a hot country where colours were more vivid & sounds more loud. A world apart . . .[34]

A writer with such power at his command could not but wish to exercise it, and Paton was filled with ambition to do something better with his life than spend it slowly rising through the ranks as a schoolmaster. His illness had given him time to think, and among the things that became clear to him was the fact that he had been losing his keenness for teaching for some time, probably since his marriage. In his autobiography he was to say that the illness had created his desire for a change of life, but in his letters to Hofmeyr he admitted that he had 'lost his zest' for teaching even before the illness, and had for some time 'not quite played the game by the Department'.[35] But if this was so, the Department returned the compliment, for when he applied for a further three months' convalescent leave, it replied that such a privilege was only given to teachers of ten years' standing, and he had served only nine years and nine months. Paton, badly in need of more time to recuperate, was enraged.[36] And though he mentioned in a letter to Hofmeyr the injustice he felt he had suffered, the Minister of Education took no action.

By the beginning of 1935 Paton was keenly seeking a way out of Maritzburg College and off the treadmill of teaching. His plan to take a doctorate in England had not been forgotten, and in April 1935 he applied for a scholarship to study at the London Institute of Education. Some of his fellow teachers knew of this application and Paton's general restlessness, and they were sorry to feel they were going to lose him. Among them was Reg Pearse, who had now joined the staff, and who valued Paton's sense of fun, his love of a good story, and his new-found fondness for a drink. He also liked Paton's capacity for deflating self-important fellow teachers.

In later years he recalled an occasion, early in 1935, when an elderly teacher, 'Scratch' Leach, assembled the school to tell them of their role in King George V's Silver Jubilee celebrations. 'Now, boys,' Leach said ponderously, 'tomorrow we are all going down to the Show Grounds to take part in the celebrations. You will all march down, and when you arrive you will line up in front of the pavilion. Each boy will then sit—down [and here Leach became so slow that he almost stopped]—on—his—own—[he searched for the *mot juste*, while a hundred boys mentally filled in a rude one]—AREA.' The silence that followed, Pearse said later, was electric, each boy holding his breath and trying not to laugh. The teachers, ranged in a stony-faced semicircle on the stage, struggled too. Suddenly, in the silence, Paton let out a bellow of laughter, and of course brought the house down.[37]

In April 1935, before he heard the results of his application for a scholarship, Paton saw an alternative opportunity to get out of teaching. In 1934 the Education Department had taken over the running of the country's five reformatories[38] from the Prisons Department, and early in April it advertised the Wardenship of three of them, letting it be known that it wanted to recruit teachers rather than the prison staff who had previously occupied the posts. Paton knew little about reformatory work, but he jumped at the chance and went straight to the top, writing to Hofmeyr on 8 April 1935:

> I see that the Dept. of Education is taking over reformatories. Do you think there is anything for me in this line? . . . There seems as yet to be no chance of a school of my own. I feel that—with luck—I might have had a big school by now under some other authority. There are times when one is tempted to wish that one had tried a little of the flamboyance and the trumpeting, & concentrated a little less on the integrity.[39]

He ended this letter with a paragraph in good idiomatic Afrikaans, the first time he had used that language to Hofmeyr, as if trying again, as he had through *Louis Botha*, to show Hofmeyr that he was now ready to stand at his side and help heal the Anglo-Afrikaner divisions. But if this was a hint that Hofmeyr might offer him a more important job than Wardenship of a borstal, it was ignored. Hofmeyr replied by encouraging Paton to apply for one of the posts being advertised, and sang the praises of one of them, Tokai.

There were two reformatories at Tokai, a beautiful Cape farm. One was for white boys, the other for Coloureds, and Paton hoped to get the white one, though he was willing to take the Coloured one if necessary. But he now ran into opposition. Dorrie, when he put the notion to her, was horrified by the very thought of a move to the Cape. She loved Pietermaritzburg and had looked forward to being the wife of a headmaster there in due course. She did not want to move to the Cape, she feared snakes there, and she could hardly believe that Alan would propose taking her to live in a prison. What effect would it have on young David to be brought up among criminals?[40] Though some of her objections seemed trivial, her opposition to the move was deep-rooted and strongly held. Up to this point Paton had done all he could to please her at every turn, and she might have expected to win her point on a matter about which she felt so strongly. If so, she was mistaken. A shift in power was taking place within the marriage, and its effects would soon become obvious to her. He was no longer going to run after Dorrie: this decision of his to leave Pietermaritzburg and teaching for ever marks the end of her dominance in the marriage, and the beginning of his.

Against her will he applied for not one post but three, on Hofmeyr's advice: the Assistant Wardenship of Tokai, the Wardenship of Houtpoort, a white borstal near the Transvaal town of Heidelberg, and, lastly, the Wardenship of Diepkloof, a black reformatory near Johannesburg. He profoundly hoped he would not get Diepkloof, about which Hofmeyr had written to him, 'It is hard to know what can be done with it'. He listed his experience, stressing his work for Toc H and his teaching, and among his reasons for application very honestly included 'the desire to escape from a service where prospects of advancement seem limited'.[41]

His application was not on the face of it a very strong one, for he admitted that he had no experience of social or reformatory work, and he appended a list of conditions, among them that he should be allowed to take accrued leave, that his salary should not be less than his teacher's pay

(£610 a year)[42] and, extraordinarily, that if he got the London Institute of Education scholarship, he should be allowed to take it up. How he hoped to do this while acting as Warden of a reformatory is far from clear, and under normal circumstances it might have been enough to rule him out as an applicant for the positions. But at the end of his letter came a sentence which he must have penned with a good deal of satisfaction: 'May I give as immediate reference, if etiquette so allows, the name of the Hon. J. H. Hofmeyr?'[43] Not many applicants to the Education Department could use as reference the Minister of Education, and Paton must have awaited the outcome with some confidence.

Yet he remained in doubt, partly because of Dorrie's continuing opposition. As he wrote to the stolid Hofmeyr,

> I don't know whether it will have occurred to you that my feelings are in a turmoil, but it is so . . . It is hard to sleep these days; grandiose schemes float thro' the brain. You needn't trouble to warn me against these with a timely reminder of reality. I've always had 'em, always will. Great novels go to the W.P.B, speeches that electrify a reverent House echo round my humble plot, & still one lives, and happily.[44]

In a later letter, while not specifically referring to his political ambitions, he made it plain to Hofmeyr that he saw the reformatory job as a stepping stone to larger things: 'the thought is certainly at the back of my mind that this may lead to a job calling for the use of more of my powers.'[45]

When the reply came, on 7 May 1935, Paton was taken aback. He had been appointed to the Wardenship of Diepkloof, and was to take up his duties there in two months, at the beginning of July 1935. Dorrie's opposition to the application had grown fiercer as Paton's determination grew; now he had to tell her that he was to take over a black borstal, in Johannesburg, the great gold-mining city for which she had all the Natalian's dislike. On getting the letter, Paton remembered years later, he crept out of the Howick Road house and hid in the garden, trying to summon up the courage to tell her. When at last he did, she was devastated by the news, and finding Paton determined, she cried bitterly and would not be comforted. 'What do you know about reformatories?' she asked him through her tears, unanswerably. 'What do you know about African delinquents? Why do we have to leave Natal, where all our family and friends are? I don't want to go.'[46]

But Paton was adamant: he was going in two months, and she could follow if she liked. For those two months she tried everything she knew to dissuade him, in vain. Her resistance lasted to the end. On 22 June 1935

Paton resigned as Natal Pilot of Toc H, being saluted in the organization's magazine *Compass*;[47] on 27 June 1935 he was presented with a silver tray as a farewell gift by the staff of Maritzburg College, and on 30 June he set off for the station, in a new suit and coat, and a pair of vivid yellow gloves to protect him from the highveld winter of which he had a Natalian's dread. He was accompanied by a big crowd of his friends to see him off, but everyone must have noticed that he was not accompanied by his wife and child. Dorrie and David had remained behind, at 10 Howick Road.

Paton hoped Dorrie would join him in Johannesburg, but to tell the truth he did not greatly trouble his mind about the matter. He was, he admitted years later, neither sober nor regretful, but light-hearted.[48] For one thing, this was an adventure that would free him from teaching. And for another, he had fallen deeply in love with a beautiful girl of nineteen, and he expected to see her in the Transvaal. 'I, guardian of the wayward and inconstant, was at that time wayward and inconstant too.'[49]

8

DIEPKLOOF
1935–1936

Freedom, not captivity, should be the keynote of the reformatory.

SIR CYRIL BURT

Paton had got to know the girl he called Joan Montgomery (though that was not her real name) six months earlier, at the beginning of 1935. She was a student at Natal University College, training to be a teacher. She was also the daughter of one of Dorrie's friends in Ixopo, and during February 1935, the month before the College opened, she had stayed with the Patons at Howick Road. Paton found her very beautiful, and she was attracted to him. She must have perceived the coolness that existed between his wife and himself. He had for years been emotionally deprived in an important sense; now Dorrie's continued reserve was contrasted for him with the warmth, youth, and enthusiasm of a girl living under his very roof. When the College opened she moved out of the Paton home, but not out of his life; she and Paton continued to see each other. They were soon deeply in love, though neither told the other.

There is little reason to doubt that it was his love for Joan Montgomery, combined with the effects of his long illness, that gave Paton the strength to take a stand on his desire to leave teaching; it also gave him the strength to cross Dorrie in this, and even to leave her in Pietermaritzburg when he went lightheartedly off to his new post in Johannesburg. For Joan was planning to come to the Transvaal with a student hockey team in August.

In the event, Dorrie arrived in Johannesburg first, a month after Paton. Paton implies, in the account of the delay he gives in his autobiography, that it was time needed for her to 'pack and follow', and that there was never any doubt that she would come; but in an earlier account he attributes the month's delay more directly to her resentment of his having turned her world upside down—not to speak of his having compelled her to follow his lead or be left.[1] Little David remembered this trip, and his mother's sisters' opposition to it, though he was all unconscious of the tensions playing beneath the surface of his parents' relationship:

> I recall the move to Diepkloof. My dad went up first, then my mother and I went up by train. I remember my aunts talking about how cold it was in Johannesburg. 'I wouldn't go to Johannesburg, it's cold there.'
>
> In the train we had a coupé. I remember waking up in the morning and there was a koppie [stony hill]. My mother said 'That's a koppie'. That was the first time I'd seen one, first time I'd been to the Transvaal.[2]

Shortly after he and his mother arrived at Diepkloof, in August 1935, Joan Montgomery came to stay too. Each morning Paton would drive her to Johannesburg, 7 miles away from Diepkloof, to join the other members of the hockey team, and each evening he would drive in again to fetch her. His own repression, less now that he was more experienced, was rapidly eroded by her closeness to him in the intimacy of the car. On one of these journeys, when they were returning to Diepkloof and Dorrie, he told Joan he loved her, and she said she loved him. 'We did not discuss the consequence of this event,' Paton was to say wryly.[3]

There was no opportunity for them to consummate their love in Johannesburg. But when Joan went back to Pietermaritzburg they wrote to each other every day, and presently Paton, making some excuse to Dorrie, went down to Pietermaritzburg for a long weekend, and he and Joan became lovers. He was not good at deceit, and it is likely that he did not make much effort to deceive his wife. At any rate, she was very quickly aware of what was afoot. Not long after his return Dorrie asked him abruptly, 'Are you in love with Joan?' and he said, 'Yes.'

The question, with its coldly positive response that was really a negative, reminds one of his asking, on their wedding night, if she was going to continue to wear Bernard Lusted's ring. When, a little later, Dorrie asked him, 'What did I do wrong?' he did not reply, for she knew the answer as well as he did. But once the affair was out in the open, he determined to end it. The action is characteristic of his rigorous moral standards: he might do wrong, but he could not live with it.

He now took some days' leave, drove down to Natal, and there told Joan that they would have to part with each other. It is clear from his account that, puritan as he was, he had found the moral discomfort of the affair balanced its pleasures:

> It was both relief and pain to say goodbye to each other. I watched her for the last time open the gate of the house where she stayed, and walk up the path, and let

herself in at the door. She turned and gave, hardly a wave, but something more like a sad salute. So it came to an end. It was a brief encounter, but I dreamed about her two or three times a year till the year that she died, and on occasions I still do, though it all happened forty years ago.[4]

When he got back to Johannesburg Dorrie met him at the door of their Diepkloof house, took him in her arms in what he described as 'that strange fierce way she had when she meant something intensely,' and said, 'I am going to make it all up to you.'[5]

This was one of the times Paton was to remember, when Dorrie had shown herself passionately responsive to him, and their second son, Jonathan, was conceived on this afternoon. And there was another tangible sign that Dorrie had learned a lesson from this affair: that night when they went to bed Bernard Lusted's ring had gone from her right hand, and Paton never saw it again.[6] If, as seems likely, his first brief extra-marital affair was a cry of distress on his part, it had been answered for the moment.

And having got his personal life into better order, Paton was free to put his whole heart into the task he had taken on at Diepkloof. The more he looked about him, the more clearly he saw that it was going to be a Herculean labour. Hofmeyr had not been exaggerating when he said it was not easy to know what to do with the place.

Diepkloof, a beautiful farm of 900 acres set 7 miles south-west of Johannesburg, had been bought by the Prisons Department in 1906, and until the First World War had been a prison for adults. The most striking historical fact to be learned about it was that Mahatma Gandhi had been briefly imprisoned there in 1913 during the protest march he led across the Natal border in an effort to improve the lot of Indians in South Africa. Between the wars the decision had been taken to transform the place into a borstal for black boys below the age of 18 years, but in fact the only change had been in the prison population: the grim, dilapidated buildings, the prison staff, the harsh discipline had all remained unchanged until the Department of Education took it over from the Prisons Department in 1934.

Well before the planned construction of the system of apartheid began in 1948, blacks had been subject to a range of discriminatory legislation, and institutions designed for them were often very poorly equipped even by the standards of the time. Paton knew in advance that he was not going to find himself at the head of a model institution. But when he first set eyes on his new charge, his heart must have sunk.

Diepkloof reformatory in 1935 looked like the ramshackle prison it was. The huge main building, of wood and corrugated iron with earthen floors, was painted a hideous yellow-brown, had heavy iron bars on the high windows, and was enclosed in a 12-foot-high barbed-wire fence supported by great iron stanchions set in concrete. An official inspection report on the place, made some months before Paton's arrival, remarked on the extreme primitiveness of the buildings, which it compared to temporary buildings in the road camps used for hard-labour prisoners, and it added that 'the high fence surrounding this [central] block gives a real prison atmosphere'.[7] This prison atmosphere was intensified by the presence of armed guards.

There were 360 inmates,[8] referred to as 'boys', though many of them were men in their twenties, and they ranged from children who had committed quite trivial thefts to young men who were experienced and extremely dangerous criminals, rapists, and murderers. Some of them would not hesitate to kill if need and opportunity arose. They were sent to Diepkloof because their offences had been committed before they reached the age of eighteen. Inspectors called for the smallest boys to be separated from the older ones,[9] but they called in vain until Paton's arrival.

The institution was used to house the young offenders during their period of committal; a well-behaved inmate would be released, after half his time had been served, into the 'care' of a farmer who would use him as unpaid labour (in effect a slave) until his time expired or he ran away.[10] This 'placement' system was much abused by brutal and exploitative farmers. One inspector remarked with astonishment that some of the boys even returned to the reformatory for shelter, and commented, 'That anyone should willingly return to such a place is a startling commentary on the conditions prevailing at some of the places to which they are sent.'[11] There were frequent escapes (and even more frequent attempts to escape) from the institution itself in spite of the bars, the barbed wire, and the uniformed guards. The reasons were not far to seek.

On the day of his arrival in Johannesburg, 1 July 1935, Paton had been driven to Diepkloof and ceremoniously welcomed with tremendous stampings and salutings by the warders, janglings of keys as the great gates in the outer fence were opened and locked behind him, and the equally large gates in the main building opened and relocked in their turn. He did a tour of inspection of the rusting, bug-infested buildings, and was keenly regarded by the 360 boys, who must have wondered how this slight, stern-looking man of 32 was going to treat them. He inspected his office, which was dark and cramped, and the other rooms, including the kitchens, primitive almost beyond belief, where porridge was cooked on antique wood-burning stoves, and dishes washed under cold water outside.[12]

Discipline at Diepkloof when Paton arrived was extremely harsh, and physical violence common. The warders, black and white, carried heavy sticks and used them frequently, severe beatings being given at will. For inmates who did not respond well to this treatment, severe floggings with a sjambok or hippo-hide whip were administered, and for the most recalcitrant cases whippings were combined with spells of solitary confinement.

The food for the prisoners was coarse and unvarying, consisting mostly of maize-meal porridge, and the inspectress of Domestic Science described their diet as 'monotonous and inadequate'.[13] Only after Paton arrived, for instance, were the inmates first given bread, at the rate of 6 ounces a day.[14] They had no dining-room; having collected their porridge in tin dishes they stood or squatted on white lines in the parade ground to eat. In rainy weather they were permitted to eat in the dormitories, which they can scarcely have thought of as a privilege.

The dormitories were corrugated iron rooms, 15 feet by 24 feet, and in each of these 22 prisoners were locked for more than 12 hours each night, from 6 p.m. to 6.30 a.m. One smaller room measured 12 feet by 12 feet, and fourteen boys slept in it, packed as tight as space allowed. The place was in fact terribly overcrowded, its population having risen by over 100 in the previous year alone, as the Depression combined with the rapid rise in urban black numbers[15] to produce a crime wave. The inmates were provided with three blankets each, most of these being worn threadbare,[16] and with these and a thin mat, in the freezing highveld winter, they slept on the floors, which were of cattle dung mixed with clay, under lights that burned all night. A grille in the door allowed for frequent inspections in search of such offences as buggery or the sharing of blankets. The temptation to share blankets in the bitter cold must have been overwhelming. Each room was provided with a single bucket of water, and another bucket to serve as a toilet. This latrine bucket was often overflowing in the morning, soaking into the earthen floor, and the smell when the rooms were opened after the night defied description, according to an official inspector.[17]

The main outdoor toilet for prisoners was described by another inspector as 'nauseating';[18] it consisted of a cross-bar on which the prisoners squatted over open buckets, practically in public, and the receptacles, of which there were only four, quickly overflowed when 360 inmates queued to use them in the morning. 'The present conditions', wrote the inspector, 'would shock even natives living in kraals. As a method of punishment they cannot be regarded as effective, as the native will probably leave the reformatory with a grudge against us.'[19]

Punishment was the admitted aim of the institution; 'reformatory' was a complete misnomer. There was a schoolhouse, consisting of a single room, 30 feet by 30 feet, with a hopelessly overworked black head teacher, Ben Moloi, who was expected to deal with a class of 112 boys.[20] An armed guard sat at the back of the room to prevent anyone from climbing out of the windows. The teaching itself was considered a punishment to be inflicted on refractory inmates.[21] Few of them seem to have learned anything, and there is no record of any passing school examinations before Paton's arrival. The main work done at Diepkloof was farm labouring and market gardening, carried out on the farm which surrounded the main buildings, and a good deal of food was produced, including quantities of vegetables; but little of this was used for the inmates, and that little was boiled up unappetizingly and served with meat twice a week.[22]

The inmates wore regulation prison garb of a thin shirt and a pair of shorts; they wore nothing else, neither a sweater nor shoes, even in the coldest weather. Sandals were, in theory, provided on medical orders,[23] but none of the boys seems to have worn them before Paton's arrival. Outbreaks of dangerous and highly contagious diseases such as typhoid fever spread rapidly, since there was nowhere to isolate such patients.[24]

The reformatory 'hospital' was a rusting iron structure, earth-floored like the rest of the institution, run by a warder without basic medical training, and generally full to overflowing. A high percentage of the inmates had venereal diseases, and in addition the sleeping arrangements ensured that many of them developed chest infections, from which a number died each year.[25] 'It is impossible', wrote a despairing inspector prior to Paton's arrival, 'to give adequate attention to all the cases of illness occurring at present.'[26]

The accommodation for the staff was better, but not much. There were two houses for the Warden and his Deputy, and they were small, dark, and dilapidated. They were placed in a dense eucalypt plantation, which added to the darkness and made gardening almost impossible. Dorrie, when she arrived, was horrified to see where she was to live, and Paton must have been glad to tell her that plans to erect a better house were already in train.[27] Two further houses, in even worse condition, housed the two married storeman warders and several single men, who slept two to a room entirely without furniture, their beds being mattresses placed on the floor.[28] And this was the accommodation for the white warders: that for blacks consisted of rondavels, circular huts, whose condition was not even described in the official reports, but can be imagined.

After having inspected the buildings of this terrible institution on this first day, Paton walked over the farm itself, and found his spirits gradually rising again. Running through the farm was a small stream which fed a dam, and in the valley, on both sides of the stream, were elaborately constructed terraces covered with vegetable beds of every kind, and with huge and most fruitful orchards. The peaches that grew here, Paton was to claim,[29] were the best in the world, and certainly the highveld climate produces unparalleled stone-fruit. The less fertile portions of the farm were given over to fields of maize. Paton had loved gardening ever since being introduced to it by his father; here was the biggest garden he had ever seen, and it was part of his job to make it bigger and better still, with virtually limitless labour to draw on. That garden, some supportive staff members, and the inmates themselves, saved him from despair.

Such was the institution which Paton took control of on 1 July 1935, and at which he was to stay until 30 June 1948. If someone had predicted to him that he would spend the happiest thirteen years of his life at Diepkloof he would have been incredulous, but in later years that was the way he was to think back on his time there. The truth was that Diepkloof was an immense challenge, and he was to find that responding to the challenge would prove immensely rewarding.

He had been told by Hofmeyr that it was the aim of the Education Department to change the place from a prison into a school. Just how this was to be done was up to Paton. He was now the lord and master of 360 boys, and of a staff of only 30, consisting of a deputy warden, Mr J. H. Laas, three clerks, a farm manager, James Barry, and 25 supervisory staff, 12 black and 13 white.[30]

The senior supervisor was Captain Dunkley, an ex-Indian Army man, very English and apparently very superior, and Paton, who calls him 'Stewart-Dunkley' in his autobiography, disliked him on sight and was glad when he left Diepkloof. Both Dunkley and Laas were notable for the elegance of their dress, whereas once Paton had settled in at Diepkloof he habitually wore khaki shorts and shirt, so that, as one of his staff was to remark, 'he looked more like the handyman than the principal of Diepkloof Reformatory.'[31] Most of the staff members on his arrival were prison officers by training, and their attitude to the inmates was one of unrelenting severity. Among the most severe were the black warders, whose heavy sticks were in frequent use. If Paton was to change Diepkloof, as he was determined to do, he would also have to change the mentality of his staff.

This he proceeded to do by weeding out staff who could not be converted from the prisons mentality, and by employing a number of new junior supervisors, who were drawn from the Special Service Battalion. This was a military unit brought into being in the mid-1930s by a right-wing government minister, Oswald Pirow, with the aim of giving discipline and training to unemployed school leavers. These new junior supervisors naturally imparted a military colour to the reformatory; the pupils received regular military drill, and were marched in groups called *spans* wherever they went.

One of the most important talents required of a leader is the ability to recognize and select a group of subordinates who will carry out the tasks needed, and Paton showed his leadership to the full in the highly talented and supportive team he gradually built up. Of these, his deputy warden, J. H. Laas, the black head warder, Ben Moloi, and a particularly talented supervisor, I. Z. Engelbrecht, were the most important, together with a young man recruited from the Special Services Battalion and named 'Lanky' de Lange. De Lange, like Paton, wore an habitual frown which concealed a genuine concern for his charges. De Lange would appear in *Cry, the Beloved Country* as the fierce-looking but idealistic young man, though this character is almost certainly also partly a self-portrait on Paton's part.

It is worth stressing that Paton did not have to convert his superiors to the idea of changing Diepkloof from a prison into a school. That battle had already been won; it was they who told him to accomplish the task. The writers he had come across while doing his M.Ed., Cyril Burt, Sheldon and Eleanor Glueck, and Homer Lane,[32] had established ideas of child-centred pedagogy and penology which by the 1930s were becoming dominant among theorists of prison and borstal reform. These ideas have been summarized in the following terms:

> Many of these writers stressed a new relationship between pupil and teacher. They emphasised individual psychological study, house-fathers and house-mothers to take care of children in institutions, and the creation of a community and family home in miniature complete with 'home-like cottages' to displace larger hostels.

These theorists also argued for a prefect system, individual instruction and supervision of pupils by university-trained professionals. Self-discipline and not external restraint should be the source of control, they argued.[33]

In particular, Sir Cyril Burt had written in glowing terms of the success of rural hostels in reforming delinquents, citing the George Junior Republic at Freeville near New York, and British institutions patterned after it, notably the Little Commonwealth in Dorset, the Sysonby Village Colony near Melton Mowbray, and the Training Colony in Berkshire.[34] All of these English experiments had collapsed, but Burt and other writers continued to support the theory that lay behind them, which held that freedom, not captivity, was the keynote of reforming the young.[35] The theoretical battle, then, had already been won. Paton's challenge was to find some way of putting these theories into practice in a terribly crowded, run-down and underfunded institution, and to carry his staff with him in doing so.

Paton came to think of Diepkloof as a microcosm of South African society. That such a comparison could be drawn in 1935 shows how false is the popular notion that apartheid dates from 1948. In particular he came to think of Diepkloof's reform as a pattern for change in South Africa as a whole. That reform, though it took several years to effect, can be readily described.

A semantic change, significantly enough, came first. Some months after Paton's arrival the Warden's title was changed to 'Principal'; the inmates were to be referred to henceforth as 'pupils'; the black head warder, Ben Moloi, was to be the 'head teacher'; the other warders became 'supervisors'. And the official name of the institution, in Afrikaans, became *Verbeteringskool*: 'Reformatory School'. Behind the semantic changes lay a conceptual alteration; Diepkloof was, like South Africa in the 1990s, on the verge of dramatic practical change, which no one had any clear idea how to effect without violent unrest.

Paton's answer was to introduce a series of rapid incremental changes, all designed to increase the freedom and the responsibility of the pupils. He saw it as vital that freedom and responsibility should go hand in hand. Aided by a supportive superior, M. C. Botha, and sympathetic inspectors such as Miss Chattey, Inspectress of Domestic Science, he began by relaxing what seemed unnecessary prohibitions on the smoking of tobacco; he had bucket latrines built, he enlarged the hospital, he revolutionized the diet by introducing bread, fresh fruit, vegetables, and more meat, he built a laundry, he introduced the wearing of jerseys and sandals in winter, he took on new staff (19 within his first year)[36] who had not been trained for prison work and who therefore were open to new and liberal ideas, he enlarged the school, and he gave the head teacher helpers.

Above all, he began to break down the punitive discipline and replace it with something approaching a contract system. If the inmates would co-operate, for instance in keeping silence after the nine-o'clock bell rang, he would respond by giving them previously unheard-of privileges, such as leaving their dormitories open from the 5 p.m. roll-call until 9 p.m. He began with the dormitory for the smallest boys, the youngest of whom was 9, and gave them the freedom of the yard inside the main building each evening. Then, step by cautious step, he opened the others, and presently began leaving them open all night. Over the next months he gave more freedom: he began marching the entire body of pupils outside the wire fence for parades.

His aim was, in essence, to break down the physical barriers separating the inmates, or pupils as they had now become, from the outside world, and he was impatient to begin as soon as possible. As early as 16 July 1935, within a fortnight of his arrival at Diepkloof, he was writing to the Secretary for Education asking permission to begin building the first 'out-side settlement', a rondavel outside the wire fence.[37] To save the Department money he proposed to build it with labour, timber, clay, and thatch from the farm. In these rondavels he proposed to house twenty to twenty-five of the most trusted inmates who would act as domestic servants for the white staff. 'Mr Paton,' wrote an inspector, 'is very anxious to have one of these cottages this year, as he has about 50 boys whom he would like to try out in a cottage conducted on free lines.'[38] The Department supported this request, though with reservations;[39] it considered the scheme 'frankly experimental'[40] but was willing to try it. Building work began within months, and was to continue for years.

The innovations of Paton's reforms lay in the nature and construction of the free hostels, and in the way he tried to substitute an internal discipline for external restraints such as bars and guards. His idea was to set out the free hostels as a little village of individual huts. In his mind's eye he saw the grim, decaying central block transformed into an idyllic rural settlement set in a beautiful garden.[41] Official Reformatory Visitors and members of the Diepkloof Board, such as the liberal-minded J. D. Rheinallt Jones and his wife, Edith, supported the village plan,[42] and Paton was authorized to begin putting it into practice.

He quickly found that his plans were much too optimistic. As one of the people who visited him at this time later remarked, 'What happened was just what people had said would happen: the boys escaped in every direction.'[43] In his autobiography Paton skates over the difficulties he had, and makes light of the absconding that followed the start of his reforms, but at the time he was severely depressed by them. In the middle of October 1935 there began a rush of escapes from Diepkloof, encouraged by Paton's early attempts to ease discipline and give more freedom to trusted boys. Two hundred and fifty boys had been given leave to spend part of their time beyond the reach of the warders' supervision when the flurry of absconding began. Paton wrote dolefully to Hofmeyr:

> Our escapes for November stand at nine so far, which will not be out of the way if there is no increase. But even then the nightmare of the end of October & the beginning of November will not be easily forgotten. In my letter of the 4th I said I was not optimistic enough to say that the epidemic was over; in fact it lasted till the 11th, & the return and punishment of some of them has had a salutary effect. We lost about 16 in as many days. I suppose it's wrong to look on escapes as the sole criterion by which to judge one's work; but one can't help it. And I look back on those two weeks as the two hardest I have ever lived through.[44]

Those who knew Paton during this first rush of absconding saw that he did not exaggerate in calling the period a nightmare. One of these was a young man with an extraordinary name, Gonville Aubey ffrench-Beytagh. ffrench-Beytagh had been born in Shanghai of English parents, and in 1929 had gone to New Zealand for four years before coming to South Africa, young, lonely, and footloose. He had made contact with Toc H in Johannesburg in 1933, and through the organization had met Paton. He often visited Diepkloof, and in November 1935 found him tremendously agitated because of the news that a whole group of boys, perhaps a dozen, had escaped. ffrench-Beytagh agreed to accompany Paton in his car, and helped him search through black townships, looking in the known haunts of the boys who had absconded. Paton, he was to remember, had been terribly drawn and anxious as they rushed through the narrow slum streets trying to locate and round up the escapees.[45]

In the first shock of these escapes Paton had, as he later recalled, stopped in a pine plantation which lay between the main block and the Principal's house, and said (rather than prayed) to God, 'Don't You want this to happen? Because if You don't want it to happen, I'm going back to a safer job.'[46] His fear, of course, was that one or more of the absconders would commit some terrible crime which would get into the newspapers, with publicity that might end at a stroke the career of this idealist in his thirties who imagined you could take down the fences of a prison without murder, rape, and mayhem resulting.

This fear he conveyed most vividly in a letter he wrote years later to an American penologist, to whom Paton described the state of mind of a reformatory superintendent who has lost some of his charges:

> Very often such boys make their mentors sadists & perverts; for above their mentors are governors, superintendents, politicians, wanting results; such boys often threaten the whole careers, reputations, self-respect, inner peace, of their mentors. People don't understand this situation. You treat a fellow reasonably, decently according to your lights, and he absconds; worse, he takes nine others; the governor rings up & says, is it true, ten absconders, my God. You can't sleep, you hear bells & whistles in your sleep. One of the ten gets caught, one you befriended, perhaps the ringleader. You let it be known—privately, quietly—that you won't be around if his fellow-pupils decide to punish him. They punish him; you get a fright when you find how much they've punished him.[47]

Paton in this letter is describing a hypothetical situation, but there is no mistaking the personal experience of the stress that lies behind this account. He was to give another and even more powerful account of it in his short story, 'The Worst Thing of His Life'.

On this and the many succeeding occasions when Diepkloof suffered a rush of absconders, Paton was lucky, as he frankly acknowledged years later: 'I was never struck the kind of blow from which I would never be able to recover, the kind of blow that exposes the vain ambitious young man with the high purpose as nothing more than a self-deluded fool.'[48] All the same, the abscondings had given him a bad fright: in his first annual report he had to admit that no fewer than 87 of the 360 inmates (that is, 24 per cent) had absconded in his first six months, and at least 33 of these were still at large.[49] A weaker man would have abandoned the programme he had set in train, and fallen back on the old methods. But Paton now showed the courage of his convictions. The absconding made him slow down the timetable for his reforms, but he did not abandon them. Instead, he redoubled his efforts to introduce a system of self-discipline that would bind the boys to their duty with chains not made of iron.

He reinforced the military discipline under which the boys marched with great precision as they moved about the farm in their *spans*. He introduced a prefect system of particularly trusted boys, and divided the whole reformatory into four 'houses' which competed against each other in sport and discipline.[50] He increased the number of religious services at Diepkloof: these had previously been conducted by the black head teacher, Ben Moloi, and by a prisons chaplain on occasional visits, but Paton now often took them himself, and he rejoiced in the beauty of the boys' singing:

> The boys sang a hymn and the beauty of it, and the earnestness and innocence and reverence of those four hundred delinquent voices, captured me then and held me captive for thirteen years. I may add that it captured many others also, and sometimes visitors to the reformatory could hardly hold back their tears.[51]

He also introduced a half-hour scripture lesson for all the boys each day.[52] Because he believed that morality inculcated through religion was a stronger and much more effective restraint than bars and guns, he used religion as a central element in the chief means of control he came up with. It was a scheme all his own, and it was what he called the *vakasha* badge, 'vakasha' being Zulu for 'to go for a walk'.

Boys were carefully observed during their first nine months in Diepkloof, and were then selected for their trustworthiness and given limited freedom in return for a promise not to abuse it. An elaborate system of observing them and keeping detailed records on their background and behaviour was put into practice by one of Paton's most valuable staff members, I. Z. Engelbrecht, a giant 6 foot 3 inches tall and with a compelling personality. And Paton devised a ceremony that was quasi-religious rather than paramilitary, to bring home to the boys the significance of what they were agreeing to do:

> On Fridays at evensong these chosen boys would be paraded before the whole congregation, and facing me. As the names were called out, each boy in turn would come and stand in front of me. I would say to him, 'Today you are receiving your vakasha badge. What do you have to say?' The boy would then turn to face the congregation and say:
>
> Today I receive my vakasha badge.
> I promise not to go beyond the boundaries of the farm.
> I promise not to touch anything that is not mine.
> I promise to obey the rules of the school.
>
> He would then turn to me again and be given a shirt with the green badge. When all the badges had been given, I would say to the congregation, 'Today these boys you see before you have received the vakasha badge,' and the congregation would applaud.[53]

The vakasha badge system, after much fine tuning, worked well. 'When I made my promise,' said one boy who had absconded repeatedly until given his badge, 'it was like a chain on my leg'. Another youth absconded after getting his freedom, fled to Durban, more than 400 miles away, and there turned himself in to the police. 'Why did you do that?' asked Paton when he was returned. 'Because of my promise,' he said.[54]

Others showed great cunning and psychological insight in misusing the vakasha system, and then wriggling out of the consequences of their actions. One of these was a boy who accompanied a black supervisor known as Baba Scotch on a trip to the nearby suburb of Booysens, where they made a visit to a fish shop. As the two of them were re-entering Diepkloof's main gate Paton received a phone-call from the owner of the fish shop, who alleged that the boy, or Baba Scotch, had stolen some of his fish. Paton sent for them both and questioned them in vain. He then dismissed the trusted Baba Scotch and questioned the pupil further, but got nothing but earnest denials of guilt. The Vice-Principal, J. H. Laas, on being asked for advice, suggested that they beat the truth out of the boy, and this Paton proceeded to do. After a few cuts the boy jumped up and accused Paton of 'crucifying me': Baba Scotch was Judas who had betrayed him, he cried out, Laas was Herod, and Paton Pontius Pilate. Much shaken by this most unexpected comparison, Paton apologized humbly to the boy, and so, very reluctantly, did Laas. That same afternoon Paton, still haunted by what had happened, went looking for the pupil to apologize again—and found him eating a large piece of fish in the dormitory.[55]

Paton found the smaller boys in particular, perhaps a hundred in number, a great pleasure to deal with; he was to say in fact that he found them 'very attractive'.[56] They responded to the smallest sign of human affection, perhaps the only signs of affection they had ever met with, with great warmth. Paton would stand by one of them during the morning parade:

> He would look straight in front of him with a little frown of concentration that expressed both childish awareness of my nearness and manly indifference to it. Sometime I would tweak his ear, and he would give me a brief smile of acknowledgment, and frown with still greater concentration. It was as though I had tweaked the ear of the whole reformatory. These small outward expressions of affection were taken as symbolic, and many of the older boys would take themselves to be included. These were the irrefutable proofs that the aim of the reformatory was not punitive.[57]

And Paton found himself affected by the deaths of these children in his care; two died of typhoid fever in his first six months,[58] and he, who had suffered so badly from typhoid himself, was deeply moved. Though he was to write little during his years at Diepkloof, one of the poems he wrote shortly after he left the institution reflected this experience, 'To a Small Boy who Died at Diepkloof Reformatory'. Its third stanza runs,

> Here is the last certificate of Death;
> Forestalling authority he sets you free,
> You that did once arrive have now departed
> And are enfolded in the sole embrace
> Of kindness that earth ever gave to you.
> So negligent in life, in death belatedly
> She pours her generous abundance on you
> And rains her bounty on the quivering wood
> And swaddles you about, where neither hail nor tempest,
> Neither wind nor snow nor any heat of sun
> Shall now offend you, and the thin cold spears
> Of the highveld rain that once so pierced you
> In falling on your grave shall press you closer
> To the deep repentant heart.[59]

Considering it important to know as much as possible about the background of his pupils, he began keeping detailed files on each of them, recording such matters as the kind of home they came from, the relatives they had living, and so on. He noted with pity pupils who had no family, but who pretended to get loving letters telling news about fictitious siblings, as with the small boy he called Ha'penny, about whom he wrote one of his most moving short stories. He tried to meet the relatives and visitors of his charges, and was much struck by some of them, notably an old black priest. 'I remember he brought the boy food and gifts,' Paton said years later. 'It impressed me that the fact his son had been arrested for stealing didn't lessen his love for the boy in the slightest.'[60] When he left the old man would say a prayer for his son, and the boy would close his eyes and listen humbly. This black priest, figure of spiritual power yet of no account in the eyes of many whites, stuck in Paton's mind, and was to be the prototype of Stephen Kumalo in *Cry, the Beloved Country*.[61]

Late in 1935 Paton gained the confidence to have the huge and forbidding main gates removed, an event which he thought of as a real milestone; then the wire from the front of the building came down, on 1 January 1936, and one year later the entire fence came down. A fine bed of flowers was planted where the fence had been, and as news of what he was doing spread, he became known, not always approvingly, as the man who had torn down the barbed wire and planted geraniums.[62] Next, boys who had served nine months with good behaviour were given the freedom of the entire large farm on Sunday afternoons, after promising not to abscond. The widespread beatings were curbed, and the Principal reserved to himself alone the right to cane or whip the pupils.

The staff took time to accept these changes. One of them, C. J. W. Kriel, remembered years later,

> While waiting [for the pupils] to gather at the gate the supervisors paraded the quad to maintain the necessary discipline and it was during this duty that I came across a pupil who required strict admonishment, but instead of doing it verbally I decided to save my breath and gave him a good wallop with the cane on the buttock, not knowing that Paton was watching through the window of his office. Needless to say, I was immediately summoned to his office where I was threatened with suspension from duty pending an investigation by the Department of Education with a view to my dismissal.[63]

But having given Kriel a fright in this way, Paton considered he had gone far enough, and he quietly let the matter drop.

The effect of these reforms, though it took time to show itself fully, was remarkable. Before Paton's arrival, escapes from the institution were frequent. In 1935, with 360 inmates, there were 13 absconders per month on average, or 43 per cent over a full year. Once the reforms were more cautiously engaged in, from the start of 1936, a decline in the numbers of absconders began too. In 1936 there were 83 absconders out of 391 pupils, or 21 per cent; in 1937 the figure was 18.5 per cent; in 1938, 15.8 per cent; in 1939, 13 per cent.[64] By 1948, with 600 inmates, many more of whom were now in a position to escape easily, there were only 3 absconders per month, or 6 per cent.[65]

Paton made many small changes, such as encouraging the pupils to paint colourful murals on the dormitory walls,[66] something which had previously been a whipping offence, and he made big ones, such as ending the apprenticeship system of effectively enslaving pupils to farmers, and building a series of trade-training shops for teaching shoemaking, tailoring, and so on. In 1937 he began allowing home leave for boys who had served a year, and who had homes to go to. And in 1938 he began building a series of free hostels, consisting of a house occupied by a black house-mother and house-father, and surrounded by huts for twenty-five boys, who would have their meals with the house-mother and live permanently outside the main block.

All these moves tended in the same direction, to replace external, enforced discipline of fences, guards, and brutality, with internal discipline, a self-discipline of trust and mutual respect, encouraged by a firm but enlightened Principal. They also tended towards the integration into a common society of a group which had been despised and set apart, as in a sense the whole black community in South Africa was. The changes worked. And in Paton's view, they offered a model of social change that could be applied on a national scale. Increasingly, he was keen to make that application himself.

9

PROVING
1936–1938

> [The passive resistance in South Africa] is the most important activity the world can at present take part in, and in which not Christendom alone but all the peoples of the earth will participate.
>
> LEO TOLSTOY, *private letter to Gandhi*, 1910.

Paton drew the comparison between Diepkloof and South Africa most clearly in a letter congratulating Hofmeyr on an unusually liberal speech in which Hofmeyr, in November 1935, had appeared to call for the unity, not just of white South Africans, but of all South Africans, black and white. Paton wrote, both encouraging Hofmeyr and warning him:

> You are in for a big fight, & I have no doubt whatever that it is a fight which God willing will change your career from that of a brilliant administrator into that of a man who definitely tried & was called to try to change the course of the history of humankind, & may even have lost what we call 'power' in making the attempt.[1]

And having urged Hofmeyr on in this way, he made it plain that he saw himself and Hofmeyr as being engaged in essentially the same task. Hofmeyr was talking of removing political barriers between white and black, and Paton was removing physical ones; but, Paton implied, the spirit that moved them was one and the same. His language becomes a curious mixture of prophecy and rousing political speech:

> I here—God helping me—am going to do all that I can do to make of this Diepkloof a place that can gladden the hearts of all who long for justice and a place in the sun for the children of God.

Realizing how sentimental this sounds, he then adds,

> I am not sentimental about it—I want a place of lawns & trees, with a village & its playing fields, with occupations & unskilled trades, with good buildings, with water & light, with a proper placement system, with a happy & understanding staff,

& happy & understood children. For these things I shall work, not till I drop—for I no longer think that wise, but with every talent that is in me devoted to that end.

But he also took care, even in this mood of exaltation, to make it plain to Hofmeyr that Diepkloof was not the end of his ambitions, but a stepping stone to a political career: 'I want to prove myself, & to stand by the proof unflinchingly. But I make it no secret from you that I intend—if it is God's plan for me—to stand one day by you in the fight for what is good & wise.'[2]

This letter, written less than five months after Paton's arrival at Diepkloof, shows what a tremendous learning experience the reformatory had been for him, and how rapidly it was opening his eyes to the plight of blacks in South Africa. He had taken on the job with little knowledge of Africans, unable to speak more than a few words of a single black language, Zulu, and having little sympathy for black problems. Diepkloof changed that, and rapidly.

Cyril Burt had argued, in *The Young Delinquent*, that the root cause of a great part of youthful crime was social deprivation. The more Paton learned about the backgrounds of his charges, the more obvious that became. Fully a third of the pupils had no father into whose care they could be released; a significant proportion could claim no relatives at all; and almost without exception they came from backgrounds of the most desperate poverty and deprivation.[3]

In addition, as Paton came to see, his pupils were the product of that interregnum of which Gramsci speaks, between an old world dying, and a new one powerless to be born.[4] The old tribal order was being destroyed by the forces of industrialization and urbanization; but blacks were increasingly being prevented from entering the new South Africa by being deprived of access to higher education, skilled jobs, and the vote. Speaking of his Afrikaans staff, Paton was to say that they, 'working in an institution for young native delinquents, have been compelled by the sheer logic of fact to recognise the forces that are on the one hand destroying the old Bantu society, and on the other hand hindering the growth of the new'.[5] This compulsion 'by the sheer logic of fact' had been Paton's own experience at Diepkloof, and it was brought home to him not just through contact with the pupils, but with the black staff too.

This was the first time he had had close and prolonged contact with blacks who were not labourers or servants. Railton Dent had introduced him to black teachers, but though he had greeted them with respect he had not had to work with them. Now that he did, he did not always find it easy to get on with them. Those of his staff, both black and white, who had found the change from a prisons regime to an educational one uncongenial had left, but among those that remained were several with whom he clashed, both black and white.

Among the whites was the farm manager, Mr James Barry, who considered Paton anti-Afrikaans, but whom he gradually won over. Among the blacks there was at least one, James Gubevu, who had repeated clashes with Paton and gradually became convinced that Paton was an irredeemable racist. There exists an undated letter of complaint about Paton from Gubevu, entitled 'Misadministration at Diepkloof', which seems to have been sparked off by a scolding Gubevu got after two boys in his charge absconded. In this document, which Gubevu seems to have sent to Margaret Ballinger, who after 1938 was one of the three representatives of black interests in the House of Assembly, he alleges that Paton had called a meeting of the black staff and told them they were inferior to whites, and he refers to 'the Principal's anti-native policy' of not letting the black staff know the rules by which the institution was run.[6] Gubevu's complaints sound overstated and fanciful (it was not in Paton's interest to keep black staff in the dark as to the rules they were administering, quite the opposite), but it seems likely that the harshness which Paton's classes at Ixopo and Maritzburg had detected in him did not disappear overnight. He demanded implicit obedience from his staff, and was angered when he did not get it. This was particularly true when it came to matters such as absconding, which affected the outside world and could lead to publicity bad enough to stop Paton's work entirely.

During one clash with Paton, Gubevu called him a racist to his face, and Paton was in such a rage that he could not speak. He went away in anger, and Gubevu must have wondered what the consequences were to be. The next morning Paton called for him, and to his surprise told him that he, Paton, had 'looked into himself, and was not happy with what he had found'. Paton then performed one of the dramatic and symbolic acts which helped to make him such a good leader and administrator. He took a shovel, led Gubevu out to the reformatory farm, and there with his own hands dug a hole. In this he planted a young tree, inviting Gubevu to help firm it into the earth. This tree, he told Gubevu, would stand to mark the most important turning point in his life. And, according to Gubevu, Paton's attitude to the black staff really was different thereafter.[7]

At least two of the pupils complained about Paton in letters to the Governor General in November 1938, Jacobus Leodwyk saying that Paton had made him work even when he was sick,[8] and Cecil Somdakakazi saying Paton had verbally abused him for asking to be released: 'it was very bad his speech'. Somdakakazi further complained that he had learned nothing in the reformatory, other than how to rape other boys, bribe them with food 'so the[y] can be my wives', and to 'teach other boy bad spirits'.[9] A second letter from each of the pupils made it plain that what they hoped to achieve by their complaint was not action against Paton, but their own immediate release: 'please let me know when I will get out from these place'.[10] These two plaintiffs had written on the same day, one clearly influencing the other, and after this isolated case in 1938 there were no complaints about Paton's rule. If he was a despot, as every prison governor must be, he was an increasingly enlightened one.

There is clear evidence that in incidents like the tree-planting with Gubevu Paton's life was being changed, and it was capable of being changed because he put his heart and soul into the work. 'Does Marais at Tokai feel the struggle as I do?' he asked. 'Or Pretorius at Houtpoort? Or am I a fool? Will Diepkloof be the same Diepkloof without my tears & my blood?'[11] But he knew the answers before he asked the questions.

As his methods produced results, Paton was able to reduce the use of the cane and the sjambok, though he never dispensed with them entirely. 'To retain the regard of the staff I do not hesitate to inflict heavy punishment for insolence, if the officer seems to expect it,' he explained in the official annual report for 1936. 'But whenever possible one attempts to dispense with corporal punishment. I myself have no faith in it, and I inflict it only where I feel that the members of the staff, feeling as they do, have a right to expect it. The attitude of the staff towards punishment shows considerable change.'[12]

However the change in practice was slow: in 1937 he inflicted 2,000 strokes,[13] and though the number fell year by year (from 5.1 strokes per pupil in 1936 to 2.4 strokes per pupil in 1941),[14] even in later years his younger son, Jonathan, who was born on 6 July 1936 and brought up at Diepkloof, was to retain most vivid memories of his father energetically wielding the sjambok as he corrected some particularly recalcitrant pupil.[15] One of his Afrikaans staff remembered years later that 'even at the reformatory he was heavy-handed in administering corporal punishment, and if my memory serves me right, he discarded the cane to use a sjambok'.[16]

He had reason to retain the use of the lash, for it has to be remembered that he was not dealing only with schoolboys. Although no pupil older than 22 should have been at Diepkloof, there were several considerably older than that. On one occasion, late in 1937, a pupil whose age Paton estimated at 26, about to be caned for threatening to assault a white supervisor, seized the cane and threatened Paton with it. Enraged, Paton punched him, but the middle finger of his left fist was cut by the man's teeth. Infection set in, and on 1 April 1938 the finger had to be amputated, leaving Paton with a mutilated left hand, about whose unsightliness he was for a time very sensitive. Particularly when being photographed, he would clasp his left hand in his right to conceal the cleft.

There can be little doubt that in making the move to Diepkloof, Paton was also consciously trying to move away from his former excessive use of physical punishment, which had got him such a reputation at Ixopo School and Maritzburg College. He was in some sense trying to reform himself, as his subordinate C. J. W. Kriel was to remark perceptively: 'It is possible that Paton was aware of his own weakness in corporal punishment and probably feared that free use of the cane might get out of hand to foil the new trend in institutional treatment, let alone the shadow it might cast on his own reputation.'[17]

His reforms continued as his confidence grew. In 1937 he introduced cobbling, tinsmithing, carpentry, and other trade training,[18] involved 100 boys in gardening rather than farming, and succeeded in having the much-abused farm 'placement' scheme abandoned. The reformatory's school was reinforced with extra teachers, and boys began taking and passing the Education Department examinations. Paton also began encouraging contact with other schools, notably St Peter's College, an Anglican school run by the Community of the Resurrection. Sports teams from Diepkloof visited St Peter's, and the St Peter's boys visited Diepkloof several times each year.[19] So did boys from an expensive Anglican school for whites, St John's, where Paton's elder son David, seven years old in 1938, was now a pupil.

The vakasha badge system was now working relatively well: by 1938 only 10.3 per cent of the boys given their badges absconded, a rate that Paton described as 'a fairly stable percentage',[20] but one that was to improve over the next few years until it stood at 3 per cent.[21] In 1937 he introduced an extension of the system, which was that carefully chosen boys who had homes to go to were given home leave once a month. After an adjustment period he found that the 3 per cent rule applied here also; of every hundred boys given home leave, 'one would return late, one would return drunk, and one would not return at all. Ninety-seven would return at the given time, and sober.'[22] Paton was rightly proud of the fact that only 1 per cent did not return at all. As he put it in his annual report in 1938, 'The record of punctuality in the matter of return was amazing. Of each 100 boys going home without supervision, 99 could be safely expected to return.'[23] And he justified his methods by advancing his view of what a reformatory should aim at:

> A reformatory where all opportunities for evil-doing are removed, is no reformatory at all. To keep a delinquent unspotted from the world for his period of detention is to do nothing for him at all. He must learn to move in a world where both good and evil are offered him, and must learn to choose; the real education lies in the choice offered, and not in the creation of an artificially crimeless environment.[24]

And he persisted with these views even when one of the pupils on vakasha leave broke into the house of a white woman and killed her, in circumstances very similar to those in which Arthur Jarvis is killed by Absolom Kumalo in *Cry, the Beloved Country*. Fortunately for Paton, the newspapers did not link the murder with Diepkloof even when the young man was apprehended, tried, and hanged.[25]

Under Paton's new regime the health of the pupils improved sharply, the decrease in the number of typhoid and pneumonia cases being the most dramatic. The district surgeon attributed the improvement to the improved sanitary conditions, the fact that the pupils were no longer locked in at night, the improved diet-scale, and the increased interest taken by staff members in their pupils.[26]

And in 1938 Paton achieved his ambition of beginning the building of his village, when four free hostels were built on a piece of open land below the main block. A fifth hostel was to follow in 1942.[27] They were placed round a playground, like a village green. Each free hostel consisted of a large building, faced with a colonial Cape-Dutch gable, containing accommodation for an African teacher and his wife, and the communal dining room for their twenty-five charges. These twenty-five boys lived in five huts placed round the larger building. The little cluster of buildings looked both African and aesthetically pleasing, and as unlike the hideous central block as could be conceived. To Paton there was nothing more beautiful on earth, and his pride speaks through the account he wrote years later:

'The spacing of the buildings, the simplicity of their design, the white walls and the thatched roofs made the village something to be proud of. In the moonlight it was—in my opinion—as beautiful as the Taj Mahal.'[28]

The village did not please everyone, though, and at least one member of Diepkloof's board, J. A. Herholdt, criticised the new buildings on the grounds that they were 'unnatural' for blacks (by which he clearly meant too good for them), and opined that 'any thinking white man' would realize that such comfortable accommodation ['gemaklike uitrusting'] would have a bad effect on them when they had to return from this 'ultra modern' accommodation to the shanty ['pandok'].[29] This racist view, which was held by more than the one man who expressed it in writing,[30] Paton rejected firmly though with restraint.[31]

He continued to believe that his life was being shaped by a power greater than himself, writing to Hofmeyr in August 1936:

> I feel myself more & more called upon by God, not in any vague but in a very real way, to let Him do this work . . . And feeling this as I do, I can say to you without fear of being thought presumptuous, that I cannot but believe that I am in some way being 'prepared'.[32]

This sense of 'preparation' and change showed itself not just in his attitude to his own black staff, but in his repeated letters urging Hofmeyr to speak more plainly for the blacks, and warmly pressing on him 'a philosophy that has no place for fear, that insists on cooperation as the essential mark of all fruitful human relationships, that in fact always insists that the other party in such a relationship is a "person" & not merely a native . . .'[33] The context of this letter was the nationwide debate about the series of Native Representation Bills which the United Party was now introducing. Paton was beginning to take a more active interest in South African politics.

Since the end of the Boer War in 1902, the main struggle for power in South Africa had been waged between the former Boers (Afrikaners, as they were increasingly called) and the English speakers. Since Afrikaners greatly outnumbered English speakers, any party which could claim their allegiance on linguistic or racial lines was bound to come to power in a system which denied almost all blacks the vote. It was for this reason among others that English speakers like Paton saw a vital importance in linking the interests of both white groups. This the Fusion government of 1933 had done, and the United Party seemed the embodiment of the hope of a united South Africa. Blacks remained a troubling side-issue which could be, and usually was, ignored before the Second World War. But among

Afrikaner nationalists in particular fear of black numbers steadily grew, for blacks seemed the only possible threat to long-term Afrikaner dominance of political power. As the thirties wore on Afrikaner nationalists increasingly focused on ways of containing that threat.

By 1935 the Nationalists under Malan were exerting increased pressure on prime minister Hertzog (who in truth did not need much urging) to introduce bills which would stifle any further growth in the tiny black vote. Under South African law, almost all whites could vote if they were over 21. Under a qualified franchise system granted by the British government in 1853, so could any African, Coloured, or Indian—if he were male, if he lived in the Cape Province, and if he had certain qualifications of education and property. These restrictions meant that Africans in 1936 constituted only 1.4 per cent of the total electorate,[34] but many whites, and Afrikaner nationalists in particular, wanted to ensure that the rights of even this tiny percentage were curtailed.

The Native Representation Bills, long debated and introduced in amended form in 1936, were aimed at rejecting African representation in the lower house, and removing African voters from the common roll, while supposedly compensating them with land rights. In the end, opposition from Hofmeyr and a handful of like-minded members persuaded Hertzog to allow the Africans three white 'communal representatives' in the House of Assembly and four in the Senate, though they would still be removed from the common roll. Hofmeyr's fighting speech at the third reading, early in 1936, attacked the principle of black 'communal representation' as implying inferior citizenship. He accused Hertzog of abandoning even the theory of a common society, and predicted, correctly as it turned out, that this retreat would not end in 1936.[35] But Hofmeyr had taken care to say in advance that his opposition to the Bills did not amount to an issue of confidence, and he carried only 10 other members with him in voting against the measures. White South Africa was overwhelmingly in favour of shutting blacks out of power, and the Afrikaner Nationalists began to realize that they had found an issue that would, in time, bring them to office. That it was also the issue that would, in time, sweep them out of power they could not foresee.

The Native Representation Act was accompanied by other equally oppressive measures aimed at controlling pressure from the rapidly growing black population. Territorial segregation was achieved by the 1936 Native Trust and Land Act, which began the process of confining Africans by law to what were later to be called the 'homelands' (or, scornfully, 'bantustans'), and influx control, which prevented Africans from leaving the desolate rural areas to seek work in the towns, was instituted by the 1937 Native Laws Amendment Act. In the case of Indians in Natal and the Transvaal, Mahatma Gandhi had marched and organized in vain; there were in place strict measures to control land and trading rights, they were forbidden to reside in the Orange Free State at all, and there was no thought of giving them the vote. The Coloured people of the Cape were slightly better off.

In this important phase of what was to become the central struggle of modern South African history Paton stood decisively on the side of the tiny minority among whites who believed that there should be gradual progress towards treating blacks like white citizens. His experiences at Diepkloof had rapidly converted him from a theoretical, Christian-based belief in the brotherhood of all men, to the realization that if Africans were not given fair and equal treatment in South African society the results would be grim indeed. At this stage he was thinking particularly of the increased crime that was sure to result; he had yet to see clearly what the impact of what would become known as apartheid would be on society as a whole. But though he saw only dimly, he was far in advance of the great majority of South Africans, who were entirely blind to the growing problem. Diepkloof had begun to peel away the scales from his eyes. But it was not only within Diepkloof that changes were taking place.

By the beginning of 1938 Paton was beginning to feel that his task at Diepkloof was approaching completion, and that his experiment was a proven success. He wrote triumphantly to Hofmeyr, 'With the building of the half of the village, to accommodate 100 pupils, the experiment reaches its logical conclusion. There remains only the consolidation of that work . . . the village is the real finish.'[36] And again he hinted broadly that it was time Hofmeyr thought of using him as a political ally, and implied that the two of them could be prime minister and deputy in a future government:

> Someday an Afrikaner who speaks English like an Englishman [Hofmeyr], and an Englishman who speaks Afrikaans like an Afrikaner (he doesn't yet)—some day, well, it might be vainglorious to finish the sentence. But one cannot escape the fact that such a combination is long overdue, & I believe would be invincible.[37]

And just in case Hofmeyr had missed the point, Paton finished his letter by telling of a dream he had had a day or two previously, in which he was a Minister, and Hofmeyr his Prime Minister. Hofmeyr ignored this account of the dream, as he ignored the hint it reinforced, and though Paton in his next letter remarked 'I'm sorry you didn't comment on my dream,' he appears to have received no response.

These hints were urgent, because South Africa was once more approaching a general election, and Paton seems to have been hoping that Hofmeyr might invite him to stand for a winnable seat. But Hofmeyr had concerns of his own which occupied him fully. His liberal views, though he held or at least expressed them very inconsistently, meant that he was in an uncomfortable position in the largely conservative United Party government. He was particularly uneasy about the Native Representation Act: although he had opposed it openly, he had stayed in the cabinet once it was passed, and had therefore to take responsibility for the implementation of an Act of which he disapproved. The Nationalists in their speeches increased his discomfort by implying that he was either a hypocrite or a fifth-columnist within the government, whose dangerous liberal views would be put into practice at the opportune time. Many of that small band of liberal thinkers who supported his stand, on the other hand, felt that he should have resigned over the Native Representation issue.

Paton saw Hofmeyr at the annual SCA camp at Anerley in July 1938, from which he wrote to Dorrie the first letter which has been preserved, itself a sign that the worst crises in their marriage were now past. 'Hoffie is well & sends his regards,' was all the news he had to tell her, other than his travel plans and official details of staff-transfers at Diepkloof. It appears that he got no more out of Hofmeyr face to face than he did by letter.

In the general election of 1938, though the United Party retained a comfortable hold on power, the Nationalists' number of seats in Parliament increased. Hofmeyr's crisis of conscience came to a head in September 1938, when a certain Senator Fourie, a former Minister of Labour who had lost his seat in the election, was appointed as one of the African representatives in the Senate on the blatantly spurious grounds that he was specially qualified for the position. Hofmeyr, who had failed to resign from the cabinet over the major issue of the Native Representation Bill, now resigned over this much more minor matter, and thereby showed, not for the first or last time, how erratic his behaviour could be.

Paton greeted the news of the fall of his mentor and patron with consternation. '*Die wêreld dws. my wêreld het baie verander,*' he wrote to Hofmeyr a week after hearing the news, his first letter in Afrikaans to the fallen leader. '*Ek kan nog nie iets skrywe omtrent jou bedanking nie. Neem dit nie kwalik nie, want my enigste gedagte is dat Diepkloof sy (haar?) skepper verloor het.'* ['The world, that is, *my* world, has very much changed . . . I can't yet write about your resignation. Don't take offence at this, for my only thought is that Diepkloof has lost its (her?) creator.'][38] And he went on to discuss his fears: that the United Party might now fall apart, that the right-wing ideologue Oswald Pirow might take over the Education Department, and that Paton would have to learn to be independent of a friend in high places. All the same he continued to hope that Hofmeyr would rise again. 'From that day on,' he wrote in his autobiography, 'I looked for the day when Hofmeyr would become Prime Minister. He, so to speak, became my politics.'[39]

And to Hofmeyr in later months he was to write that the resignation had been quixotic, but justified. 'We shall have to say "Our country did this & that",' he wrote, 'but we can at least add, "but even in those days men revolted".'[40] In the later, and much darker, days that were to come, this was to be his credo too; even if he could not change the course of events he would not submit to calling them inevitable, much less calling them right. 'In other words,' he explained to Hofmeyr, 'there are for me two kinds of greatness, that of the idealist who condones no evil & removes none; & that of the realist who condones one that he may remove another.'[41] He himself, when obliged to choose between these two kinds of greatness, would choose the first early in his political career, and increasingly, as he aged, the second.

And as he had done once or twice in the past to Hofmeyr, he sketched his own future in half-joking terms, those terms in which he found it easiest to tell Hofmeyr of his ambitions:

> I don't mind telling you now that the possibility seems remote, that at the end of five years (you once asked me how long I wanted) you might have elevated me to some place from which I could have stepped off into politics. I feel as convinced as any stage-struck girl that goes to Hollywood that I could become a star. Something tells me that I could move men, administrate efficiently, command respect, face danger, retreat in good order. Something tells me that I shall never be readier for it than now. But now no J. H. Hofmeyr stands behind. Good—I've accepted that.

And having, as he said, accepted that Hofmeyr would not be able to do anything for him, he thought about how he could do it for himself, and came up with another of his uncanny prophecies: 'I shall probably in a year or so write a world-conquering book about Diepkloof, make £10,000, resign, & heigh-ho! me for the caucus.'[42] The war would throw his timetable out, the caucus would never see him, but he made no mistake as to the world-conquering book.

In the same letter he clearly stated the terms in which he saw Hofmeyr, as 'a guarantee of several great things, racial co-operation, liberal native policy, Commonwealth connection, democratic institutions'.[43] In point of fact Hofmeyr was to return to the Cabinet with increased power on the outbreak of war almost exactly a year after his resignation, so that the guarantee he represented to Paton was not removed for long. On the other hand, Paton over-estimated Hofmeyr's liberal convictions, which were waveringly and inconsistently held. Paton's attribution of his own values to Hofmeyr was in part wish-fulfilment. All the same, these were the chief values which Paton saw as embodied in the liberal members of the United Party, and which he rightly feared would be swept away if the interests of Afrikaner nationalists should be uncoupled from those of English speakers.

He himself continued to hope that this uncoupling could be avoided, and grew more enthusiastically pro-Afrikaner. It was probably in the second half of 1938 that he wrote his dramatic portrait of his parents as Mr and Mrs Kingsley. In the last part of the fragment that survives, the Kingsleys' daughter Jennie enters with her young man, a young and courteous Afrikaner named Marthinus Jacobus Petrus de Wet, who gets very favourable treatment in the play and who effortlessly and politely shows his superiority to Mr Kingsley.[44] Paton's enthusiasm for all things Afrikaans continued to increase until it suffered a rude shock during 1938.

The event was the Centenary celebrations of the Great Trek, that pioneering emigration in 1838 of Boers from the Cape, who moved northwards in their great covered ox-waggons to escape the influence of British rule. Their motivations included a desire for independence and living space, and a desire to escape the doctrines of racial equality which British government officials and churchmen preached. The Purified National party under Malan saw in this centenary in 1938 a perfect opportunity to revive enthusiasm for Afrikaner nationalism, desire for independence from the British Empire, and doctrines of racial purity. They came up with an inspired plan, which was that the Trek should be re-enacted, with waggons coming from all over the country to a hilltop outside Pretoria, where the foundation stone of a commemorative monument in the form of a great square tower would be laid.

Paton was swept up in the enthusiasm for these celebrations and determined to take part in them. He had written to Hofmeyr while the latter was still a minister, asking permission to take the Diepkloof waggon and oxen to Pretoria, and declaring his own intention to accompany them. And he was as good as his word; he grew a small beard ('scraggy' was the adjective one of his Afrikaner staff used of it),[45] as burning Afrikaner zealots all over the country were doing—the only crop ever grown in South Africa without a subsidy, as opponents of the Trek celebrations joked.

Paton took the whole business with the utmost seriousness. He set off for Pretoria on the waggon in December 1938, incongruously attired in the striped blazer of the NUC.[46] He found that other bearded travellers greeted him warmly, while one clean-shaven motorist called out aggressively, 'Look at the bloody old Boer'. Paton was delighted, and told his staff the story.[47] He was determined to show that English-speaking South Africans supported these celebrations as fully as Afrikaners. The waggon flew the *vierkleur*, the flag of the old Boer Republics, and on it sat almost all the Afrikaans staff of Diepkloof. Only two blacks from the institution accompanied the waggon, and they were the two boys who had been brought along to look after the oxen.[48]

But even before he reached Pretoria Paton saw ominous signs that Afrikaner nationalists rejected the support of English speakers. In various towns along the way mayoral delegations came out to make welcoming speeches and provide civic receptions for the Trekkers. If these mayors were Afrikaners themselves the welcoming gestures were well-received, and there would be speeches in Afrikaans and much singing of the nationalist song *Die Stem van Suid Afrika* [The Voice of South Africa], drinking of coffee and eating of traditional rusks. But similar gestures of welcome and respect from several non-Afrikaner mayors were rebuffed. The radicals who had organized the Trek commemoration saw it not as a event to unite South Africa, but as an event to unite Afrikanerdom, and to accomplish their aim they were quite willing to split further a country already divided.

The Diepkloof waggon reached the appointed meeting place, already known as *Monumentkoppie* [monument hill], to find 250,000 people milling about there in terrific heat and dust. Paton, hot and tired from the journey, headed for the temporary washrooms, stripped and had a shower. Another naked, bearded man under the next shower said to him with the utmost enthusiasm, '*Het jy die skare gesien?*' [Have you seen the crowds?] '*Ja*', said Paton. And the man, with what Paton later called 'unspeakable comradeliness', said to him, '*Nou gaan ons die Engelse opdonder*' [Now we'll knock the stuffing out of the English].[49] And Paton, as one of the 'English' who would be '*opgedonder*' if it came to that, felt a chill.

There were more such chills to come, for at the celebrations that followed English speakers who brought messages of goodwill were drowned out by the singing of Afrikaans songs until they agreed to give their messages in Afrikaans instead, when they were loudly applauded. And when one of Paton's staff members, Nonnie Pienaar, tried to buy a book in English as a gift for Paton's son David, the girl in the bookshop told him in Afrikaans, 'This is not an English celebration. You'll find nothing in English here.'[50]

'What I had done in good faith and such good will', Paton wrote bitterly, 'turned to ashes. I wanted only that the celebration should come to an end. It was a lonely and terrible experience for any English-speaking South African who had gone there to rejoice in this Afrikaner festival.'[51] He left Monumentkoppie early and left his white staff, convinced of the great advance Afrikanerdom had made through this gathering, to return on their own, drawn on their emblematic waggon by the oxen, and led home by two young black miscreants. On reaching home on 17 December 1938 he said to his wife, 'I'm taking off this beard and I'll never grow another'.[52]

And two days later he wrote his first poem in Afrikaans, 'Lied van die Verworpenes', [Song of the Outcasts],[53] a sad repudiation of the Trek celebrations. In it an excited child sees the waggons coming, and asks permission to go out and welcome them; when he does, the waggons run him down and crush him under their great wheels:

,Kyk ma, daar kom die ossewaens,
Hoe trots waai daar die mooi vlae!'
,Gaan uit my kind en groet'nis sê
Vir helde van die vroeëre dae.'

Hy hardloop uit, my skat, my kind,
Ek hoor sy stem ,hip hip hoera!'
Dierbare God, hy loop reg voor
Die osse van die ossewa.

Rustig die edel jukgediert
Stap aan, sy arme lyfie oor.
Onkeerb're wiele kraak en dreun
Dis al wat nou die skare hoor.

O drome van die toekoms skoon
Wat ek vir jou my kleintjie had,
O Pad van ons Suid-Afrika,
Was dit nie ook my kind se pad?[54]

The poem may be literally translated thus:

'Look ma, here come the ox-waggons
How bravely the bright flags wave!'
'Go out my child and greet
The heroes of earlier days.'

He runs out, my treasure, my child,
I hear his voice: 'Hip, hip, hooray!'
Dear God, he walks right in front
Of the oxen of the waggon.

Tranquilly the prize beast
Treads on, over his poor body.
Unstoppable wheels creak and groan
That is all the crowds hear now.

O dreams of the bright tomorrow
That I had for my little one,
O road of our South Africa,
Was it not my child's road too?

The last stanza, with its question spoken by the crushed child's mother, suggests Paton's mingled sorrow and anger that his own dreams of a united South Africa had been destroyed by the Trek celebrations.

This highly competent Afrikaans poem, the first of only two he is known to have written,[55] also marks the high point of his efforts to master Afrikaans, and shows how successful he had been. From the date of this poem, 19 December 1938, although he continued to seek and cultivate friendships with Afrikaners, he regarded Afrikaner nationalism as something to be countered with all his strength. And he regarded it with fear, for it was plain to him that the nationalism represented by Malan and his party was on the rise. The liberal forces in the United Party were now in retreat, a retreat symbolised for Paton by Hofmeyr's resignation. In fact, he came to believe, it would take some world-shaking event to prevent Malan, who at this stage controlled only 27 seats of the 150 in parliament, from coming to power sooner rather than later. As it turned out, world-shaking events were on the way.

10

WAR
1939–1943

They also serve who only stand and wait.

MILTON

By the beginning of 1939 it was clear to many observers that the world was moving rapidly towards war. Mussolini had been in power in Italy since 1922, Hitler in Germany since 1933. The two dictators, singly and in tandem, had tested the will of the democracies repeatedly, and had found them wanting in resolve. Italy's invasion of Abyssinia in October 1935, the dictators' involvement in the Spanish Civil War on Franco's side from 1936 to 1939, Hitler's militarization of the Rhineland, each of these events was met by the democracies with appeasing language and an apparent inability to act. Britain and France felt too weak militarily to oppose Germany, and the dictators increasingly showed that they were not going to be stopped by anything short of force.

The slide towards war gathered pace all through 1938. In March 1938 Hitler annexed Austria while the British and French governments stood helplessly by. There was another crisis in May 1938, when a German invasion of Czechoslovakia seemed imminent; when it did not come the British were temporarily elated. Hitler renewed the pressure, however, and in September Neville Chamberlain met him at Berchtesgaden, and again at Godesberg, to try to avert a German invasion of Czechoslovakia. War seemed inevitable, and on 27 September 1938 the British Fleet was mobilized. The next day Hitler agreed to a conference in Munich, at which he obliged Chamberlain and the French Prime Minister, Daladier, to agree to almost all his demands of the Czechs, who were not even consulted. Chamberlain flew back to London on 30 September 1938, waving his agreement with Hitler and proclaiming 'peace with honour' to relieved and jubilant crowds. Both peace and honour were illusory. When, in

March 1939, Hitler invaded Czechoslovakia, Britain and France could do nothing other than to try to ward off an attack on Poland by committing themselves to war on the Polish side if a German invasion should come.

English-speaking South Africans, as members of the Empire, shared at long range successive British feelings of fear, relief, and humiliation. But for many Afrikaners, their defeat in the Boer War still rankling, threats to British power were welcome from whatever side they came. And while British rearmament went on apace, and the world waited to see in which direction the dictators would strike next, ugly echoes of Nazism's martial music began to be heard in South Africa.

These echoes were in part the outcome of the Trek commemoration in which Paton had taken part. One movement which flowed directly from it, as the name suggested, was the Ossewa Brandwag (Oxwaggon Sentinels). Beginning as an ostensibly cultural movement in February 1939, the Ossewa Brandwag rapidly became a paramilitary body like the German SA, with its own storm troopers under a commandant-general, J. F. J. Van Rensburg, an ex-Administrator of the Orange Free State. Its numbers, and with them its power, grew rapidly as war drew closer, and the increasingly large gangs of paramilitary thugs it could muster at rallies made its potential for violent intimidation of its opponents considerable. Before long it was boasting of larger numbers than the South African army could muster.

The Ossewa Brandwag had many links with an older movement, again cultural in origin, the Broederbond (League of Brothers). More insidious than the Ossewa Brandwag, the Broederbond infiltrated members into every aspect of South African life. Secretive as to its membership and methods, the Broederbond tried particularly to influence the young, enlisting teachers wherever possible, starting an Afrikaner alternative to the Boy Scouts called the Voortrekker movement, and promoting Afrikaner festivals such as the Day of the Covenant, December 16, in commemoration of a Boer victory over the Zulus at Blood River. The Broederbond also moved to seize some economic power by launching a bank (Volkskas), an insurance group (SANTAM), an investment firm (SANLAM), and even a retail chain (Uniewinkels). These two movements between them seemed to be developing a state within the South African state.

In political terms both the Broederbond and the Ossewa Brandwag supported Malan's Nationalists and opposed the United Party of Hertzog and Smuts. As the war neared, Malan's group of MPs did what they could to support the dictators, rejecting demands for sanctions against Italy over Abyssinia, and asking for a ban on Jewish immigrants from Germany and other countries.[1] This anti-Semitism was reflected also in the leader pages of the newspapers which supported the Nationalist line, *Die Burger*, *Volksblad*, and *Die Transvaler*. Extreme nationalists gained confidence from the rise of Hitler as they had from the Great Trek centenary celebrations. Diepkloof itself reflected this growing division in the country. Most of Paton's white staff rejoiced in the idea of England's defeat; like some Irishmen at this time, they were not so much pro-German as anti-British. Even Paton's enlightened supervisor, 'Lanky' de Lange, a particular favourite of his, told him, 'If there's a war I don't want England to lose it, but I want her to suffer.'[2]

When Hitler attacked Poland on 1 September 1939, then, and Britain's declaration of war followed on 3 September 1939, South Africa was in a terrible dilemma, her white population divided between British loyalists who wished to come to the aid of the mother country without question, and a sizeable portion of the population which was fiercely opposed to joining in a war on the side of the power which had defeated them in 1902 and dominated them ever since. Parliament in Cape Town fully reflected this cleavage: in a decisive debate on 4 September 1939 the ruling United Party showed itself united no longer, dividing largely along linguistic lines, into Afrikaners who supported the prime minister Hertzog's motion that South Africa remain neutral in the war, and English speakers who supported Smuts's amendment that South Africa declare war. The vote was close, Smuts winning by 80 votes to 67; Hertzog resigned and joined Malan's Nationalists in opposition, and Smuts, whose support was bolstered by the small Labour and Dominion parties, became prime minister in his place with Hofmeyr as his deputy and minister of finance. South Africa, troubled, divided, and excited, was at war.

In the first year of war, as Nazism triumphed across the Continent, violence seemed just below the surface in South Africa. Hitler conquered Poland in less than a month, in September 1939; in April 1940 the Germans took Norway and Denmark; and in May 1940, in a brilliant campaign, they invaded Belgium and Holland, struck through the Ardennes to the Channel, paralyzed the French, and isolated the British Expeditionary Force. The evacuation from Dunkirk began on 27 May and ended on 3 June; the Germans entered Paris in triumph on 14 June, and by 17 June 1940 the French were suing for an armistice. Britain now stood alone, and during the next few months one invasion scare followed another.

In South Africa nationalistic Afrikaners, outraged to find the imperialist Smuts in control of their nation's destiny, and believing that South Africa had joined, not only the wrong side, but also the losing side, flocked to the

Ossewa Brandwag in huge numbers. They marched and manœuvred on farms, their leaders held huge stop-the-war rallies in the cities, and they made speeches attacking 'British-Jewish capitalism',[3] predicting that Germany was sure to win, and looking forward rejoicingly to this outcome. A coup attempt seemed on the cards, and violent groups, including the openly anti-Semitic Greyshirts and another paramilitary movement called *Stormjaers*, were on the increase.[4]

Many of Diepkloof's staff were Ossewa Brandwag members, proudly wearing the badge of the organization on their lapels. When in 1940 Smuts felt confident enough to forbid civil servants to wear these badges, they defiantly wore the crown caps of mineral water bottles instead. Paton told them to remove these, and with much foot-dragging they obeyed.[5] His relations with his staff became increasingly difficult, and he confined himself to official contacts with them. Previously he had occasionally invited individuals among them to his house for a drink in the evening, or to play 'reformatory bridge', a card game he and Dorrie had invented, and of which they became very fond; but now these contacts largely stopped. Those of his staff who were in favour of the war effort had left Diepkloof to enlist, and had been replaced by diehard nationalists.

The war touched him personally very early on, when his brother Atholl became the first South African soldier to die in action. Paton had had little contact with Atholl from the time he had taken his first teaching post at Ixopo, and the two had scarcely seen each other since Paton had come to Diepkloof. There was in any case little love lost between them, and as men they had become utterly different characters. Atholl, who worked for the Pietermaritzburg municipal electricity department as a clerk, was a handsome and uninhibited pursuer of pleasure, in a hearty, beer-drinking, rugby-playing, woman-chasing way: it was his form of rejection of his father's values. On deciding to join up he had at the age of 36 married his current girl-friend, Grace Allison, and had quickly found that the army was just the place for him, being rapidly promoted to sergeant as the South African army moved north. On 16 December 1940 he was killed in a minor skirmish at El Wak on the Kenya border with Somaliland.[6] Rather to his own surprise and puzzlement, Paton broke down and sobbed when on 21 December 1940 he received a telegram from his mother: JUST HEARD ATHOL KILLED IN ACTION—MUM.[7] His white staff, in Afrikaans, and his black staff, in English, sent him letters expressing their condolence.[8]

Paton himself, in spite of the pacifist beliefs with which he had been brought up, had decided early on in the conflict to enlist, although he was 38 in January 1941. In the event he had been rejected, his superiors ruling that he was doing work of national importance and must stay at Diepkloof.[9] Late in 1942, after the heavy losses sustained by the South African army in the fall of Tobruk, he again requested leave to enlist, and was again refused.[10] He had to content himself with joining the National Reserve Volunteers, a home guard, and serving at weekends in it as a private from 1940 to 1941, and then as a sergeant until 1944. He considered that this service, for which he later received the Africa Service medal and ribbon, and a citation from King George VI, had been entirely futile, except that going off in uniform each Saturday had been a demonstration to his staff, white and black, of his willingness to do something tangible in defence of his beliefs.[11]

Essentially, though, he found himself thrown back into the routine of running an institution which he had already transformed beyond all recognition, and which had now settled into a pattern which was not to change substantially until the end of his time there in 1948. It was a task, in other words, which he felt was finished, and from which he wished to move on. The remaining changes he had envisaged and planned, including the building of a new central block and the establishment of a school in the nearby black township, where Diepkloof boys could receive aftercare once they had been released, were suddenly beyond his grasp, for the funds that were to have been set aside for these projects were now devoted to the war effort. In addition the quality of his staff declined as some of his best men left to join up. The progress of change at Diepkloof became glacial after 1939. Paton, though he had been prevented from joining the forces, was marking time. And in these circumstances he looked around him for other worlds to conquer.

In later years, when he had left Diepkloof, he would look back with clear eyes and analyse the queer, complex mixture of motives that drove him on at this time:

> I was ambitious, compassionate, desiring approval, capable of deep pity & capable too of expressing it in practical action; my duties pressed in upon me, & while I could be lazy, I discharged most of them with some distinction; & this in its turn led to respect & prestige, so that one felt one was building up something, so that one grew still more ambitious & still more compassionate & still more dutiful. Such a mixture indeed are we all.[12]

Perhaps so, but few of us have the honesty or the clear-sightedness to describe ourselves as Paton did here, in a private letter.

It might have been thought that the tremendous effort he had made in cleansing the Augean stables of Diepkloof would have exhausted his energies, but in fact even before the outbreak of war he had been flinging himself with equal vigour into a range of other interests, all tending in the same direction: the service of others. And to begin with, he continued and greatly expanded his work for Toc H during the war years.

Having given up the position of Natal Pilot of the organization when he left Pietermaritzburg, he made his appearance at a Toc H meeting very soon after arriving in Johannesburg. A very considerable reputation had preceded him, and not all the Johannesburg Toc H men were unequivocal in their welcome. The young Gonville ffrench-Beytagh, for one, was prepared to dislike the newcomer:

> The first time I met him was in a Toc H committee. I think we were preparing for Tubby Clayton's visit. We heard that Paton was coming up from Pietermaritzburg to Johannesburg, in 1935 I think, and would be part of Toc H in Joburg. Toc H had district teams, and I was on one of them by this time; and this unknown chap from Natal was coming up, and he would be made chairman, either of the area or the whole district of the Transvaal. A great reputation came before him, everyone was talking as if he were God almighty. He was apparently hugely popular.
>
> I was prepared not to like this chap. I think Alan could see that, and he completely took the wind out of my sails by turning to me and saying. Now what's your opinion about this? That completely won me over, and he had a tremendous effect on me.[13]

Paton's energy in working for Toc H found ample release on the outbreak of war, for Toc H and the Young Men's Christian Association (YMCA) had taken on the project of organizing a recreation centre for South African servicemen in every camp in South Africa and beyond. The President of Toc H was J. H. Hofmeyr, who had been recalled to the cabinet by Smuts on the outbreak of war. Hofmeyr needed a national chairman of the YMCA-Toc H work, and he called on his protégé, Paton.

As National Chairman Paton rapidly got to know a range of powerful businessmen who could raise what to Paton seemed astronomical sums of money for the Toc H war work, and he was to admit in later years that he had deeply envied them their wealth and their possessions.[14] He might wish to be a saint and give his all to serve his fellows, but he also longed for material success. His ambitions were manifold and, in part, contradictory.

In addition to the tycoons, he came in contact with a range of deeply moral men, many though not all of them committed Christians. Gonville ffrench-Beytagh was one of these. Not long after his first meeting with Paton, ffrench-Beytagh was set upon by thugs in Braamfontein subway in

1936 and badly beaten, spending many weeks in Johannesburg General Hospital with a broken jaw and other severe injuries. When he recovered consciousness in the hospital bed the first person he set eyes on was Paton, to his surprise, and Paton returned to visit him several times. As ffrench-Beytagh recalled at the end of his life,

> Around about that time, in hospital, I was about 20, I began to take stock of my life. I'd done a lot of different things, travelling round the world. And I remember thinking, as I lay in bed in that hospital, if I ever got married, what would I like my sons to be like? I'd always had a great admiration for people like Bulldog Drummond, and I suddenly discovered I didn't want them to be like Bulldog Drummond, I wanted them to be like Alan Paton. And this shook me because I didn't like the church.
>
> On one of his visits I got talking to Alan about religion, I had the usual objections about what hypocrites Christians are, and I remember Alan saying to me why don't you read the Gospels? I did read them about that time. That started me inquiring then. I came back to the church—I had been confirmed as a youth, but it hadn't taken—through Toc H.[15]

Under Paton's influence, though he was to say that Paton never 'preached' and in fact hardly discussed religion at all, ffrench-Beytagh's own faith grew rapidly, and in 1939 he was ordained an Anglican priest. In 1954 he would become Dean of the cathedral in Salisbury, Rhodesia, and in 1965 Dean of St Mary's, Johannesburg, one of the most prominent churchmen in southern Africa, and a thorn in the side of the Nationalist government, who would harass him, imprison him, put him on trial for sedition, and eventually succeed in forcing him out of the country. This was the road on which he had been started by Alan Paton's influence, and he never regretted it. The reality of the impact Paton made on those around him, without any conscious effort on his own part, is most clear in the way he changed the lives of people such as ffrench-Beytagh, who many years later, dying in a London hospital, took pleasure in witnessing to Paton's influence on him.[16]

As an offshoot of his work with Toc H, Paton helped to run clubs for black boys, a practical effort aimed at keeping them off the streets and out of crime. Paton's energy was such that in a short time he was appointed Chairman of the Transvaal Association of Non-European Boys' Clubs,[17] a post he continued to hold for a number of years.

But even at this period of his closest involvement with Toc H, shortly before the war, he was a problem to the organization, for the reason that he seemed too idealistic to other members. In particular, he argued that though Toc H saw its main duty at this time as giving aid to the armed forces, it must not identify itself with the fight against Hitler. In no sense, Paton argued, should it join the struggle against him, though it should continue to give all help to those who fought him. This self-contradictory argument caused confusion in Toc H ranks, and was adopted by Paton out of deference to the Afrikaner members of the organization, many of whom of course passionately opposed the war. Nothing more clearly shows how closely he still tried to identify with Afrikaner sentiments.[18] Over time he had to abandon this untenable argument, but it was the first sign of a tendency that would, in the end, destroy his links with Toc H: his willingness to take a stand on an issue which he regarded as one of principle, and which the rest of the organization regarded as wildly unrealistic.

But even more than his work for Toc H, Paton threw himself into the life and work of his local Anglican church, and it was this that had the most lasting and profound effect on him of any experience at this time, other than being Principal of Diepkloof. Throughout the years he was at Diepkloof, he worshipped at the church of All Saints, Booysens, about 5 miles from Diepkloof, and served as a churchwarden there. In addition, during those thirteen years he was also one of the parish's representatives on the annual Anglican diocesan synod. And it was through these synod meetings that he got to know the Bishop of Johannesburg, Geoffrey Clayton.

Clayton was in some ways a man in the mould of J. H. Hofmeyr, possessed of a brilliant intellect and powerful personality, but to many people a most unattractive man. Clayton had the same sloppiness in dress and eating habits as Hofmeyr, the same rudeness, the same physical ugliness. In face and figure he strongly resembled John Tenniel's conception of the Frog Footman in *Alice*. He had Hofmeyr's clumsiness towards women, although in Clayton's case this resulted not from the influence of a tyrannical and jealous mother, but from the fact that he was (Paton believed) a homosexual, though a man of very chaste life.[19] 'It's not that I dislike women,' he once told Paton, 'it's just that I can't tell one from another.'[20]

He was also a man of a very strange sense of humour. When he later became Archbishop of Cape Town and occupied Bishopscourt, a house famous for its garden, a visiting Archbishop's wife remarked to him, 'Your Grace, the garden is beautiful.' Clayton considered for a while, and then said to her with vehemence, 'I *hate* beauty.' He then laughed so that his great belly shook.[21] Some of his jokes were funnier, but they remained the kind at which only he laughed. At one Episcopal Synod, when the visiting bishops and their wives had finished lunch, Clayton asked what their plans were for the afternoon. Mrs Vernon Inman, wife of the Bishop of Natal, said, 'I'm going to bed with a thriller.' A satirical gleam appeared in Clayton's eye and he looked round that impeccable company: 'Which one?' he asked.[22]

And yet Paton could see past this mask, as he could see past Hofmeyr's, to the soul beneath, a soul with the courage to fight for what it believed to be right, and a personality big enough to inspire others to do the same. As Hofmeyr gradually faded as an influence on Paton, Clayton took his place. Paton came to admire his bishop unreservedly:

> The Bishop of Johannesburg was an extraordinary man, one of the few persons in my life of whom I would have used the adjective 'great' . . . To sit under him at synod was something never to be forgotten. He was a chairman without peer . . . Synod was very like a big class, sitting under a formidable master . . .[23]

In that final phrase one has one of the keys to Paton's love of both Clayton and Hofmeyr. Hofmeyr appealed to him most as a fellow schoolboy, giggling and playing rough games at the SCA camps; Clayton made Paton feel like a schoolboy in a different way, by behaving towards him as a schoolmaster might have done. Both allowed him to shrug off for a time the authority he had assumed as schoolteacher and reformatory Principal and become a gleeful, carefree boy again.

Clayton seems to have become aware of Paton's talents as an administrator and committee member very early on. Once the war broke out, Clayton realized that an entirely new order would come into being as a result of the conflict, and it was his view that the church had the duty to map out a moral path for South Africa to follow after the war. In particular, he took the view that a country which would not allow its black citizens to live and work, freely, side by side with all other citizens, was heading towards catastrophe.[24] The church should do what it could to show the way forward. At the annual synod of 1940, therefore, he set up a Bishop's Commission to 'define what it believes to be the mind of Christ for this land'.[25] Clayton chose 33 committee members, of whom two were black, to begin the work: one of them was Paton.

In later years Paton was to say that there were five seminal events in his life. They were, in order, a Christian upbringing; joining the Student Christian Association, at the age of 16; becoming Principal of Diepkloof, at

32; joining the South African Institute of Race Relations at 33; and serving on the Bishop's Commission, aged 38. After 1953 he would add a sixth, his membership of the Liberal Party.[26]

The Bishop's Commission sat from 1940 to 1942, and during that time Paton gave up at least one evening every week to its meetings. 'There were nine committees,' he wrote, 'and I was a member of two of them, Education and Social Welfare. It was one of the seminal events of my life, after which I was never the same again. I had to open my eyes and look at South Africa as I had never looked at it before.'[27] And what he saw is best understood by looking at the clauses of the Commission's final report.

There were seven of these, labelled A to G. Clause A was merely introductory. Clause B, written largely by Clayton, was a theological examination of the duties of a Christian, in which Clayton affirmed that though it was not the church's mission to reform society, it was the mission and the duty of Christians to do so. Paton already believed this: Clayton's clause B stated it so clearly that it could not be dodged, and its implications, Paton came to realize, were revolutionary in the South African context. In particular, as he said later, 'white supremacy and the principles of Clause B were irreconcilable'.[28] You could not believe that a black man was your brother in Christ, and go on treating him in a way in which you would not have wished to be treated yourself. Coming to accept the clear statement of Clause B changed Paton's life by focusing for him much of the teaching he had absorbed during the rest of his life to this point. If a black man is accepted as a brother, a common society becomes inevitable.

The rest of the Commission's report spelled out the practical implications of Clause B's argument, concentrating on race relations. Clause C criticized South Africa's refusal to apply within her borders the principles of freedom and justice her army was fighting for outside them.[29] Clause D condemned the inequalities of income between black and white, the system of migratory labour, and the control of South Africa's industrial resources by whites alone. Clause E condemned racial segregation, particularly the reservation of certain occupations to whites alone, and the total inadequacy of funds and equipment in schools for blacks, an inadequacy which was to continue into the 1990s.

And Clause E had a further important section, on political representation, which was, of course, the nub of the problem. If blacks had a share of political power consonant with their numbers, other problems of inequity would rapidly be swept away. As we have seen, the Hertzog-Smuts fusion government had in 1936 removed those few black voters who qualified for the franchise in the Cape to a separate roll. The Bishop's Commission now recommended the following three steps:

1. The extension of the Cape franchise for Coloured, Indian and African men to all Coloured, Indian and African men and women in South Africa.
2. A corresponding increase in the number of white M.P.s and white Senators who represented African voters.
3. A common roll for all citizens, so that through a common election, M.P.s should represent *all* qualified voters.[30]

Clause F dealt with the need for Christian teaching in the schools to prepare children for this new order, and Clause G concluded this revolutionary document, rightly stating that the implementation of the Commission's recommendations would require 'a change of heart within the nation', and proclaiming a hope that 'the nation be called back to God'. Few of the Commission members can have been under any illusions about the likelihood of such a national conversion. But clearly a majority of them, and Paton was certainly one of them, believed that such a conversion was vitally necessary and should be striven for. Better to light a small candle than curse the darkness.

The report was presented to synod in November 1943, and came under heavy fire, on the one hand from pedants who criticized its language, and on the other from activist priests, most of them English migrants to South Africa, who wanted a crusade against racism, with the church at its head. Chief among these were Father Trevor Huddleston and the Revd. Michael Scott. In the end Clayton, exerting his authority and the force of his personality, pushed the report through without major changes, but Huddleston and Scott were clearly going to exert an influence on the Anglican church in South Africa.

Paton knew them both already from his work at Diepkloof. In many ways they were strikingly similar: they were both English, tall and ascetic, both burningly committed to the fight against the racial system in South Africa, both capable of fanatical dedication to their chosen cause. Scott was an assistant priest at St Alban's Mission for Coloureds, and he and Paton had taken an interest in each other's work. He was a tormented man who had had a number of nervous breakdowns before coming to South Africa in 1943; he was plagued by such matters as his inability to understand why God permitted the existence of evil, by his desire (and failure) to reconcile his beliefs in Communism and Christianity, and by his own sexuality, which he could not accept. He was brave, gentle, and diffident, and though his attempts to right injustices were often ineffectual, he did not stop trying. Paton sympathized with him, and was to pay tribute to him by putting him into *Cry, the Beloved Country*, and by publishing a thoughtful assessment of him in 1958.[31]

Huddleston, a member of the Anglican Community of the Resurrection, an English Dominican group based at Mirfield in Yorkshire, was Prior of the Church of Christ the King in the sprawling black slum of Sophiatown, a huge job for a man who was only 29 in 1943. Among his many duties was the oversight of a black school, St Peter's, whose pupils played sport against the teams from Diepkloof, and the two institutions exchanged annual fraternal visits. Huddleston had immense energy, a genius for organization, a burning moral zeal, and an unshakeable will akin to Paton's own. But he also had a conviction of the rightness of his own views which Paton, for all his ambitions, was too humble to match. This conviction led him not only to pursue what he saw as the right, but to condemn and accuse those he saw as being in the wrong, tactics which did not always further his own cause. He also had an egotism and a love of publicity which Paton found uncomfortable. He and Paton, however, for all their differences, developed a friendship which endured for many years, and Huddleston came also to know and befriend Dorrie.

Paton, during these years of war, was being educated to an awareness of the problems of his complex and divided country, and educated into a sympathy with those who were not of his own group. Starting from a position of regional chauvinism in which he thought of Natal with love, and considered South Africa as a rather distant abstraction, he had first developed a profound sympathy for the Afrikaners, and now was extending that fellow feeling to Africans, Indians, Coloureds—extending it, in fact, further than any but a very few of his white fellow-countrymen. He was also, through his work at Diepkloof, through the Bishop's Commission and through his membership of the South African Institute of Race Relations, building up a network of like-minded liberals, few in number but all the more closely united for that. Once having had his eyes opened, he could never close them again. 'As for myself,' he was to write, 'having lived for 38 years in the dark, the Commission opened for me a door, and I went through into the light and I shut it against myself, and entered a new country whose very joys and adversities were made resplendent by the light.'[32] The teleological imagery shows the extent to which he thought of the change that had come over him as being as important as a conversion to Christianity. It was in fact an affirmation of the consequences of all his religious beliefs, and in the long run it would alter his life utterly.

One consequence was that just as he had made his approach to sympathy with the Afrikaners through learning their language, he now began to learn an African language, and to cultivate friendships among blacks. This was by no means easy, for the gulf that separated South African whites from most blacks was wide, compounded of linguistic and cultural differences so profound that they could only be overcome at the risk of many false steps, false starts, and embarrassments. To do so required a certain courage, but courage was something Paton had in abundance. Coming from Natal, in which the dominant group is the Zulu, he naturally spoke a certain amount of Zulu, and he began taking lessons in that language from the black head teacher at Diepkloof, Ben Moloi. Moloi was a handsome, open-faced man with a broad grin and a good sense of humour. He had also been well educated, at a Mission school, and wrote a better hand than many of Paton's white staff. After the lessons Paton and Moloi would have tea together with Dorrie and the children. On a couple of occasions Paton also invited some of his Afrikaans staff members along to the lesson. They accepted gladly, but on finding that they were expected to have tea afterwards with Moloi an icy constraint fell on them. They behaved with scrupulous politeness, but they did not come again.

Paton greatly admired those among his friends who could achieve real friendships with Africans. The most striking of these was Edith Rheinallt Jones, who apart from her work on the Diepkloof Board of Management ran an organization called the Wayfarers (a sort of Girl Guide movement for black and coloured children), a hostel for black girls in Johannesburg called the Helping Hand club, and numerous similar good works. Edith Jones was a very large woman with a bad heart, who consistently ignored her doctor's warnings to slow down. On several occasions Paton drove her deep into the countryside to visit one of her Wayfarers groups at rural schools in the northern Transvaal, in the Venda 'homeland'. He found her friendship with the black teachers extraordinarily natural and easy. Many years later he wrote a piece entitled 'Case History of a Pinky', in which he described her conversation with one teacher, a Mrs Takalani:

Mrs Takalani was in a state of spiritual intoxication.

'You must bring her again,' she said to me. 'When she comes she makes things new.'

She turned to Mrs Jones. 'Did you hear what I said?' she asked. 'I said when you come you make things new.'

'Don't talk nonsense,' said Mrs Jones.

'Don't you say I am talking nonsense,' said Mrs Takalani. 'Why do you think all these people come here? They come here to see you.'

'That is nonsense,' said Mrs Jones. 'They come here to see what their children are learning in the Wayfarers.'

As for me, I listened fascinated. I had never before heard a white woman and a black woman talk to each other in this fashion. It was something new for me. At that time my own relations with black people were extremely polite, but I realized that these two had long passed that stage.[33]

And when Mrs Jones died in April 1944,[34] a month after Paton had witnessed this scene, her funeral was attended by great crowds of blacks and whites, coloureds and Indians, rich and poor, Jew, Christian, Hindu, and Moslem. 'I was overwhelmed,' Paton was to write of this sight. 'I was seeing a vision, which was never to leave me.'[35] It was a vision of a South Africa undivided, a South Africa moving with painful ascent, as he put it, to that self-fulfilment no human being may with justice be denied. This visit, and the vision of the future it embodied for Paton, also found expression in a poem he was to write in 1948, 'Black Woman Teacher'.[36] And in due course his description of this funeral would find many echoes in the funeral of Arthur Jarvis in *Cry, the Beloved Country*, in which Arthur Jarvis's character owes much to Paton's observation of Mrs Jones.

From now on he would work for a single, united country, and for a nationalism founded on one thing only, and that was a common consciousness of sharing the land of South Africa. It is a fragile basis for a country's foundation: but lacking unity of race, colour, language, religion, history, or tradition, it is all South Africa has to build on. The enemies of this fragile vision were group loyalties—racism and Afrikaner nationalism on the one hand, and complex black nationalisms on the other. Finding a way of reconciling them was to be the task of the rest of Paton's life, and it was during the war that he dedicated himself to it. The prize for success would be the peace of his country; the price of failure he would not even contemplate. The task might prove impossible, but he was going to give his whole heart to it.

11

PEACE 1944–1946

There is scarcely less bother in the running of a family than in that of an entire state.

MONTAIGNE

'Our family life', wrote Paton of the war years, 'was a happy one.'[1] And he then goes on to talk about his relations with his sons. What he does not say is that the birth of his younger son had greatly complicated his relations with Dorrie. Jonathan was born on 6 July 1936 in Durban, where Dorrie had gone to give birth in reach of the support of her sisters. Paton and little David, now nearly 6, were camping together in a tent at Park Rynie (the coastal village where Paton had recuperated after his typhoid in 1934) when the birth took place. Paton had german measles, and they were not allowed into the wards to see the new arrival: instead, Paton secured a ladder and climbed up to the first-floor window to peer in at his new son.[2]

Whether because of the circumstances of the child's conception, or because of Jonathan's personality, or simply because he preferred younger children to older ones, Paton came to love Jonathan more than his brother, and was unable to conceal this fact from either boy. As a result, and in spite (or because) of the names their father had given them, the sibling rivalry between David and Jonathan was equal to that which had existed between Paton and his brother Atholl. 'David and Jonathan—a risk to take—but it turned out all right,' Paton wrote optimistically.[3]

But early on, David, perhaps inspired by the notion of having to climb up to see his brother, fabricated a tale that Jonathan was really a monkey, an animal common in Natal. As he put it when they were both grown men:

My brother and I had this thing, that he's not really my brother, that he was an ape, there was a band of apes going round Durban at this time, they came across the nursing home, saw this baby, snatched it, the nurse was desperate, no baby, she grabbed a small ape by the tail, it came off, she wrapped him up and put him in the bed, and that's Jonathan.[4]

Many children fabricate such insulting stories about their siblings and then forget about them in half an hour. This ape joke (but it was not really a joke) David told many times throughout their childhood, and he was still keeping the story alive when he and Jonathan were successful professional men in middle age.[5]

Paton could see what was happening (he wrote that David 'ruled Jonathan with the elder brother's rod'),[6] but he could do little about it. Having been so mercilessly beaten by his own father, he was determined not to beat his sons, and in fact beat David only twice, and Jonathan once. Physical coercion was left to Dorrie:

> David was five and a half years older than Jonathan, and he played the part of the lordly elder brother. He would drive Jonathan to distraction, and finally Dorrie would lose her temper and smack them both as hard as she could. I was able to observe the way in which a mother could rebuke and chastise a son, and a father could not.[7]

For all that they were rarely beaten by him, the boys at times feared and disliked their father. David's relationship with Paton was the simpler: after the birth of Jonathan, David and his father were never very close, though they shared such experiences as cycling together from Johannesburg to Durban in the December holidays of both 1943 and 1944, with a detour each time to spend a few days in the Drakensberg with Paton's old friend Reg Pearse. These were marathon rides of more than 400 hilly miles each for a boy who had just turned 13 when they did the first of them. The fact that they were both strong and determined no doubt put distance between Paton and his elder son, whom he described as 'self-willed';[8] more important, in later years, was the fact that David, in spite of having been sent to a very expensive Anglican school, St John's in Johannesburg, was not a Christian, to the lasting regret of his father.

With his younger son Paton's relationship was much more complex. Jonathan's feelings towards his father alternated between almost excessive affection and dependence on the one hand, and hostility on the other: and hostility was in the end to predominate. Jonathan longed for demonstrations of affection from his father, kisses and hugs, but these Paton thought inappropriate once the boy was no longer a small child. Instead he showed increasing sternness. On their annual holidays to Natal there would be regular disputes between David and Jonathan as to whether the car windows were to be up or down. This squabble would go on until Paton's patience was exhausted and he would fly into a rage which Jonathan, at least, found terrifying. Then it was enough for him to growl, 'David, let Jonno keep his window open,' or 'Jonno, put your window up,' and there would be silence in the car for the next half hour.[9]

From Paton's own side, it is plain that although relations with David were always slightly constrained after Jonathan's birth, he delighted in Jonathan. In *Cry, the Beloved Country* he was to draw a vivid portrait of a small boy 'with a brightness in him': this small boy was modelled on Jonathan, and there is no doubt that Paton loved him deeply. Jonathan was also the inspiration for one of Paton's best poems, 'Meditation for a Young Boy Confirmed'. Even towards the end of his life, Paton would write of Jonathan to a mutual friend, 'I am vulnerable to the young man';[10] and it was true.

Jonathan, on the other hand, was often deeply jealous of his father and would fly into fits of rage if rebuked by him. When the Principal's new house at Diepkloof was nearing completion, Paton led Jonathan, then three, to a newly painted wall on which Jonathan had scrawled, and said, 'Who did this?' Jonathan's reaction was to contort his face with fury and resentment, and to come at his father with his small fists flying, so that Paton, startled and wondering, let the matter drop.[11] Paton tells this one story in his autobiography, but there were many more such incidents, and they continued through Jonathan's boyhood.

Jonathan loved music, and Paton did not; though he enjoyed classical masterpieces such as Beethoven's Moonlight Sonata, he was ignorant of modern music and largely indifferent to it. To him anything not classical was 'jazz'. If Paton heard the gramophone he would come and tell Jonathan to turn it off, and on at least one occasion Jonathan, by then a teenager, was so enraged that he smashed the record against a wall. Then Paton would demand an apology for this behaviour, and all his own father's inflexibility and dominance would come out. 'Say sorry,' he would insist, as tears of impotent rage began to course down Jonathan's cheeks: 'Say sorry, say you're sorry that you threw that!' And he would go on for as long as necessary until the sobbing apology had been received. 'That was terrible treatment,' Jonathan was to say after his father's death, 'almost worse than a beating.'[12] It is a curiosity that Paton, who had done so much to treat delinquent black boys with loving care rather than harsh discipline, should have inspired such bitter and lifelong resentment in one of his sons, and near-indifference in the other.

Paton with his siblings: Eunice ('Dorrie') in the foreground, Ailsa behind, and Atholl, *c.*1911

Paton with his father James at the front steps of 551 Bulwer Street

The boy student at Natal University College, 1920

Railton Dent, the greatest influence on Paton's life, at Natal University College

Reg Pearse, Paton's closest friend at university, *c.*1920

Neville Nuttall, with whom Paton shared the first bottle of port, *c.*1921

Tramping from Pietermaritzburg to Ladysmith with Reg Pearse, the trip Paton said changed his life, July 1922

Paton, behind the boy holding the sign, dwarfed by his Form V class at Maritzburg College, 1928

J. H. Hofmeyr, Paton's friend and patron; this was the photo that Paton hung on his study wall all his life.

Dorrie Paton in 1935, the year they moved to Diepkloof

One tiny sign of their emotional distance from him was their disinclination to refer to him, in his absence, as 'Dad' or any such term reflecting a warm familial relationship. Instead they called him 'Mr P', and by extension came to call their mother 'Mrs P'. Their friends in due course adopted this strangely distanced term too, and so, over time, did Dorrie, except when she was talking to her sons, when she would call him 'Dad'—'Dad says this'. But 'Dad' was a term his sons would not use except to Paton's face. 'My early memories of my father,' Jonathan would say, 'are indeed of a stern and rather angry man, and strangely enough, these have also become my memories of him in old age. But there were also times of considerable affection.'[13]

The children's warmer memories of Paton were to involve their holidays together, or his skill in teaching them. He had taught David to read at the age of three; he did the same for Jonathan, so that by the time he went to school he was far ahead of his peers. The boys remembered his pleasure in planning their annual holidays, his love of mapping out the route elaborately in advance, taking in as many mountain passes as possible, his joy in being on the road early, his delight in roadside picnics, the way he seemed to notice a host of things about the countryside that no one else did. To travel by car with him was an endless education. They also loved going for walks with Paton, as he had with his own father, and he would display his encyclopedic knowledge of South Africa's flora and fauna. And once he had identified a bird or wild flower for them, using both the common and the Latin names, he expected them to remember. 'What's that tree up ahead?' he would ask, or 'What's that bird calling?' and he would be very disappointed if they could not answer correctly.

Jonathan's feelings about Paton are (in 1992) more bitter than David's, and the reason is not far to seek: Dorrie and Jonathan had a very close relationship, and she used it to keep Paton at arm's length. To be precise, Jonathan slept in her bed. Many women adopt this practice with the newborn, as it makes night feeds easier. But it is a rare mother who will voluntarily keep the child in her bed after it is weaned. Jonathan slept in his mother's bed at least until he turned 10, and the result was an oedipal bonding between himself and his mother that showed itself in episodes of passionate resentment of his father.

Paton and Dorrie shared a room, but had separate beds, though Paton all through his married life wished they shared a double bed.[14] Dorrie's control over their sex-life had always been close and miserly. In Stoppard's *Jumpers*, a character complains that his wife only just stops short of the issuing of ration coupons, and Paton might have made the same joke if he had thought this a joking matter.

After the arrival of Jonathan things grew worse, and Dorrie's promise to 'make it all up to you' seems soon to have been forgotten. Although Paton longed for a third child, Dorrie steadfastly refused to have another. Years later he would write to her, rather sadly, 'I am only sorry that we didn't have at least one more child, but that was your choice, wasn't it?'[15] It was her choice, and she used her second child to make sure there was no third. In his maturity Jonathan was to reflect on this strange situation:

> I had a very close relationship with my mother. I actually slept in my mother's bed, I would say until I was at least ten or eleven. When you think about the implications of that it's quite something to think about. My parents had separate beds, but shared a room. And so it was a very strange situation.
>
> I have a theory about that situation, and I see it almost in Oedipal terms. I was in a sense taken by my mother. There was a rift between my parents . . . if you look at my father's writing, my mother used to wear the ring of her first husband, until it caused a rift between them, and finally it was discarded and he feels that the rift was made up. But in a sense I feel it was never really made up, though in a way they were very fond of each other. I felt that my sleeping in her bed was a kind of protection.

And in Jonathan's view, the 'protection' was generally effective.

> Were my parents sexually active? I can hardly think so. Incidentally sex was never something that was discussed in our house at all. I really learned about it when I finally went to boarding school. I really was incredibly ignorant.
>
> I just remember waking up sometimes and finding my mother wasn't in the bed. I don't know about my parents' sexual life, and I don't think one would ever know. But I have a theory about why I was sent to boarding school, which incidentally was the unhappiest time of my life.[16]

Jonathan, while sharing his parents' room, was witness to some intimate scenes. Some were amusing, as when his father used the chamber pot kept under his bed, while his mother made little jokes about her husband 'having a tinkle again'. Some were not, as when Jonathan, then about 6, saw Paton, in a rage, wrestle his wife through the door of their room and lock her out. Afterwards Jonathan sat with her on a sofa, asking 'Why don't you leave him and we'll go away?' And his mother replied, 'I might.'[17] Denied the love his passionate nature craved, Paton threw himself into his many activities, and made for himself a life of social service that took him out of the house on many evenings after the work at Diepkloof had finished.

Even when he did not go out after work, some part of his work tended to come home with him. Because the Patons lived on the Diepkloof farm, less than half a mile from the main building, their family life and the life of the institution were closely intertwined. Both David, as soon as he came to Diepkloof, and Jonathan, as soon as he was old enough, were given command of a *span* of the smallest pupils and were taught to drill and march them about the property, rather as Frederick the Great supplied his small son with a real army to play with. Jonathan's childhood companions were black delinquents, and he was fond, in later life, of telling his friends that he had been brought up in a reformatory.

But although he and his brother marched the reformatory boys about and gave them orders, for friendship they turned to the children of the white staff. The Vice-Principal's younger son, Tommy Laas, was Jonathan's friend, but, even in childhood play, politics and the war intruded. On one occasion the boys were playing war games, speaking Afrikaans, and Jonathan said, 'We're going to attack Hitler'. Tommy Laas flatly disagreed. 'No, we're not going to attack Hitler, Hitler's a good man,' he said, and Jonathan was taken aback. After much discussion they reached a compromise: there were two Hitlers, one bad, one good, and they would attack the bad one. J. H. Laas might be Paton's right-hand man, but he and Paton were politically at opposite ends of the spectrum.

For all that, he and Paton were friends as their sons were friends. Laas and his wife were regularly invited to the Paton's home after dinner for drinks and a game of 'reformatory bridge', and so, until the war, were the young white supervisors. 'Reformatory bridge' was based on a game called 'haul',[18] and had been invented by Paton and Dorrie together. In essence it was a simplified form of bridge, played for points. Paton would infuriate Dorrie by appearing to throw away strong cards deliberately. 'Alan, you are the meanest, dirtiest player I have ever met,' she would say, and when he continued to annoy her, she would fling down her cards and refuse to play. Then there would be laughter and cajoling, and sulkily she would resume the game until the next explosion.[19] Visitors never knew quite how seriously to take all this, but they enjoyed the games enough to come whenever invited, and such conviviality became an important source of the group spirit that linked some of the Diepkloof staff in spite of their political differences.

As the tide of war turned these differences diminished. Once Germany invaded the Soviet Union in mid-1941, and the United States entered the war after the Japanese bombing of Pearl Harbour in December the same year, it became clear that Hitler was going to lose the war, and, as a direct consequence, in South Africa the threat of violence from the Ossewa Brandwag gradually receded. Paton, like everyone else, began to turn his thoughts to what would happen once the war ended. Smuts's star, and with it Hofmeyr's, appeared to rise as victory grew more certain; in the general elections of 1943 Smuts's United Party increased its majority, the Nationalists being thrown into disarray by infighting between Malan and Hertzog, and by the defection of 16 National MPs to Oswald Pirow's Nazi-style New Order group, which refused to stand for election and so was politically annihilated. It seemed certain that Hofmeyr was destined to take over the leadership of his country in due course.

Paton still nursed political ambitions, but he had begun to realize that it was not enough to convince Hofmeyr of his worth: he must make the public aware of what he had accomplished at Diepkloof. His gradual loss of faith in Hofmeyr as a patron did not reflect any diminution of admiration for him; it was the result of repeated disappointment of his hopes that Hofmeyr might advance him, and a slowly growing conviction that Hofmeyr was psychologically incapable of breaking away from Smuts and starting a new and more liberal party of his own. Such a party would embody Paton's views on the need for a common society and, just as importantly, would have a place for Paton and those who thought like him: J. D. Rheinallt Jones (who since the election had been one of the four Senators representing black voters), Professor Hoernlé, Edgar Brookes, and others. But if Hofmeyr would not found and lead such a party, who would?

By 1941 Paton was eagerly seeking publicity for Diepkloof. The minutes of meetings of the board of management of 5 August 1941 reflect this, with Paton seeking suggestions from members. It was pointed out that the public 'was in general quite ignorant of the function and aims of the reformatory, and knew more of its failures than its successes'.[20] The board suggested an open day for the public to inspect displays of manufactures produced in the reformatory workshops, together with flowers and fruits, and performances of drill, singing, and tribal dancing. And, in fact, Paton made such open days annual events, made a point of inviting journalists from the Johannesburg papers,[21] and in addition invited along to see Diepkloof anyone who at any time expressed interest in the work he was doing there.

But he aimed at a wider audience still. If he were going to make a move into politics, he needed a national reputation, and the best way to achieve that was by writing. He had been too busy at Diepkloof to produce anything but an occasional poem since 1935; now, in August 1941, he took a month's 'rest' and wrote the greater part of a book he entitled *The Afrikaner.* All but ten pages of this has been lost, but its inspiration seems to have been the dying kicks of Paton's enthusiasm for Afrikanerdom, and it was at least partly historical. 'I have written it without great difficulty, & indeed have avoided topics into which research would have been needed,' he wrote to Hofmeyr. 'But for all that I believe it to be anything but superficial; I even believe it to be a unique contribution. At times I am moved to passages which are I believe of some beauty.'[22] But *The Afrikaner* joined *Ship of Truth*, *John Henry Dane*, *Brother Death*, *Religion and My Generation*, his play *Louis Botha* and a slowly growing pile of poetry, in unpublished limbo, at least in part because he was doubtful if a public servant should publish such a book.[23]

And after finishing the book, in November 1941, he penned his most revealing letter to Hofmeyr, an almost desperate appeal for 'Dear Hoffie' to procure him a more important job. He was as tired of Diepkloof as he had been of teaching:

> I have always been loyal, obedient, modest even if self-confident; but I dread an endless succession of such obedient and modest days. I fear that the latent fires may die down, beyond all hopes of ever blazing again. I have never dared to say to you 'Have you any plans for me, or is it your plan that I should stand on my own feet & paddle my own canoe & solve my own problems?' But I should like to say that to you, & know at least that you do not or do intend to say 'Come this way'. I have an ambition that I have never revealed to you; I believe in it, & I believe that I would do you credit if the chance were given me. But I have not dared to tell you what it is; for in spite of my self-confidence, it would cast me down to the depths if you thought I was presumptuous.

And he now made a bold self-analysis:

> I believe that I have a contribution to make to South Africa greater than that of any English-speaking South African I know or read of; I cannot but believe that God sent me to Diepkloof to prepare me for that. I can go out on the lawn in front of the house, & in the dark put into words (both English & Afrikaans words) what I feel & believe & would die for. But I ask myself whether I am to go on doing this, a magnificent apprenticeship for a job I shall never do. I cannot expect you to believe in me as I believe in myself; I have a deep unalterable conviction that I am ready, & a fear, of which I am not proud, that I may instead join the ranks of those who missed the boat & have a grievance against life.[24]

In the face of this renewed and almost desperate appeal, one has to ask once again, what exactly was Paton hoping for? In the short run, the answer is almost certainly that he wanted a job in the higher ranks of the civil service, most likely the post of Director of Prisons. In the longer run, though, he was continuing to dream that when Smuts retired (and the distinguished statesman was now 71) Hofmeyr would become Prime Minister, and that he would summon Paton to be his deputy, as Smuts had summoned Hofmeyr. But time was passing, and Paton felt the best years of his life were passing too: 'I'm 38, & never again will I have the energy & the enthusiasm that I have now.'[25] But Hofmeyr, even in the face of this almost desperate appeal, seems to have remained vague and noncommittal as he had so often in the past.

Paton did not intend to be ignored for ever: he had not devoted his prodigious talent and energies to Diepkloof for so long to continue as an invisible civil servant. It was not enough merely to achieve transformation of the institution, triumphantly effective though that transformation had been. Paton intended that it should be widely known and understood, partly because he was proud of his achievements and driven by ambition, partly because he thought Diepkloof a lesson South Africa needed to learn. 'I have been a teacher all my life,' he recorded in old age. 'In the first half I taught boys and girls, in the second half I tried to teach white South African adults the facts of life, but they are a tough proposition.'[26]

This teaching involved making of Diepkloof an act of communication, and Paton set about it with his usual energy. On several occasions during the year he would invite his superiors, Hofmeyr, the Secretary of the Department or various other bureaucrats, to take the salute at a grand parade at Diepkloof, when the boys would be marched out with military precision, and the visitor taken on a tour of inspection and then given a good tea. Paton's many contacts in Anglican lay-person's organizations, and groups such as Toc H, served him well now, and he made sure they knew what Diepkloof was doing and why. Such was his success as a publicizer of his reforms that by 1944 a leading liberal journal, *Forum*, for which Hofmeyr himself often wrote, could describe him as 'Principal of the Diepkloof Reformatory, famed far and wide as the Union's Boys' Town'.[27]

He valued this reputation, for he now knew that he must have a larger audience than the Minister of Education and the senior members of his department. By 1942 he had turned to the press in earnest, and began publishing a series of articles, a steady drumbeat designed to publicize his work at Diepkloof, and more importantly, the views on racial issues that lay behind that work.

His first article of this type, 'New Schools for South Africans',[28] focused on his old concern, relations between Afrikaners and English speakers. But he quickly began concentrating on the even more menacing divide between black and white. Few white South Africans thought much about this issue at the time. If they considered the 'native problem' at all they concentrated on the crime wave for which they blamed blacks. Paton seized upon this fear and used it as a spur to change South African society in the direction he was moving Diepkloof. 'Society Aims to Protect Itself'[29] was the first article in which he began approaching this issue, in October 1943, but his attack became clearest when, in January 1944, he published in *Forum* an article entitled 'Real Way to Cure Crime: Our Society must Reform Itself'.[30]

In this important piece he argued that the cause of crime among blacks was only partially the breakup of tribal order, inadequate education, slum conditions, and the rest. The root cause, in his view, was that blacks in South African society lacked what he called 'social significance'. They were denied equality, dignity, and the vote, and the resulting violence and lawlessness were, in his view, inevitable results. He put forward a detailed programme for reform, a programme that amounted to the beginnings of radical reform for the whole of South African society:

> Education: Compulsory and free primary education. Technical and secondary schools to prepare for new employments.
>
> Economic: Raising of wages and progressive removal of restrictions on opportunity.

Innocent though these two points look, they clearly implied movement towards abolition of the system of reserving skilled jobs for whites, a politically explosive issue. But there was more to come:

> Political: Increased opportunity to share in local government, beginning in purely Native areas, urban and rural.

The word 'beginning' there carries the same implicit threat or promise of further movement to come: power sharing in areas not 'purely Native' is clearly in Paton's mind.

> Judicial: Local courts. Native assessors and juries. Removal of barriers and difficulties in the way of student-lawyers. More Native officials; Native probation officers. More Native police, of higher standard, and recruited from urban population.
>
> Municipal: Ownership of land and houses. Townships to be limited in size. Right to build. More Native officials; recreation and social centres, under local councils and voluntary bodies.
>
> And lastly: Abolition of the compound system for married labour.[31]

This was an extraordinarily detailed and far-reaching programme for the Principal of a black Reformatory to put forward. It looks much more like the manifesto of a political party, and Paton hinted plainly that it was just the start of what should be an ongoing process leading to a common society in South Africa.

But he was not going to lay his programme out in full now. 'Such a problem . . . fills the enfranchised South African with fear, even when he is convinced of the ultimate necessity,' he wrote. 'For this reason I shall not dwell on the full implications. It is a mistake—as it is with children—to present a remote goal when that presentation may paralyse action and delay progress.'[32] This slyly condescending remark does not disguise the fact that the progressive nature of his programme for South Africa is strongly reminiscent of his reforms at Diepkloof, with small freedoms and privileges leading on to greater ones, with the final goal being the complete integration of the black inmates into a wider society. This, of course, was precisely what Afrikaner Nationalists and other supporters of what would come to be called apartheid were afraid of.

This seminal article was followed by a stream of others, whose titles convey their thrust very well: 'Who is Really to Blame for the Crime Wave in South Africa?',[33] 'Prevention of Crime',[34] 'Behandeling van die Oortreder' ('Treatment of the Offender'),[35] 'Freedom as a Reformatory Instrument',[36] and so on. In several of these articles he hinted that he saw himself as moving towards a political career: 'I, if I had power, would not rest till society had launched a concerted campaign on crime . . .'.[37] It was not just Hofmeyr he wanted to impress now with the ideas he had developed at Diepkloof and put into practice there: it was the whole of South Africa.

His series of articles was a success in attracting attention, but not always in the way he had hoped. One sour observer of his changes was the Afrikaans paper *Die Transvaler*, which was to become the Transvaal mouthpiece of the Nationalist government when it came to power. Its editor was the brilliant, fanatical man who was to achieve notoriety as the architect of apartheid, Dr Hendrik Verwoerd. On 24 June 1945, after Paton had published eleven of his articles implicitly calling for change in South Africa, Verwoerd struck at him with a leading article sneeringly entitled *Diepkloof Wysheid* [Diepkloof Wisdom]. In this article Verwoerd, attacking Paton by name, made a frontal assault on Paton's view that South African society needed changing, charging him with blaming the police, the government, society, anything but the real cause of the problem, which in Verwoerd's view was softness towards black criminals.

He sneered at Paton's predictions that the racial situation in South Africa would deteriorate further unless fundamental reform of society were attempted. Blacks at Diepkloof (*klonkies*, he called them, a contemptuous term akin to 'sambos') were being petted instead of punished, and it was about time taxpayers' money stopped being wasted in this way. And the worst example of this waste was Diepkloof itself. The institution, Verwoerd announced in bold type, was a colossal failure (*kolossale mislukking*). Mr Paton's foolish pampering theories (*vertroetelingsteorieë*) were the reason that his management and his work were so futile. Was this to be endured any longer? Verwoerd asked with anger and contempt. This demand that taxpayers' money no longer be wasted on Diepkloof was an attack, not just on Paton, but on Hofmeyr and Paton's other superiors, and his blood must have run cold when he read it.

And there was one paragraph in which Verwoerd's mask of concerned citizen fell away. Members of the public, he wrote, saw nothing of the earth-shaking reforms at Diepkloof; all they knew of was constant absconding, and the loafing about of the 'little black ladies and gentlemen' on the farm, whose white officials were told by the Principal to plead with them, with a 'won't you please do a little work'. The Afrikaans is vitriolic: '*Leeglêery van swart dametjies and heertjies op die plaas wie se blanke amptenare deur die hoof eenkeer angesê is om hulle met 'n "asseblief tog" te smeek om ietsie te doen . . .*'[38]

Paton was horrified by this public attack on himself and his work: not just because the allegations of loafing and increased absconding were false, but because of the contempt Verwoerd clearly had for the very idea of treating black people with politeness. Nor was the editorial Verwoerd's only comment on Paton's work. The editorial page of the paper every day carried a box containing a biblical quotation. On this day the quotation was from Proverbs: *Die weg van 'n dwaas is reg in sy eie oë*: 'The way of a fool is right in his own eyes'. In this too one saw the nature of the man who would not only found a system on racism, but affect to support it with theology. Pain, anger, and resentment were Paton's chief emotions as he read, but fear must have been there too.[39] In deep trouble of mind he took the newspaper and drove at once to Pretoria to show it to his Minister, Hofmeyr, who merely snorted derisively: 'What else did you expect?'[40]

Paton had clearly expected a sympathetic public response to his humane theories and their successful application; Verwoerd had shown him how wrong he was. It must have been obvious to him that the more successfully he turned that act of signification which was his work at Diepkloof into an act of communication, the more contemptuous would be the rejection from racists. This was his first direct experience of Verwoerd's attitude: it would not be his last, and it would confirm him in his resistance to Afrikaner nationalism. Certainly he was not going to give up, or be frightened from his course. Some different means of conveying the same message would have to be found, and a still wider audience sought.

Paton had for years been torn between seeing himself as a man of action, reformer, politician, theorist, and seeing himself as an artist. Whenever his work at Diepkloof got him down he would long for the life of the writer. In October 1944 he heard that the Education Department was planning to move Diepkloof to an entirely new site, 16 miles from Johannesburg. The new site was a fine one, but it would mean the end of all Paton's work of beautification of Diepkloof, the abandonment of the village he so loved, and a completely fresh start. 'It must inevitably bring me to a parting of the ways,' he wrote to Hofmeyr, '& I expect to withdraw from Johannesburg & to write. As you probably know I have always been torn between the two—the man of affairs & the poet & writer; and I think this change will make decision inevitable.'[41]

As usual, this is in part an appeal to Hofmeyr to do something for him or he would retire, a technique he had used earlier in 1944 in threatening to leave Diepkloof for a bigger job in Rhodesia, and would again in 1945, when offered the headship of Adam's College in Natal.[42] The year 1944 represented a crisis year for Paton, in which he was most tormented by incompatible ambitions; and it was to produce, in another letter to Hofmeyr, the plainest statement of those ambitions even while he pretended that he had abandoned them:

> I find that personal ambitions seem less & less important. I once thought I would succeed you as Prime Minister, but as that seems less & less probable, and indeed more and more fantastic, it seems less & less important that I should occupy any distinguished position. I am finding greater satisfaction in writing, & think I shall do something about South Africa very soon . . . In your last letter you said that you would avoid the Prime Ministership if you could—with honour, that I took for granted. I could not help reflecting that I, like Marshal Foch,[43] had been 'hungry for responsibility', but that it had avoided me. You must not think I am dissatisfied.[44]

Hofmeyr cannot have been deceived for a moment by the last sentence, but his mother's training had made him a follower, not a leader. 'I wish I could see more clearly than I at present do,' he would write to Paton after one of these appeals, how you could turn your gifts . . .'[45] How this appearance of powerlessness, in a man who had the power to transform Paton's life by offering him any of the important jobs in his gift, must have made Paton grind his teeth! Yet in a sense Hofmeyr's protestations of impotence were no more than the truth. He was unable to summon the will to make use of Paton's undoubted brilliance in organization and administration.

In the biography Paton was to write of Hofmeyr, he tells the story of his friend, when he was a man in his thirties and the most powerful civil servant in South Africa's richest and most populous province, being taken by his mother to have a cut treated in a chemist shop. 'Jantjie, don't jump like that,' barked Deborah Hofmeyr when the iodine was being applied. 'How can the lady help you if you jump like that?' 'But it hurts, Ma, it hurts,' whined the mighty Administrator of the Transvaal.[46] In an important sense he never grew up. Academically brilliant he might be, but he had virtually no understanding of or sympathy with other human beings, none of that mature insight and strength that Paton himself displayed to such a marked degree. Above all, Hofmeyr was psychologically incapable of breaking away from Smuts and taking the risk of forming his own party, in which Paton would have found a natural place. Paton had been slow to realize this, hero-worshipping Hofmeyr as he did, and wishfully investing him with Paton's own strengths.

By 1945 Paton was feeling deeply depressed with Diepkloof. He was now coming to the end of his war work for Toc H and the YMCA, resigning from the national executive of those bodies in July 1945 when the war ended. 'I am anxious to withdraw from what public life I possess,' he told Hofmeyr with irony, '& to turn my attention to my work, my home, & my pen.'[47] His home life, though stable and outwardly happy, left him unfulfilled; his work at Diepkloof seemed to him to have gone nowhere during the six years of war, and he gloomily anticipated that the pent-up demand for housing would make any further building at Diepkloof, in particularly the longed-for replacement of the main block, unattainable for another six years.

His pen remained a refuge, and he planned, in mid-1945, a novel, set, like the opening of *Louis Botha*, during the Boer War. As he told Hofmeyr,

> My story is simple. A party of young Boers, during the siege of Ladysmith, find their way down as far as Mooi River, blowing up bridges (if there are any, I'll find out) & interrupting communications. Then comes the news of the relief of Ladysmith, & their retreat is cut off. But a young Natalian of Boer sympathies offers to guide them over the Berg, which as you know is at its grandest here. And up there, with freedom near, they are caught by British scouts. The hero, a young bitter Free Stater, is wounded in the leg. Others are wounded too; & willy-nilly, captors & captives settle into a cave under the great peaks, a cave looking to the

East & out over Natal. The leg gets worse, & a doctor must be sent for. He brings his daughter too (perhaps that is inevitable). The leg must be taken off. And so the young bitter man has to spend much time in the company of the chief captor, an older man, wise & experienced. The young man learns that the deeper things of life are the same for all good men, & there could be no better place to learn that than so high above the world.[48]

Paton was never to write this novel, but even the plot summary marks an advance on the episodic and over-coloured manuscripts he had produced in the past. It is true that he clearly planned to pile on the local colour in the Walpole manner, but it is also clear that his story had a well-constructed plot, and was, in its nature, its setting, and its chief theme, based on his own experience. It seems likely that, had he written it, this would have been a fully South African novel, as *Cry, the Beloved Country* is, rather than a largely derivative book of the *Ship of Truth* variety. He was clearly bubbling with ideas for the serious novel he had for years been predicting he would write. What he needed was the right circumstances in which to get to work. And in 1946, though he could never have predicted it, those circumstances would arise. Though Hofmeyr would continue to fail him, his life was about to change in the most extraordinary way.

12

CRY, THE BELOVED COUNTRY 1946

By their fruits shall ye know them.
MATTHEW 7: 16

At intervals since his student trip to Britain in 1924 Paton had planned to return there to study. Sir Cyril Burt's accounts of English experimental reformatories along Diepkloof lines seem to have given him the notion of undertaking a study tour of such places once the war ended, and early in 1946 these plans began to take clearer shape. He made a preliminary trip to Rhodesia, travelling by train, to see Rhodesian prisons between 13 and 22 March, 1946,[1] and found that there was much to be learned through such visits.

With his usual energy he applied for six months' paid leave, gathered funds (Dorrie self-sacrificingly cashed in the life insurance policies in her name),[2] arranged cheap accommodation overseas for himself through friends in Toc H, and rapidly extended his proposed area of investigation to include the United States and even the Soviet Union. When the Friends of the Soviet Union, the wartime group through which he had made his contacts, fell asunder with the start of the Cold War, he crossed the USSR off his list and substituted Scandinavia.[3] His aim was to prepare himself for the post of Director of Prisons, though he continued to think of this as a jumping-off point into politics. 'If the invitation came to go to Prisons, as I think it might', he wrote to Hofmeyr (who in 1946 was acting Prime Minister in Smuts's absence abroad, and therefore might well be the person from whom the 'invitation' came), 'it would be a very difficult choice, for it seems as tho' Diepkloof may be shifted & re-built after my return. I still think my strongest suit was politics, but dooty is dooty.'[4]

Leaving his deputy, J. H. Laas, in charge of Diepkloof, Paton left Johannesburg late at night on 8 June 1946 by one of the first of the newly instituted South African Airways flights to London.[5] It was the first time he had flown, and he was both excited and terrified. Ten years before it had been a rare and wonderful thing to fly from London to Paris; to fly from the southern end of Africa to Europe in 1946 was wildly glamorous. Paton thought the 36-hour journey in the four-engined Douglas DC4 a wonderful experience, though he was too frightened to rest. 'I hardly slept the first night, & felt a bit nervous up there in the air', he wrote to Dorrie after the trip was safely over,[6] with a show of casualness; but he was later to recall that during his first take off, 'it was some minutes before I realized that we were not going to crash, at least not yet'.[7] His fear returned whenever the rhythm of the engines altered, which they did repeatedly, and he stayed glued to his window watching for possible emergency landing places below.

The flight progressed up Africa in what now seem absurdly short hops: they landed at Salisbury, then at Kisumu, a dirty town where they breakfasted in a dirty hotel, then high across the savannah of east Africa until they made a three-hour stop at Khartoum, where Paton posted cards to his sons. Then he foolishly walked hatless in baking heat along the banks of the Blue Nile, looking for birds and wondering why he was the only soul stirring. Presently a sensation that his head was on fire told him the answer, and he went very slowly back to his hotel to take three cold showers without effect, and thereafter stayed out of the sun.[8]

At midnight on the second night they were in Tripoli, where 'Itis', as Paton called Italians, served him a most un-Italian meal of eggs and baked beans. At dawn he was flying over the Mediterranean, there was a brief stop in hazy Paris, and at 9 a.m. on 10 June 1946 he landed at Heathrow. By bus and taxi he made his way to the home in Seven Kings of Mrs Dench, mother one of his Toc H friends in Johannesburg, and then, exhausted as he was, he went into the city to assure himself that it was still there.

He found London dirty, and bomb damage was everywhere to be seen; Londoners, not surprisingly, looked drawn and slightly shabby. 'It was this first day that it seemed so incredible to think that 48 hours before we had been packing', he wrote in wonder to Dorrie. 'I have since been to Toc H, South Africa House [the embassy building], found after much searching the Allens with whom I stayed in 1924. But times have changed. Mrs Allen has a flat, no servant, & took me into the kitchen to help with the tea. Mr Allen is dead, & the select Kensington Places & Gates & Gardens are dirty & converted into flats.'[9] He did not dislike this change; the grandeur of London in 1924 had given him a sense of inferiority, and he enjoyed staying with Mrs Dench, a working-class woman whose courage and sense of humour he admired. Sugar rationing he, Natal born, found the worst deprivation: he ingeniously tried sugaring only every second cup of tea, in the hope that he would be able to taste the sweetened ones.[10] Finding this did not work, he collected his sweet ration, 400 grams of coconut ice and groundnut toffee, and wolfed the lot in company with Mrs Dench. 'I believe I could even eat fudge,' he wrote wryly.[11]

He did a good deal of sightseeing in his first weeks, not only in London but outside it. On Box Hill, near Stoneleigh, he walked in 'the loveliest of woods', and at Epsom he walked round the famous course. He spent a weekend in Sussex with the oddly named administrator of Toc H, Lake Lake, seeing Bodiam Castle and Battle Abbey, visiting Hastings and St Leonards, and shooting rabbits at Sevenoaks. Before he had been in Britain a month, though, he was becoming homesick. 'Am missing you all,' he wrote to Dorrie, '& while walking in the English countryside, find myself thinking back to South Africa.'[12]

Toc H took up a lot of his time, and he saw a good deal of its founder, Tubby Clayton. He was unimpressed by Clayton, who revealed a streak of egotism and phoniness. He particularly disliked Clayton's dominance in conversation. 'I listened, we all listened,' he wrote to his wife, in terms reminiscent of Keats's irritated reaction to Coleridge. 'There is no chance even for me when Tubby is about.'[13] And in his autobiography he tells of Tubby Clayton's habit of ejaculating 'Good, good' or 'Grand, grand' after having failed to pay attention to what someone said. On one occasion he advanced on a near stranger, vociferating loudly. 'Jack, my dearest fellow, how are you, my dear Jack?' 'Well, thank you,' said Jack. 'Good, good,' said Clayton heartily. 'And Jack, how is your dear wife?' Jack, deeply affected, said, 'Tubby, she died last Friday', and Tubby Clayton boomed 'Grand, grand'.[14] Paton's enthusiasm for Toc H was cooling fast, as his telling of this story shows.

Before embarking on his study tour of European borstals, he had planned to attend an international gathering organized by the Society for Christians and Jews, a group which had sprung up in the 1930s in an attempt to combat the anti-Semitism so evidently spreading across Europe. The Society had spread to South Africa in 1937, and among its founder members had been Paton's bishop, Geoffrey Clayton, and Professor Hoernlé. Paton had joined just before the war as a means of opposing the anti-Semitism which he saw growing in extreme Afrikaner nationalists. The Conference organized by the Society opened in London at the end of July 1946, with the Archbishop of Canterbury in the chair and with introductory addresses by the former Chief Rabbi of Berlin, Dr Leo Baeck, and the American theologian Reinhold Niebuhr. It then moved to Oxford, where it divided into six commissions, Paton being elected chairman of one, the Commission on Mutual Responsibility in the Community. 'I was the best of a poor lot,' he wrote modestly to his wife.[15]

Paton was moved by Baeck, who spoke about the help which Christians, at the risk of their lives and safety, had given to Jews in their darkest hour.[16] And he was immensely impressed by Reinhold Niebuhr, whom he thought the most enthralling speaker he had ever heard, talking without notes for more than an hour, and holding his audience in the palm of his hand. Niebuhr was a moral philosopher, and the gist of his remarks on this occasion, as Paton later recalled them, was that individual man could become a saint, but that collective man was a tough proposition.

But though Niebuhr was the single person to impress Paton most at this conference, by articulating what Paton already knew, it was the survivors of the Nazi concentration camps, both Jews and Christians, who most moved him. Of Leo Baeck he wrote, 'I have never seen a human face more full of suffering; but it lights up when he smiles, & he is full of kindness and thoughtfulness and gentleness.'[17] The stories of the concentration camps filled him with horror:

> We were told by Pastor Fricker, a Protestant in the Dachau Concentration Camp, of a trainload of Jews who arrived in Dachau on a Saturday. The SS guards were not on duty for the weekend, so the doors were not opened till the Monday, & no food or water was given. Many were dead, & three corpses had been eaten to the very bones.
>
> Pastor Gruber spoke in German, but with such power that one was moved even when one did not know what was being said. Pastor Fricker speaking in German told us of Catholic Father Schneider,[18] who ministered to dying Jews from the Old Testament and dying Christians from the New. And each day he was beaten, & each day he persisted, till he was beaten to death.
>
> The terrible horror of Nazism is brought home, & it is just as terrible to hear that anti-semitism is not dead in Europe, but that it is being increased by the tens of thousands of homeless Jews drifting across Europe. It seems as if man cannot learn, & it makes one all the more conscious of one's own responsibilities in these matters. But it is moving to note that most of those who did not succumb to the poison were practicing Xians [Christians], who suffered death & untold suffering because they would not yield . . . I should like David to see at least the first page of this letter; I hope when he grows up he will fight all injustice & hatred.[19]

And in an article he wrote for the South African journal *Outspan* and mailed home, Paton applied the lessons of the conference to South Africa directly, in the terms in which he would soon be crying out against apartheid:

> We are not a cruel or unjust people, but we are criminally careless in our entertainment of anti-Jewish prejudice, which, seized upon by those ready to exploit it, could bring even us to disaster. Consent to evil a little, and it will grow. Consent to evil a little, and a great evil will overwhelm us, till no tongue crying out against it will be heard any more. That of which we should have been master will be the master of us. No shame or remorse will save us then. Therefore consent not to this evil at all.[20]

Many of the delegates were Americans, and Paton found some of them rather overbearing. Many of them, on hearing that he planned to visit their country in November and December, promptly invited him to stay with them, and though he was a little sceptical, he wrote down their names and addresses. In due course he was to find that these offers were perfectly sincere.

On 6 August 1946, after the conclusion of the conference, Paton returned to London and Mrs Dench, and then began the series of visits to prisons and reformatories which was the real reason for his trip. His first tour was of Wormwood Scrubs, on 9 August 1946, and he kept a record of the many subsequent inspection tours which he made, working his way gradually through Britain: Wandsworth Prison on 12 August 1946; an approved school at Kemble in the Cotswolds on 13 August; back to London for Brixton Prison on 14 August; Chelmsford Prison on 15 August; Rochester Prison on 22 August. In due course he was to submit a detailed and carefully thought out official report of 55 closely typed pages to his superiors, by far the most important of his theoretical writings on penal reform and penal practice, though it has never been published.[21] His pocket diary, however, gives more immediate responses to individual institutions, and shows that he considered his work at Diepkloof compared well with what he was being shown. 'Leicester Prison—7/10/46—Not very impressed. Sherwood, Nottingham—8/10/46—15–16 absconders out of 70 in 8 months [that is, more than 30% per annum compared to Diepkloof's 12% in the same year]—Not v. impressed.'[22] He found that the effects of the war were not confined to London: 'Very often I spent a day visiting a prison or a borstal, but sometimes the hosts had not sufficient food to give me lunch, and were both embarrassed and ashamed that they could not.'[23]

The homesickness he had revealed in telling his wife he thought much of South Africa becomes a constant theme in his letters from September 1946 onwards: 'Everywhere I go here, I think always back to South Africa, & to the places we have been to & seen.'[24] This homesickness grew when, after a Toc H conference in London at which he delivered a paper, he sailed for Sweden, embarking on the SS *Saga* from Tilbury to Gothenburg on 5 September 1946. The British passengers could scarcely believe the abundance of food at the first smorgasbord, and gorged themselves until the sea grew rough. Paton, after two months of British rations, fully shared their hunger.[25] Fortunately for him he did not share their seasickness.

Homesickness was another matter, and everything he saw and heard seemed to remind him of South Africa. The Swedish countryside made him long for his family. From Gothenburg to Stockholm was, he wrote to Dorrie, 'a seven-hour electric train journey, thro' country of small farms & birch & pine forests. I wished only that we all could have been together.'[26] The magnificent cake-shops he found in Stockholm made him reflect, 'I give DFP [his son David] two days here, & he would be sick for a month.' He could not afford the cakes, however, for he was saving every penny, and his letters to Dorrie are full of schemes for getting some small extra allowance from the Education Department. No doubt these money worries also inclined him to think longingly of home.

More strikingly, the Swedish language reminded him powerfully of another Germanic language, his beloved Afrikaans:

> I am sure I could learn this language in a month or two. Its roots are very similar to Afrikaans, & there is hardly a shop name that I cannot make out after some puzzling. A Konst shop is an art shop, Afr. 'Kuns'. 'Kladderi' is outfitter—Afr. 'Kledere'. 'Högtids klader' puzzled me, but evening-suits were in the window, & I concluded it was 'High-time clothes', which is right.[27]

His homesickness was made worse by the fact that Dorrie had decided not to write to him in Scandinavia, believing that her letters would take 20 days *en route*, and not reach him before he left for Britain again. He felt very cut off as a result. Already he was writing of his plans to return to South Africa from the United States by a small, cheap, one-class boat: 'That will be a good day, won't it?'[28]

And soon, an omen, his homesickness began to take literary form. He dined on 15 September 1946 with an eminent Swedish penologist, Torgayi

Lindberg, and other Swedes including an official named Uno Eng, a desk-bound poet who acted as Paton's interpreter. Eng cared much more for poetry than for prisons, and the talk around the dinner table was almost exclusively literary. It was a bibulous meal ('we drank schnapps and beer and red wine and liqueurs' he reported rather woozily to Dorrie afterwards)[29] and it cost 26 crowns, about £2, which Paton thought alarmingly expensive. Spurred on by drink and the company, he wrote a sonnet which he gallantly presented to Mrs Fredrickson, the crippled wife of one of his Swedish hosts. Given the circumstances of its composition, it is a remarkably polished piece of work, but what is most striking about it is the evidence it gives that South Africa was filling his mind and his imagination:

> From where the sun pours on the southern sand
> From break of day to the declining hour
> His fierce & unabating southern power
> I come, to this austere & northern land.
> The child of sun is driven by the rain
> Along the streets, to welcome with a cry
> The first thin breaking of the northern sky,
> That breaks, that closes, till it rains again!
> Yet these are things outside us. Talk of friends
> For Nature's cruelty makes high amends.
> The sun is in the mind, the heart, the mouth
> And in the rooms where friends are met together,
> It shines, regardless of the northern weather,
> With all the power of the golden South.[30]

This poem is a portent, for the 'golden South' and its sun were increasingly in Paton's mind, heart, and mouth. It is hard to know if his hosts understood the poem, though, as he told Dorrie, it 'brought me great fame, & they think I am a great South African poet!'[31] Probably even Uno Eng never knew that, as Paton was to put it, 'his modest guest from that faraway country was pregnant with child.'[32] However it was not Sweden that would see the beginning of the birth, but Norway.

He had worked hard in Sweden, travelling to the various Swedish borstals at Langholmen (11 September 1946), Hall (12 September), Uppsala (13 September), Skrubba (14 September), and Skenäs (17 September). They seemed to him munificently funded by comparison with Diepkloof, and they were dealing with a class of young offender utterly different from the hardened young adult criminals who were his worst burden. He also envied the Swedish staff-inmate ratios which were 1:2 or 1:3 compared with Diepkloof's 1:10 or worse.[33] Yet he found he had less to learn from the Swedes than they from him; in spite of their funding, manpower, magnificent buildings, and supportive public, their best absconding rates were worse than those at Diepkloof.[34]

But he was about to take a brief holiday, the fulfilment of a long-term ambition, by visiting Norway. This was the only period of free time he would have on his trip,[35] and he was to make good use of it. He travelled to Norway by train on 24 September 1946, and as the train crossed the border at Storlien and then began to run alongside a green foaming river of a kind he had never seen before, he was filled with excitement and a strange exultation. 'I was in a strange mood,' he wrote later. 'I spoke a great deal to myself, composing sentences which seemed to me to be very beautiful.'[36]

He had longed to see Norway ever since reading the novels of Knut Hamsun, probably soon after Hamsun won the Nobel Prize for Literature in 1920, while Paton was still at Natal University College. Hamsun, whose real name was Knut Petersen, had been born in 1859 in central Norway and was still alive; he would die in 1952. His best-known work is *Hunger* (1890), but he had written many others, including a book that had a major impact on Paton, *Growth of the Soil*.

On his arrival by train in Trondheim, Paton made his way to an hotel, the Hotel Bristol, and had difficulty in making himself understood to the girl at the desk, who spoke no English. Here he was approached by a stranger who introduced himself as Mr Jensen, an engineer, who acted as an interpreter for Paton, and having secured him a room, asked him if he would like to see the city's superb Nidaros Cathedral. Paton never forgot what happened next:

> It was now almost dark, and the cathedral itself was in darkness. The lights had been turned off at four o'clock to save electricity, for Norway in 1946 was bitterly poor, having been stripped of almost everything by the Germans. Mr Jensen showed me round the famous building with the aid of a torch. It is the most beautiful cathedral in Norway, and kings are crowned there. It has also one of the most beautiful rose windows in the world, and when we had finished our tour we sat down in two of the front pews and looked at it. There was still enough light in the sky to see its magnificent design and its colours. We did not speak, and I do not know how long we sat there. I was in the grip of powerful emotion, not directly to do with the cathedral and the rose window, but certainly occasioned by them. I was filled with an intense homesickness, for home and wife and sons, and for my far-off country.

'Let us go back to the hotel,' said Mr Jensen, 'and at seven o'clock I shall call for you and take you to dinner.' It must have been about six o'clock when I reached my room, and I sat down and wrote these words: 'There is a lovely road that runs from Ixopo into the hills. These hills are grass-covered and rolling, and they are lovely beyond any singing of it.'

That is how the story started. I do not remember if I knew what the story was to be. But the first chapter of *Cry, the Beloved Country* was written in my room at the hotel Bristol, while I was waiting for Mr Jensen to come to take me to dinner.[37]

Paton was to tell the story of how he came to begin *Cry, the Beloved Country* many times, and always he mentioned the rose window, as if it were self-evident what gazing on a stained-glass window in a Norwegian cathedral had to do with a novel about the racial conflict in South Africa. When an American interviewer in later years showed scepticism, 'You mean a rose window had that much effect?' Paton had little to add: 'Not just the rose window, I suppose. I had been reading John Steinbeck's *The Grapes of Wrath* and that was mixed up in my feelings. Also, I was feeling a great homesickness.'[38]

He had read the Steinbeck novel in Stockholm on 22 September 1946[39] and had been powerfully affected by it, but homesickness was the dominant emotion evoked by everything he experienced in Norway. The language was part of this homesickness, as it had been in Sweden; Norwegian is even easier than Swedish for someone with a good knowledge of Afrikaans to make out, and if Paton opened a prayer book in the cathedral he would have found himself able to follow the service without much difficulty. And the window itself reminded him powerfully of home. In St Mary's Cathedral, Johannesburg, where Paton had so often worshipped during his years as a synod representative and member of Bishop Clayton's Commission, there is a strikingly beautiful window, older and smaller than the Trondheim one (which was only finished in 1930, a gift from Norwegian women) but not dissimilar to it. The sight of that window would naturally have increased his homesickness, even though Paton seems to have been unaware of the precise nature of its effect on him, and it helps to explain why gazing at that rose window should have sent him back to his hotel room to write such a paean of love for the hills over which he and Dorrie had walked as they courted.

A strong desire for Dorrie was no doubt also at the heart of the novel's inspiration: Paton saw his writing as a way of escaping sexual temptation of the sort he experienced in any large city with a red-light district. Once he had started writing it, his book not only filled every spare moment of his time, but it kept him from temptation. 'I have no roving spirit,' he had written to Dorrie on 18 August 1946, but that was early in his trip. Later on, from the United States, he would admit more frankly, 'This book has probably saved me . . . What I should have done with so many evenings I don't know. Probably gone on the razzle.'[40] Many years later he was to tell a friend, the journalist Ian Wyllie, '*Cry, the Beloved Country* kept me off sex,' a delphic remark which Wyllie did not understand.[41]

Trondheim cathedral had yet another effect on Paton, and this was to provide him with a central theme of *Cry, the Beloved Country*. It reminded him of Hamsun's *Growth of the Soil*, the book that had brought him to Norway in the first place. *Growth of the Soil* (originally entitled *Markens Grøde*) is a novel of great lyrical power, with the nervous, hallucinatory quality for which Hamsun has been praised.[42] The plot tells of a peasant, Isak, settling in the far northern wilderness and establishing a farm, Sellanraa. His wife Inger has a harelip of which she is desperately ashamed. She insists on giving birth alone each time; her first two children are born sound, but then a daughter is born with a harelip and Inger kills it. She is sent to prison in Trondheim, and after some years returns with big-city ideas that spoil her: vanity, materialism, loss of love, disinclination to work on the farm. Isak gradually reforms her with a mixture of love and firmness. The story is one of great power and sensuality, and Hamsun's evocation of the beauty and desolation of northern Norway struck Paton profoundly; his side trip to Norway was the direct result.[43]

He was never to reach northern Norway, the setting of the farm on which most of the novel takes place. But he could not have failed to remember the vivid account Inger gives on her return from imprisonment:

> She told them about the cathedral at Trondhjem, and began like this: 'You haven't seen the cathedral at Trondhjem, maybe? No, you haven't been there!' And it might have been her own cathedral, from the way she praised, boasted of it, told them height and breadth; it was a marvel! Seven priests could stand there preaching all at once and never hear one another. 'And then I suppose you've never seen St Olaf's Well? Right in the middle of the cathedral itself, it is, on one side, and it's a bottomless well. When we went there, we took each a little stone with us, and dropped it in, but it never reached the bottom.'
>
> 'Never reached the bottom?' whispered the two women, shaking their heads.
>
> 'And there's a thousand other things beside in that cathedral,' exclaimed Inger delightedly. 'There's the silver chest to begin with. It's Holy St Olaf his own silver chest that he had . . .'[44]

Paton could hardly have sat in that cathedral without thinking of this passage, and it is no surprise that his novel, begun under these circumstances, should have taken as a major theme the corruption of peasants drawn to the city's bright lights and febrile society. His experiences at Diepkloof had given him an urgent need to say something about social and moral disintegration in South African society, and Hamsun's novel showed him how it might be done. Homesickness and Hamsun's theme, then, were brought together for him by the sight of that rose window, and the result was the beginning of *Cry, the Beloved Country*.

As he had done for his earlier, unpublished novels, he sketched the plot-outline in a column of brief notes, though it is not clear quite at which stage of the writing he did this. *Cry, the Beloved Country*, as it rapidly developed, concerns the search of a humble black priest, the Reverend Stephen Kumalo, for his son Absalom and his sister Gertrude, who have gone to work in Johannesburg and have not returned. In the course of his search Kumalo leaves his impoverished rural parish and travels to Johannesburg, where he finds that his sister has turned to prostitution and Absalom has murdered the son of a white farmer, James Jarvis. Absalom is convicted and sentenced to death, and Kumalo returns home with Gertrude's son and Absalom's pregnant wife. The novel ends with the reconciliation of Kumalo and Jarvis, and Jarvis's determination to rise above tragedy by helping the impoverished black community.

On the surface, this plot could scarcely be further from Hamsun's tale of the frozen north, but in fact his struggling Nordic peasantry, battling isolation and the elements, are not far removed from Paton's rural Zulus, faced with drought and a dwindling population. In both Hamsun's and Paton's tales, there is a strong conviction that country life has a spiritually beneficial effect, a tradition running back through the Romantic writers to Classical times. Movement to the city, for Hamsun's Inger and Paton's Gertrude alike, is morally extremely dangerous, and the best hope for recovery is a return to rural virtues.

The effect of the Steinbeck novel was more general: *The Grapes of Wrath*, with its dirt-poor farming family fleeing the dust-bowl for the promised land of California, seems to have appealed to him as a profound myth of that journey and that search which are central to *Cry, the Beloved Country*. And Steinbeck had another more minor influence too: Paton was taken by Steinbeck's habit of indicating speech by a preliminary dash, rather than by inverted commas, a trick Steinbeck had imitated from Joyce. After using inverted commas in the first few pages of his manuscript,[45] Paton switched to the Joycean dash, and used it thereafter not just in *Cry, the Beloved Country* but in all his subsequent fiction.

At least two other books had some peripheral impact on the writing of *Cry, the Beloved Country*. One was *Black Boy*, the recently published novel by the black American author Richard Wright.[46] Though this violent story of oppression and poverty among black Americans had no direct effect on the form of *Cry, the Beloved Country*, Wright's pronounced left-wing views seem to lie behind the protests of the black bus-boycotters to some extent, and behind the writings of the murder victim, young Jarvis. The other book Paton had read in Sweden was an old friend, *Pilgrim's Progress*.[47] Bunyan's story, like *The Grapes of Wrath*, no doubt reinforced Paton's choice of the journey and the search as the major element in *Cry, the Beloved Country*'s structure.

But it was Paton's own experience, and his own nature, that contributed most to his novel's themes as they developed: as one critic was to summarize them, they include such universal notions as 'division and reconciliation between fathers and sons; crime, punishment, and the law; hope in circumstances of personal desolation; the movement of rural populations to industrial centres; and the plight of squatters and the homeless in cities that are also the focus of power and wealth.'[48]

Paton's novel opened with the lyrical description of Ixopo's hills, but it turned rapidly from the beauty of the white farmlands to the impoverishment of the black, drawing the contrast as Paton had drawn it in his unpublished story 'Secret for Seven' in 1934.

> The great red hills stand desolate, and the earth has torn away like flesh. The lightning flashes over them, the clouds pour down upon them, the dead streams come to life, full of the red blood of the earth. Down in the valleys women scratch the soil that is left, and the maize hardly reaches the height of a man. They are valleys of old men and old women, of mothers and children. The men are away, the young men and the girls are away. The soil cannot keep them any more.[49]

This was a vivid recollection of the terribly impoverished black hamlet Nokweja, not far from Carisbrooke, which Paton had often visited on his walks around Ixopo.[50] He had been particularly impressed by the simplicity and bareness of a little Anglican church in Nokweja, and the church rapidly found its way into his story. His book was launched. His next letter to his wife, after telling of his visit to Trondheim cathedral, contained the restrained sentence, 'It was raining heavily & I returned to my room & began my novel about South Africa.'[51]

Once he had begun writing, on 24 September 1946, though he went on with his tour, he snatched every spare moment to work on his novel.

From Trondheim he took the spectacular train ride to Oslo through the Gudbrandsdal, with the autumn foliage flaming on both sides of a pistachio-green river, and did a little dutiful sightseeing of Viking ships and Vigeland sculptures; but when the rain began coming down again he went eagerly back to his hotel and wrote hard, beginning the story proper now, on 26 September 1946, with the arrival of the letter summoning Kumalo to Johannesburg.

The central theme, the clash between black and white cultures and their possible reconciliation, was not at first as clear to him as it subsequently became. An examination of the manuscript shows that among the details he cut out as he went along were references to Kumalo's quarrel with his chief, and many small references to the Boer War as Kumalo travels by train through towns in which battles had been fought at the turn of the century; it was not quarrels within racial groups that finally interested him, but inter-racial conflict, and he made changes where necessary to keep this central conflict in clear focus.

Even when minor disasters overtook him he went on writing like a man possessed. On 27 September he tried to return by train to Gothenburg to catch his boat back to Britain, but at the border town of Halden was put off the train by Swedish border guards on the grounds that his visa, obtained in England, had allowed him only one entry to Sweden. Horrified and almost in tears, he told the Swedish officials that they were as bad as Nazis. 'SS!' he shouted at them, 'Just like the SS!' This abuse did nothing to endear him to them, and he had to stand in the dark on an icy platform watching the lights of his train disappearing down the line. With it went his chances of getting back to England on schedule, and continuing with his prearranged round of visits to more reformatories.[52]

From this situation he was saved by the wife of the British Consul in Halden, Mrs Thompson, who worked out a plan whereby he could return to Oslo for a visa, borrow money from the British Ambassador there, and catch the next boat to England in a few days' time from Gothenburg. He slept that night in an empty railway carriage, and the rest of the plan went smoothly; he spent a night in a cheap hotel in Halden, writing about Kumalo's journey to Johannesburg and his first adventures there, and three nights in a pensionatet in Gothenburg waiting for the boat, and here too he wrote hard.

Before long Diepkloof reformatory had entered the story, and with it the idealistic young man with the fierce and frowning looks, whom Paton was to say was based on 'Lanky' de Lange, the supervisor, but which is just as much a self-portrait. Several of the other characters who were entering the book at this time were also based directly on people Paton had known in Johannesburg. Father Beresford was based on Michael Scott, the activist priest Paton had met while working on the Bishop's Commission;[53] Father Vincent, the pink-cheeked English priest who helps Kumalo, was based on Scott's fellow-activist Trevor Huddleston,[54] the name Vincent being taken from Father Vincent Wall of Johannesburg, who was murdered by a black intruder in similar circumstances to the younger Jarvis.[55] Msimangu was based on the first black Anglican to become a monk, Father Leo Rakale,[56] whom Paton had met through Huddleston.

He took the SS *Saga* back from Gothenburg to Tilbury on 30 September, and wrote sedulously on the voyage, the story now coming to him with great speed. The manuscript reveals the rapidity and ease with which he wrote; the handwriting is fluent and excited, and except for occasional passages which Paton rewrote on the versos of the many odd pieces of writing paper he was picking up in hotels and on ships, it shows few signs of major hesitation or correction.

In England he went back to Mrs Dench, and resumed his borstal inspections, travelling as far as Wakefield, Sherwood, and Gringley, and each time returning on the same day to London. He went to Usk in south Wales, and to Leicester.[57] He got sick of rail travel. He caught a bad cold, having just shaken off one he had caught in Sweden, and this held his writing up for nearly a week. When he resumed it he was dealing with the bus boycotts by black commuters, perhaps as a result of hearing that there was more industrial unrest in South Africa.

Black mine workers had struck in August 1946 against their risible wage of two shillings a shift, and the government's response, as it had been on former occasions,[58] was to send in armed police to put the strike down by force. Several lives were lost and many miners suffered severe injuries; Hofmeyr, acting Prime Minister during Smuts's absence at the United Nations Conference in Paris, took the tough and uncompromising line of a fundamentally weak man who finds himself faced with hard decisions. He not only denied black miners more pay, but rewarded white miners who had not struck, thus increasing the already large pay disparity between the two groups. Paton, in England, feared a blood bath unless the black Miners' Union was recognized by the government, recognition Hofmeyr refused to accord. 'The unrest is inevitable,' Paton wrote to Dorrie, '& the latest decision to give European miners 1/6 per day extra while native miners get nothing, will not make it any less. In the meantime I want to write my book.'[59]

With this background, his book naturally focused increasingly on the uncompromising attitude of authorities who saw blacks suffering as they protest against price rises they could not afford, but refused to bend and ease the injustice. The bus boycotts had actually started in 1940, and had continued intermittently until 1945; Paton focused on them partly because he had witnessed them himself, partly because to focus on the miners' strike which was so much on his mind while he wrote would have been interpreted as a criticism of his friend, and Minister, Hofmeyr.

At the end of October he said an exhausting round of goodbyes to all the friends he had made in England, gave final talks to groups such as Toc H, the Howard League, and the Society for Christians and Jews, and fitted in a few more prisons such as Wandsworth. 'All I have seen of Sweden and England makes me realize how progressive Diepkloof is,' he wrote with quiet pride to Dorrie. 'I believe I could honestly say—privately of course—that I have not found one place that I think better, but that may of course be only conceit.'[60]

On 6 November 1946 he left Southampton for America, travelling in style on the *Queen Elizabeth*, in a second-class cabin that cost him £60 and seemed to him luxurious,[61] and all the way across the Atlantic he wrote his novel. It was probably on the voyage that he had the idea of the murder of a man who was to be the son of a farmer from Kumalo's own part of the country. This development drove into the background the figure of Gertrude, the prostitute, who was to have played a more important role, and whose last-minute turning back from a new life was based on a Diepkloof boy, Jacky.[62] As America came closer, and as his novel took form and fire, he felt an increasing sense of excitement. His life, he must have realized dimly, would never be the same again after this journey. But even he could not yet sense the extent of the change that was about to overtake him.

13

CRY, THE BELOVED COUNTRY (CONTINUED) 1946–1947

There is no test of literary merit except survival.

GEORGE ORWELL

At 3 a.m. on the morning of 11 November 1946 Paton went on deck to see the Statue of Liberty, and, even more impressive to him, what he called 'the incredible skyline of New York'. In the late twentieth century, when every city in the world has a surplus of skyscrapers, it is hard to conceive the degree to which New York seemed, to a wondering traveller in 1946, a vision of the future—as indeed it was. He was filled with excitement at this first sight of the country which had come out of the crucible of the war so much strengthened as to bestride half the world. Britain had seemed to him bled white, exhausted, and hungry; America by contrast struck him as almost overpoweringly purposeful, energetic, and optimistic. He was at first intimidated by the air of hustle, but soon found himself enjoying it.

He had asked a Jewish friend he had made at the Conference of Christians and Jews in Oxford, Audrey Kanter, to confirm a hotel booking in New York for him. She was still in Europe when he arrived in America, but with characteristic generosity she had arranged with her mother for Paton to be taken on arrival to a better hotel than he had asked for, the Granada in Brooklyn, and he had hardly settled in when the Kanters arrived to invite him to their apartment for dinner. 'The Kanters are at the top of a 20-floor apartment house, & the night view is stupendous,' he reported excitedly to Dorrie. 'New York is an incredible place. I could hardly walk along the streets without looking up to these enormous heights. The State Empire building [sic] is 1,100 ft. I eat with negroes at these snack-bars, & they—but I shall tell you about that later.'[1]

He felt, he said 'like a hick from the country',[2] gawping at the buildings, stunned by the constant roar of traffic, marvelling at the size of railway stations, learning about hot-dog stands, breakfasting on hash browns, grits, and eggs sunny side up. His experiences were not unlike those of Kumalo on first coming to Johannesburg, though he was too shrewd and suspicious to be cheated like the simple black priest. In spite of this, money seemed to run through his fingers like water; the Granada cost $9 a day, at which his Scots ancestry rebelled, and the Kanters found him a cheaper hotel, the Gramercy Park, nearer the heart of the city. 'You can't get the simplest meal for under a dollar, which is 5/-,' he reported, scandalized. 'And an ordinary seat at the cinema is 85 or 95c. A meal at a restaurant would be 2 or 2 and a half dollars.'[3] It was now 18 November, and he found the cold outdoors as oppressive as the heat indoors; he was obliged, 'with what reluctance I cannot tell you,'[4] to spend $32 on an overcoat.

His letters of introduction from Hofmeyr and the Education Department at once got him entry to the world of courts and reformatories, as they had in England and Scandinavia, and he was swept into another round of visits. Every day he would set out by train from the vast concourse of Central Station, and visit one of the many New York State reformatories which, even more than the Swedish institutions he had seen, bowled him over with their wealth of staff and equipment. 'I visited Children's Village at Dobb's Ferry, with buildings more substantial than the new hostels at Diepkloof but not so beautiful, with every kind of facility that one could dream of, with thirty percent of the staff engaged on aftercare work compared with seven percent at Diepkloof, and a principal's house that made my own look like a cottage.'[5] But even here he felt that Diepkloof was not altogether inferior.

> They have nothing like our Placement Office, Lanky's work I mean. [Lanky de Lange had the task of preparing each boy for his release from Diepkloof.] I explained our system to the Asst. Dir. & the psychologist, how we start from complete security, & how I was sure that a building could be built which would be secure but not prison-like, nor would it give the children the idea primarily that it was secure. And the psychologist said, quite seriously, I would like to work for you. I suppose this Children's Village is one of the richest & most up-to-date places in the world. It would cost $3,000,000 dollars [sic] to rebuild, but the more I see, the more I realize just how much we have done at DK with just the staff, the money, the facilities that we had.[6]

And each evening, on his return to the Gramercy Park Hotel, he would write. 'I find I am not anxious to go to Coney Island or Bronx Park or go to the top of the Empire State Bldg, or see plays—in any case I can't afford to, & any way I prefer writing my book, & am feeling already that this is going to be the most worth-while part of my trip,' he told Dorrie prophetically.[7]

He had arranged to attend a National Conference on Juvenile Delinquency in Washington, and he spent five days from 20 to 25 November 1946 there among over a thousand fellow penologists, hearing papers without number. He did some sightseeing too: the Washington Monument, the Capitol (where he was moved by the great bust of Lincoln in the Rotunda),[8] Mount Vernon, the Library of Congress, art galleries, and Arlington National Cemetery, where he heard a fat man, at the Tomb of the Unknown Soldier, drawl to his wife, throwing a large cigar to one side of his mouth with a practised movement of the tongue, 'Say Mamie, what do you know, they don't even know who he was.'[9]

But it was the Lincoln Memorial that particularly moved him. He climbed the steps from which one gets such a splendid view, and spent a long time standing in silence before that vast statue of the seated President, whom Paton called 'surely the greatest of all the rulers of nations, the man who would spend a sleepless night because he had been asked to order the execution of a young soldier. He certainly knew that in pardoning we are pardoned.'[10] It is characteristic of Paton that it was the quality of mercy in Lincoln that attracted him. His book made steady progress, night by night, in his room at the Statler Hotel, and Lincoln got into the story when the older Jarvis goes through the papers of his murdered son, finds a collection of books on Lincoln, and reads the Gettysburg Address, thus preparing him and the reader for his eventual reconciliation with the father of his son's murderer. It was Railton Dent who had first introduced Paton to the writings of Lincoln, and Paton was moved to send Dent a postcard of the Memorial in which he had stood with such awe.[11] Dent was, in fact, the model for the younger Jarvis; like Jarvis, Dent had a small library of books on Lincoln,[12] and was determined to do something practical for his black countrymen; and he was much in Paton's mind as he wrote this section of the novel.

Back in New York he moved to a still cheaper hotel, the Henry Hudson, wrote hard each night, and continued his round of visits, seeing Coxsackie Prison, a Catholic borstal called Lincolndale, and the famous Sing Sing jail. It was now early December, and winter had begun to set in, a winter such as Paton had never experienced.

On Monday I went to Sing Sing. A bitter day, the coldest Dec. 2 for 71 years. My ears & nose were numb, & I was afraid of frost-bite. The temperature was 17°F below freezing. I learned nothing at Sing Sing. It's a huge place, much cracked-up, but too big to be good. I saw the Death House which was a solemn experience. They don't show visitors such things in England. I came away feeling that we in S.A. must solve our own prison problems with our own intelligence.[13]

He now bought a railway ticket for the rest of his journey around America, to Atlanta, New Orleans, Houston, the Grand Canyon (for two days' holiday), Los Angeles, San Fransisco, Sacramento, Denver, Omaha, Chicago, Toronto, Ottawa, and back to New York. He later recalled that his ticket for this marathon journey was a yard long and had cost him $120: it actually cost $130, and his money worries fill every letter to Dorrie, to whom he continued to write every Monday without fail.

He left New York for Georgia on 9 December 1946, and spent 5 days in Atlanta. Here he stayed with a family named Groseclose, who had been fellow travellers with him from New York and spontaneously asked him to spend his week in their home. The experience he had had with the Kanters in New York was to prove typical of his experiences all over America; like other visitors before and since, he found that Americans in America fully lived up to their reputation for warm-hearted generosity. As he worked his way through his rail ticket, he worked his way through the list of addresses he had collected at the Conference in Oxford, and he never found either to fail him. The energy and warmth of Americans impressed him ever more deeply. He also found their country extraordinarily beautiful. Thus began his love-affair with the United States, which was to last to the end of his life.

Atlanta he found, on 10 December 1946, shocked by a terrible fire in the Hotel Winecoff, in which 120 people, including many children, had died, some by jumping 16 storeys into the street. The tragedy affected Paton too; he went to see the hotel, still surrounded by staring crowds, and wrote about it several times to Dorrie. He found the south a curious change from the northern states. Georgia, like so much he had seen, reminded him of home: 'It is strange after New York to ride in buses where whites sit in one place & negroes in another. It is just like being in South Africa.'[14] Unfortunately he said something very like this in an interview published in the *Atlanta Constitution*, in which he criticised an industrial farm for black boys he had inspected, and his host Frank Groseclose took grave offence at this slur on the good name of Georgia, so that he was glad to be leaving the next day.

On 14 December 1946 he left for New Orleans, and then went straight on through Texas, where he spent a day in Houston between trains, and wrote hard in the Hotel Tennison. Texas, because it reminded him of the Orange Free State or the Karoo, he loved. 'What a country, what a country!' he said to a Texan in his carriage, and the man warmed to him at once.[15]

And on he plunged, through New Mexico to the Grand Canyon, where he stayed in a friendly hotel overlooking that stupendous chasm. 'The altitude is 7,000 ft', he told Dorrie, '& it is cold; snow lies in patches, & deer move about in the trees, & sparks shoot out whenever you touch the electric light switches.'[16] He was determined to walk down to the floor of the canyon, which reminded him of the Drakensberg, 'except that instead of looking at it from below, you look at it from the other side, from its own level.'[17] On 19 December 1946 he started out, walked down 3,000 feet, and then hurt his ankle on a stone and took four hours to limp painfully back up. That night he was too tired and sore to write, but in fact his book was now nearing completion, and he knew it was good. 'I've written a book about South Africa which, good or bad, satisfies me in my depths,' he told Dorrie with sober pride.[18]

The chapter in which Kumalo, towards the end of the novel, returns to Ndotsheni, seemed to him the true climax of the novel, and he considered that it had produced his best writing. 'You say your favourite scene is Kumalo's return to Ndotsheni,' he was to write to one appreciative reader some months later. 'That is chapter 30. Well, you are right. It is the best in the book. When I wrote it I knew it was the best, & as you yourself can see, it evoked the best & deepest language of the book.'[19]

More trains: to Los Angeles, where he took a bus tour of Hollywood, seeing the homes of Tom Mix, Will Rogers, Charlie Chaplin, and others, the guide at intervals making the stock joke, 'Of course I don't know who's living with him now.'[20] After just a day there he took the night train on to San Francisco: he had learned that sleeping on the trains, though uncomfortable, was the cheapest way of spending a night.

He loved California, where the grass-covered hills reminded him of his own Natal hills.[21] In San Francisco he stayed at the Hotel Somerton in Geary Street, nursed his badly swollen ankle, and wrote. Here he had more contacts made at the Conference of Christians and Jews; one of them, Sonia Davur, invited him to a party on Christmas Eve, at the home of her

The prison reformer: Paton, soon after his arrival at Diepkloof, still in the blazer of Natal University College, proudly posing with his first Chevrolet, 1935

Diepkloof's main gate, before Paton tore it down: the first Headboys outside the inner wire, 1935

Famous author in the bosom of his family: Dorrie, Paton, Jonathan, David in 1949, after the success of *Cry, the Beloved Country*

'Out with racial discrimination — Paton sweeps South Africa clean', Jack Leyden cartoon, *Sunday Tribune*, 1976

Paton in the United States in 1954

Paton and Anne on his 80th birthday, 11 January 1983

employer, and here he shared a table with a cheerful couple, Aubrey and Marigold Burns, who heard that this small limping man with the odd accent would be alone in San Francisco on Christmas Day. The next day Paton went to a Christmas service with Mrs Lindberg, another of his Oxford contacts, and on returning to his room was telephoned by Aubrey Burns, who invited him to his house in Fairfax, in Marin County, north of the city, for a Christmas meal. This seemed to Paton the culmination of all the kindness Americans had shown him, and the friendship he now began with the Burnses was to last until Aubrey's death in 1983.

Burns was a handsome, cheerful man rather younger than Paton, and a poet of some merit; his wife Marigold, energetic and indomitable, also had a love of literature. In the course of that Christmas day in their unusual house, 127 Cypress Drive, the chief feature of which was four redwood trees growing up through the floor and out through the ceiling of its living room, so that water coursed down their trunks whenever it rained, Paton told them that he had written a novel, and they secured from him a promise that when it was finished he would let them read it. 'They are poor & happy, and both witty,' he reported to Dorrie. 'I played with the children, and enjoyed myself. But they are teetotallers, he is a Methodist parson, so my Xmas Day was dry.'[22] Nor in truth did he much like the children, Hal and Christopher, aged 7 and 3, who seemed to him much in need of Diepkloof discipline, so that he itched to smack them but restrained himself.

He wrote the last words of his still-untitled novel four days later. 'I finished the book today', he wrote triumphantly to Dorrie on 29 December 1946. 'I went to church in the early morning [as he had done every Sunday throughout the trip], & then wrote in my room all day. The room is now in a terrible fug, but I've finished my book, & I can now use my innumerable hotel evenings to patch it up.'[23]

He spent another fortnight in San Francisco, a city he came to love, staying in the Hotel Somerton and going out almost every day to visit institutions. Ray Studt and Roy Votaw of the California Youth Authority, and its Director, Pearl West, were his guides. Votaw in particular took to Paton, years later remembering him as 'one of the most thoughtful concerned people I met in my years with the Youth Authority'.[24] Like everyone else to whom Paton described the Diepkloof system, he was much impressed by what Paton had managed to do with pitifully few resources in South Africa. Votaw invited him to dinner, where he delighted the Votaws' two children with his apparently endless stories about

South Africa and its animals.[25] In return the Votaws showed him the redwood forests, which he found resembled the Knysna forest in South Africa, 'only, I regret to say, better'.[26]

It was the institutions near San Francisco that Paton had most wanted to see, particularly one named Preston School, which had been one of the models he had had in mind when he began his work at Diepkloof. But he found that the reforms at Preston had been reversed, and was horrified by the fences, the unspeakable detention cells, the inhumanly rigid discipline and the air of apathy and despair among the inmates. He was taken to a girls' school called Los Guilucos, and a boys' camp in the mountains near Sacramento.[27] He also, out of interest, visited Alcatraz, the famous maximum-security prison on an island off San Francisco, which had suffered a prison riot some months before. He returned from Alcatraz, as he had from Preston, in silent depression, 'having been a witness to the ultimate conflict between man and society'.[28] He wrote humorously to Dorrie, 'On Wednesday Jan 8 I went to the famous Alcatraz, & I have felt sober ever since.'[29]

Shortly after finishing his book on 29 December 1946 he had delivered the first half of his only copy to Aubrey and Marigold Burns. On his last two days in California, the second of which was his birthday, 11 January 1947, he stayed in the Burns's house and gave them the second half of the book, writing to Dorrie while they finished it '—with great interest, I may say'.[30] They read it with more than mere interest: they were profoundly moved by it. 'I found it difficult to read,' Burns was to say, 'chiefly because it is difficult to read small script through water—tears rose up as from a mountain spring, from one phrase to another, and from one emotion to another.'[31] The Burnses knew they were dealing with a work of genius. 'This book will go on living long after you are dead', said Aubrey.[32] Even as they spoke to him of it, they wept: 'Aubrey blew his nose a great deal, and Marigold had recourse to her handkerchief. It was clear that they had been through an emotional experience which if not as deep as mine was akin to it. No writer in the world could have asked for more from his first two readers.'[33] Paton wrote to Dorrie,

> I thought that book was good, but not so good as they thought it. In fact I don't care to repeat what Burns said, it would seem foolish. But he has written off to friends in New York and is determined to get it published here. I am telling you this in confidence, because nothing may come of it. But I am not allowing myself to get wound up by it, & if nothing happens, I shall bring it back to S.A. to have it published humbly and obscurely. Should however the book be what Burns

thinks it is, that will be what I made this trip for. And it might conceivably alter all our plans for the future. You needn't be afraid that I have gone batty. If, by the way, the book is accepted, you will get a cable saying I am coming back by air, & you will know what it means.[34]

His excitement is already palpable, and it is plain he did not think he would have to publish the book 'humbly and obscurely'.

He and the Burnses discussed possible titles, among them *The Tribe that is Broken*,[35] but in the end agreed on *Cry, the Beloved Country*, a phrase which occurs several times in the novel.

> Cry, the beloved country, for the unborn child that is the inheritor of our fear. Let him not love the earth too deeply. Let him not laugh too gladly when the water runs through his fingers, nor stand too silent when the setting sun makes red the veld with fire. Let him not be too moved when the birds of his land are singing, nor give too much of his heart to a mountain or a valley. For fear will rob him of all if he gives too much.

And the Burnses also suggested numerous small changes to the manuscript, many of which Paton put into effect: in particular he reduced the role of the chief at their suggestion, shortened the Biblical quotations, cut out references to race-riots, cut out much of the preaching, and wrote a new ending.[36]

With characteristic American energy the Burnses set about finding Paton a publisher before he left the United States. They listed fifteen American publishers to whom chapters should be sent, and Marigold found friends who would type fifteen copies of six chapters, including the first and last, and dispatch them to the publishers on the list. Interested publishers would be asked to write at once to Paton using the address of the South African High Commissioner in Ottawa, where he was scheduled to arrive in twelve days' time.

Paton meanwhile took a train through the High Sierras to Denver, Omaha, Chicago, and Ottawa. As he crossed the Canadian border the heating in his carriage failed, and Paton, who had not been told about the failure, put on all the clothes he owned, hugged himself hard, and concluded that Canada was very cold. When he got out in Ottawa he found that he was right.

> I took my two suitcases and struggled into the street, but I soon had to take refuge in a café because I was blinded by tears, my nose was running, my ears and hands were threatening to drop off. I was afraid that my tears would freeze and that I would be blinded. The temperature was -30°F, and it was colder than anything I

had ever known. Eventually I found my modest hotel. To get off the freezing street and into the warm hotel partakes of the nature of ecstasy.[37]

But almost at once, having bathed and shaved, he forced himself back outside into the cold, for he was alive with curiosity and excitement to see if there was any mail for him at the High Commission. In his autobiography[38] he was to say that there were letters from nine publishers asking for more of the book, but this is an understandable exaggeration; his letters to Dorrie show that 'there were four publishers' letters waiting me, one in particular (from Scribner's) being very promising'.[39] The thought of being published by Scribner's, the American publisher whom he most respected, thrilled him, and by telephone he arranged with the Burnses to air-mail them the rest of the manuscript so that they could have it typed and sent to Scribner in New York.

He would have done better to have taken it to New York himself, for his precious manuscript, the greater part of it his only copy, was sent by rail, the aircraft being unable to fly because of the snow-storms that now descended on Ottawa. It arrived in California, at the small post-office in Fairfax, just before the place closed for the weekend; Marigold Burns, by some feat of organization and pleading known only to herself,[40] contacted the postmaster and got him to open his office on a Sunday and hand her the parcel, and she and her friends then divided up the manuscript between them and hammered away at their typewriters night and day, posting the resulting copy to Scribner sequentially.

Meanwhile Paton, holding his excitement in check, was going dutifully through a few more prisons: in Kingston, Ontario, where he saw a women's penitentiary, and was prodigiously ogled by the sex-starved inmates, and in Guelph, where he inspected a productive prison dairy-farm. And at last he came to New York once again, and went straight to the Scribner's offices on 4 February 1947 to talk to Scribner's famous editor Maxwell Perkins. Perkins, well known for such eccentricities as wearing a hat at almost all times, indoors and out, was the discoverer of a range of writers Paton venerated, and he was tremendously excited at this contact with the great man, and drank in every detail of his office. As he wrote to Dorrie,

there was a drawer of a filing-cabinet open. You know how the little tabs project slightly beyond the cards, & there on the cards were Ernest Hemingway, Ben Hecht, Sholem Asch, & many other names that you know. Do you think I am going to join that select company. I said to Miss Wyhoff the secretary, well, this is an experience. She said, yes, this is the greatest publishing house in the world. But of course your book is good.[41]

Perkins had not received any more of *Cry, the Beloved Country* than the original six chapters, but he took Paton to lunch, and while they were lunching a large parcel of typescript from the Burnses arrived, seventeen chapters. Perkins would read it and give Paton the verdict in three days' time. Perkins himself Paton found very difficult to converse with, and he could not decide whether or not Perkins liked his book. In particular he found it unsettling that Perkins simply did not respond to many of Paton's questions and comments: what he did not realize was that Perkins was very deaf.[42]

Over the next three days the rest of the typescript arrived, and when Paton returned on 7 February 1947 Perkins and Charles Scribner, the firm's founder, implied that the answer was yes.

I went on Friday afternoon & Mr Maxwell Perkins the chief editor threw up his hands & said, I've only read parts of it, but we can't refuse it. He had just read the last chapter to see how it ended. He said I must send the final copy from SA as soon as possible. He said also Scribner's can't publish in SA so I could get it published there myself. The contract will be arranged by post. He introduced me to 'Charles' & I learned later that this was Charles Scribner himself. He said, Charles, we must take it. He said I must come & have a drink.[43]

Paton wrote describing to Burns how Perkins had said quietly to Scribner, 'We must take this book,' and Scribner replied, 'Well, take it,' and how the three of them then went for a celebratory drink across the street from the Scribner offices, hardly talking at all.[44] Paton told Dorrie, however, that they had talked of Thomas Wolfe, 'the great mad genius whose works I have been reading here', and that Scribner had warned Paton, 'Don't expect money, you can never tell how these things go.' Paton had replied, 'It wasn't written for money.'[45] Scribner lifted his glass in a toast, but, as Paton told Aubrey Burns in some bafflement, he did not say what the toast was. When Paton subsequently proposed a toast—'Here's to our association'—neither Scribner nor Perkins responded.[46] Soon after that, Perkins said that South Africa must be a sad country. 'Why?' asked Paton, but he got no reply.

Two days before this bizarre little scene, Paton had written to Hofmeyr to let him know what was happening:

I came away really to discover what to do with my declining years [he had just turned 44], & to see South Africa from far away. Now while I was in Norway &

> Sweden, living in hotel rooms & shut off from the people by reason of the language, I began a novel. And I have been writing it in all the hotel rooms of the world, a story of South Africa. Others must decide whether it is good, but my heart went into it. It is first a novel, but second & secondarily, a book with a message. And whether that message will be popular with the authorities, especially as South Africa is the hot news all over the world, I strongly doubt. And I strongly doubt too if I should stay in the Public Service.[47]

Paton clearly saw his novel as a work with direct political significance, and in fact it could hardly have come at a more opportune time. South Africa was, as he said, 'hot news', and it would get steadily hotter. The issue, of course, was racial discrimination, though at this stage it was not Africans who were in the limelight. South Africans of Indian descent had begun a campaign of passive resistance, along Gandhian lines, in response to new legislation, introduced by Smuts, aimed at restricting the rights of Indians to buy land or businesses. The Indian Congress in South Africa, of which militants under Dr G. M. Naicker and Y. Dadoo had seized control late in 1945, began a highly visible publicity campaign through the United Nations, enlisting the help of the Indian Government. Paton, who had found these matters splashed all over the New York papers, knew well that his book might prove a severe embarrassment to his political masters.

The Scribner acceptance was not the only remarkable offer he received that day in New York. He had, on his prison visits, impressed a great many American penologists. Nothing shows this more clearly than a visit he paid to the Penal and Correctional Division of the United Nations at Lake Success, New York, when the Director, whom Paton had probably met in Washington at the conference he attended there, offered Paton a post. 'When I went there to say goodbye it was the last thing in the world I expected,' he told Dorrie. 'That was on Friday morning but on Friday afternoon Scribner's took the book, & I confess I've hardly thought of the other. How would you like to bring all the family here to NY?'[48] He knew very well the answer to that question: Dorrie had not wanted to leave Pietermaritzburg for Johannesburg, and the thought of New York would have terrified her.

But Paton had already determined to turn the American offer down: he was just enjoying peeping through the doors that seemed to be opening invitingly on every side, after his years of Diepkloof drudgery and disappointed hopes that 'Hoffie' would wake up. And he ended this letter with excited postscripts for his sons, to whom he had written occasional affectionate letters throughout his trip. David had been planning to sell his stamp-collection to buy a horse, and Jonathan had longed for a new electric train set:

> Dear DFP,
> Some news eh? What about that horse? Dad.
> Dear Jonno,
> Some news eh? What about that train? Dad.[49]

These were days of joy for him, when he found he could not get to sleep until 2 a.m. and woke again at 4 a.m.

And although Scribner had warned him not to count on making money, of course he was doing so, having asked Perkins what the normal contract would stipulate. 'The contract would be $500 down, 10 per cent on retail price for first 3000, 12 and a half per cent for next 2000, 15 per cent thereafter,' he told Dorrie. 'Most books are $2 & $2.50. Now my dear, a book published in this country of 130,000,000, by the great house of Scribner, even if it doesn't sell well, should sell how many? You can work it out.'[50] Who could grudge him these dreams, when he had for years scrimped on a civil servant's salary which in 1947 had risen to just £720 a year,[51] and for the last six months had felt he was being dangerously extravagant if he spent more than $1 on a meal?

He showed his sense of new-found wealth by tipping generously as he left the Henry Hudson Hotel for the last time on 10 February 1947, and tipped again in the taxi, and in the train, feeling like Maecenas. He even took a Pullman from New York through Boston to Saint John, New Brunswick, from where he was to embark for South Africa.[52] It was a delightful change after all those nights of sitting upright and trying to sleep in the trains across the continent and back.

He sailed from Saint Johns, the only passenger on a small cargo ship, the SS *Fort Connolly*, after a delay of several days; Aubrey Burns, ever-faithful friend, sent him $40 with a telegram, 'Delayed sailing means unexpected expense'. It was not the first such gift from the Burnses, who were not rich, but Paton was to repay them amply. When his contract with Scribner was drawn up, some months later, he asked that the Burnses should be named as his agents, meaning that they would get 10 per cent of the royalties on the original edition. Perkins remonstrated with him, but he held firm. Though Paton's opinion of his book was high, he could not have foreseen that it would sell considerably more than a million copies in his lifetime, but he never regretted this princely repayment of the Burns's kindness.[53]

When the *Fort Connolly* finally sailed on 18 February 1947, Paton, the writing habit strongly ingrained by now, settled down to spend the four weeks of the voyage writing the elaborate 55-page report to his superiors about his trip. This work, he was to say, 'kept me out of the intrigues of this unhappy ship'.[54] The crew seemed to be at daggers drawn with each other; the master, Captain Proctor, was a drunkard who spent the entire voyage in his cabin, and he was rumoured to have a range of schemes for cheating the ship's owners. Paton found himself involved in one of these, when Proctor asked him to sign a receipt for one shilling for his work as a member of the crew; Paton also, and this is more to the point, signed a Certificate of Discharge showing himself to have worked as a purser on the *Fort Connolly*, having ostensibly been engaged at Hull; no doubt this allowed Proctor to indent for the wages of a crew member he had never paid. His copy of this dubious document Paton kept all his life.[55]

On 21 March 1947 the SS *Fort Connolly* docked at Cape Town: Paton had been abroad for eight months. In January he had written to Dorrie about this homecoming, 'I hope you will arrange things better than you sometimes do (joke, of course), & that you will show what you feel, & say what you think, & let us make it a bit of a honeymoon.'[56] He had very much hoped that she would come down to Cape Town to meet him, and was bitterly disappointed to find at the docks that she had not.[57] He took the train to Johannesburg, to be met there by Dorrie and his sons, and when he saw her on the platform of the Johannesburg station on 24 March 1947, her face was alight with joy. Lanky de Lange drove them home to Diepkloof, and, as Paton recalled after Dorrie's death, 'the moment that he left us, we were in each other's arms, loving and wordless. I think I whispered to you, *shall we have a bath?* and you nodded your head. And so we made love, after all those many months. I remember it now, not with sorrow, but with joy.'[58] He was back at Diepkloof again, but not for long.

14

LOTUS-EATING 1947–1948

And deep asleep he seem'd, yet all awake,
And music in his ears his beating heart did make.

TENNYSON

Cry, the Beloved Country was published in New York on 1 February 1948, which meant that after his return to Diepkloof Paton had nearly a year to wait before he knew whether the book would be the success he hoped for. During this period he lived on supportive letters from Perkins and Scribner.[1] Perkins, after arguing that the novel reaches a climax at the trial, and thereafter runs on too long, refused Paton's offer to rewrite the last third of the book on the grounds that to do so would slow up the production of the volume.[2] In the end Perkins did virtually no editing of *Cry, the Beloved Country*.

Paton also sent his typescript to friends whom he hoped would like it. Among these was Hofmeyr, who criticized the title, which he said was meaningful only when the novel had been read, and the construction, which he said suffered from the anticlimax of having Kumalo part from his son and go back to Ndotsheni several chapters before the boy is hanged in Johannesburg. Hofmeyr also complained that Gertrude is built up for an important role, and then simply drops out of the book.

Paton was hurt by this criticism, and delayed his reply by nearly a fortnight, something very rare for him when corresponding with Hofmeyr. When he did write back, he defended his book rather feebly:

I wish I could have given more attention to your criticism; if I had the time I would reconstruct the last six chapters & shorten the anti-climax. But Scribner's have written to say that they anxiously await the final copy. You see, it is a hotel-room book. Its virtues lie rather in the compulsion under which it was written than in the construction. But if I write another, I assure you it will be as perfectly constructed as I am able to do it. And I quite agree with you too about the characterisation of Gertrude. The book rather got out of hand at that stage, & it was only then that I conceived the idea of the murder of a man who was to be the son of a farmer from Ndotsheni itself. The shock of this discovery made me rather forget poor Gertrude, who was to have played a more important role, quite what I don't know now.[3]

He showed his continued regard for Hofmeyr by asking if he might dedicate *Cry, the Beloved Country* to him, and Hofmeyr acceded. Paton then had second thoughts, considering that people might think it odd that he had not dedicated the book to Dorrie,[4] and the English edition of *Cry, the Beloved Country* was eventually to appear with the dedication,

To
MY WIFE
and to my friend of many years
JAN HENDRIK HOFMEYR.[5]

The truth was that Paton's regard for Hofmeyr was somewhat diminished; he could not but reflect that if Hofmeyr's respect for him were anything like his for Hofmeyr, the great man would have found him some more important job by now.

But with his hopes for the reception of *Cry, the Beloved Country* so high, it scarcely seemed to matter. Even the renewed threat to move Diepkloof to a new site deep in the countryside, so destroying all his work on the farm, gardens, and village, left him curiously unmoved. 'It would have been a terrible blow to me, had not I, by virtue of this new but quite uncertain change of events, had the strange feeling of being in some sense a spectator,' he wrote to Hofmeyr,[6] and the fact that he could tell his Minister that he was now almost indifferent to what happened to Diepkloof shows the gap that was opening between the two men.

Meanwhile he published a shortened version of his official report on his trip in two articles in the Prisons magazine, *Nongqai*, 'The Penal Practice of Some Other Countries'.[7] He also published a pamphlet based on his official report, entitled *Freedom as a Reformatory Instrument*.[8] These publications, together with Paton's official report, give a clear picture of his view of Diepkloof as compared with what he had seen round the world, and since Paton was now preparing to resign from Diepkloof should *Cry, the Beloved Country* prove a success, it is worth pausing for a moment to evaluate the nature of his achievement as a prison reformer.

One scholar, summing up his work at Diepkloof, has argued that 'what was novel about Paton's work was that he was given the freedom to apply to blacks the diversified system of juvenile justice and welfare in existence for whites'.[9] But Paton's report makes it plain that while the theories he applied at Diepkloof were commonly held, the practical achievements he had arrived at, using resources that were pitifully few, were in his view unexcelled anywhere in the world. Those institutions that most closely approximated to the open-hostel system he had instituted at Diepkloof had either closed down after a few years, such as the Little Commonwealth in Dorset, the Sysonby Village Colony near Melton Mowbray, and the Training Colony in Berkshire, or had almost entirely abandoned the experiment with freedom, such as Preston School in California.

The closest rivals to Diepkloof's village system he came across on his tour were Swedish institutions, the success of which he attributed partly to resources that left him speechless with envy, partly to a much lower level of crime in the community, and partly to a public highly supportive of experiments in freedom, and tolerant of absconding. Certainly none of his Swedish counterparts had had to deal with a Verwoerd. And yet nowhere did he find an institution willing to give the degree of freedom he gave the pupils in the Village at Diepkloof, who slept in rondavels on their own, with unlocked doors and windows. 'In no single instance did I find anything approaching the complete freedom of the Diepkloof hostel, where boys do not sleep under the same roof as the cottage master and matron . . . I am satisfied that our method of granting freedom is equal to any that I saw,' he wrote with restraint. Nor did he find anywhere an institution that could boast the low absconding rates he had achieved at Diepkloof. As he was to put it a private letter, 'our absconding . . . is the lowest in the world. Last year [i.e. 1947] 800 leaves were granted for boys to visit their homes; & of these two failed to return, & two tried to fail to return, & two got into trouble.'[10] He justly claimed that his use of freedom as a reformatory instrument was, in practice, unparalleled outside Scandinavia.[11]

Paton's achievement should be seen, not in terms of theoretical advances (he was a popularizer rather than an originator of penological theory, in his many articles) but in terms of his success as a practical demonstrator of the value of freedom as a reformatory instrument. In his transformation of Diepkloof from prison into reformatory school and farm, with tiny resources and an original staff at best uncomprehending, and at worse antagonistic, he showed that he had a genius for organization. This genius was demonstrated in the speed and smoothness with which he produced his changes, in the great skill by which he built up a team of loyal subordinates convinced of the worth of what they were doing, and it extended to the clever dodges with which he squeezed extra resources out of an Education Department still wrestling with the effects of the Great Depression and a society that regarded Africans as worthy only of what they could pay for themselves. At Diepkloof he made a purse, not silk but perhaps satin, out of a sow's ear.

It is true that he himself believed that his reforms did not go far enough. The freedom of his Village, of which he was rightly so proud, was restricted to about 125 pupils out of the total of more than 600 in 1947. The grimly forbidding Main block, which he had tried repeatedly to have replaced, was still home to the majority of pupils, though many of these enjoyed the freedom to walk unsupervised at certain times anywhere on the large farm. And this failure of the authorities to allow the needed changes to buildings, or to provide such facilities as a library or a swimming pool, was paralleled by a failure to build the aftercare hostel in Johannesburg for which Paton pleaded repeatedly but in vain, and which he saw as necessary to bring his reforms to a logical conclusion. Just before his resignation from Diepkloof it was agreed by Hofmeyr that such a hostel should be built, and named the Alan Paton School, but Hofmeyr's fall from power caused the cancellation of the plan before it could be put into operation.[12]

Just as importantly, little heed was paid to Paton's repeated requests for magistrates to distinguish those pupils who could benefit from the reformatory from those who could not. Although freedom was the aim at Diepkloof, strict discipline continued to be used, in part because Paton, unlike many of his counterparts in Sweden and the United States, had no say as to which pupils were sent to him, and consequently continued to receive a proportion of incorrigibles with whom the reformatory system could achieve nothing.

But there was an even larger area which Paton felt was outside his control, and that concerned the environment from which his pupils came. In his two articles in *Nongqai*, he argued that the reforms that can be achieved inside a reformatory system were relatively minor compared with the changes that should be made to the Courts, the diagnostic authority, the Prisons authority, and (not least important) public opinion itself. But all the signs were that these changes were not about to happen. On the contrary, as Verwoerd's attack on him had shown, public opinion was, if anything, hardening towards the problem of black crime. Paton had argued, in his stream of articles during the war, that the real cause of crime was a society that shut blacks out of power and robbed them of pride in their own worth. As the year 1947 drew to a close, there were ominous signs that the determination of white South Africans to block the social advance of blacks was hardening, and this hardening process was accelerated by the robust development in 1946 and 1947 of opposition from black and Indian groups.[13] Portents of the Nationalist victory which would come in the 1948 elections were becoming clearer. Paton became increasingly convinced that his work at Diepkloof was ultimately being negated by changes in South African society.

In this he was absolutely right. Within a year H. F. Verwoerd, who had publicly savaged Diepkloof and all it stood for, would be the minister in charge of reformatories. He would ensure that the aftercare school in Johannesburg was never built, and one by one he would reverse the advances Paton had made. In 1958 he would close Diepkloof altogether, in pursuit of his view that African delinquents should be punished in the homelands to which they notionally belonged, and the pupils were ethnically grouped and dispatched to 'youth camps' in desolate rural areas. According to one of Paton's Afrikaner staff members, C. J. W. Kriel, 'The anti-Paton feeling [emanating from Verwoerd] was so strong that only one of the Diepkloof staff was transferred to the youth camps, and he was a storeman.'[14]

Paton was bitter about the 'penal reform' the Nationalist government instituted. In 1955, when the consequences of Verwoerd's moves had become clear, Paton wrote angrily to a friend,

> They [the Nationalist Party] will make what changes suit them, within the rigid framework of the sequence Offence-Punishment. They do however pay a certain attention to the Christian sequence Offence-Punishment-Forgiveness-Restoration, but they do this in a kind of apologetic way, as though they were being a little unfaithful to the Awful Sovereignty of God & a little too sensitive to the Softness of Man. The sequence in their hands becomes Offence-Punishment-Unctuousness, which makes me sick.[15]

Not surprisingly, Verwoerd's camps proved a failure, but Diepkloof had been destroyed. Today the beautiful farm is gone, and on its site stands an army camp, Doornkop Military Base. The house in which Paton lived for the best part of thirteen years there still stands, occupied now by the camp commander, and on the lawn can still be seen the bird-bath Paton made with his own hands. But of the greater work he did there, of his dreams of straightening twisted lives and a twisted society, and of the remarkable reality into which he transformed some of those dreams, nothing remains. Instead the desolation Verwoerd made lies all around.

Paton could scarcely have borne to stay and watch the destruction of Diepkloof: it would have broken his heart. Fortunately for him, he did not have to witness it, for his novel provided him with the way out. On 1 February 1948 *Cry, the Beloved Country* was published in New York, and on 5 February the first reviews reached its tense author. He must have been immensely relieved to read them. The first two he read[16] set the tone of the many hundreds that were to follow. Writing in the *New York Times* of 1 February 1948, Richard Sullivan called *Cry, the Beloved Country* 'a remarkable book, . . . a beautiful novel, a rich, firm and moving piece of prose,' and said, 'There is not much current writing that goes deeper than this. There is not much with a lovelier verbal sheen.' And that same day, in the *New York Herald Tribune*, Margaret Hubbard compared *Cry, the Beloved Country* to Olive Schreiner's *Story of an African Farm*, and commented that Paton's book stood by any standard. 'But above all,' she wrote, 'the quality of the style is a new experience. Here, in English, is a new cadence, derived from the native tongues. English words are used with the limpid rhythm of Zulu or Xosa [sic], as picturesque, as simple in expression, yet as delicately suggestive, as those languages.' Margaret Hubbard spoke neither Zulu nor Xhosa, and Paton, as we have seen, spoke little Zulu; before sending the typescript back to New York in 1947, he had asked Ben Moloi, the head teacher at Diepkloof, to go through the Zulu and correct it for him. His 'dialect' speech is a poetic invention akin to Synge's evocation of Irish speech-patterns,[17] and the beauty of his language derives not so much from his knowledge of Zulu as from his early immersion in the language of the Authorized Version of the Bible. But such pedantic criticism of his critics never occurred to him: he was being praised, and praised with all the generosity he could have wished for.

And after these first reviews reached him, they fluttered in as fast as homing pigeons, and each critic seemed intent on surpassing the last in praising the new author who had burst upon the world. 'A novel of absorbing interest told by a passionate humanitarian with a complete command of his subject . . . a book to read and remember';[18] 'Paton's prose is almost Biblical in its simplicity; his passion for humankind is Biblical, too';[19] 'Beside the sprawling verbosity, the tawdry cleverness of currently touted novels, *Cry, the Beloved Country* shines with a quiet radiance';[20] 'Exquisite, tragic, yet hopeful story';[21] 'The finest novel I have ever read . . . amazingly deft fusion of realistic detail and a symbolical synthesis of various points of view . . . it is brilliant.'[22] Many of these adulatory reviews were illustrated with a studio photograph of Paton, distributed by

Scribner's, in which he wears a pin-stripe suit, puts his chin out determinedly, and peers through his round spectacles with an expression of courage and deep intelligence.

Paton was bowled over by the reception of his book, which exceeded even his ambitions for it. 'The American reviews were extraordinary; you would think I had written them myself,'[23] he joked to Hofmeyr, and it was true. If this reception took him by surprise, it overwhelmed his publishers. For the truth seems to be that in spite of their rapid acceptance of *Cry, the Beloved Country*, Scribner's did not expect this unusual novel, about a country which at this time many Americans could scarcely locate on a map, to be a huge success. Certainly Scribner's had done nothing to prepare the way. As one of the leading American critics noted, the book was given 'no preliminary ballyhoo or build-up. It was ignored by every book club. It was not scheduled for discussion on any of the radio programs dealing with books. *Publisher's Weekly*, organ of the book trade, did not tip off buyers to its qualities. Its publisher, Scribner, did not greet it with pre-publication puffs . . . There were no ads in the papers announcing it on publication day.'[24]

As a result, Scribner's were totally unprepared for the reaction. *Cry, the Beloved Country* sold out on the day of its publication, and was hastily reprinted. The second printing was snatched up as quickly, and was followed by a third which met the same happy fate. By April 1948, just three months after its first appearance, the book was in its sixth printing, and the demand showed no sign of slackening. 'The sales on your book are still relatively small to the furor that it has produced,' Scribner wrote to Paton. 'Friends of mine who rarely read a book are enthralled by it. The Church (including R.C.s) are preaching about it—& capitalists & leftists embrace it equally. Its fame is spreading by word of mouth which is the only advertising that really counts. All that we can do is to back this up. You should be here—it is most exciting but exhausting. In my 35 years of publishing I have never known the like and if the book does not eventually sell into the hundreds of thousands I shall be equally at a loss.'[25] For all Scribner's excitement he was continuing to underestimate Paton's novel—over time it would sell into the millions.

Nor were reviews the only signs of success. On 2 March 1948 Scribner's cabled Paton that the British film director Alexander Korda wished to buy the film rights to the book (Korda's cable began HAVE JUST READ YOUR GREAT BOOK),[26] and a few days later came another cable, telling him that the American playwright Maxwell Anderson[27] wished to buy the rights to turn

Cry, the Beloved Country into a musical, in the form of a Greek tragedy complete with Chorus, with the help of the composer Kurt Weill.[28] What more could a new author ask for? Paton's cup of happiness overflowed.

To cope with these new demands on him, he engaged a literary agent, Annie Laurie Williams, on Scribner's recommendation. Annie Laurie Williams, who in spite of her musical name was an elderly and diminutive Texan, closed with Korda for an advance of £1,000 and 2 per cent of gross receipts on the film, a deal which Paton later came to think a very bad one. 'At that time the reputation of the book was so high that Annie Laurie could have sold it outright to Korda for a sum greatly exceeding one thousand pounds. The deal with Maxwell Anderson was much more successful'.[29] Even £1,000, of course, was more than a year's salary to Paton at this time. This large payment was rapidly followed by a second: in April he received the first royalty cheque from Scribner for £1,239. These sums, handsome though they seemed to Paton, were to prove just the first drops of the miraculous stream that would flow without ceasing for the rest of his life. By the time he died in 1988 *Cry, the Beloved Country* would have sold over 15 million copies in twenty languages.[30]

It brought about a remarkable change in Paton's fortunes. There exists a very incomplete list, in Dorrie Paton's handwriting, of the receipts from Paton's writing after 1947. In 1948 she recorded that the receipts from *Cry, the Beloved Country* amounted to £1,348, but this figure certainly excluded the film rights payment. In 1949 she recorded earnings totalling £4,343. In 1950 the total recorded income was £11,815, more than 15 times Paton's salary at Diepkloof three years before. Thereafter the figure appears to drop sharply, to £2,898 in 1951 and £1,809 in 1952, but Dorrie was not a methodical book-keeper, and her interest in keeping the accounts seems to have tailed off after 1950. One of the more complete years hereafter is 1955, in which she recorded an income totalling £8,850; another is 1957, in which she recorded £9,576. It is probably safe to say that Paton's income from *Cry, the Beloved Country* did not fall below £3,000 a year after 1949, and even after the war this was a very comfortable income indeed in South Africa.[31] In many years in the mid-1950s his income approached £10,000.

To put this in context, the Patons' expenses for 1949, when they had begun to live up to their new-found wealth, were to average £98 a month, or £1,176 for the year, a sum which included such items as Jonathan's school fees, David's university fees, servants' wages, insurance, and the costs of holidays.[32] 'We were rich, very, very rich,' Paton wrote happily.[33] He gave at least a tenth of his income away annually, to Toc H, his local church, the Institute of Race Relations, and the Christian Council, as well as to many individuals, mostly Africans.[34] This habit, probably begun as a young man, he was to continue all his life, though the recipients would change; among his papers are many letters from people who asked him for money, and, if they were evidently in need, they got it.

Annie Laurie Williams sold British publication rights in *Cry, the Beloved Country* to Jonathan Cape, whose South-African-born reader, William Plomer, one of the outstanding publisher's readers of the period and an important literary figure in his own right, was immediately enthusiastic about it.[35] Cape planned to publish only in 1949 because of the post-war paper shortage; in the event the novel's huge success in America made Cape hurry it out in September 1948. Even before it appeared, the South African press was paying attention to the rave reviews in America, and in March 1948 a photographer arrived from the liberal magazine *Forum* to photograph Paton, dressed in a tweed jacket and intently reading one of their magazines, surrounded by his adoring family: Dorrie, rather plain and thin, David, a burly 17-year-old, and, hanging on to his father from behind, 12-year-old Jonathan in the blazer of St John's College preparatory school.

As soon as he was assured that his book was a financial success, Paton sent in his resignation to the Department on 31 March, with effect from 30 June 1948. He had wanted to resign as of 31 March, but Dorrie had nervously resisted the move, and in deference to her he postponed matters.[36] With the American reviews, and the fan mail that now began to pour in on Diepkloof from America (as it would continue to do for the rest of his life), his self-esteem rose steadily. Among the early letters that came in were appreciations from Laurence Housman (novelist, critic, and brother of the poet), Sybil Thorndike, who had read Paton's unpublished *Louis Botha* in 1933, and Pauline Smith, the South African short story writer, whose work Paton much admired.[37] 'Recognition is pleasant,' he wrote to Hofmeyr, 'but I am also profoundly grateful that the talent that I buried in the ground, out of a sense of duty I believe, has, as it did not in the parable [Matthew 25], grown while in abeyance. It is hard to say to a man, I have a talent buried in the ground; he doesn't really believe you.'[38] This was, of course, a veiled reproof of Hofmeyr, who for so long had ignored Paton's offers to use his talents at Hofmeyr's side.

And Paton now gave Hofmeyr a clear signal that he was no longer dependent on him:

I am also deeply grateful that my sense of individual significance, which had been waning as I told you, has returned to me. I feel that I am going to do something for South Africa after all. You probably never knew it, but for a long time I was very dependent, too dependent, on you. My weaning commenced some years ago, but it was an unhealthy process, & contained some elements that were unworthy. This success in my own right has removed these elements . . .[39]

And in a subsequent letter he added, to make it plain that no promotion could now keep him in the civil service,

It is a strange thought that I, who would have been transported to the seventh heaven by some success in my old direction, have suddenly found that no offer of elevation would interest me now.[40]

His friendship with Hofmeyr would continue, but it was no longer that of subordinate appealing to superior, in any sense.

And as Paton's star rose, Hofmeyr's fell. On 26 May 1948 the General Election took place, and it demonstrated that the successful prosecution of war no more guaranteed Smuts a hold on power than it had guaranteed Churchill. Smuts was swept to defeat, the Nationalists and their allies under Malan winning 79 seats to Smuts's United Party's 71. Smuts himself lost his seat of Standerton, which he had held for twenty-five years; the wreckage seemed complete. Malan took over as Prime Minister, and Hofmeyr, exhausted and depressed, became part of an Opposition which was destined never again to exercise power. When Paton went to commiserate with Hofmeyr after his defeat, his former mentor said gloomily, 'There's no hope for this country.'[41]

The long slow rise of Afrikaner Nationalism had at last achieved that success of which Smuts and the war had deprived it for so long. From this point on the National Party would increase its majority in election after election, with one small setback, for the next forty years, until the rise of another political group, predominantly black this time and even stronger, would begin to shoulder it from power. What Paton had dreaded and fought against so quietly and stoutly had come to pass: the fragmentation of his beloved country. As he wrote years later, looking back ruefully on the ruin of his hopes,

Malan came to power pledged to separate the different races of South Africa in every conceivable sphere of life, in schools, universities, trains, buses, hospitals, sports fields and sporting events, and above all, residential areas. Marriage between whites and persons not white would be forbidden, as well as sexual relations between whites and others, outside marriage. And at last the Cape franchise . . . would be abolished. There would be only one franchise for the whole country, and it would be for white people only.

. . . After the election Nationalist Afrikanerdom was filled with a jubilation that knew no bounds, while people like myself were struck dumb by this change in our fortunes.[42]

It was certain that the changes to come would rapidly affect Diepkloof, undoing the greater part of what Paton had achieved.

But he was preparing to leave, with a mixture of sadness and relief. He entertained his staff, black and white, in a series of tea-parties at his own house, and was quite worn out by the effort, as he told Hofmeyr.[43] He also was given an elaborate and affectionate farewell by his parish church of All Saints in Booysens, and Toc H gave a dinner in his honour.[44] He agreed to continue to serve as the Toc H Vice-Chairman for Southern Africa for a further three years.[45]

On 30 June 1948 he completed his last day at Diepkloof, being sent off by his staff and pupils with a grand parade followed by a tea. As he sat at his office table on that last day, shortly before the parade, he experienced one of those strange fits of sudden emotion that had overtaken him when he heard of the death of his brother Atholl, whom he had not loved. Now, though he had often longed for release from Diepkloof, and knew he was going just in time to avoid seeing the destruction of all he had worked for, he burst into a fit of desperate, uncontrollable sobbing at his table, and cried so loudly that his chief supervisor, Koos Verwey, came and shut the door so that he and the clerks would not have to hear.[46] Paton, deeply emotional and unembarrassed by his emotions, would not have thought of shutting the door himself.

He might more appropriately have wept at the tea that followed the parade, for his Vice-Principal, J. H. Laas, had made it a completely segregated affair, with the black guests seated at separate tea tables a hundred yards away from the whites. Laas was a convinced Nationalist, and this was the first sign of what was to come. His party had won, and he was making it plain that from now on South Africa, and Diepkloof with it, were going to be different places. Paton, filled with anger, went over to join the blacks, and Ben Moloi said to him, 'Well, at least we know how things are going to be.'[47]

The black staff, headed by Moloi, presented Paton with a beautifully lettered and coloured scroll, signed by six of their number:

To Alan Stewart Paton

On the occasion of your departure from Johannesburg we, the African members of the staff of Diepkloof Reformatory, wish to express our deep appreciation of the great work which you have done during the thirteen years that you have been with us.

This work, and the reforms which you have wrought here have been the salvation of many thousands of African delinquents who have passed through this Institution, and will remain as a lasting monument to your memory.

We wish you every success in the future and sincerely hope that you will maintain connections with this Institution and keep alive the many friendships you have made here.[48]

Paton had this scroll framed, and it hung on the wall of his study all the rest of his life. If his white staff gave him a similar scroll, he chose not to preserve it.

This was a time of change and changing allegiances for Paton in more ways than one. As his fame grew, with it came an ebullient self-confidence that for a time did not brook criticism easily, and this led to a straining of relationships with some of his oldest friends. One of these was Railton Dent, whom Paton had had so much in mind as he wrote *Cry, the Beloved Country*; the younger Jarvis's idealism and his liberal values were those of Dent. Paton sent him a copy of *Cry, the Beloved Country* to read in May 1948, and Dent responded almost at once with a long letter that begins with two pages of generous praise, calling *Cry, the Beloved Country* a 'beautiful and moving book . . . one of the most important pieces of writing yet produced in this land'.[49] But he then went on to criticise the novel, as being too much a novel of ideas:

Perhaps my main critical reaction is that your book would have been a finer work of art had you refrained from attempting to show so many facets of our so-called Native Problem. It seems to me that you have tried to bring in something of everything . . .

My next criticism, if criticism it be, is that humour and touches of light-heartedness would have relieved the tension, and by contrast would have heightened the tragedy . . .

I like your direct translation of Zulu forms of speech. 'Stay well', 'Go well', 'In the side of the hand you eat with', and particularly effective, 'Now now'. But why oh why did you reject the beautiful Zulu 'Umkulunkulu' [God] for the ugly Xhosa 'Tixo'? 'Umkulunkulu' would have toned in so well with your simple and beautiful language, where as I found 'Tixo' always a discord.

A possible criticism is that you have tried to make your characters too faultless. Kumalo, Msimangu, Mrs Lithebe, both Jarvises and even the agricultural demonstrator are practically without fault. . . .

And that final mountain scene again, some critics may consider it very improbable. They may claim that spiritual experiences on mountain tops are sought only by people who understand the poetry and the cultivated solace of nature and so on. . . . But I have said more than enough. Mabel has read the book too, and we have put our opinions about it together in this letter. We agree on nearly all points.[50]

Paton was stung by this criticism; more than that, he was shocked and deeply hurt by it. 'This book was myself,' he wrote to Reg Pearse, adding, 'The book was, in a sense, written for them [Dent and his other friends]; they were often in my mind while I wrote it. In a sense, therefore, criticism of the book may have seemed like a rejection of what I had become.'[51] His response to Dent was delayed so long that Dent wrote again to ask if his first letter had been received. Paton reluctantly replied,

I have already written some seven or eight letters to you, some even closed and stamped. The relations between us are quite other than those between me & some reviewer I have never seen nor known. I am emotionally involved, & to write an objective reply is almost impossible. This letter however must be written, & I am now almost recovered from the emotional shock administered by your intellectual and analytical response to a work which I thought would elicit from you of all people an emotional one. I suppose I thought, this is Joe's book in a way, his child, his cause; in fact, as you knew, you were much in my mind when I was writing it. All this is quite irrelevant critically, but very relevant emotionally.[52]

And he then, rather emotionally, defended himself against the charge that he had produced a novel of ideas rather than a work of 'pure art':

I must say also that I had in this book no desire to produce art which would have no social reference. But I would judge myself to fail if that were all it had . . . I do not know what I shall do next; but it may well be another book with a purpose. I would rather write a book that would help our cause than write all the plays of Shakespeare, as I feel now.[53]

If he had stopped there all might still have been well, but he did not:

And perhaps you will forgive a parting shot. Mabel's agreement on 'nearly all points' I take to be evidence of marital bliss rather than of critical confirmation.[54]

This remark seems to have cut Dent to the quick; he replied with real hurt and anger, and the two old friends drew away from each other and were never as close again, though they maintained contact through mutual friends such as Reg Pearse.

The truth was that Paton, who was being praised to the skies by every noted American critic (and the paean was soon to be swelled by the British and South African press, when the Jonathan Cape edition of *Cry, the Beloved Country* appeared on 27 September 1948), neither wanted nor appreciated adverse criticism from those he had hoped would most appreciate his success and the blow he had struck for the liberal political cause in South Africa. In his excited letters he boasted, quite understandably, about his reviews:

> *The Star* called it a 'great novel', & the *Rand Daily Mail* one of the most significant documents to come out of Africa. The *South Coast Herald*, our local [Anerley-Southport] paper . . . says that 'a good book is the precious life-blood of a master-spirit.' (Ahem!) *Femina* calls it a 'wonderful book'. *Milady* calls it a 'powerful story'. *The Sunday Tribune* says it is 'a milestone in literature'.
>
> The English reviews I thought much more restrained than the American, but the latest ones made me change my opinion. Punch—& isn't that exciting?—says 'it is very difficult to describe the book's great quality'. *The Daily Herald* says 'better than *Uncle Tom's Cabin*'. Francis Brett Young calls it 'a work of art' & the *Daily Mirror* calls it flatly 'a great book'. The *Observer* says that to read it is an 'absorbing & deeply moving experience'. *The Friend* says it is 'enthralling'. And the plum I keep to the last; the *Manchester Evening News* says 'Genius tells story of negro'. Isn't that nice now?[55]

'Did all this excitement go to our heads?' he was to ask years later and to answer with his characteristic honesty, 'To some extent, undoubtedly.'[56] It would have been astonishing had it not.

The world seemed open to him, and he played at this time with many plans. The novelist Sarah Gertrude Millin, friend of Hofmeyr, wrote to him saying that the tax he would have to pay on books published abroad would amount to nearly 100 per cent, and until he discovered that she was wrong he seriously considered living abroad, probably in America. 'This is a year of destiny for us, & one can't say yet how it will end. It may even be that if I am to write, we shall have to leave South Africa,' he told Railton Dent.[57]

In the event he did nothing so drastic. Instead he took a long holiday with his family in the Kruger Park, the magnificent eastern Transvaal game reserve a little larger than Wales. He had visited the park several times before, and anticipated every detail of the trip with delight. The long drive from Johannesburg to the eastern Transvaal over the great rolling hills of the highveld, and then the precipitous drop down that magnificent, mountainous escarpment to the bushveld below; the rapid increase in temperature, and the subtropical plants everywhere; then the arrival at the great gates of the park, and the stern injunctions from the rangers not to get out of the car on pain of being eaten by lions; and at last the entry into the strange world of dense bush, swarming with every kind of big game, a world which even in 1946 had long disappeared in the rest of South Africa, and which in the 1990s exists nowhere in Africa in its pristine state. For Paton the particular joy was not the big game, but the birds about which he already knew so much, and which could be observed here in bewildering variety. This was the Kruger Park which he had come to love, and which he would revisit annually most years for the rest of his life.

He finished this holiday with a slow journey down to Natal (through Swaziland and Zululand)[58] to revisit his old haunts, and to look for a house he might rent while he decided what he was going to do. He found one in the village near where the SCA camps were held, Anerley, and rented it, initially for a few months, at £15. 15*s*. a month.[59] He also revisited his old friends in the world of education, and was flattered to be offered the job of Director of Education for Natal. Ten years ago, even five, he would have been delighted with the thought that he might rise to such heights; now he turned the offer down without hesitation. 'Was I going to sacrifice my future Bohemian career for the sake of sitting in a big office . . . ? Was I going to sacrifice the excitement of London and New York for the sober delights of Pietermaritzburg?'[60] He was not. Nor did he ever regret his decision.

Having rented the house in Anerley, he drove back to Johannesburg to pack up, and to return his children to St John's, which they had attended for years, Jonathan as a day boy at the preparatory school and David as a boarder at the College itself.[61] The school was and is one of the most expensive in South Africa, and Paton had only been able to send them there because the school had waived part of the fees for him. Now both his sons became boarders at full fees. Paton's hope was the school would make his sons good Christians. With David this hope came to nothing, and for his part Jonathan bitterly resented this enforced parting from his mother. He wrote to her almost daily, and she to him, and he was mercilessly teased for this flow of maternal missives. He begged to be taken away and allowed to attend the Port Shepstone school near Anerley, but he begged in vain.[62] There can be little doubt that Paton intended that his married life also would be transformed by the success of *Cry, the Beloved Country*, and getting his younger son away from Dorrie was a high priority. Nor would he ever believe how much Jonathan hated St John's. In old age Paton was to write that Jonathan had enjoyed the school,[63] and for all Jonathan's angry protests that that was quite untrue, he refused to change it.[64]

The Patons moved into the rented cottage at Anerley at the beginning of August 1948; they would find a second, more suitable house nearby after a year. Paton now settled down to the life of a writer of independent means. In part he was taking an extended holiday, and after his exertions of the past thirteen years no one could have begrudged him that. But in part he was hoping to give free rein to that creative spirit in himself that he had been unable to release while he was tied to Diepkloof. While he had been the Principal there he had replied to his friends' question on why he did not write by saying, 'You cannot do work like this and write as well, for both draw on the same source of creative energy.'[65] He saw his life, as he had once explained it to Hofmeyr, in terms of a struggle for dominance of the artist and the man of affairs, and he planned to give the artist his chance.[66]

He had very little idea of what he was to do next, and was content to sit and wait for the next book to come to him as *Cry, the Beloved Country* had done. 'I approach this whole matter of a second book with much hope, but a quite incredible ignorance of how & when & where it will come,' he wrote to Maxwell Anderson in New York, with whom he was now corresponding about the proposed dramatic version of *Cry, the Beloved Country*. 'My mind is full of ideas but they don't add up to anything. I don't at this moment at least, intend to produce a second 'Cry,' but would like to be released from the necessity to propound anything . . . I never intended to inflict on you the thoughts of the novice who fell in & found he could swim, & now stands on the brink of a deliberate plunge. But it is exciting, is it not?'[67]

The same note of excitement mingled with apprehension appears in his first letter to a young woman who had written him a fan-letter from England, Mary Benson, known to her friends as Pixie. Having compared *Cry, the Beloved Country* with *War and Peace*, she asked him about his other books, and he replied, 'This is my only book. The life-work mentioned above [Diepkloof] took all my energy before. But I am leaving it to write another, not without fear. For the first one happened in spite of me, & here I am setting out to write one deliberately.'[68]

Cry, the Beloved Country continued its extraordinary sales. Following its publication by Jonathan Cape in England in September 1948 it became a Book Society Choice there, and almost simultaneously it was made a co-selection of the American Book of the Month Club, a selection which immediately sold 40,000 copies. Within six months of first publication it had sold 100,000 copies in the United States, and the acceleration continued. *Omnibook Magazine* published it in abridged form, selling 250,000 copies, and the *Reader's Digest* published an abridgment of 200,000 copies.[69] Translations of the complete novel rapidly followed, and within two years there were editions in French, German, Dutch, Norwegian, Swedish, Danish, Finnish, Italian, and Zulu.[70]

'My wife and I have no desire to own anything that we do not own now,' Paton said in a radio interview broadcast in South Africa late in 1948, 'except perhaps a farm on which we could help the Diepkloof Reformatory to find a place for those pupils who by reason of temperament or physical defect are quite unable to make any kind of way in the world.'[71] But the more distant Diepkloof grew from him, the less inclined he was to set up a private Diepkloof of his own in this way. In the end he was to take on only one Diepkloof boy as a gardener. This was Sponono, the boy Paton was to write of in the moving story of that title.

His correspondence with Maxwell Anderson made him very keen to go to New York for the Broadway opening of the Anderson–Kurt Weill adaptation of his novel, for which the title *Lost in the Stars* had been selected. Paton sent Anderson lists of Zulu names for possible use in the adaptation,[72] and he also offered to send Weill examples of African music, particularly the beautiful hymn 'Nkosi Sikelel' iAfrika',[73] which had already become an unofficial anthem of the African National Congress in South Africa. And he gave Anderson further information about the elder Jarvis which shows that he had a clear idea about the characters in *Cry, the Beloved Country*, and could supply details that go well beyond the confines of the book:

> You want to make Jarvis less sympathetic to the African native and his son's efforts. He was originally of course never very sympathetic; we have thousands of such men in this country. They are not violent or abusive or cruel. They are upright and honest. They do not see the black man as a person at all. If suffering is put under their noses they will do much to alleviate it but they will not think about it when it is not there. Jarvis I think was like that . . .
>
> There is a second point. Jarvis's relationship with Kumalo was stumbling and awkward. You are quite right in supposing that he came to feel guilt and make reparation but if he has to talk to Kumalo about these things it must always be stumblingly and awkwardly.[74]

Paton knew what he was talking about, for in spite of his intention to cultivate black friends, he had made very few himself. At this stage it is probable that he could only have named Ben Moloi as a friend, and his relationship with Moloi had inevitably been that of Principal with subordinate. Though he tried to cultivate black friends in Anerley, he did not find the task easy, partly because of the awkwardness he speaks of, which resulted from the gulfs of language, culture, education, and income that separated white from black at this stage, but also because there were very few people of any kind whom he could easily befriend at Anerley.

His days there were on the surface idyllic, and that is the word he used to describe them:

After an early-morning cup of tea we would walk to the Southport beach, where we could swim in the Indian Ocean or in the Southport pool. Back to Anerley through a field of flowers, to the local shop to buy the newspaper, and home to the cottage for breakfast [prepared by the black maid they had engaged]. After breakfast we would go to the post office to collect the mail, the size of which brought me local fame. Then home to deal with the requests, the beggings, the invitations, the praises, the condemnations . . .

After lunch we would have a siesta, enjoy afternoon tea at four o'clock, and then take the dogs for a walk through Anerley-Southport. After sundown, drinks and dinner with some wine, we would read, and go early to bed.[75]

To this lotus-eating life the magnificent climate and the subtropical vegetation lent themselves. The village streets, unpaved then, were lined with orange and scarlet milkwood, jacaranda, and other flowering trees, which put on an apparently endless pyrotechnic display; even in August, when the highveld grass at Diepkloof would have been bleached and crackling with frost, the Natal air was silky in its mildness, and the rolling hills inland green as an English spring.

The idleness of his days here meant that he welcomed the arrival of visitors, particularly journalists and photographers, as an excuse for taking several days off to show them around. One of these was Rebecca Reyher, an American journalist, whose women readers were chiefly interested in Dorrie. Of her Rebecca Reyher was to write,

She makes no effort to shine or impress; she welcomes you, draws you in, but remains rooted in the land, despite the world acclaim of her husband. She is deeply religious . . . The house in which they are living is small and shabby, compared to the big house in Ixopo in which Doree [sic] Francis grew up. It was here I spent my day with Mrs Paton . . .

The Patons have a large dog, part alsatian, named Timmo. Every morning Doree has a bathe in a pool among the jagged rocks that rise from the silver-sand beach. She walks Timmo then back to the house, perhaps five minutes away. She returns to the beach for a walk as the late afternoon sun sinks.[76]

Another visitor was a brilliant photographer, Constance Stuart, who was to become widely known in America as Constance Stuart Larrabee. She drove down to Anerley to photograph Paton for *Harper's Bazaar*, and he, finding her petite and sympathetic, took to her at once and spent two days driving her up to Ixopo to photograph many of the scenes which had found their way into *Cry, the Beloved Country*: the rolling hills, the little African chapel at Nokweja (Ndotsheni), Carisbrooke siding, the small train which puffs out of the valley there, and so on. She also photographed the boy Sponono, naked to the waist and polishing a pair of Paton's boots, and the large black cook about whom he would write that fine poem 'A Discardment'. Sponono was named as 'Spinono' in Dorrie Paton's accounts, which also record that he was paid a shilling and twopence a day, and the cook £3 a month, with board and lodging free.[77] Years later Constance Larrabee was to recall,

> Alan took me to Ixopo, and Carisbrooke, and among the mountains; and we went to the little church there [Nokweja], and I photographed the minister, who was so like the minister he describes in the book. The minister knew him, and lots of the people too; I'm sure he modelled the characters on life . . . He drove me all about showing me all the places mentioned in the book, and I took photos of them and had this complete record.[78]

Neither she nor Paton could know that towards the end of his life they were to renew in America the friendship begun on this trip.

Being photographed and interviewed was pleasant, but Paton was soon troubled to find that he was writing very little. 'I haven't a single idea about a book. At present freedom has gone to my head. I do almost nothing.'[79] 'Almost nothing' was not nothing, for his expectations were high. He was slowly returning to the writing of poetry, after a period when his only poem had been the sonnet he wrote in Sweden. Now, in August 1948, he produced a steady stream of poems, of which the first, 'Only the Child is No More', was the result of his memory of the sea-side holidays on which his father had taken him at rare intervals:

> The sea roars as ever it did
> The great green walls travel landwards
> Rearing up with magnificence
> Their wind-blown manes.
>
> His wonderment I recapture here
> I remember his eyes shining
> I remember his ears hearing
> Unbelievable music.

I hear it now, but the high notes
Of excitement are gone
I hear now deeper
More sorrowful notes.

All is the same as ever it was
The river, the reed lagoon
The white birds, the rocks on the shore
Only the child is no more.[80]

This comparison of what he had been with what he was now, a theme which is common to poets from Wordsworth to Larkin, reappeared in another poem written a few days later, on 8 August 1948, 'To Walt Whitman'. In this he refers to his fears that his power to write, what he calls

. . . the great living host of tumbling words
Was a delusion, a brood of children
Locked within a womb that ne'er would open.

But these fears had proved groundless when *Cry, the Beloved Country* came to birth, and now he is ready to give tongue.

So now, great brother, I am ready to sing now.
These songs shall be presented to you
And what does it matter if they are unworthy,
If they are not so gay and sorrowful as I thought . . . ?[81]

This poem shows the fear Paton had shown in his letters to Anderson and Mary Benson, the fear that the miracle of *Cry, the Beloved Country* might be beyond his powers to repeat.

But he had a fall-back position prepared: if it should prove that he could not make a career as a writer, if the sales of *Cry, the Beloved Country* dried up and he proved unable to produce a successful second novel, he intended to go into politics, as he had intended to for many years. As an internationally known writer he would have little difficulty finding a winnable seat for the United Party, and once in Parliament he was confident that his name and his powers of oratory, his energy and vision, would take him to the top. He was determined to fight the Nationalist Party's rapidly emerging plans to segregate the entire country and confine blacks to positions of permanent powerlessness. One of the first poems he wrote at Anerley was a bitterly ironic attack on the emerging policy of apartheid, and the fact that he wrote it on 11 August 1948, when few South Africans outside the inner circles of the Nationalist Party could see the implications of the policy clearly, shows how much understanding of the mind of Verwoerd and his ilk Paton had. The poem is entitled 'We Mean Nothing Evil Towards You':

> Black man, we are going to shut you off
> We are going to set you apart, now and forever.
> We mean nothing evil towards you
> You shall have your own place, your own institutions.
> Your tribal customs shall flourish unhindered
> You shall lie all day long in the sun if you wish it
> All the things that civilisation has stolen
> Shall be restored . . .

Then irony gives way to most biting sarcasm:

> A fresh new wind shall blow through your territory
> Under your hands freed from our commandment
> You shall build what shall astonish you.
> The ravished land shall take on virginity
> The rocks and the shales of the desolate country
> Shall acquire the fertility of the fruitful earth.
> Chance-gotten children shall return to the womb . . .

And having pointed out the ludicrousness of Malan's and Verwoerd's dreams of turning back the clock and sending the urbanized African back to the 'homeland' and the tribe, Paton has his deluded narrator ask the black man for something in return for his largesse:

> Can you not make a magic that will silence conscience,
> Put peace behind the frowning vigilant eyes,
> That will regardless of Space and Time
> Wipe you from the face of the earth?
> But without pain . . .
> For we mean nothing evil towards you.
> Our resolve is immutable, our hands tremble
> Only with the greatness of our resolution.
> We are going to set you apart, now and forever,
> We mean nothing evil towards you.[82]

Paton caught brilliantly, in this poem, the queer mixture of idealism, blindness, and cruelty that lay at the heart of the apartheid doctrines. His intense study over many years of the minds of men like Verwoerd had not been in vain.

The steady stream of poems continued through these early months at Anerley. On 17 August 1948 he wrote 'The Laughing Girls',[83] another exploration of what one might call the apartheid state of mind, and 'Could you not write otherwise?', a reply to criticism of his book;[84] on September 14, 'Black Woman Teacher', a memory of Mrs Takalani, whom he had met in company with Mrs Rheinhallt Jones;[85] on September 19, 'Maria Lee', a poem about a woman hanged for murder, who had gone to the gallows singing hymns. This poem he chose not to publish subsequently. On October 8 he wrote a superb and moving poem, 'A Discardment', in which the gift of an old garment to the woman who cooked for him and Dorrie came to symbolize the whole tragic impact of western culture on tribal life.[86]

We gave her a discardment
A trifle, a thing no longer to be worn,
Its purpose served, its life done.
She put it on with exclamations
Her eyes shone, she called and cried,
The great bulk of her pirouetted
She danced and mimed, sang snatches of a song.
She called out blessings in her native tongue
Called to her fellow-servants
To strangers and to passers-by
To all the continent of Africa
To see this wonder, to participate
In this intolerable joy.

And so for nothing
Is purchased loyalty and trust
And the unquestioning obedience
Of the earth's most rare simplicity
So for nothing
The destruction of a world.

Of this poem Paton was to write to an inquirer, many years later,

This woman came from the hinterland because of economic necessity and had to leave her children at home. I remember she used to tell me of the great difficulties of bringing up her children, and she ascribed their failures to naughtiness, whereas in fact it was the absence of the mother that led to their difficulties, and so her family world had begun to fall to pieces. My wife had just made the observation that the destruction of a world may also mean the destruction of her pride in herself. We live as you know in a parasitical society which affects deeply both whites and blacks, though in different ways.[87]

And the extraordinary stream continued. On 10 October 1948, 'Indian Woman';[88] on October 11, 'My Sense of Humour';[89] on November 4,

'Dancing Boy';[90] on November 18, 'The Chief', a trivial piece about sitting on the toilet, which he was right never to publish; on November 20, 'The Stock Exchange';[91] and on November 21, 'Singer of Childhood', a poem about the rainbird, a variation on the theme of 'Only the Child is No More'. In January 1949 there followed 'Anxiety Song of an Englishman';[92] and on 3 February 1949 'Sanna', a fine protest against the hypocrisy of the Nationalist government's laws against miscegenation:

> The village lies in Sabbath heat
> The dog lies in the sun
> But stern and strict the elders go
> They pass me one by one . . .
>
> And stern and strict the sabbath clothes
> And stern the eyes above
> And stern and strict the elders go
> To hear the words of love.
>
> And Sanna follows all demure
> And plays her little part
> The child of love moves in her womb
> And terror in her heart.[93]

In March 1949 he wrote 'To a Small Boy who Died at Diepkloof Reformatory',[94] and, probably in the same month, 'In the Umtwalumi Valley'.[95] Most of these poems were to remain unpublished in his desk drawer for many years, though at the time he had rapid publication in mind: 'I've written 20 pieces of verse, & hope to publish them soon,' he wrote to Reg Pearse. 'They are not poetry, but deal with the problems of our country. Something gets itself said, & I have to be satisfied with that.'[96]

But he was not satisfied with his writing, though he worked tremendously hard at it; what he wanted was a new novel. He was also reading hard ('One must study novels now,' he told his fan Mary Benson).[97] In the course of two weeks in August 1948, for instance, he read Whitman, whose poetry he loved, and current fiction from Britain and America: Graham Greene's *The Heart of the Matter*, which he thought 'a sad, hopeless, magnificent book'; Ross Lockridge's *Raintree County* (noting gloomily that Lockridge, unable to follow up this best seller, had gassed himself),[98] and other very recently published books: Nigel Balchin's *Borgia Testament*, Norman Mailer's *The Naked and the Dead*, and Thornton Wilder's *The Ides of March*. He also burrowed through four volumes of Ford Madox Ford.[99] He struggled with Proust, in vain, and read Eliot with renewed delight. With the advice of the Afrikaans writer Uys Krige, whom he seems to have met at this time, he began an intensive study of Afrikaans literature.[100] He was in fact catching up with a good deal of the reading he had missed during the thirteen Diepkloof years, and from now until the end of his life he would be a voracious and omnivorous reader. In particular his knowledge of the literature of South Africa and other countries of the British Commonwealth became encyclopaedic.

But literature was not the only thing on his mind. On 4 December 1948 Paton opened his paper at the breakfast table in Anerley to read that his friend J. H. Hofmeyr had died of a heart attack the day before. He had known that Hofmeyr's health was not good, but the shock of this news was profound. 'For days I could think of nothing else,' he wrote.[101] It was not just the death of a friend that so moved him, but the fact that Hofmeyr had been the leader of the liberal wing of the United Party, and he had no obvious successor. Just a few weeks before Paton had written an article for a British paper, 'From Van Riebeeck to Hofmeyr', subtitled 'The concise history of South Africa, written for a stranger', in which he had argued that the difference between the policies of the Nationalist and United Parties was small: 'It might be said that the Nationalist programme is "We will ensure survival, but we shall try to be just", and the United Party programme is "We shall try to be just, but we will ensure survival".'[102] And this small difference was largely the work of Hofmeyr in the United Party, Paton believed. With Hofmeyr gone, he feared that over time even that small difference would disappear. His genuine grief at the loss of Hofmeyr, and the cause which he identified (only partly accurately) with him, emerges clearly in the poem he wrote for Hofmeyr on 4 December 1948, the day he got the news. It seems to draw inspiration from Whitman's famous elegy on the death of Lincoln, with its reference to the 'tolling, tolling bells' perpetual clang'.[103] Paton's lines run:

Toll iron bell toll extolling bell
The toll is taken from the brave and the broken
Consoling bell toll
But toll the brave soul
Where no brave words are spoken.

Strike iron bell strike ironic bell
Strike the bright name
From the dark scrolls
Of the blind nation.
Strike sorrow strike shame
Into the blind souls . . .

His grief at Hofmeyr's going also had a longer-term consequence for his writing: less than a month after his friend's death Paton was writing to Mary Benson, 'I'd like to do a Life of Hofmeyr, but that will be a long job, & I shall have to debate with myself whether I can give so much of what is left to me to such a task.'[104] Even during Hofmeyr's lifetime Paton had half-jokingly told him on several occasions that he intended to write his life. Now, within a few years, he would embark on the huge task, the first of his biographies.

But Hofmeyr's death laid yet another burden on him, though he was to evade it for some years more. It was clear to him now that the Liberal cause in South Africa, by which he meant the cause of a common society, open to South Africans of all races, could not longer look to the United Party as its champion. Yet what party, other than the Communists, whom Paton intensely distrusted, would champion non-racialism? And if it were to be a new party, who would be its leader? He had often hinted to Hofmeyr that he would like to enter politics by his side; now he began to sense a call to enter politics in his place.

The death of Hofmeyr was itself one element of that call; the rioting that broke out in Durban a month later, on 13 January 1949, was another. An assault by an Indian shopkeeper on a Zulu boy set off a wave of anti-Indian violence, crowds of Zulus hunting down and killing Indians wherever they could be found. Armed car-loads of Indians retaliated. The violence lasted nearly a week, and spread down the coast towards Anerley; by the time it was contained by the police and the army, 55 Indians and 87 Zulus had died, and more than a thousand people had been injured, many severely. About this time, Paton wrote an untitled poem which seems to explore some of the causes of this violence:

> The world is changing too fast for me.
> I remember the valley of the Umzimkulwana
> When I was a boy, how it was my kingdom
> Shared only with redwing sprews & the oribi
> And the iguana crashing away startled
> Through the undergrowth: but the undergrowth is gone
> And the huts of the Indians stand nakedly
> Where the oribi stood silent as stone
> Under my eyes; and the kingdom of childhood,
> The sacred inviolable places, the glades & the glories
> Are desecrated by the alien encroachments
> Of uncountable Indians, much more than I ever remember
> When I was a boy.[105]

Paton in later years chose not to publish this poem, probably because on reflection he came to believe it told only part of the truth about the violence between Indian and Zulu. Drawing on his deep experience at Diepkloof of the causes of violence and crime, he had no doubt that the root cause of the violence could be traced in the political, economic, and social structures of South Africa.[106] He knew well that matters would grow worse if there were no leadership urging the country to turn away from the road of apartheid. Yet who would provide that leadership now Hofmeyr was gone? On 1 February 1949 he wrote a poem entitled 'To Edgar Brookes',[107] in which he salutes the Liberal Senator (clearly thinking of him as inheriting Hofmeyr's mantle) and offers to stand by his side:

> . . . when you report to the lord
> Offer this pen with your sword.[108]

The call of South African politics, that dark and sibilant snake-pit, was growing louder in his ears.

15

BLEEDING ONTO THE PAGE 1948–1950

Writing is easy. You sit down with pen and paper, open a vein and just bleed onto the page.

ANONYMOUS

There is plenty of evidence that from 1948 to 1952 Paton was tormented by an inability to write fiction. The reasons are not hard to find. After years of desultory apprentice work, none of which he had thought worthy of publication, he had under extraordinary circumstances produced a hugely successful novel. He thought of *Cry, the Beloved Country* as the product of inspiration, as a gift from on high rather than the result of his own workmanship, and he had no idea how to work the miracle again. At the same time, the pressure on him to prove (if only to himself) that he was not a one-book man grew month by month as he waited for another marvellous novel to come to him, and nothing happened.

His publishers, and many of his readers, urged to him follow up the success of *Cry, the Beloved Country* with another book in the same style, or a sequel using the same characters; Paton refused. 'Something told me not to do it. Something told me that *Cry, the Beloved Country* was a book that should not be repeated or imitated.'[1] He also was to say, years later, that he did not want to write a novel before he went to England and America, a trip which was planned for August 1949. Yet this was eight months away, and he had written *Cry, the Beloved Country* in the evenings of just three months, and while travelling hard. The truth is he very much wanted to write another novel, but found himself barren of ideas. One of the poems he wrote at this time is entitled 'I have approached':

I have approached a moment of sterility
I shall not write any more awhile
For there is nothing more meretricious

Than to play with words.
Yet they are all there within me . . .
Therefore, words, stay where you are awhile
Till I am able to call you out,
Till I am able to call you with authentic voice
So that the great living host of you
Tumble out and form immediately
Into parties, commandos, and battalions . . .[2]

The battalions, however, appeared to be mutinous. He tried to write a novel: there exists a single page of one, written on the verso of a draft of 'Black Woman Teacher',[3] the poem he wrote on 14 September 1948. This single page, headed as if it were the beginning of a first chapter, is a section of dialogue between a fierce old patriarch named Jan van Vlaanderen, and his son just back from university in Stellenbosch. The name, and something of the situation, would be reused when Paton came to write his next novel, but of this attempt there survives only one page. He also sent one correspondent at this time 'an outline of a novel I have in my mind,' adding, 'You will observe that its tragedy [is] peculiarly South African; it is so in a far narrower sense that [sic] Kumalo's story was peculiarly South African.'[4] Unhappily the outline itself did not survive with the letter, but it and the single page of manuscript are likely to have been connected, and they show that as early as September 1948 he was trying and failing to write the book that would become *Too Late the Phalarope*.

By April 1949 he seems to have abandoned it, and gone off on another tack, perhaps suggested by his reading at this time of Carlo Levi's *Christ Stopped At Eboli*, which had just been published.[5] 'I've started on a novel about a visit by Christ to Johannesburg, but it hasn't gripped me yet. It might, I'm hoping,'[6] he told Mary Benson; but he hoped in vain, and in time abandoned this unpromising notion.

Nor was it just the novel form he was struggling with. On 5 November 1947 the American director Luther Greene had written to him, through his friend Aubrey Burns, asking him to write a play for Greene's wife Judith Anderson, the Australian-born dramatic actress whose reputation in the United States was at its zenith at this time. Paton agreed to produce a play for her: 'Would I like to do such a thing? Very much.'[7] Subsequently he tried hard, but found himself utterly barren of possible plots. He even appealed, only half-jokingly, to his admirer Mary Benson in London, who was working as the secretary of the British film director David Lean, and with whom his correspondence was by now regular and warm, to provide him with a plot, offering her a handsome fee. She tried, and failed to come up with anything,[8] and when she wrote rather gloomily about this, he replied, 'I'm depressed myself (about ever writing another word, of course).'[9]

He must have been relieved to get away from this constant unsatisfiable pressure to produce when, in August 1949, he set off for London, where he planned to meet Zoltan Korda to discuss the film of *Cry, the Beloved Country*, and for New York to attend the opening of the Maxwell Anderson–Kurt Weill adaptation of the book, entitled *Lost in the Stars.* 'SCRIPT AND MUSIC FINISHED,' Anderson and Weill had cabled in March 1949, 'GREAT ENCOURAGEMENT FROM PLAYWRIGHTS COMPANY. EXPECT START REHEARSALS EARLY SEPTEMBER. HOPE YOU CAN BE IN NEW YORK THEN.'[10]

He would have liked to have taken Dorrie with him to share his glory. Sitting on the south coast of Natal, 70 miles south of Durban, he had felt terribly isolated from the great literary world in which he was causing such a stir.[11] So far he had been able to share in the huge excitement generated by the success of *Cry, the Beloved Country* only through reviews and letters; now he was going to be fêted and lionized in the capitals of the world, and what more natural than that he should want his wife to be there too? But Dorrie refused to come with him. Instead, as Paton put it with the restraint he showed in discussing such matters after her death, 'she felt she should be near our twelve-year-old son Jonathan'.[12] The transformation he had hoped for in his married life had not come about, and he must have known by now that the marriage would never offer him part of what he had hoped for from it. Even before he left for London he left Dorrie, for he made a trip to the Cape alone in March 1949,[13] for reasons that remain unclear, and then attended one last SCA camp near Anerley for ten days in July 1949. He did not enjoy it, and never went to another.[14]

He flew to London with KLM (an airline he came to prefer to any other) on 30 August 1949, to work with Zoltan Korda on the film script of *Cry, the Beloved Country.* Zoltan Korda, handsome as a film star himself and bubbling with ideas, was much in awe of his brother Alexander, the head of London Films. The Kordas, Hungarian Jews of brilliance, were trying to do in London what Sam Goldwyn, Louis Mayer, and others had done in Hollywood, and for a time it seemed that they might succeed. Certainly Alexander Korda had grand ideas and the grand manner; he put Paton up in the luxurious Piccadilly Hotel, granted him a single interview in which he praised *Cry, the Beloved Country* and gave Paton the clear impression that it would be dangerous to cross him, and then dismissed him and Zoltan to get on with their work.[15]

It did not make rapid progress, partly because Zoltan Korda's ideas of what he wanted changed continually, partly because Paton had not the slightest notion of how to write a film script, and partly because Paton was thoroughly enjoying being back in London again, and Korda was happy to wine and dine him in expensive restaurants. On some of these evenings Korda would talk about his wife whom he had left behind in America, and on others he would console himself with the presence of a beautiful young Hungarian girl named Zuzhi who was an employee (in a vague capacity) of London Films.[16] And there was another source of distraction.

In the first volume of his autobiography, *Towards the Mountain*, Paton, after telling about his affair with Joan Montgomery, says that in later years he had a second love affair, but gives no details. It began in London, in 1949, after he went to have dinner with Mary Benson on 3 September. They had been corresponding since she sent him a fan letter about *Cry, the Beloved Country* in May 1948. Mary Benson was a South African twenty years younger than Paton. She had had a comfortable and conventional Pretoria upbringing ('the country club life', she was to call it)[17] and had as a teenager been determined to become a film star. Slim, fair-haired, and very pretty, she had travelled to Hollywood after finishing school, but like many another aspiring starlet had failed to break into films. She returned to South Africa on the outbreak of war in 1939, and after serving as a secretary to the British High Commissioner had joined the Women's Auxiliary Army Service and served in North Africa and Europe. After the war she had again pursued her desire to work in films, and had landed the job of secretary to David Lean in London; it wasn't acting, but it allowed her to move on the borders of the longed-for world.

It was from here that she had begun writing to Paton, having been deeply moved by *Cry, the Beloved Country*, which had opened her eyes to the injustices of South African society. She was eager to learn about the country and to get to know Africans. She turned to correspondence with Paton himself in the hope of discovering some purpose for her life, or giving herself some direction. She felt that she was drifting rudderless—she had not recovered from a wartime love affair which had ended unhappily and was finding the secretarial work for Lean unfulfilling. Paton was naturally flattered by this, and amused by her witty letters, and he responded by telling her of his own problems with writing, and his own fears of failure. Of all his many correspondents at this time, Mary Benson learned most about his intimate inner life.

He told her in May 1949 of his wish to join a community of some sort, perhaps a tuberculosis community for blacks; of his desire to create a trust for his children and put all his money into it, so that he and Dorrie could be 'poor & perhaps useful';[18] he even told her that a major reason for this strange ambition was his hope that identification with a community would allow him to write again: 'Of such base stuff are my highest and noblest intentions.'[19] Paton sent Mary Benson occasional poems he had just written, including one, 'To a Black Man who Lost a Child thro' Starvation' which survives only in her unique copy. She sent him news of the latest books in England, asked him to use his link with Maxwell Anderson to get her some sort of job in the theatre world in America, (Paton tried), and told him of the turmoil in her mind. She had become bored with the work for Lean and was planning to go to America. 'I want to have lived for some "purpose"!' she cried out to Paton, and he, who in certain moods longed for sanctity, knew exactly what she meant.[20]

Soon the two of them were joking like old friends. 'Do you really read *Ulysses*?' he asked her. 'You are cleverer than I thought.'[21] And at the end of one of his wittier letters (he had mentioned seeing a photograph of David Lean, 'burning with inner fire, although he just appears to be smoking') he told her, 'This is a feeble letter, isn't it? Well it matches your last two to a hair.'[22] Benson, whom Paton had begun to call by her nickname 'Pixie', sent him some of her verse, and then pretended to think his response to it too generous. He wrote in mock irritation, 'Glad you were *amused* by my reception of your verse; send me some more, I'll give you something to amuse you. (Of course it was doggerel, but I was quite right to call it verse. If it hadn't been so bad, I'd have asked who wrote it for you. Now send me more if you dare.)'[23] The correspondence had become a long-range flirtation which each of them enjoyed.

As the time grew closer for his arrival in London she wondered what he thought she looked like. 'You can't be ugly,' Paton replied, 'or you'd never dream of asking such a question, & that's a comfort!' And he admitted he was curious about her appearance, adding, 'As for mine, it is entirely undistinguished. Of medium height, too thin in the face, uninterested in clothes, not a brilliant talker, sometimes won't talk at all.'[24]

As soon as he arrived in London he phoned Benson at Pinewood Studios, and two days after his arrival he dined with her in a flat she had borrowed for the purpose in St John's Wood, thinking the Kensington attic where she rented a room too small for the entertainment of such a guest. She was very excited all day, fussing about the cooking and cleaning. Perhaps in consequence, she was disappointed in his appearance when he arrived. In her engaging memoir, *A Far Cry*, she gave a most vivid description of Paton in this, his forty-seventh year:

> He arrived with a signed copy of his novel and a box of chocolates, a spiky man, shorter and a good deal older than I. Yes, he was undistinguished, with glued-down hair and pale-blue eyes that peered at me over his spectacles, and yes, his off-the-peg double-breasted suit showed a certain failure in interest . . .[25]

But the disappointment wore off when Paton opened his mouth:

> When he began to talk he was immensely entertaining—his eyes sparkled, his lemon-sour mouth curled around witticisms that were engagingly accompanied by chuckles. We talked non-stop until he left at midnight.[26]

Subsequently Mary Benson was to say that Paton had not even kissed her goodbye on this occasion. But a few days later, when the woman from whom she rented a little attic room was away, Paton had stayed the night and they had become lovers.

This was the beginning of a strange, desultory affair that began as a physical relationship but soon became a warm friendship, continued over many years, whenever Paton and Mary Benson had the opportunity of a meeting. She was later to say that she presently realized that she was not in love with him, and on a subsequent occasion she rejected his advances, and then felt that her rejection had showed a failure of generosity. The relationship (to her at least) never became a serious one, and in later years she wrote the following description of it:

> While Alan was in London we met repeatedly, seeing films and visiting art galleries, mutually stimulated and amused and, briefly, becoming lovers. But I soon realized I was not in love with him, nor he with me, I'm sure. Looking back I think the success of his book, the movie and the musical, were tremendously exciting for him and I was part of the adventure. Clearly, his buttoned-up expression concealed a deeply passionate nature. But I think our relationship was a kind of reassuring—we felt very comfortable, after all we'd come to know each other's weaknesses and faults—reassuring each other that we could be sexually attractive.[27]

Paton struck her as never having had an opportunity of expressing his passionate self before, and there is little doubt that he was intensely grateful to this warm-hearted and beautiful woman. For him at least it was a month of intense happiness, never to be forgotten.

They talked about South Africa, and the political and moral struggles which Paton saw as inevitable there; they talked of those few white fighters for the black cause, and Paton mentioned Michael Scott, the English priest who had shown such courage in working for black rights. This was the first time Mary Benson had heard of Scott, for whom she would presently work, and with whom she would fall hopelessly in love. Yet Paton told her that unlike Scott, he himself was 'not political'; words were his province now, not actions. And instead he talked of his plans for a new novel: the Christ-comes-to-Johannesburg theme seemed to have lost its attraction, but he was increasingly inclined to write about 'a young white policeman obsessed by desire for a black woman'.[28] *Too Late the Phalarope* was by now definitely gestating.

And because Paton was at least fifty per cent a puritan, the month of joy was also a month of guilt. If he wrote letters to Dorrie at this time, they have not survived. What have survived are two poems he wrote during this month, 'The Prison House' and 'Samuel'. 'The Prison House' in particular shows the divided feelings of a man who has been living by the strictest rule for years, in prison metaphorically, and who has escaped it only to find that freedom no longer satisfies him. The images of the sexual wasteland of his marriage are drawn from his experiences at Diepkloof, and the still small voice is his keeper:

> I ran from the prison house but they captured me
> And he met me there at the door with a face of doom
> And motioned me to go to his private room
> And he took my rank from me, and gave me the hell
> Of his tongue, and ordered me to the runaway cell
> With the chains and the walls, and the long night days,[29] and the gloom.
>
> And once on leave that goes to the well-behaved
> I jumped in fright from the very brothel bed
> And through the midnight streets like a mad thing fled
> Sobbing with fear lest the door be closed on me
> And in silence he let me pass, he let me be,
> No word but your clothing's disarranged, he said . . .

Paton, like the vakasha boys of Diepkloof, was chained with fetters not made of iron:

> He can take the hide from my back, the sight from my eyes,
> The lust of my loins and the sounds of the earth from me,

Fruit's taste, and the scent of the flowers and the salt of the sea,
The thoughts of the mind, the words of music and fire
That comforted me, so long as he does not require
These chains that now are become as garments to me.[30]

The magical, troubled month in London came to an end when Paton had to leave by air for New York in October 1949: the rehearsals of *Lost in the Stars* were by now well advanced, and Maxwell Anderson and Kurt Weill wanted Paton's opinion of it before the first night. How he felt on parting from Mary Benson will never be known; for her part she accelerated the plans she had been making for some time to go to America herself.

In New York he found himself the centre of a whirl of publicity, parties, dinners, and rehearsals, dizzying and overwhelming to a man who had spent months in the seclusion of a small Natal coastal resort, and who before that had been a humble civil servant. On his first night he met Maxwell Anderson and his family, Kurt Weill and his wife Lottie Lenya, Rouben Mamoulian, who was the director of *Lost in the Stars*, and members of the cast, including Todd Duncan, the sophisticated singer who would be taking the part of Paton's simple priest Kumalo.[31]

Also on that first eventful night, he attended one of the last rehearsals of the musical, and found to his dismay that he hated it. The problem was partly that it was nothing like his book: it was played by people who had no idea of the simplicity or spirituality of Kumalo and his kind, and placed amid scenes constructed entirely without knowledge of South Africa or South African society. Paton's teeth were set on edge by scene after scene which rang hollow, and by dialogue which seemed to him entirely artificial or dead.

Worse still, both Anderson and Weill were agnostics at best, and in place of the deeply religious and hopeful theme of *Cry, the Beloved Country*, they had filled their musical with gloomy outcries against the desertion of man by God: to use the words of their title (which Paton hated as much as anything else) man had been abandoned in space, 'lost in the stars'. There is a scrupulous and very attractive honesty about Paton, in spite of his occasional lapses from it, and he loathed being put into situations where he had to compromise it. Now, at the end of the rehearsal, everyone crowded round him to hear him praise the show, and in particular Anderson wanted to know how he liked the script. 'This terrible evening', Paton was to call it later.[32] Another man would have lied his head off, quieting his conscience with the reflection that it was his duty to encourage the cast and retain his friendship with Anderson, but Paton could not. He did his best; he said what was true, that he had liked Weill's music, and that he had been deeply moved by the chorus 'Cry, the beloved country', which used his own words. But the script he hated from end to end, and the best he could do was to say nothing about it. The friendship between himself and Maxwell Anderson was nipped in the bud, as he later put it, and he crept to bed that night in the Hotel Dorset wishing that he were back in Anerley with Dorrie.[33]

Things improved. There followed a feverish round of dinners and speeches. There was a grand lunch given on 19 October 1949 by the *New York Herald Tribune* Women's Book Club, at which Paton and Alfred Bertram Guthrie (author of *The Big Sky*) made speeches. Guthrie in his address told several dirty jokes, to which Paton and the assembled women gave a frosty reception.[34] The time was coming when Paton would learn to tell a *risqué* joke himself, but it had not come yet, and never would he tell one to a woman not his wife.

In his own speech (as in an interview which he had given to the *Herald Tribune* a few days earlier) Paton gave the women a brief history of South Africa, in which he compared the race problems of his country with the racial problems of America. The Malan government, he said, had come to power because white South Africans were afraid of being engulfed by black South Africans. Americans, he suggested, were beginning to have the same fears, though on a much smaller scale. And both countries, he said, should fear and resist any attempt to counter the power of blacks with the power of whites. And he ended with a peroration that stands as a statement of his deepest beliefs, and a prophecy that was to be fulfilled by the years:

> I might conclude with an affirmation of my own belief, that the only power that can resist the power of fear is the power of love. It's a weak thing and a tender thing; men despise and deride it. But I look for the day when in South Africa we shall realize that the only lasting and worth-while solution of our grave and profound problems lies not in the use of power, but in that understanding and compassion without which human life is an intolerable bondage, condemning us all to an existence of violence and misery and fear.[35]

This speech, for all its moderation and morality, attracted the grave disapproval of the Nationalist government of South Africa. Malan's Nationalists did not want South Africans drawing attention in foreign cities to the problems of their country. The South African Information Office in New York monitored Paton's talk, and the response was rapid. The South African embassy in Washington telegraphed details of Paton's speech to Pretoria, and was instructed to bring Paton to heel. A few days after his speech, he received a visit from two men, Norton and Hahn, who advised him to be careful of what he said abroad. Norton was the acting director of the South African Government Information Office in New York; Hahn was probably his subordinate. On 7 November 1949 the ambassador reported to Pretoria, 'While I cannot say that we have succeeded in persuading Mr Paton to refrain completely from being critical of conditions in the Union [i.e. South Africa], I do believe that Messrs. Norton and Hahn succeeded in ensuring a more responsible approach by the author to Union problems.'[36] This was Paton's first experience of being closely watched by his own government and of its attitude to free speech: it was not to be his last.

More lionizing followed this sobering episode: a *Time-Life* lunch in the Rockefeller Center, a huge Press Club lunch to receive the annual 'Page One' award, and more contact with a government, this time a meeting with Pandit Nehru, the Indian Prime Minister, who invited him to another New York hotel to discuss with him the problem of Indians in Natal. The two of them discussed a topic which, if they had known of it, would have set Hahn, Norton, their ambassador, and for that matter prime minister Malan to frothing at the mouth: economic sanctions against South Africa, to compel her to treat her Indian citizens better. In particular they discussed oil sanctions, and Paton never forgot Nehru's remark, 'Mr Paton, oil companies are more powerful than governments.'[37]

This was heady stuff, and Paton longed again for Dorrie to share it with him. But now an opportunity arose for him to try once more to get her to come over. While the rehearsals were going on in New York, the London *Sunday Times* wrote that he had won its annual £1,000 book prize, shared with Winston Churchill (for *The Gathering Storm*) and Frederick Spencer Chapman (for *The Jungle is Neutral*). He decided to use his part of the prize to fly Dorrie to London, and she could then come on to New York to see the opening night of *Lost in the Stars*.

Why she agreed to do what she had previously refused to do is not clear, and it is possible that Paton told her about his affair with Mary Benson. The Patons' letters for this period, both sides, have been destroyed. At all events Dorrie acceded, and flew to London on 5 November 1949, Paton having arrived there the previous day. Shortly before he left New York for London, such is the complexity of human nature, he had resumed seeing Mary Benson, who had resigned her job with David Lean and flown over to New York. In his scrappy diary for this period, the unelaborated entry 'Pixie Benson' occurs on Sunday 2 October 1949, and again on Friday 7

October and 30 October 1949, though there were probably other meetings too.[38] In later years, however, Benson was to say the affair was over by this time.

When Dorrie arrived in London on 5 November Paton was uncharacteristically late at the airport in meeting her, but in his autobiography he records his pleasure at making love with her again in the flat Zoltan Korda had arranged for them in St James's.[39] He rejoiced in introducing her to the exciting new world in which he was now moving. The grand *Sunday Times* function was held at the Dorchester, and Paton made a speech afterwards about the art of writing. Then, or a little later, Dorrie said to Paton, 'In my family I was the lucky one',[40] and he, to whom she seldom said any appreciative thing, was moved.

They flew to New York together for the opening night of *Lost in the Stars*. Though Paton disliked it, he wanted the show to be a success. That morning, 30 October 1949, he cabled good wishes to Anderson and Weill. The cable to Weill read, I SEND YOU MY BEST WISHES FOR TONIGHT AND HOPE THAT YOUR GREAT MUSIC WILL SATISFY EVERYONE AS DEEPLY AS IT SATISFIES YOUR FRIEND ALAN PATON'.[41] His wishes were amply fulfilled: *Lost in the Stars* had an ecstatic reception. 'People wept and shouted and clapped,' Paton wrote.[42] The reviews were all the producers could have asked for. The *New York Times* set the tone:

> With the production of *Lost in the Stars* the kind of theatre we all respect and admire has been restored to Broadway . . . When *Lost in the Stars* gets down to its climactic scenes, concentrating all the terror and violence of the story in the spiritual tragedy of two broken old men, it has the power of an epic . . .[43]

And *Life* magazine followed suit:

> No matter how many pretty good shows have already opened, Broadway's season never really gets rolling until, soon or late, the first big hit comes along. Last month Broadway finally had a winner in *Lost in the Stars*, a musical play adapted from Alan Paton's distinguished novel. . . . *Lost in the Stars* is a credit to its creators because they have retained the spirit of a powerful and reverent book.[44]

Since this was exactly what the show did not do, in Paton's opinion, he could not whole-heartedly join in the rejoicing, though he tried. There was perhaps a personal reason why he felt more than ordinarily uneasy at the first night: not only was Dorrie present, but Mary Benson too.[45] They did not meet. *Lost in the Stars* went on to enjoy a run of over a year, made a lot of money for Anderson, Weill, and Paton, and has been several times revived in the years since, with renewed success.[46]

But Paton did not wish to stay for more. After a fortnight, on 14 November 1949, he and Dorrie left New York by air for San Francisco, where Paton introduced her to his good friends Aubrey and Marigold Burns, who since his previous visit had basked in the reflected glory of the book they had helped to bring to birth. Paton found the Burns children, particularly Christopher, now six, as annoying as he had three years before, and Dorrie presently gave the boy what Paton called 'a sharp crack', whereupon his parents sent him off to stay with friends until the Patons had come to the end of their visit.[47]

It lasted only four days, which they spent sightseeing in and around San Francisco, before Dorrie flew back to South Africa (via New York and London) to join Jonathan for his school holidays, having been with Paton for just six weeks. On 19 November 1949 Paton left the Burns's house too, but not to return to Africa. His intention was to find somewhere quiet to write. There can be little doubt that, having failed to write a new novel in the seclusion of Anerley, with Dorrie by his side, he hoped that total isolation in a foreign country would work again the magic his lonely trip in 1946 and 1947 had done.

Aubrey Burns recommended a quiet motel 200 miles north of San Francisco, just off Highway 101, between Legett and Eureka, and very near the hamlet of Piercy. Lane's Flat Cabins, as the motel was called, catered mainly to the trucking crews thundering north towards Canada and returning. Paton's cabin was out of earshot of the traffic, very near the Eel river, and in the shadow of the giant redwood trees which are a feature of California's forests. Here he proposed to write all day in his cabin, from which he could hear the salmon jumping in the river, and he would take his meals in the Lane's Flat cafeteria. Anyone else would have thought an American trucking motel an odd place in which to write, but to Paton it seemed perfectly logical: *Cry, the Beloved Country*, after all, had been written in the hotel rooms of the world.

He settled into this cabin on 20 November 1949, equipped, as he told Mary Benson (now in Pennsylvania, staying with married friends on their farm), with 'cigarettes & a bottle of whisky & two packs of cards, an anthology of verse, an English & a Zulu bible, an Afrikaans New Testament, Jeffers's poems, [Henry] Green's 'Loving', Thoreau's 'Walden', & half a dozen Pocket books, paper, pen, ink.'[48] Among the Pocket books was a volume of Whitman, whose poetry Paton particularly admired.

He now began one of the most miserable periods of his life, and it lasted until he abandoned Lane's Flat two months later, on 21 January 1950. The problem was not that he disliked living on his own (Thoreau's *Walden* was a good choice for him), nor that he grew bored with truck-drivers' food and truck-drivers' company. The problem was that he could not write.

He had come with two ideas, that of the young policeman obsessed with lust for a black girl, and that of a novel set during the Great Trek, apparently centring on the figure of Piet Retief,[49] the Boer leader who was treacherously murdered by Zulus. At intervals he considered a third, the previously abandoned notion of Christ's visit to Johannesburg. Every morning he would sit himself down in front of the blank page, pray for inspiration, take pen in hand, and try to write: but nothing came. Then he would go out for an enormous American breakfast in the 'joint', as he called the motel restaurant, and come back to stare at his blank page again until he could no longer bear it, and would burst out of the cabin to go for a walk. After one attempt he gave up trying to walk in the redwood forests, which were too damp, and too full of fallen branches, for his liking. 'Redwoods are fine, but no sun gets through them; it's nature all right, & nature at its grandest, but a trifle raw,' he told Mary Benson. Instead he had to walk along Highway 101, then a narrow and dangerous road, where the roaring of the trucks never stopped.

At night he would feel gloomy about the day's lack of production, and would have a whisky to cheer himself up, and then another, and another. And while he drank he would play endless games of patience, and smoke endless cigarettes. He would go to bed very late, often early in the morning; and then, finding himself unable to sleep, would lie miserably listening to the fish jumping in the river until dawn, when he would fall asleep and wake late in the morning, guilty. And then the whole round would begin again. The worst part of this cycle was the inability to write; a close second was insomnia.

He does not seem to have suffered from loneliness, though he must have been sorely tempted when Mary Benson—still trying to find her 'purpose'—wrote suggesting that she might join him at Lane's Flat. 'It would have been very pleasant for you to come here,' Paton wrote to her, '& I really thought about it from every conceivable angle, but it was suddenly made impossible by Aubrey Burns's decision to come here & finish his novel. That arrangement was so long standing that I had to accept it, but I may tell you that he comes here to Lane's Flat on the strict understanding that he does not come to my cabin unless by invitation, except of course at the times when we relax.'[50] But Burns's presence was an excuse. Paton went on to admit, 'the truth is that even if Aubrey Burns hadn't decided to come up, I was against your coming here, not just because I was writing (which is a factor I considered), but because I did not think that was the right way for you to start the new & purposeful existence which you seem to be benevolently considering.' And he ended with a warning that her search for a purpose in life would be painful: 'If what you want is marriage, I wish you all fortune; but if it's something else, I don't think you realize yet that you'll have to pay for it.'[51]

His strictness with himself paid off gradually. As had happened in Anerley when he strove to write a novel, and could not, poems came to him instead. His scrappy diary records that he had begun to write a series of poems he thought of as psalms, and in fact in their imagery many of the poems he produced at Lane's Flat owe a great deal to the Bible. The first of them, though not particularly psalmodic, is plainly drawing on his thoughts of a novel about Christ in Johannesburg, Eloff Street being one of the main streets of that city. He wrote it on 3 December 1949, and the influence of Whitman, whom he had been reading steadily, is evident in the long free-verse lines and the prophetic voice, as it would be in all the poems he produced at Lane's Flat:[52]

> I saw the famous gust of wind in Eloff Street
> It came without notice, shaking the blinds and awnings
> Ten thousand people backed to the wall to let it pass
> And all Johannesburg was awed and silent,
> Save for an old prostitute woman, her body long past pleasure
> Who ran into the halted traffic, holding up hands to heaven
> And crying my Lord and my God, so that the whole city laughed
> This being no place for adoration.[53]

And after this beginning, as his diary records, the poems came quickly. On 3 December, the same day as 'I saw the famous gust', he wrote a poem he subsequently referred to as 'Psalm 5', and which cannot now be identified with certainty; on December 6 he produced two 'Sonnets to Sleep', with his insomnia very much in mind:

> Sleep is the mocking thing when the mind's knot
> Won't yield and loosen, when the mind's eye fears
> To close completely, when the mind's ear hears
> Rumours of danger, plot and counter-plot,
> And threats and intrigues, and the long night's shot
> All through and through with restlessness and tears . . .[54]

On December 12 he wrote two poems which appear to have been lost, 'The Lord is the Great Judge' and 'The Lord is my Neighbour'; on

December 13, 'My Lord has a Great Attraction' and 'My Lord, Hearken to the Cries', the latter of which appears to have been lost; on 15 December 1949 he wrote 'My Lord has a Magic with Animals'; and on December 16, 'The Monument'. Suffering from insomnia on December 18, he wrote a particularly beautiful psalm, 'My Lord in the Forest' (later retitled 'A Psalm of the Forest'), setting it down at 3 a.m. It is noticeable that even while he was writing in an American forest, what comes to his mind is a South African forest, the yellow-woods of Tsitsikamma rather than the redwoods of California:

I have seen my Lord in the forest, He walks from tree to tree laying His hands upon them.
The yellowwoods stand upright and proud that He comes amongst them, the chestnut throws down blooms at His feet.
The thorns withdraw their branches before Him, they will not again be used shamefully against Him.
The wild fig makes a shade for Him, and no more denies Him.
The monkeys chatter and skip about in the branches, they peer at Him from behind their fingers,
They shower Him with berries and fruits . . .
And the winds move in the upper branches, they dash them like cymbals together,
They gather from all the four corners, and the waterfalls shout and thunder,
The whole forest is filled with roaring, with an acknowledgment, an exaltation.[55]

And the flow continued. On 19 December 1949 came 'I'll Sting the Conscience of the World'; the next day he recorded writing 'Psalm about Existentialism', but this appears to have been lost, together with another poem written on 22 December, 'I Thought to Myself O Lord'. On the last day of December 1949 he wrote 'Oh Lord, my Enemies Overwhelm me' and perhaps another poem, 'For the Earth is Corrupted', though his diary is vague about this.

The most beautiful of the poems he produced at this time was 'Meditation for a Young Boy Confirmed', which he wrote early in the morning of 3 January 1950. It had been sparked off by Jonathan's confirmation at his school, St John's in Johannesburg, on 26 November 1949. Paton regarded this as a step of great importance in his younger son's Christian life, and not only wrote to him beforehand, counselling him to take the ceremony seriously and with humility, but sent him a cable of congratulations. Jonathan responded with a letter on the day of his confirmation:

Dear Dad,

Thank you very much for your letter.

I was confirmed today with 57 others. We receive our first communion tomorrow morning . . . Before I forget, thank you very much as well for your cable received this afternoon (at lunch time).

Sunday: We received our first communion to-day. I was rather nervous at first but everything went well. [He then fills up the remaining space on the sheet by copying out his school time-table for the entire week.]

Well, that's our time-table but I must close now and get ready to go out with Carrot [his nick-name for his mother] for the day.

Much love,
Write soon,
Jonathan.

On 3 January 1950, soon after returning to Lane's Flat, having spent Christmas and the New Year with the Burnses, Paton woke at 2 a.m. and wrote 'Meditation for a Young Boy Confirmed', which opens,

I rise from my dream, and take suddenly this pen and this paper
For I have seen with my eyes a certain beloved person, who lives in a distant country
I have seen hands laid upon him, I have heard the Lord asked to defend him,
I have seen him kneel with trust and reverence, and the innocence of him smote me in the inward parts.
I remembered him with most deep affection, I regarded him with fear with trembling,
For life is waiting for him . . .

And the poem wryly predicts that among the hurts waiting for Jonathan and all like him is religious doubt:

You will observe that virgins do not bear children, and that dead men are not resurrected;
You will read in the newspapers of wars and disasters, but they will report miracles with impatience.
You will be distressed . . .

The advice that follows shows the toughness of Paton's own faith:

Do not hastily concede this territory, do not retreat immediately,
Pass over the slender bridges, pick your road quickly through the marshes,
Observe the frail planks left by your predecessors, the stones gained only by leaping,
Press on to the higher ground . . .
But do not lie to yourself, admit this is the journey of the heart.

And his mixed feelings about the his love affairs show in stanza XIII:

If you should fall into sin, innocent one, that is the way of this pilgrimage;
Struggle against it, not for one fraction of a moment concede its dominion.
It will occasion you grief and sorrow, it will torment you,
But hate not God, nor turn from him in shame or self-reproach;
He has seen many such, his compassion is as great as his Creation. Be
tempted and fall and return, return and be tempted and fall
A thousand times . . .[56]

He knew this poem was good even as he wrote it, and the next morning (4 January 1950) he showed it to Aubrey Burns, who was much moved by it and called it 'the best ever'.[57]

But Paton was not satisfied with poems, even poems as good as 'Meditation for a Young Boy Confirmed'. What he had come to Lane's Flat to produce was a novel, and he thought of his poems as a mere exercise. His diary reveals that his gloom continued in spite of the poems:

Tuesday 13 December [1949]: Rose 11 a.m.—must stop this. Wrote A[ubrey] & M[arigold] re presents. Wrote psalms 8 & 9. The Lord has a great attraction, My Lord, hearken to the cries. Later changed to 7, 8. Had biscuits and ginger 3 p.m. . . . Walked till 6, had bath and played patience till 1.30 a.m.

Fri 16: Rose 10.30. 11.30, did Ps 14 and started on 24 Xmas cards. Wrote many letters—Tried to read script—felt faint & went walking—Came back disinclined for work—Played too much patience—Went to bed 11.30 and read Walden—A bad day—Homesick maybe.

Sun 18 Dec 1949: 3 a.m. Still awake. *Wrote Lord in the Forest.* B'fast 12 noon—Wrote DOP [Dorrie] & Xmas letters. Revised psalms. Wrote letters. Bathed & exercised 6 p.m.—Ate 9 p.m., bed to read [William Faulkner's] Sanctuary 10 p.m.—Put off light 12—Again woke 2 p.m.—Tried to write Psalm [about] Man's knowledge—No lights—No good!

The recurrent electricity failures seem to have become symbolic of his own lack of illumination: 'NO LIGHTS,' intones his diary in doleful capitals, and again, 'NO LIGHTS—NO LIGHTS'.[58]

For his own peace of mind he needed to write a novel. And he did in fact drive himself to write one, producing chapter 1 on 8 December 1949, and recording in his diary that chapters 2 to 5 were already written. At intervals thereafter one can trace his progress with the book. On 18 December 1949 he had reached chapter 8; on 31 December 1949 chapter 12, 'not with great satisfaction'; on 2 January 1950 he wrote chapters 13 and 14: 'They have quite a swing but don't strike any wonderful notes.' Shortly after this, however, he began to feel that the thing was taking fire. On 4

January 1950, there is a real note of excitement and exultation for the first time: 'After supper wrote chapter 15: Swinging along!' At the same time his insomnia ceased to torment him: 'sleeping at last becoming normal'.[59] By the time he celebrated his 47th birthday on 11 January 1950, receiving cables from Dorrie and his sons, and enjoying a party put on for him by the staff at Lane's Flat with a bottle of whisky as a gift, he felt he was aloft again on the wings of his muse, as he had been while writing *Cry, the Beloved Country*.

But the happy change was not to last, and Paton suffered terribly from the let-down. By the middle of January 1950 it was raining continuously after a snowfall during which the electric power failed once more; the Eel river came down in spate until, on 17 January 1950, he recorded, 'Eel river is working at my steps—it has probably risen.'[60] His confidence was being undermined too: the next day he recorded writing chapter 18 of his novel, but the excitement has gone: his hero, he records, is 'still meandering'. A meandering novel was not what he wanted; after his own criticism of *Cry, the Beloved Country*'s structure, he had promised himself a novel 'as perfectly constructed as I am able to do it',[61] and this loose-knit manuscript did not satisfy him. On the contrary, he grew snappish and irritable, and on some occasions when Aubrey Burns (whose own novel was never to see the light of day) came to his cabin, the two friends did not get on well:

> He [Burns] brought me a poem, 'Birthday Celebration' for my birthday—We had 4 whiskies instead of 3, & he got more & more talkative & I more & more remote—A[ubrey] talkative is stupid—He made very stupid remarks about FDR[oosevelt]—When he noted that I was remote he spent the rest of the evening explaining his remarks—I find him an excellent, at times a touching companion, but he does chatter too much . . .[62]

It was not Burns's criticism of Roosevelt, one of Paton's political heroes, that annoyed him, but his own lack of productivity.

On 21 January 1950, unable to bear the grind and failure of Lane's Flat any longer, he left the cabins and seems to have abandoned his novel at the same time. He may well have flung it into the Eel river.[63] In later years, in his autobiography, he would pretend that he had completely forgotten his purpose in going to Lane's Flat in the first place. 'What did I go there for? To write a novel? I must confess that I cannot remember.'[64] It was pain he did not care to remember, the pain that rings clearly through a letter he wrote Mary Benson in April 1950: 'God took me aloft, & the moment I said "I'm aloft" He let me drop, & smashed every bone in my body.'[65]

16

RECOVERY
1950–1952

Real generosity towards the future lies in giving all the present.

ALBERT CAMUS

Not broken but badly bruised, he made his way back to South Africa via Los Angeles, New York, and London. In Santa Barbara on 24 January 1950, on his way to Los Angeles, he spent a day with Luther Greene, the producer who had not given up trying to draw a play out of Paton, and met his wife Judith Anderson. He got on well with the Greenes, remembering years later that they had taken him to a Spanish-style restaurant in Santa Barbara, where he drank so much tequila that he neither knew nor cared where he was.[1] Greene also took him to meet the poet Robinson Jeffers, who lived at Big Sur. Paton, who admired Jeffers's work, could not draw one word from him. Even direct questions failed: 'What's that bird?' Paton enquired when one perched on the window sill, and Jeffers opened a bird book at a particular page, and handed it to Paton—still in silence.[2]

On 25 January 1950 he spent an afternoon and evening with Aldous Huxley, and here the difficulty was quite the reverse: Huxley, peering blindly before him, long fingers intertwined, never stopped talking. In spite of this, Paton, humbled by his experiences at Lane's Flat, was to say, 'My few hours with him were too short.' He considered Huxley a genuinely wise man, and was to quote him to great effect in the most spiritual of his own writings, *Instrument of Thy Peace*.[3]

Returning to the Greenes, he spent a few days roughing out a first act of a play, but by 1 February had to admit defeat. 'LG [Luther Greene] . . . read Act I—No good—But cannot explain—obviously must study technique elsewhere.'[4]

On 7 February 1950 he flew to New York for another round of parties of the sort that had greeted the opening of *Lost in the Stars*, which was still running to full houses. He did not go to see the show again, and contacted Maxwell Anderson only to learn that the playwright, still smarting from Paton's inability to praise the script, was too busy to see him.[5] But there was compensation in elaborate lunches with Charles Scribner and Maxwell Perkins, for *Cry, the Beloved Country* was proving a financial locomotive for the house of Scribner, and showing no signs of losing its momentum. Scribner's also, on 21 February 1950, arranged a cocktail party in honour of Paton and John Steinbeck, at which Steinbeck spent all evening at the bar talking to the barman, so that Paton, nothing loath, had the attention of the guests all to himself.[6]

He also had a pleasurable meeting with Edmund Fuller,[7] an American reader who had written him a fan letter in May 1948 after the publication of *Cry, the Beloved Country*, and with whom he was to have a lifelong friendship pursued mostly through correspondence. Fuller was a writer of remarkable power, and he had sent Paton a copy of his novel *A Star Pointed North*, which concerns the slave Frederick Douglass, an historical figure who escaped from Baltimore, published an abolitionist newspaper, and was several times consulted by Paton's hero, Abraham Lincoln. Fuller was living in Hand's Cove, Vermont, the name of which for some reason appealed to Paton powerfully, so that he longed to go there. To his disappointment he found he had too many appointments to do so, but Fuller travelled to New York and had a brief and rather awkward conversation with Paton, their first, in the offices of Paton's agent Annie Laurie Williams.[8] Williams, for whom Paton developed great affection, was described by Fuller as 'a tiny, deceptively frail-looking woman who resembled a gnome or gargoyle. She had such severe scoliosis that she was bent at nearly a forty-five degree angle and had to twist her head, like a bird, to look up at you. She was very knowledgeable, especially of Broadway and Hollywood.'[9]

On 25 February 1950 Paton flew to London, and spent nearly a month there working on the script of the film of *Cry, the Beloved Country* with Zoltan Korda. It was now beginning to take clear shape, and Korda planned to do much of the filming on location near Ixopo later in the year. Paton also resumed his contacts with Toc H, and on 5 March 1950 was accorded the signal honour of preaching in St Paul's Cathedral, only the second layman ever to do so, as he proudly recorded.[10] This invitation had come to him through Canon John Collins, who had read and been deeply impressed by *Cry, the Beloved Country*, and the friendship initiated in 1950 was to run a most interesting course in the years to come. Nor was the sermon in London the only one; he was now in great demand as a speaker, and he travelled to St John's College, Oxford (1 March 1950), to Cambridge (6 March), and to Birmingham (9 March) to give similar addresses.[11]

But writing scripts and sermons did not fill all his time. As he had done while working with Zoltan Korda before, Paton began to see Mary Benson, who was back in London after her American visit, and they continued their friendship.[12] In point of fact he went straight to St Paul's Cathedral from lunch with her, at which she served him kippers and then was afraid that he would belch during the sermon as she was doing in the congregation.[13] 'Be tempted and fall and return,' he had written in 'Meditation for a Young Boy Confirmed'; and it is one of the most interesting aspects of this complex man that his puritanism warred always against the side of his nature that longed for human warmth and love. In his sermon in St Paul's, which Mary Benson had typed out for him, he used a striking image which sums up the dilemma:

> You know that if a magnet is brought near to iron filings, they will, if they are free to move, arrange themselves immediately in the field of its attraction. They take upon themselves a pattern whose unity is clearly to be seen. But some may be unable to respond; there may be roughnesses and obstruction in the surface on which they lie, which prevent them from responding to the greater and more powerful attraction; others may be prevented by those that lie across them, weigh down upon them, and otherwise hinder them; yet others may contain impurities which prevent them from fully responding.
>
> So indeed with ourselves; for us also there is a pattern into which we may move under the influence of God's attraction, a pattern whose unity is dimly apprehended by us, but whose realization eludes us. Some of us too may be unable to respond because we are prevented by some roughness or obstruction in the plane on which we live, which, small though it may be, appears to us like a mountain that no faith of ours has ever been able to move. Some of us are prevented by others that lie across our path, that weigh down upon us, and hinder us; and we ourselves prevent others, because we obstruct them and weigh down upon them, and hinder them. Some of us respond but feebly to the great attraction, because our metal is impure, and our being divided.[14]

It is a precise and deeply insightful piece of self-analysis, and only a profoundly spiritual man could have made it.

Benson, having been guided by her long discussions with Paton, had begun working with Michael Scott, the priest about whom Paton had spoken on their first evening together. In the course of helping Scott with his work on behalf of tribespeople in South West Africa, Benson had fallen deeply in love with him. Over the years, as she was forced to accept that

Scott wanted no more than a working partnership, Paton proved a sensitive and understanding friend.

On 20 March 1950 he left London for Amsterdam, his first visit to the Dutch capital, and from there took the KLM flight to Johannesburg, as usual enjoying the service. On this flight, probably in the glow of his post-prandial whisky, he produced an epigrammatic tribute:

> The Dutch may be a stolid race
> And noted for their phlegm,
> But with what master touch & grace
> They made the K.L.M.[15]

Until the end of his life he was to love almost every aspect of air travel. Having finished his dinner with relish he would drink a good deal before falling easily into a profound sleep, and he seemed never to suffer from the jet lag that so plagues many international travellers.

From Johannesburg he went straight on to Durban for one of those rare passionate meetings with Dorrie, which he was to describe with his characteristic sly humour. 'When we reached the hotel, I said, *I'm going to have a bath.* You said *I'll come too.* Jonathan said, *I'm coming too.* David, who at that time ruled Jonathan with the elder brother's rod, said most emphatically, *You are not going too.*'[16]

Zoltan Korda arrived in South Africa in April 1950, and he and Paton worked in Johannesburg for some weeks, putting finishing touches to the script of the film. In May 1950 the filming began at Carisbrooke near Ixopo. Most of the cast were accommodated near Carisbrooke at Rayfield, the farm belonging to Dorrie Paton's brother, Garry Francis. This arrangement was made necessary because neither of the hotels in Ixopo would accept as guests the black stars, Canada Lee (who was to play Kumalo), Charles McRae (playing the old priest's friend), and Vivien Clinton (playing the wife of Absalom). Sidney Poitier, who played the young priest Msimangu, was not needed in Ixopo. In the event the Plough Hotel, showing some courage, agreed to accommodate Vivien Clinton in a caravan in its grounds.[17]

The filming caused enormous excitement in Ixopo, and brought a good deal of money into the town, helping to fund the Memorial Hall.[18] The townspeople queued to work as extras at £5 a day, and those not actively employed came to watch in fascination. Among the throngs of onlookers was a young man named Guy Butler, who would later become known as a South African literary figure in his own right. Butler was to remember talking to Paton, and hearing that he planned a new story about a young Afrikaans policeman who had interfered with a black girl in the cells.[19] 'Yes,' Paton was to write years later to Butler, 'the story that I mentioned was the germ of *Too Late the Phalarope*, but of course as you know when the creative genius (!!) takes over the story changes.'[20] When the filming at Nokweja was finished, London Films demolished the small, ramshackle church where so many scenes of film and novel had taken place, and built a new bigger one in its place. Thus art replaced life.

Then, in June and July 1950, there was more filming in Johannesburg, where the reformatory scenes were made at Diepkloof itself, with Ben Moloi appearing as an extra, and being photographed together with Paton, Sidney Poitier, and Canada Lee.[21] Paton was aware from his conversations with Moloi, and with his own successor as Principal, W. W. J. Kieser, that his achievements at Diepkloof had already been substantially reversed. Among those whom the new hand at the tiller most directly affected was Moloi, who would be sacked a year later, in May 1951. Paton would remain his faithful friend, getting him a job at a Natal school, educating his children, and even supplying him with the bride-price when he wanted a new wife.[22]

Paton must have suspected that Diepkloof would soon be closed altogether. He felt the tug of the institution into which he had put thirteen of the best years of his life, so much energy, and so many dreams. On 4 May 1950 he noted about his visit to the place, with the self-mockery that in his letters often conceals deep emotion, 'My heart melted and my bones became water.'[23]

He did not much enjoy the filming of *Cry, the Beloved Country*, in part because there was little role for him to play once the script had been completed, in part because he disliked Zoltan Korda's arrogant harshness with the actors, particularly with Canada Lee, an ex-boxer whose intelligence was not high and for whom Korda showed increasing contempt.[24] Paton did enjoy getting to know the few black South African actors involved in the film, including Lionel Ngakane and Reggie Ngcobo, with both of whom he kept up his friendship in later years.

He broke away from the filming at least twice to take extended holidays, once travelling to the Kruger Park with Korda, and in June 1950 taking Dorrie and his sons for a long motoring holiday in the Rhodesias (now Zambia and Zimbabwe) and Portuguese East Africa (now Mozambique).

The filming moved to London at the end of 1950, and Paton was flown over by London Films for it, but had hardly got there when Canada Lee fell ill and everything stopped until he could recuperate. Paton returned to South Africa complaining that 'the film has been a bally distraction',[25] and more determined than ever to give his whole mind to his writing. 'You shouldn't be doing this,' Zoltan Korda's brother Vincent had said bluntly to Paton of the film-making. 'You were meant to be a writer of books, not film scripts.'[26] Paton knew he was right, and when he returned to London in 1951 for the final period of filming there, he came ready to write.

The Nationalist Party had now been in power in South Africa for nearly three years, and its grip on the country was steadily strengthening. It was becoming increasingly clear that the rightward lurch of the white electorate in 1948 was not an aberration, but the beginning of an era. And that apartheid era was marked by a rapid strengthening of the laws dividing South African whites from other South Africans. Once Verwoerd took up the post of Minister of Native Affairs on 18 October 1950, the complex of interlocking laws designed to keep power in the hands of whites was rapidly strengthened and systematized.

In 1927 the Immorality Act had made it illegal for a white to have sexual congress outside marriage with an African. In 1950 the Nationalist government passed the Mixed Marriages Act, prohibiting inter-racial marriage and any inter-racial sexual relationship. Before this could be enforced, the racial group of every person had necessarily to be established in law: hence the Population Registration Act of 1950, which required every citizen to register his or her 'race', with humiliating tests designed to weed out persons of mixed race claiming to be white. Nor could people of different races be allowed to live in the same areas of town: the Group Areas Act of 1950 uprooted hundreds of thousands of Indian, black, and Coloured people, and moved them to new, usually underdeveloped and inferior areas.

Paton regarded these laws as not just bitterly unfair, but thoroughly un-Christian. How could people as apparently devoted to Christianity as the Afrikaners give assent to Christ's commandment that they should love their fellow human beings, and simultaneously give assent to a series of laws founded on fear, contempt, and hatred? Paton could not understand this mystery, for all his careful study of the Afrikaner mind over so many years.[27] He suggested several explanations, including white fear of being overwhelmed, racial abhorrence of Africans, its concomitant racial arrogance, and greed.

It does not seem to have struck him, or if it did he did not wish to say so, that many white South Africans were deeply proud of their attempt to build a First World country in Africa, and feared the descent into Third World conditions which in 1950 seemed likely to follow the transfer of power to an overwhelmingly more numerous black population. This fear was to increase in the second half of the century as African countries gained their independence and many slid into despotism and disorder.

But in 1951 all this lay in the future, and Paton believed that he could change the hearts and minds of white South Africans with a book showing how cruelly unfair the new laws were. The theme of the white policeman who offends against the Immorality Act which he is paid to enforce had been in his mind at least since he told Mary Benson about it in September 1949. In later years he was to say that the story had come to him when he saw a newspaper report of a policeman in a country town in the Transvaal, who was being charged under the Immorality Act. What caught his imagination was not the 'crime' itself (such reports were common enough in South African papers until the 1980s) but the fact that 'the policeman's wife sat in court throughout the trial, and by her demeanour showed that she had forgiven him'.[28] As so often with Paton, compassion was what spoke to and through him, and the policeman story began to grow in his mind.

By February 1951 he was writing to Mary Benson that this was the theme he now definitely intended to pursue, and when he went to London that year to complete the film he took an early opportunity to go off to a quiet English town, Falmouth in Cornwall, which he had probably visited by motorcycle on his first trip to England as a student delegate in 1924. Here he settled into the Green Bank Hotel, in much the way he had settled into Lane's Flat Cabins. But unlike that previous disastrous time of struggle and frustration, he found that isolation and the hotel room worked the magic they had on his 1946 trip: the book began pouring from his pen.

The novel focuses on Pieter van Vlaanderen, an heroic figure who is brought down and destroyed by the very system of which he is a part. A widely admired war hero, rugby player, and scholar, he is a lieutenant of police in a small town in the eastern Transvaal, and by his position should uphold and enforce the Immorality Act. Instead he sleeps with a black girl, Stephanie, is found out, arrested, and tried, and the effect of this shame is the destruction of his family.

Too Late the Phalarope is an elaborately and consciously constructed novel, which moves to its conclusion with all the inevitability of a Greek tragedy. And, in fact, it shares a number of links with Greek tragedy, from the heroic protagonist who is brought down by a tragic flaw, to the use of a narrator who constantly comments on the action and predicts what is to come. Paton may have been given the idea of using the Greek model by Maxwell Anderson's initial telegram proposing the transformation of *Cry, the Beloved Country* into *Lost in the Stars*:

WOULD LIKE TO TRANSLATE INTO TERMS OF THE THEATRE AND FEEL THAT CASTING IT INTO TERMS OF GREEK TRAGEDY WITH EQUIVALENT OF A CHORUS WITH MUSICAL ACCOMPANIMENT BY KURT WEILL COMPOSER OF INTERNATIONAL DISTINCTION WOULD ENABLE ME TO CARRY OVER INTO THE PLAY THE FULL SUBSTANCE OF THE BOOK AND PRESERVE THE SIMPLICITY AND BEAUTY OF THE DIALOGUE ALMOST INTACT.[29]

Yet though it is a more polished piece of work than *Cry, the Beloved Country*, *Too Late the Phalarope* lacks the power of the earlier novel, and has generally been judged inferior to it. This is partly because the details of the tragedy are less obviously universal than those of *Cry, the Beloved Country*, partly because the tension is reduced early on by the repetitive predictions of the chorus, and partly because Van Vlaanderen is a little too good to be true. In *Too Late the Phalarope* Paton was writing specifically for a South African audience, and he was determined that his message should be unmistakable not just to Railton Dent and his other friends, but to those who made and enforced the unjust laws of his country. The result is that the book seemed to non-South-African readers to lack the appeal of *Cry, the Beloved Country*, while some South Africans, even Paton's friends, felt that it nagged rather.

For all that, Paton himself was passionately engaged with the book while he wrote it, not least because in describing the subtle inadequacies of Pieter Van Vlaanderen's marriage he was describing his own relationship with Dorrie. Van Vlaanderen's wife is a decent, loving woman, and he can scarcely formulate his complaint against her; but she neither loves him as he loves her, nor, timidly conventional as she is, can she give him the complete response he longs for in love. In the last resort she cannot understand him, and when put to the test he finds himself desperately alone. One of the chief themes of the novel was one of the chief failings of Paton's marriage: the lack of communication which he felt plagued his relations with Dorrie. The very title makes this plain: the interest of Pieter Van Vlaanderen's father in the phalarope, a sea-bird which also nests inland, comes too late to bridge the widening gap between himself and his bird-watching son. *Too Late the Phalarope* is full of missed opportunities for characters to speak to one another and to avoid thereby the tragedy that stalks them. In its essentials Van Vlaanderen's situation was Paton's, and his descriptions of his protagonist's struggles against desire, and his heroic spiritual wrestlings with temptation, are wholly convincing because Paton had lived through them himself, and more than once.

Of the writing process itself little record remains. Virtually every page of *Cry, the Beloved Country* can be dated from the manuscript and Paton's copious letters of the time. But of his correspondence during the writing of *Too Late the Phalarope* little survives. The best evidence would have been his letters to Dorrie from England, but all were subsequently destroyed. The reasons are very likely to have been found in the theme of the novel itself, and in the fact that, during the writing, his friendship with Mary Benson had been resumed.

By 1951 Mary Benson was involved in a variety of initiatives to help Africans in South Africa, South West Africa (now Namibia), and Bechuanaland (now Botswana) achieve self-government. She had found her cause, and in later years would write several distinguished books about the struggle of black South Africans.[30]

For five months in the first half of 1951 she and Michael Scott were assisting Tshekedi Khama, Regent of the Bangwato in Bechuanaland, who had been exiled from his tribal area by the British government which controlled the Protectorate. While helping Tshekedi Khama campaign for his return home, Benson worked with him in the exclusive hotel run by the British government for diplomats and visiting dignitaries in London, at 2 Park Street, Mayfair, behind the Dorchester Hotel. One morning, to her astonishment, she chanced to meet Paton there. Growing tired of his Falmouth hotel, and needed for consultations by Zoltan Korda, he had decided to move to London once *Too Late the Phalarope* was approaching completion. His British publishers, Jonathan Cape, had secured him a room at 2 Park Street, and he was writing hard each day just a floor above Tshekedi Khama's suite.

Paton had not told Benson where he was, no doubt partly because of the moral struggle he describes Pieter Van Vlaanderen undergoing, but almost certainly partly because he feared that regular meetings with her might distract him from his book. After this happy chance encounter, however, they resumed their friendship, and Paton found that his writing did not suffer at all. Every evening he would bring Mary Benson the handwritten pages he had produced that day, and she would read them and comment on the novel he was writing with such apparent fluency. London was a much less gloomy place to him once she re-entered his life. They were amused and excited that, at long last, he had succeeded in breaking through the writer's block, just as she had succeeded in finding a purpose in life. He found her a most generous and intelligent companion, and they achieved that rare thing, true friendship between a man and woman not married to each other, and no longer lovers.

Paton's book flowed rapidly and smoothly with her as his reader and friend, though it was still not finished by the time he left England. On the other hand, by the middle of 1951 the filming of *Cry, the Beloved Country* was finished, and before he left for South Africa Paton attended a private viewing at the headquarters of London Films.

He was relieved to find that he did not hate the film as he had hated *Lost in the Stars*, and for a time he would defend it rather feebly in letters to acquaintances: 'Yes I am pleased about the movie; I was too close to it to be any good as a judge, & would naturally defend things that others would criticise.' But he also added, 'I don't think I'd do it again.'[31] And, more openly, he told Mary Benson, 'So you didn't like the film very much. Well you are not alone in that. But I don't see how it can be blamed entirely on Korda; I probably didn't argue enough.' And he concluded ruefully, 'In any case there it is, "Screenplay by Alan Paton".'[32]

The film of *Cry, the Beloved Country* is faithful to the original but rather plodding, and by comparison with the book it is terribly impoverished by the loss of Paton's magnificent prose. Its slow opening infuriated Sir Alexander Korda, because London Films' productions had been criticized by Graham Greene, among others, for their dreamy lack of pace. At the private viewing Alexander Korda sat in growing rage through the first ten minutes and then electrified Paton by launching a public attack on Zoltan. 'Suddenly the lights of the theatre at 146 Piccadilly are turned on. Sir Alexander is on his feet. He turns to his brother and speaks to him in tones of barely controlled ferocity. "When does your film start?" he says. "When does your film start?"'[33] Alexander Korda's rage was justified. When the film was released at the end of 1951 it flopped, and Paton made from it only the £1,000 advance on royalties secured by Annie Laurie Williams.

Too Late the Phalarope, by contrast, was a great critical success. Paton had not finished it when he returned to South Africa from England in the middle of 1951, and he is mistaken when in his autobiography he claims that it was 'written in two places only, the Green Bank Hotel in Falmouth, Cornwall, and 2 Park Street, London'.[34] He returned to London in April and May 1952, apparently for consultations with Jonathan Cape about the book. During this visit he had a meeting with a group of expatriate South African writers in London, hosted by the novelist Laurens van der Post.

There exists a striking photograph of Paton, wearing a pair of rumpled trousers and a jacket considerably too large for him, standing in a Sloane Square garden with the impeccably dressed Van der Post, the painter Enslin du Plessis, the Afrikaans writer Uys Krige, and the larger-than-life poet Roy Campbell.[35] Paton was fascinated by Campbell:

> He and Uys could not stop talking—the rest of us just listened. Krige says somewhere that when Campbell was talking, he could not get a word in. I don't believe it. It was a dead heat. I was interested to hear Roy's accent. It contained nothing of English, or Spanish, or French [Campbell had lived in each of these countries], but was purely Natal Carbineer/Durban High School, but of the lower not the higher echelons. I asked him if he had been punching any more poets on the nose. [Campbell had punched Stephen Spender at a poetry reading.] He thought that was very funny and laughed delightedly—a sort of tee-hee laugh, as I remember. I am sorry that I never met him again.[36]

In April 1952 Paton gave his old friend Railton Dent, whom he ran into in England as Dent was coming to the end of a holiday there, a copy of the incomplete manuscript of *Too Late the Phalarope* to read and comment on during his voyage back to South Africa. In response to Dent's letter of suggestions (a letter both cautious and complimentary, as well it might be after the reception of his criticisms of *Cry, the Beloved Country*), Paton wrote from his home in Anerley in July 1952, 'The book is advanced by at least two or three chapters that you haven't seen, but I think it best to finish the whole first, therefore please retain the manuscript. I am going to stay with Mrs Putterill for the best part of July to finish the work.'[37]

Mrs Putterill was an acquaintance from Ixopo days, who owned an isolated farm near Kamberg in rural Natal. According to Dent's wife, Mabel, Paton spent several weeks up there, without Dorrie, writing in complete seclusion, and it was there that he finished *Too Late the Phalarope*.[38] The novel was published in August 1953.

When an author's first novel has had rave reviews, a second book is often coldly received, as if the critics felt a need to reassert balance. Paton must have feared this would be his fate after the astonishing success of *Cry, the Beloved Country*, but he need not have worried. *Too Late the Phalarope*, like *Cry, the Beloved Country*, was first published in America, and the American reviewers sang in chorus. 'Alan Paton's Second Masterpiece', announced the New York *Commonweal*, and its critic, Harold C. Gardiner, went on to describe the new novel as 'even more impressive than *Cry, the Beloved Country*'.[39] The *Los Angeles Times* agreed: 'New Alan Paton Novel Surpasses Early Work', it proclaimed.[40] 'A Novel of Universal Meaning', said the *Christian Science Monitor*, describing the book as telling 'in terms of simple dignity and great power the ageless story of man's conflict with evil'.[41] 'Paton Novel a Mountain Among Foothills of Current Fiction', was the Washington *Star's* headline, and it went on to call the book 'beautiful' and 'heartrending'.[42] It was left to Alfred Kazin, in the *New York Times* Book Review, to inject a note of moderation. After praising Paton's ability to convey the atmosphere of South Africa and its terrible racial problems, he became magisterial:

> Mr Paton's characters, and his telling of the story in a too even biblical style, are finally much less satisfactory. Although he can give us an almost hypnotic sense of these people, they are, after all, just types—Pieter, the blond Boer hero; Kaplan, the Jewish shopkeeper with his kindliness and his raptures over Tchaikovsky; the mother, so saintly as to be invisible, and the aunt, so intense as to be unintelligible. Everything in such a story depends on the author's ability to make us believe in the heroic strength [of Van Vlaanderen] far more than in his flaw, which should *surprise* us; here the catastrophe is hinted at so steadily through the book that when it finally comes we are moved, but not enlightened.[43]

But Kazin's was very much a minority voice. Large sales were guaranteed by the book's immediate adoption by the American Book of the Month Club as its choice for August 1953. The book was published shortly afterwards in Britain by Jonathan Cape, and in general the English reviewers were adulatory, though Paton, whose expectations were high, professed himself disappointed. 'The reviews here are very mixed,' he was to write from London to his wife. 'Daily Telegraph, Times Lit[erary] Supp[lement], Manchester Guardian very good—Spectator, New Statesman very bad—others non-committal.'[44] The South African critics recognized the extent to which this was a book aimed specifically at Paton's homeland. 'It is a book to be read here and now, especially in South Africa,' wrote Geoffrey Durrant in the *Natal Witness*, 'a work of great penetration and honesty, a genuinely serious and moving book in a time when serious writing is regrettably rare.'[45]

Paton had proved, by the success of *Too Late the Phalarope*, that he was not a one-book man. But the long struggle to produce it had also shown him that he was not going to be able to turn out a series of books with the power of *Cry, the Beloved Country* in short order. One clear sign of this was his decision to press ahead with a biography of Hofmeyr. He had first, half-jokingly, suggested this project to Hofmeyr on 23 February 1938, and it had recurred in his correspondence at intervals subsequently, particularly after Hofmeyr's death. But in 1950, while still in the process of finishing off *Too Late the Phalarope*, he began actively gathering material on Hofmeyr.

His reasons for deciding on this project were complex. Many a novelist suffering a temporary failure of inspiration (Virginia Woolf, Graham Greene, William Plomer, the list could be expanded almost at will) turns to biography in the belief that it is easier to write than a novel, and then discovers that this belief is delusion. Paton at this time unquestionably believed that he could keep the writer's flywheel turning with a biography while waiting for the next great novel to visit him. And there was a second reason for his decision to embark on a life of Hofmeyr: he was by now under increasing pressure to enter politics, and the biography he planned would be a move in that direction.

On 5 February 1951 Paton wrote to Mary Benson, 'There is still pressure on me to enter politics, but I still feel a great repugnance.'[46] The pressure was undoubtedly coming from the many liberal-minded friends he had made during his period at Diepkloof, his contacts in church circles, and above all those readers who had been moved by the compassion and enlightenment of *Cry, the Beloved Country*. A small minority of white South Africans were horrified at what the Nationalist government was doing to divide whites from other groups in the country, and were casting about for a political party which would fight apartheid.

The choice was not large. The Communist Party was certainly eager to condemn racism and to press for black rights, but Paton was not alone in believing that this front concealed an ideologically-driven despotism closer in practice to Nazism than to liberalism. In any case the Nationalist Government, in an echo of the McCarthyism now sweeping the United States, had banned the Communist Party in 1950, by the Suppression of Communism Act. The United Party, chastened and subdued by its defeat in the 1948 election, by the death of Hofmeyr that same year, and the death of Smuts himself in 1950, was trying to occupy the middle ground by opposing the ruthless apartheid measures of the Nationalists. In this endeavour it was hamstrung by feeble leadership and an apparent inability to put forward any serious plans for sharing power with blacks. Where was the party that would do so?

Paton, as we saw, had for many years had a powerful attraction towards political life. Now it seemed the way was open for him. He was an internationally known figure, an effective public speaker with a clear moral message and an equally clear vision of how blacks could gradually be brought into the process of government. He had laid it out with remarkable clarity, at least in embryo, in his articles while head of Diepkloof, notably his article, written in January 1944, entitled 'Real Way to Cure Crime: Our Society must Reform Itself'.[47]

What gave him pause was his sense that he did not have the experience to lead a new political party, or even a faction of an established party. This was what had made him wait so many years and in vain for an invitation from Hofmeyr, rather than taking the initiative by working his way into the United Party and standing for preselection for a seat. The liberal wing of the United Party seemed to have died with Hofmeyr; an entirely new party was needed, and Paton did not believe he was the man to found it. This was the root of the 'great repugnance' he spoke of to Mary Benson.

The biography of Hofmeyr was a way of advancing liberal views, and paying homage to his dead friend: a way of reconciling the roles of writer and activist which he often felt were dividing his energies and interests. But he had not gone very far with the research when Hofmeyr's mother effectively brought the project to a stop. She, who had been keen that Paton should write her son's life, heard that during his stay in Britain in 1951 for the filming of *Cry, the Beloved Country* and the writing of *Too Late the Phalarope*, Paton had made a brief trip to Scotland to talk to three men who had been professors at the Johannesburg University College (now the University of the Witwatersrand) while Hofmeyr had been its youthful Principal. They had witnessed Hofmeyr's ruthless campaign to dismiss the professor of medicine, Stibbe, whom Hofmeyr's mother (probably wrongly) suspected of having an affair with a typist. This affair was wholly discreditable to Hofmeyr, whom it showed as having the most limited understanding of human nature, and it was even more discreditable to his mother, who had urged him on to destroy Stibbe.

Mrs Hofmeyr was outraged that Paton was researching the Stibbe affair. She drew herself up to her full height (though she was no taller than Paton) and said accusingly, 'But *I* told you all about the Stibbe affair.' Paton replied with courage, 'Mrs Hofmeyr, when one is writing a man's life, one must go to everyone who knew something about him.' Deborah Hofmeyr, of whom Paton was to say, 'of all the women I have known, she had the fiercest will',[48] had determined that the book would be written to her prescription and hers only. And she did not believe in 'raking up the past', as she put it, if to do so was not flattering to herself and her son. Finding Paton equally determined that he was going to tell the truth, she wrote him off with a characteristic little stammer that gave her statement finality: 'I don't think we'll m-m-meet again.'[49]

Nor did they. Paton went on with the research, but Mrs Hofmeyr controlled her son's copyright, and it gradually became clear to Paton that his book could not appear while she lived. Nor, indeed, could he easily have faced her with a book which he knew she would condemn. 'She was often likened to a gorgon, those mythical women whose looks could turn men into stone,' he wrote. 'Some other name would have had to be found for her. Her looks turned men into jelly.'[50]

By September 1952, then, when this interview with Mrs Hofmeyr took place, Paton felt he had once again reached a crossroads in his life. He had given up his school career for Diepkloof in 1935; he had given up Diepkloof in 1948 to give himself freedom to write; and now freedom seemed to have produced long periods of artistic paralysis. Though he had invested enough of the earnings from *Cry, the Beloved Country* to be confident of financial security, and shown with *Too Late the Phalarope* that part of that success could be repeated, he could not abide the thought of living a beachcomber's life at Anerley much longer. Now the oblique combination of politics and writing offered by the Hofmeyr biography had been snatched from his hands.

And the call of his country grew more urgent by the day. 'Things are not good here, as you know,' he wrote to an American admirer, Irita Van Doren. 'We are governed by fanatics, who do not understand anything of history or of the outside world & the way it is moving. They are obsessed by the idea of survival, & are therefore planning suicide.'[51] The wit of his analysis does not conceal its grim accuracy. It was time for him to do something practical about it, but he was from the start realistic rather than enthusiastic about the new course he was to embark on. 'I am . . . willy-nilly being drawn more & more into activities. I don't think that these activities will necessarily change the course of events, but they are duties to be done.'[52]

As so often at crucial points in his career, he felt he was in the hands of a power whose presence he could only dimly sense, but whose will he would obey whatever the cost. In this sober spirit he embarked on sixteen years of political activism which were to make him hated and feared by the Nationalist government of his country, and which would bring the greatest rewards and dangers of his life. As he had shown long ago in his father's house, he would not seek a fight, but if it came he would fight it on his own terms.

17

THE POLITICS OF INNOCENCE 1953–1956

> The Liberal Party was a practical operational attempt to redeem history by bringing the politics of innocence into operation.
>
> TONY MORPHET

Paton's sense that things in South Africa were rapidly going from bad to worse under the Nationalist government had begun to sharpen during the spate of repressive laws passed during 1950, particularly the Mixed Marriages Act, the Population Registration Act, the Group Areas Act, and the Suppression of Communism Act. This last Act he particularly execrated as 'our first deliberate step away from the rule of law',[1] because the Act empowered the Minister of Justice to ban from public life any person he 'deemed' to be furthering the 'aims of communism'. This vague wording gave the minister terrible powers, and it was not long before banning orders were being issued against persons who had never had contact with the Communist Party, and in fact loathed it.

Paton himself disliked communism: he told a Liberal Party colleague, Pondi Morel, that he abominated the extreme Left as much as the extreme Right. She was to remark, years later,

> He hated Communism. I don't mean he hated the original idea, we'd all been through the phase of thinking Communism might be the saving of the world, but he hated what happened as a result of it, in Russia and the other Eastern European countries, and any country in which you got Communism.[2]

Paton hated communism partly because, being a religious believer, he recognized that ersatz religion clearly for the fraud it was. And he was too rigorous and muscular a thinker to fall into the arms of a left-wing totalitarianism in fleeing a right-wing one. 'I am prejudiced against the communist,' Leonard Woolf wrote once, 'because he seems to have the same end [in mind] as I and turns out so obviously not to have it.'[3] Paton felt the same way.

But being an anti-communist was not enough to save one from being banned under the Suppression of Communism Act, which was soon being wielded with a tyrannical freedom. A person banned was commonly restricted to an area a few miles in extent, could not be present at any gathering of two people or more (including private dinners and children's birthday parties), and could not publish or be quoted. The severity of the banning order varied from case to case, but in the worst cases it amounted to house arrest with restrictions stopping just short of solitary confinement.

There had been repressive legislation in South Africa well before the Nationalist victory in 1948, but the legislation of 1950 and succeeding years seemed to Paton fundamentally different. It was not merely that the government had changed, and become much more extreme, uncaring, and ideologically driven than the Smuts and Hertzog governments had been. Africans had also changed, their hopes and aspirations having been raised by the war and the establishment of the United Nations. The world had changed fundamentally, and was not going to allow South Africa to go on doing what it pleased within its borders. Finally, Paton himself had changed: as a result of his Diepkloof experiences he was much more sensitive to social problems, and the publication of *Cry, the Beloved Country* had put him in a position to do something about them. His sense of personal responsibility had grown to the point where it could not be ignored.

He had already several times attended meetings called to consider the formation of a new political party dedicated to propagating liberal political views in opposition to the racism of the right and the authoritarianism of the left in South Africa. Among the first of these meetings was one held in Talbot House, the Toc H centre, in Durban, in mid-1948. This was soon after *Cry, the Beloved Country* had been published in the United States, and just after the general election which brought the Nationalists to power, but before the death of Hofmeyr. Present at this meeting were two or three Anglican clergymen associated with Toc H, Kenneth Kirkwood, then on the Faculty of the University of Natal and later Rhodes Professor of Race Relations at Oxford, and a friend of Kirkwood's, Edward Callan.

Callan was a young Irishman who had come out to South Africa in 1938, served in the armed forces, and then gone to Witwatersrand University on an ex-serviceman's scholarship. In later years he would go on to a distinguished academic career in the United States, establishing himself as an internationally recognized authority on Yeats, Auden—and Paton.[4] In June 1948 he was in Durban for a brief holiday from his studies at Witwatersrand University, and Kirkwood, who had known him during the Italian campaign, got him along to the meeting with the promise that Alan Paton was going to be there.

When Kirkwood and I got there the meeting consisted of two or three Anglican clergymen and us. And the Liberal Party was discussed. Kenneth had expected Alan to be there, but he was out walking. One of them said He's got a new novel working in his head. It would be *Too Late the Phalarope*.

And while we were seated in the room, the half dozen of us, he did come in, and he sat in a chair near the door he'd come through, and he said nothing. Just listened. And when the discussion was more or less over someone asked him direct, Alan, what do you think about this suggestion for a Liberal Party, and he said at once Oh no, Jan Hofmeyr says we should work with the liberal wing of the United Party. And that was that. So he certainly had no thoughts then of starting the Liberal Party himself.[5]

Paton was pessimistic about the chances of a Liberal Party in part at least because he knew very well that only a tiny minority of white South Africans would support policies that led to the handing over of power to Africans. Virtually nobody believed that power could be shared in South Africa; the enormous and growing numerical superiority of the black population meant that under a system of universal suffrage whites would be politically marginalized.

At another meeting, in Johannesburg, someone made a joke about the tiny numbers of whites of liberal convictions. On 11 November 1949, in New York for the opening night of *Lost in the Stars*, Paton wrote a poem which he called 'The Joke', recording this wisecrack, and his own wry response to it:

Let me relate a small affair.
A room. Johannesburg. Ten of us there.
One said, don't think that I presume
But if a bomb fell on this room
Why that would end the liberal cause.
Oh my the laughs! the loud guffaws!

Why, even now, I choke
Over that joke.[6]

He choked because he recognized that it was true, and knew well the extreme difficulty of trying to propound liberal views to a population which already felt besieged by its black fellow countrymen. He considered that Christian beliefs of the unity of all human beings offered the best hope for South Africa, but acknowledged that few white South African Christians practised their faith's beliefs in this regard. Even in his own local church at Port Shepstone, 6 miles from Anerley, to which many black Christians came, they were seated at the back and took communion only after the whites had done so.[7]

Though he did not consider the position hopeless, he believed that a tremendous job of educating white South Africans lay ahead. He saw signs of the difficulties everywhere. At the Durban première of the film of *Cry, the Beloved Country*, (for which the South African board of censors ordered the cutting of young Jarvis's political writings),[8] Paton sat next to the wife of the Prime Minister. Mrs Malan was taken aback by the bleak picture of Johannesburg's slums presented in the film. During the interval she asked Paton, scandalized, 'Do you really think Johannesburg looks like that?' to which he replied, '*Mevrou*, I lived thirteen years of my life among scenes like that.'[9] He did not record her response to this reply, and it is probably only coincidence that within a few years the great clearances of Johannesburg's shanty towns began. The fact was that, like Mrs Malan, many white South Africans had only the vaguest idea of the conditions in which blacks lived.

Opposition to the Malan government's repressive legislation was scattered and largely ineffective in 1950 and 1951. The main political organization representing blacks, the African National Congress (ANC), organized a series of peaceful protests which culminated in a big May Day strike in 1950. This was followed by widespread rioting in which 18 Africans were killed. 'You know about the riots of course,' Paton wrote to Mary Benson. 'I am at one moment anxious to be in the breach, at another anxious to get away & write.'[10] The government's response to this unrest was the Suppression of Communism Bill, which became law in June 1950.

In June 1951 parliament, by a simple majority, moved against the Coloured voters of the Cape by placing them on a separate roll. This Act was struck down as unconstitutional by the Appellate Division of the Supreme Court, and there was also a widespread though short-lived movement of ex-servicemen, under the banner of an organization called the Torch Commando, to defend the Coloured vote. Paton joined the Torch Commando, like hundreds of thousands of other whites, and had great hopes of it for a time.[11] But the organization was fatally flawed: while criticising apartheid, it practised it, being almost wholly segregated itself, and it was to fade away after 1953. Following mass protest meetings of blacks organized by the ANC and other organizations in April 1951, there was rioting when the Torch Commando presented a petition to Parliament on 28 May 1951. The response was another government crackdown. On 26 June 1951 the ANC began a Defiance Campaign against curfew laws, 'whites only' signs, and other manifestations of what later became known as 'petty apartheid'. Blacks sat on benches marked 'Whites Only', or read in libraries reserved for whites, waiting to be arrested. Sympathetic whites entered black townships without the required permits, and 8,326 people had been arrested by the end of 1951.

The government's response to these protests was to show even more determination to force through the programme of separating whites from other South Africans. It brought the Defiance Campaign to a sudden stop by instituting three-year prison terms for protesters. But the Nationalist Party now realized that it needed a new mandate, and a two-thirds majority in Parliament, to accomplish its aim. These it set out to win in the general election of 1953. Whereas the 1948 election had caught many people unawares, the buildup to the 1953 election made it perfectly plain that what was at stake was the future of South Africa, and that in particular the Nationalist Party was seeking a mandate to shut all but white South Africans permanently out of power.

Against his will, Paton was drawn to the conclusion that a new party must be formed to fight Nationalism, and that he must be part of it. The writer in him would much rather have crept away to hide somewhere. 'If I did what I wished, Pixie,' he wrote to Mary Benson, 'I have no doubt what it would be. It would be to buy a small place in the Lake District, & be no more heard of.'[12] But this was a passing mood: he was no hermit, and no pietist either. On the contrary, he was preparing for action. And just as when at Diepkloof he had considered a political career, he began to prepare the ground with a press campaign.

On 14 October 1950 he published his first letter to the papers on political questions since leaving Diepkloof, a letter to the Johannesburg *Rand Daily Mail*. It was an appeal for funds for the social work of the Anglican church among blacks, particularly an organization called Ekutuleni, but it drew a vivid picture of the violence and terror which Paton feared would engulf South Africa if the government did not change course. And it ended with an implicit challenge to the racist ideology that lay behind apartheid, and an appeal to South Africans of good will: 'By identifying ourselves with the work of Ekutuleni, we play some part in the army of God; and to the world, particularly the world of Africa, we testify that under God all men are of one flesh and brothers.'[13]

This was a mild enough first shot, but in his unpublished writings he had begun to produce a series of bitter little parables which showed more clearly what he thought of the course the country was taking. He wrote several of these in September 1952, while writing in a hotel just outside Bulwer in Natal: he tended now to do much of his writing away from Dorrie, a sign that his marriage had not improved. ('Come back very loving,' he warned her on one of his trips away, 'otherwise I shall soon be off again.')[14]

One of the allegories he wrote in the Bulwer hotel, entitled 'The Dangerous Doctor', is an attack on the policies of his old enemy Dr Verwoerd, now Minister of Native Affairs:

> It is a nightmare journey. The car sways from side to side, & the doctor seems to take a delight in going straight for immovable objects and other vehicles. The drivers hurl curses at us as we fly past.
>
> The doctor glares at me.
>
> 'Do you hear what they say?' he says. 'They say we are a danger to the world. Such criticism is intolerable.'
>
> He glares at me again.
>
> 'But even more intolerable', he says, 'is criticism from one's passengers.'
>
> My companion, who is an alien from some foreign country, whispers to me.
>
> 'The man's crazy,' he says. 'I shall write to my friends about it.'
>
> One would have thought that the doctor had overheard us.
>
> 'It is because one's passengers criticise,' he said, 'that aliens hate us.'
>
> Coming towards us is a titanic yellow bus, full of people. I have seen it before, it rejoices in the name of PRIDE OF INDIA. We make straight for it, hooting loudly. There is a shower of curses, from the doctor, & from the other driver, a man wearing a Gandhi cap. Somehow we avoid disaster, but I cry out, 'for God's sake, doctor, let me drive.'
>
> The doctor looks at me balefully.
>
> 'The day you drive,' he says, 'will be the end of us all.'
>
> A passenger from the Waterberg [a deeply conservative area of the northern Transvaal] turns to me.
>
> 'You wait till I drive,' he says. 'This will be nothing.'[15]

And an unpublished poem beginning 'Dr Verwoerd my boss my boss', written in the 1950s after Paton had unexpectedly got a small writer's grant from the government, ends with an ironic prayer:

> My boss, my boss, I hope and pray
> Harder than I can really say

I pray that you will live to see
The end of all your industry.[16]

By 1952 Paton was expressing his opposition to Verwoerd's apartheid doctrines, and his own desire to 'drive', more openly, writing to newspapers like the *Natal Mercury* putting forward distinctly political views in the name of the Liberal cause, and pleading for a middle way between white domination of blacks, and black domination of whites:

Now it seems clear to Liberals that the era of European domination (with or without justice) is drawing to its end. It is not a question of whether this is a good thing or a bad one . . . it just is a fact. No matter what dominations still await man, this is not one of them. Already more and more of our attention in South Africa is being given to the task of maintaining European [that is, white] domination. There will come a point when the demands on us will become unendurable; we will be so busy protecting our lives that it will no longer be possible to live them. The Liberal foresees no end to this but revolution . . .

Liberalism is supposed in South Africa to be an extreme creed; in fact it is a middle creed. It is the only alternative to two kinds of domination, both of which will be intolerable to decent men.[17]

His confident talk of the programme of Liberalism, before the formation of a Liberal Party in South Africa, was the result of his increasingly active discussions about such a party with his circle of liberal friends. During his Diepkloof years he had associated with such thinkers as Professor Hoernlé of the University of the Witwatersrand, Edgar Brookes, who served as one of the Senators representing Africans, and J. D. Rheinallt-Jones, who had been on Diepkloof's Board and now was Brookes's ally in the Senate. Through these contacts, and others he made in the South African Institute of Race Relations, Paton rapidly built up acquaintance with many of the active proponents of liberal thought in South Africa.

Among these were several Cape liberals, including Leo Marquard, Margaret Ballinger, Leslie Rubin, Donald Molteno, Colin Eglin, and others, many of whom were to be important in Paton's political career.

Leo Marquard came of a distinguished Afrikaans family, his father having been Moderator of the Dutch Reformed Church in the Orange Free State, and his mother being descended from a well-known Scots clergyman in the Cape, the Revd. Andrew Murray. He went to New College, Oxford, as a Rhodes scholar, and shortly after his return to South Africa founded the National Union of South African Students (NUSAS) and was its president for seven successive years. He early began to interest himself in building bridges between the races; in 1927 he founded the Joint Council of

Europeans and Africans, he helped to found the Institute of Citizenship in 1947, he was a life member of the South African Institute of Race Relations, and was to be its president in 1957, 1958, and 1969.

Margaret Ballinger came of an impoverished English-speaking Cape family, the Hodgsons, was educated at Oxford, where she had got to know Paton's friend Hofmeyr, worked as an academic under him at the fledgling University of the Witwatersrand, and in 1938 was elected to Parliament as a 'Natives' Representative'. She was re-elected continuously until 1960, serving with another Liberal, Walter Stanford in the House of Assembly, and with Rheinallt-Jones and Brookes in the Senate, representing Africans.[18] She was a large, dominant woman with a very keen mind, but with a tendency to deliver a lecture every time she opened her mouth. Paton never much liked her, but he recognized her worth to the Party.[19]

Leslie Rubin was of a Cape Jewish family and was trained in the law; and like Ballinger, Rheinallt-Jones, and Brookes, he became a parliamentarian, being elected a Liberal Party Senator representing the Africans of the Cape Province. He was urbane, witty, and very much a man of the world, and of the Cape Liberals he was to become Paton's closest friend. Paton seems to have been fascinated by Rubin's polish. The two of them shared a taste for whisky, and would consume large quantities of it together under the benevolent eye of Pearl, Rubin's wife. And when Pearl had gone to bed Rubin would begin telling the *risqué* stories of which he had an extraordinary fund, and Paton not only enjoyed them, but gradually learned to tell one or two himself.[20] What is more, Rubin was one of very few people to whom Paton could talk about his sex life, and Rubin probably came to know more about the difficulties of Paton's marriage than anyone else at this time.[21] Thus began what Rubin was to describe as 'a strange close friendship between a Jewish atheist and a devout Christian . . . perhaps the most enriching experience of my life'.[22]

When in Cape Town in this period Paton would stay with the Rubins at Sea Point, and it was here that many of Rubin's visitors met him. One of them, the American writer Robert St John, left a vivid record of being invited to meet Paton at one of Rubin's parties, of waiting all evening in vain for Paton to appear, imagining that he would be greeted with great fanfare, and finally discovering that Paton had been there all the time, humbly sitting on the sidelines:

> I had another scotch and soda and went to the far corner of the room to try to make conversation with a short, bespectacled, very serious-looking man who I had decided must be a professor of ancient history or perhaps a geologist at the local

university. He had been sitting on the fringe of the circle all evening, only occasionally volunteering an opinion on the matters under discussion.

'Paton,' he said without a wasted word, holding out his hand as I approached him. 'Your name?'

Paton rarely spoke unless someone asked him a direct question, and then he was always modest in answering. He preferred the dark corners. If he was forced to talk he would rather it be about politics than anything else.[23]

For all his modesty, Paton was playing a more and more important role in Liberal circles. In January 1953, with the general election approaching, Liberal-minded groups in the larger South African cities formed the South African Liberal Association at a meeting in Cape Town, to which each of the four provinces sent representatives; Paton was one of the representatives for Natal. The meeting decided to ask a former South African Chief Justice, N. J. de Wet, to be President of the Association and of the Party which it foreshadowed, but he declined. The Association had no effect on the outcome of the general election in April 1953, in which the Nationalist Party increased its majority to 94 seats against the United Party's 58 and the tiny Labour Party's 4. More apartheid legislation was obviously going to follow.

In May 1953, in the wake of the election, the Liberal Association met again in Cape Town and decided on the formation of the South African Liberal Party. Margaret Ballinger was elected President, and there were two Vice-Presidents, Leo Marquard and Paton. Oscar Wollheim was to be National Chairman. Paton was not Vice-President for long. In July 1953 the Party held a National Congress at which it was decided that the positions of National President and Vice-Presidents should fall away; Ballinger became Leader, Marquard the National Chairman and Wollheim the Deputy Chairman.[24] In 1956 Paton would replace Ballinger as Leader of the Party.

It is worth asking what Paton hoped to achieve through the formation of the Liberal Party, for to it he was to give much of his prodigious energy, and a considerable part of his income, from this point until the party's dissolution in 1968. The aim of political parties is to come to power, but Paton never seems to have entertained the delusion that the Liberal Party of South Africa might do so. He was well aware that the mere sight, in press photographs, of whites and blacks occupying the same stage during Liberal Party meetings, was enough to alienate a majority of white voters. He also knew well that the Liberal Party's inability, in these early years, to endorse the idea of universal franchise, was enough to alienate potential black supporters in the African National Congress and the Indian Congress. Nationalists on one end of the political spectrum, and Communists on the other, regarded the Liberal Party with contempt. Not only had Liberals no chance of coming to power; they were never to achieve the election of a single one of their candidates for parliamentary seats under the Liberal banner. Their official representation continued to be the Senators representing Africans, their chief competitors for these positions coming from the Congress of Democrats, a whites-only successor to the banned Communist Party.

Why then did Paton, seeing all this as clearly as he did, join the Party and work so long and so hard for it? The answer, both simple and sufficient, is that, like Luther, he believed he could do no other. 'We had to come out,' he wrote in October 1953, 'because we thought it was the right thing to do, and the sensible one. We hope to accustom to the language of justice and common sense thousands of ears that are accustomed to that of fear and prejudice. We believe that we speak the language of sanity, that it is our view alone that can save our country from a future of tragedy and violence.'[25] More directly, he wrote years later, 'I believed that I could not be true to my Christian beliefs and at the same time keep my mouth shut.'[26] The long-term educative function of Liberalism inspired in him an almost missionary zeal.

He personally had nothing to gain, and much to lose, from having entered on this new and much more public life. One immediate and most unwelcome consequence was that wherever he went from this time on he was apt to be followed by a car containing members of the South African security police. He, who had always been a most loyal citizen of his country, hated this attention and felt both menaced and shamed by it.[27] The security police were to be as constant as his shadow for the next thirteen years.

The police did not have to go to Anerley in search of him, for in May 1953, shortly after the formation of the Liberal Party, the Patons moved to the Valley of a Thousand Hills, inland from Durban. They did not move there for the wonderful climate or the spectacular natural beauty; they went to work, at a wage of £5 a month each, caring for young black patients recovering from tuberculosis in a recently-founded Toc H TB Settlement in whose development Paton had from the start taken a close interest.

Having achieved worldly success, Paton had felt very diffident about his new-found wealth, and resisted the temptation to buy a mansion and a limousine. He longed to do something worthwhile with his money. His thoughts of starting a farm on which young black delinquents could continue their rehabilitation had come to nothing in the excitement of travel, writing, and film-making. By December 1950 his thoughts had turned to founding a Gandhian settlement for black TB patients, and he planned to do this through his contacts with Toc H.

One of these contacts was a most energetic man named Don McKenzie, chairman of the Toc H organization in Natal, a good manager and a man whom Paton considered saintly in his devotion to the black cause.[28] McKenzie and Paton, with a good deal of input from other Toc H workers, built up a TB settlement on 60 acres of flat ground in the Valley of a Thousand Hills, about 10 miles from what is now the dormitory settlement of Hillcrest. McKenzie, thin and craggy, and with a way of glaring under his black brows, was the head of the place and would do most of the administration; Dorrie Paton would keep the records and accounts, and Paton (as he liked to boast wryly) would make the compost. He also ran the workshop in which the young patients learned to turn out sandals and other items, much as his pupils at Diepkloof had done. But Paton also did something else: he funded the Settlement, at least in its early days, virtually single-handed. When the film of *Cry, the Beloved Country* had its première in Durban, Paton donated the entire proceeds to the Tuberculosis Settlement (which McKenzie had founded in December 1950),[29] and he continued to give it substantial sums until McKenzie, an inspired fund-raiser, had brought in enough money from other donors to make the Settlement prosperous. It thrives to this day.

Paton was to pay tribute to the spirit that prevailed at the Settlement: it was the spirit he would have liked to inculcate at Diepkloof, though it had escaped him there because of the age and toughness of his charges. But the Settlement was quite different:

> Many of the patients were small children, in those days all boys, and full of mischief. There were no external evidences of discipline, no fear, no commands, no raised voices. A word of reproach from him [Don McKenzie] was sufficient punishment. When Don appeared in the mornings, he would be surrounded by children. Soon there was an established form of greeting—Oo la la! and in time it came to be the name for Don himself. One could without sentimentality describe the settlement as a place ruled by love.[30]

The Patons lived simply in one of the small houses scattered over the settlement farm, and devoted a good deal of their time to making the place a success. Paton, however, would spend part of each morning writing, and he gave interviews to curious journalists who dropped in to see what the author of *Cry, the Beloved Country* could be doing living in a matchbox house among African TB sufferers. 'I'm the compost maker,' he would say in response to their questions about his role, and at least one of these interviews was published with a photograph of Paton, dressed in his Diepkloof outfit of khaki shorts and shirt, sturdily turning the compost with a pitchfork.[31]

Fame and fortune had made little difference to his appearance. He was 50 years old when he moved to the Botha's Hill TB Settlement, and his fair hair was now receding and greying so that he had a pepper-and-salt appearance. He was rather thinner than he had been in the days when he took a great deal of heavy exercise. His mouth was more firmly set, his frown a little more deeply incised, in token that the unrelenting effort to subdue his desires went on, that struggle of which his severe discipline of others as a schoolteacher and borstal Principal and his writings of moral struggle were external manifestations. But his delight in the beauties of nature and his enjoyment of a good joke were what they had been when, as a brilliant youth at the University College, he had walked the Natal hills in company with Pearse and Armitage, pulling their cart and reciting their poems, with the world at their feet.

Many people were frankly puzzled at this strange turn in Paton's career; what did a famous writer with no medical training think he was doing on a farm for patients recuperating from tuberculosis? The journalists who visited him seem to have suspected that this was a smart publicity stunt. But the truth is that one strong strand in his nature was an urgent desire for sanctity. In 1936 he had written to Hofmeyr from Diepkloof an unintentionally hilarious account of a French saint, Jean-Baptiste Vianney, who lived on nothing but half a glass of milk and two small potatoes a day, heard confessions without ceasing, and slept only one hour a night, on a bed of planks. Paton had concluded, 'I begin to see that the love of God may become so compelling that one does not struggle to give up luxury, comfort, leisure, but that one has no time for luxury, comfort, leisure.' Then, realizing that this sounded a little overblown, he added, 'Do not be alarmed—I am a long way off, a terribly long way off. But my critical self is acutely aware that the vision has never been so clear or so compelling.'[32] The vision did not remain quite so clear or compelling, but there was always a side to him that longed for simplicity, poverty, obscurity, even while another facet of his being pursued power, wealth, and fame. 'You are like myself I think,' he once wrote to Mary Benson, 'a mixture of sinner & saint, earth & heaven, body & soul, desire & aspiration.'[33]

The first side was satisfied with compost-making in the Botha's Hill Settlement; the second rejoiced in interviews, political speeches on behalf of the Liberal Party, and invitations to gatherings such as the World Council of Churches conference at Bossey, near Céligny in Switzerland, in August 1953, in preparation for the second assembly of the World Council of Churches planned for Evanston, Illinois a year later. Paton was invited to serve on the Commission on Race. He enjoyed his visit to Switzerland, although he had to work extremely hard, writing the final report of the Commission on Race himself; and he had time for only one long walk in the mountains, through beautiful small farms on which stoutly built chalets stood lapped in woods, and emerald fields carpeted the valley bottoms.[34] But he was glad to return to the TB Settlement, where he told occasional American visitors that he was now planning to found a semi-religious order to work at the racial problems of his country.[35] The example of St Francis of Assisi was always before him.

It was the Liberal Party, however, rather than any overtly religious order, that took up his time and, increasingly, his income. Because it had little popular support, the Party also got very little financial support, and was always desperately poor. It had essentially two sources of income.[36] One was Alan Paton; the other was Peter Brown, who was to become Paton's closest friend during the second half of his life.

Peter McKenzie Brown was more than twenty years younger than Paton, having been born in 1924. He was the son of a wealthy Durban family, which had made a fortune through a wholesale business, W. G. Brown & Co. Brown was a schoolboy at the exclusive Natal school, Michaelhouse, before the war, when Paton, then still Principal of Diepkloof, came to give the school a talk on behalf of the Institute of Race Relations. It made little impact on Brown, who for some time after he left school was much less interested in politics than in drinking vast quantities of beer and performing such feats as driving a car across the Sahara.[37]

After serving in the forces during the war, however, Brown went to Cambridge, and there attended a meeting addressed by the South African mixed-race novelist Peter Abrahams, author of a powerful novel of black South African life, *Mine Boy*.[38] This meeting sensitized Brown to the plight of black South Africans, and on his return to South Africa he took a degree in Native Law and Administration at the University of Cape Town. By the early 1950s, then, he was one of that tiny minority of South Africans with a strong social conscience, and he formed a small non-racial group who met in his large and beautiful house in Pietermaritzburg, Shinglewood. Paton, who was then living in the hotel outside Bulwer on his own to write, heard of this group, expressed an interest in it, and was invited along. In this way he and Brown met in October 1952.[39] Brown was a broad-built, athletic young man with an extremely keen mind and a great capacity for hard work. Inheriting a comfortable fortune, he would proceed to multiply it several times. But, like Paton, he was determined to use his resources to do something to help others.

In this and subsequent meetings, Brown was profoundly impressed by Paton, who became his political mentor and inspiration, and presently was to become his friend. Under Paton's influence Brown's political views crystallized into a conviction that apartheid was morally wrong, and in its application both cruelly unjust and likely to have disastrous consequences for all South Africans. Through the Liberal Party, of which he was a founding member, he strove with Paton to turn the hearts of their fellow white South Africans. And he, much more than Paton, was to pay a terrible price for standing in the way of the Nationalist juggernaut.

Unlike Paton, Brown did believe throughout the 1950s that the Liberal Party stood a chance of coming to power. In an interview conducted in 1990 he was to remark,

> Alan said in his biography that he never thought the Liberal Party would come to power. But in the early days I did. The mood of those times internationally, the whole process of decolonization was under way at that time. Perhaps it was a naive hope, but one did think that Liberalism could triumph.
>
> From the later 1950s it was clear to the leadership of the party that the emphasis should be much more on extra-parliamentary pressures. The hope was still one of eventually attaining power, but by a different route. So I don't think one could say that at any stage the idea of getting into power was abandoned. Except by Alan, but he would never have expressed that directly I think. Certainly I can't remember him saying that to me.[40]

What made the Liberal Party unique in South African politics was that it was multi-racial. As Leslie Rubin was to argue, 'The formation of the Liberal Party in 1953 marked a basic change in the course of South African liberalism; it also transformed the pattern of white politics'.[41] After the banning of the Communist Party in 1950 only the Liberals argued and demonstrated that non-racialism was a viable course for South Africa to follow. The Indian Congress admitted only Indian members; the African National Congress, in spite of claims that would be made for it later, was from its foundation until this period essentially if not entirely a party of blacks only. The Congress of Democrats had only white members. Liberals set out to woo both white and black members, and this meant forging links with the more liberal members of the United Party, and with the Indian Congress and the African National Congress.

With the United Party the Liberals had some initial successes. The Natal section of the United Party was particularly liberal in inclination, and the Chairman of its Berea branch, Hans Meidner, was one of the prized defectors to the Liberal Party. Other notable white members in Natal included Pauline ('Pondi') Morel, who was the headmistress of an Indian school in Durban, Leo Kuper, professor of Sociology at the University of Natal, the one-time United Party MP, Ray Swart, the mathematician Ken Hill and his wife Jean, and a large number of idealistic young people. For all these people the formation of the Party provided a catalyst and stimulant: for the first time an organization untainted by communism was openly propagating views opposed to apartheid. 'I have no doubt', Leslie Rubin was to write in 1992, 'that the Party was an important factor in the process, pitifully slow though it has been, which has culminated in the new non-racial democratic South Africa which is emerging today.'[42]

The Liberals also had considerable success in attracting Indian members; in Natal by 1955 one third of their members were Indians. One of these had attended Peter Brown's small multi-racial group: he was S. R. Naidoo, a most conservative member of the Indian community. Others to join later included Pat Poovalingam, who had risen from a background of abject poverty, had taken a law degree by sheer determination, and was clearly a rising star in the Durban Indian community. E. V. Mahomed, who had links with the African National Congress, built up a strong Liberal Party branch in Stanger, and later became National Treasurer. Yet another prominent Indian Liberal was Manilal Gandhi, son of the Mahatma, who ran the Phoenix Settlement on Gandhian lines not far from Durban. But the Indian Congress itself was hostile to the Liberal Party: Yusuf Dadoo, the Transvaal leader of the Congress, called the Liberal Party 'half-baked'; and two Durban leaders, J. N. Singh and I. C. Meer, at an ill-tempered meeting which Paton attended with Margaret Ballinger, accused the Liberal leaders to their faces of 'dividing the people'.[43]

The Liberal Party had early success in attracting African members, a goal which Paton and other leading Liberals thought of the utmost importance. The Party's work on behalf of black landowners threatened with the loss of their land in the government's infamous policy of removing 'black spots' (black-owned land in areas zoned for whites) evoked much support from blacks, particularly in Natal, where within two years of the Party's foundation a third of the members were African.[44] Peter Brown was a major influence here; he spoke Zulu and Xhosa, and his multi-racial group had contained some Africans who joined the party, including Selby Mzwana; others to join early on included Elliot Mngadi, Mlahleni Njisane, and Jordan Ngubane. Mngadi, who was to play a prominent role in Paton's third novel, *Ah, But Your Land is Beautiful*, was a small, cheerful man, who showed great courage in leading opposition to the government's removal of black residents: Paton was to publish a tribute to his valour in 1964.[45] Ngubane was a leading black journalist and had been a founder member of the African National Congress's Youth League in 1943; in due course he would become the Liberal Party's National Vice-President.

But Ngubane was one of very few members of the ANC to join the Liberal Party. The ANC, then as in the 1990s, was, despite its small size at this time, an umbrella organization with a range of factions, not all in agreement on any given issue. An important section of the ANC in the early 1950s was profoundly influenced by the Communist Party, which, after its banning in 1950, had set up a cover organization deceptively named the Congress of Democrats. Not all members of the Congress of Democrats were covert Communists, but a large percentage were, and the organization's direction and agenda were controlled by the Communists. The South African Communist Party had been actively wooing the African National Congress for many years, and its influence on the ANC was long-standing. The transition from Communist Party to Congress of Democrats had not greatly disturbed this relationship. The Communists were implacably antagonistic to the Liberal Party, which had declared its opposition to 'all forms of totalitarianism such as fascism and communism'.[46]

Paton soon found, when he approached the ANC on behalf of the Liberal Party, that the ANC was unwilling to countenance formal links with the one legal non-racial party in the country. In October 1953 he and Peter Brown visited Chief Albert Lutuli, who had become President of the ANC in 1952, and had a long talk with him and the Secretary of the ANC, M. B. Yengwa. They followed this up with a second visit in February 1954.[47] Lutuli was a physically impressive man, and Paton was strongly predisposed to admire him; he was presented with a signed photograph of Lutuli, which would hang on the wall of his study for the rest of his life, among his other workplace treasures.

Over the years the two men would build up a good personal relationship, and Lutuli found Paton very useful. For one thing, Paton on several occasions wrote his speeches for him. An example is the speech Lutuli delivered when given an honorary degree by Glasgow University in 1964;[48] even more strikingly, his magnificent acceptance speech when he was awarded the Nobel Peace Prize in November 1961 was almost certainly written by Paton, whose style is everywhere evident in it.[49] Paton liked Lutuli, even though he considered that the ANC leader suffered from an 'overwhelming weakness of laziness'.[50] Lutuli was in fact a man of great charm but little drive.

In spite of the goodwill between the two on a personal level, Paton could not achieve what he sought, which was overt and practical co-operation between the Liberal Party and the ANC. He found Lutuli friendly, but emphatic that the ANC would have nothing to do with any group willing to be party to whites-only elections. Paton reported to Leslie Rubin:

> We agreed to differ on the franchise question and we agreed to co-operate where possible. Luthuli [*sic*] was, however, emphatic that Congress would take no interest in elections on the 1936 model. He would therefore not raise a finger to support any candidate, but I myself have heard that Congress may advise a different attitude.[51]

Paton was right to think that Lutuli was not his own master, but wrong as to the direction he believed Congress would move in. Personal relations between Lutuli and members of the Liberal Party would continue to be good, but in general the ANC would keep its distance from the Liberals. Nor was the reason hard to find: as Paton reported to Leslie Rubin in Cape Town, 'It is the Communist element which is bitterly opposed to the Liberal Party.'[52]

The ANC could not but recognize that the Liberal Party, in its courageous opposition to apartheid, was a valuable ally, and it was undeniable that those Senators who were Liberals (Ballinger and Rubin)[53] were an authentic voice for Africans, until the abolition of 'Native Representation' in 1960 silenced it. Lutuli himself recognized this when he addressed a Liberal Party Congress in 1958:

> It [the Liberal Party] stands for and represents lasting values, values which would make South Africa a country to be honoured. We in the ANC would particularly like to work with the Liberal Party. I must say that we do usually co-operate in those matters where we agree and as the years have gone by, we have found ourselves more and more in agreement with the Liberal Party.[54]

These were kind words, but the fact was that a gulf lay between the two organizations.

In 1953 Paton also reported to Rubin that the ANC had brought up the issue of 'unconstitutional action', that is, some kind of guerilla campaign of the type the ANC would launch in 1961.[55] If the fact that Liberals were willing to support a limited franchise was one wedge between the two organizations, the question of violence was another. Paton, with his strongly pacifist upbringing and belief in Christianity as the way of peace, was utterly opposed to acts that would cause hurt to any human being, even if attempts were made to justify such acts as 'counter-violence'. He did not believe that ends justified means. He believed in the rule of law, not the rule of the Communist Party.[56] Lutuli was a moderate too, but the pressures within the ANC were increasingly beyond his control. From this very first meeting with Lutuli, then, in October 1953, the issues that would divide Liberals from many in the ANC emerged with painful clarity, despite all the friendly intentions.

Paton liked such Communists as Bram Fischer, the distinguished Afrikaner who was the leader of the South African Communist Party, but he did not doubt that the Communists, both white and black, would turn and destroy Liberals if they gained power. The young Liberal Peter Rodda, one of Jonathan's university friends who greatly admired Paton and who visited the Paton house often during the 1950s, remembered that Dorrie told a little joke which illustrated her views on this clearly enough:

> Tonto and the Lone Ranger are riding along and they suddenly see that the whole horizon is ringed with Indians on their horses. The Lone Ranger turns even paler than usual, and he says 'Look Tonto! We're surrounded!' And Tonto says, '*We*, white man?' Psychologically a very interesting joke.[57]

But the Communists were not the only group opposed to the Liberal Party. The Nationalist Party, of course, scorned and derided them, and from the start harassed their members. This opposition, like that of the Communists, Paton could accept with equanimity. It was in a sense an affirmation of the rightness of Liberal values: these were enemies that did Liberalism credit. But there was other opposition much more unexpected and more hurtful.

Paton found to his great dismay that Toc H, the organization to which he had given so much of his time and energy since 1928, and of which he had been Honorary Commissioner for South Africa since 1949, was also opposed to non-racial politics being in any way associated with itself. Paton had believed that Toc H's message about the brotherhood of all human beings applied to black ones too, in spite of the fact that Toc H branches in South Africa had always declined to have black members. Now, after a quarter of a century of dedication, Paton found that Toc H objected to his engaging in Liberal Party work while he lived at the Toc H TB Settlement. One of his co-workers there, Bill Evans, another Liberal, was obliged to leave the Settlement for having helped Paton call and advertise a Liberal Association meeting at the Settlement.[58] Paton, angered by this, determined to leave the Settlement too, and did so at the end of 1953. He became increasingly convinced that Toc H members were satisfied to have no black members, and when the organization showed its opposition to the Liberal Party openly Paton broke with it for good in 1956. 'To put it very bluntly', he was to write in 1976, 'I think that Toc H S. A. has been false to the ideals which it proclaims.'[59]

After leaving the TB settlement at the end of 1953 he and Dorrie moved into the Rydal Mount Hotel in Durban for three months while they hunted for a house, and were very tired of hotel life by the time they had found and bought one, at 23 Lynton Road, Kloof. Kloof, unlike Hillcrest in those days, was a well-developed, plush dormitory town, with wide tree-lined streets and big houses in huge leafy gardens. The Patons' new house was in fact old by South African standards, an odd rambling structure which gave the strong impression of having been designed without the help of an architect. It was essentially L-shaped, but with a round, rondavel-like appendage in the front, where there was a big living room. The large front door led straight into this room, from which bedrooms opened off, and a long passage led past other bedrooms and via a covered way, half passage and half verandah, to a big kitchen, added by the Patons, and a dining-room.

In this dining-room stood an enormous, kidney-shaped table which Paton had had specially made, and round it a great many informal Liberal party gatherings were to take place. Peter Rodda remembered in later years how Pat Poovalingam and his wife Sakhunthalay were often there when he visited, and Jordan Ngubane sometimes, though he lived further away; frequently Peter Brown would turn up, and many younger liberals too. Rodda recalled, too, that it was round that table that Walter Saunders, a dilettante poet and painter, and Tom Sharpe, the novelist, who was teaching in Pietermaritzburg at the time, almost came to blows on one occasion.[60]

There was a huge garden, 5 acres of it, and at the beginning of 1956 the Patons made it even larger by buying part of a neighbour's property. In this garden, but nestled near the house, Paton built himself a work room in the shape of a rondavel with large windows and thatch picturesquely arched, a cross between an African hut and a small English cottage. The garden was the real glory of the place, or rather it gradually became so as Paton landscaped it, planned and built driveways, paths, and ponds, planted literally dozens of indigenous trees, and laid out a huge croquet lawn. Croquet was from henceforth to be one of his passions, and he played it with barely controlled aggression, delighting in victory and being downcast by a rare defeat.

23 Lynton Road rapidly became a place where members of the Liberal Party felt free to drop in at almost any time. The Patons kept open house. Dorrie Paton had supported Paton's decision to join the Liberal Party without apparent hesitation. Whatever else might have divided them, in this they were at one; Dorrie had became deeply religious since her marriage to Paton, and she needed no lessons on her duty to her neighbour. She was quiet and seemed timid, and few who got to know her at this time would have guessed at the way she had dictated terms to Paton early in their relationship. But she had a character of her own, and many who came to hear Paton stayed to befriend Dorrie. Pondi Morel was to say of her,

> At first you thought of her as Alan Paton's wife. But she soon came across very clearly as a woman in her own right. And she seemed to be closer to all of us than he was, and she was more accessible. He was a little frightening sometimes, you know. It was partly his way of peering over his glasses in a rather judgmental way. And the cut of his mouth. With Dorrie it was always a much easier relationship. And she was very friendly with all the people he was closely associated with.[61]

Nor was she as strait-laced as she appeared. Leslie Rubin, seated at the kidney-shaped table for breakfast, listened to Walter Stanford and Paton talking about birds. Stanford, who represented the Africans of the Transkei in the house of Assembly, seemed to know as much about birds as Paton, and Rubin determined to deflate him. 'Walter's favourite bird, of course', announced Rubin to the company, 'is the Brown Transkei Tit', and Dorrie shook with laughter.[62] She remained, until near the end of her life, the zestful, amusing and life-loving woman Paton had first been attracted to in Ixopo, capable at any moment of telling an off-colour joke or doing a Zulu war dance.[63]

As the grip of apartheid tightened on the country, 23 Lynton Road became a non-racial meeting-place, where people of liberal views could call in unannounced at any time after mid-morning, be offered endless cups of tea round the dining-room table (or glasses of whisky, later in the day) and find themselves in a circle that included not just their famous host and his wife, but intelligent and politically informed people of all races. For many of them it was their first experience of such a gathering, and it served as a vision of what South Africa could become, as indeed the Liberal Party itself did.

It was at Kloof that Paton's huge birthday parties, later such an institution, began. Paton had always liked to have his birthdays commemorated by a family party, but now he began inviting his political friends as well. The first of these semi-public parties, held in the big living room, seems to have taken place on 11 January 1955, and they grew year by year until they constituted large gatherings of the Liberal Party faithful. They quickly began to follow a formula, with a toast elaborately proposed by one of Paton's friends, a poem which had been written for the occasion read out, and a lengthy and usually very funny reply from Paton. Dorrie would uncomplainingly perform prodigies of catering. As the 1950s wore on and the pressure on the Party, in the form of harassment and bannings of prominent Liberals, increased, these parties were a way of rallying and encouraging the remnants of the Party, and reaffirming the worth of what it stood for.

This apparently endless socializing came to a stop only when Paton was travelling overseas, which he did a good deal after 1950. In most years he would make at least one trip, in some he made two or three, and his love of travel never waned. Early in 1954, for instance, he travelled to America at the invitation of *Collier's Magazine*, and there undertook an extensive tour of the country in preparation for writing two articles on the position of blacks in the United States. His travelling companion was an unusually talented young photographer, Dan Weiner, who photographed Paton talking to blacks and whites all over the United States.

This was the point at which Paton, who had always admired the energy and the freedom of the United States, came to recognize clearly the major difference between the rights of the individual in his own country, and those same rights in the United States. The rights of American citizens, he realized, could not be abrogated by even the most powerful and determined administration, because they were protected by the American Constitution, and defended by a separation of powers within government. No American government could do what the South African government had done and intended to do: disenfranchise a large group of its citizens, and then proceed to ride rough-shod over them. Paton came to envy the United States Constitution and system of checks and balances very deeply, and often expressed the longing for his own country to have such a system. In this he found himself challenged by the radical views of Weiner, who believed that the gap between constitutional theory and practice in American life was wide. Perhaps in consequence, Paton's articles very clearly brought out the disadvantaged position of black people in the United States, and did so nearly ten years before the civil rights movement gained momentum there.

Paton and Weiner got on very well together. It was Weiner, whose energy was equal to Paton's own, who proposed that the two of them should also do a book on South Africa together. Once the *Collier's Magazine* articles, 'The Negro in America Today', had been published, to great acclaim, in October 1954 (they won Paton the Brotherhood Award of the National Conference of Christians and Jews in 1955), Paton and Weiner did a lengthy tour of South Africa, and the resulting coffee-table book, *South Africa in Transition*, was published by Scribner's in 1956. Weiner died in an air crash the following year.

Following his tour for *Collier's Magazine*, Paton returned to South Africa briefly before leaving again, this time for Britain, taking a very excited Jonathan with him on 1 July 1954. Jonathan was now 18 and in his first year at Natal University. This was his first trip to Europe, and he never forgot the excitement of the flight, and going with Paton to meet Dorrie, who, with a niece, Jill Richburn (daughter of her sister Rad), had travelled by sea from South Africa, so that Jonathan was reunited with his mother on Southampton dock. To judge by the detailed diary Dorrie kept throughout this trip, all three Patons enjoyed the stay in London followed by a tour by car to northern England and through Scotland, returning to London on 30 July 1954. From here Jonathan flew back alone to South Africa, while Dorrie and Jill Richburn toured Scandinavia from August 9 to 21, and then visited France and Italy until September 15.

Paton, meanwhile, had flown to the United States again in August 1954, to attend the second council of the World Council of Churches in Evanston, Illinois. He had been appointed the scribe of Section V of the assembly, which was detailed to prepare a report on 'The Church Amid Racial and Ethnic Tensions'. Here he had the interesting experience of hearing the representatives of the South African Dutch Reformed churches, known by their initials as the NGK and NHK,[64] try (and fail) to justify apartheid in theological terms to a huge gathering which considered segregation utterly un-Christian.

Paton made two friends at Evanston: one was the German pastor Martin Niemöller, who had opposed Nazi ideology and had been sent to Sachsenhausen in 1937, and later to Dachau, where he spent the war and miraculously survived. He and Paton took to each other at first sight: Paton had heard of Niemöller at the Conference of Christians and Jews he had attended in London in 1946, and greatly admired his courage; while Niemöller had read and been moved by *Cry, the Beloved Country*. The two of them took all their meals together, and parted from each other with great regret when the meeting came to an end.

The other friend was Liston Pope, Paton's room-mate at Evanston. Pope was a distinguished churchman and academic, Dean of the Divinity School at Yale. He and Paton collaborated on a paper, 'The Novelist and Christ', a detailed analysis of the problems of presenting Christ in works of literature.[65] It was Pope, long an admirer of Paton's writing, who arranged for Paton to receive the first of what were to be many honorary degrees, that of the Doctorate of Humane Letters, awarded him by Yale Divinity School on 28 September 1954 once the World Council of Churches meeting came to an end. Three bishops received the honour with him, including the Archbishop of Canterbury.

He returned to South Africa on 6 October 1954, curiously nervous to be getting back to his own country, and was received at Jan Smuts airport in Johannesburg in the way he would come to know well: the security police waiting at the entrance, picking him out with stares of implacable hostility; the immigration official recognizing his name in the passport, and handling it as if it belonged to Satan himself; the customs official turning his case inside out with brutal thoroughness and repacking it, Paton thought, as roughly as possible. And then the carload of secret policemen to follow him as he left the airport. He was home again. He pretended not to mind, but he did mind, very much. An honourable man striving to do his best for his country does not easily get used to being treated like a criminal.

He threw himself anew into working for the Liberal Party, which was now thrashing out, at its annual congresses, the divisive issue of the franchise. The Conservative wing, made up chiefly of Cape Liberals, notably the brilliant lawyer Donald Molteno, argued for a limited franchise along the lines that applied to the Cape Coloured people. No one should have the vote unless they had passed the eighth year of schooling—except whites. More radical Liberals, mostly from the Transvaal and led by Marion Friedmann, the highly intelligent wife of a Johannesburg ophthalmic surgeon, opposed this as contemptible racism of the sort the Liberal Party was trying to fight, and said that if it became Party policy there would soon be no black Liberals at all. They would all have gone over to the ANC, which Friedmann considered was increasingly controlled by the Communists of the Congress of Democrats.

On various sides of this central and complex debate were strong characters such as Hans Meidner and Leo Marquard,[66] and the Liberals were a very fractious party indeed. The Party also seemed to attract more than its fair share of extraordinary eccentrics, such as the gentle mathematician Ken Hill, who wore an overcoat in the heat of Durban's summer, and who would stand for many minutes admiring a scene he found beautiful, inclining his head at an angle to get a new view of it, or turning his back on it and looking over his shoulder in a way that aroused a good deal of interest among passers-by.[67]

It was Paton who held the party together, with Peter Brown as an increasingly respected ally. 'Alan had so much influence on all of us, he was the leader, there's no doubt about it,' Pondi Morel was to say after his death. 'It was a small group, but he certainly was the leader. I don't mean that we always agreed with him, all of us, not even Dorrie. But we had an absolute sympathy with what he so fervently believed in.'[68] In a real sense the Liberal Party was Paton's party, for all that many others made great contributions of energy and intellect to it. 'The Party is badly split over the franchise resolution,' Paton wrote to Dorrie from New York on 4 September 1954, after getting his honorary degree from Yale. 'Much depends on what I decide.'[69] It was true, and not just of the franchise issue.

Although he had initially supported the limited franchise proposal, he came down in the end, and with some reluctance,[70] on the side of a universal franchise. 'I frankly do not think that the present state of the world any longer permits us to think in any terms but those of complete equality,' he wrote. His was the view that prevailed, though the Cape Liberals thought him crazily idealistic. They predicted, correctly, that this move would sink any chances the Party might have had of winning parliamentary seats. Paton did not budge from his view that it was the right decision morally, and hang the consequences. 'The trouble with you, Paton', said Molteno angrily, 'is that you think the Liberal Party is a church.' Paton thought this remark very funny because he knew it was half true, and it was to find its way into his third novel, *Ah, But Your Land is Beautiful.*

Molteno, Ballinger, and the other Cape Liberals were proved right: the Liberal Party's decision, at its congress in 1954, to embrace the concept of universal franchise destroyed whatever chance there might have been of Liberal successes in the polls. The overwhelming majority of white South Africans regarded the Party's programme as a formula for handing over power to blacks, and they believed the consequences of such a handover would be disastrous for the country. They had little difficulty in persuading themselves that continued white rule was in the best interests of whites and black alike.

The Party continued to contest seats at provincial and national elections until its forced dissolution in 1968, but won none of them. In most cases its candidates lost their deposits. In some Natal constituencies even the three registered voters required to nominate a candidate for election could be found only with difficulty. Finding Party members willing to stand for election became very difficult, and the joke went round in Party circles that a number of candidates had agreed to stand only on condition that they should not be required to sit.[71] Lamentably for the Party and for South Africa, there was no likelihood that any of them would win seats, nor did they.

An eventual result of their loss of this crucial debate in 1954 was the quiet and gradual withdrawal of Molteno, Ballinger, and other Cape Liberals from active participation in Party affairs. Ballinger herself was to resign the Presidency of the Party at the end of 1955, but well before she did so she was playing a diminishing role in the organization. Accordingly Paton and others felt impelled to step into the breach. Paton spent more and more time travelling to make speeches at Party meetings, often in company with Leslie Rubin. They were a good team.

Paton's gifts as an orator, honed by his debating experience at university, were very remarkable, and they rapidly developed during his public career until he struck many people as an electrifying speaker. He had superb timing and delivery, and his range of speaking talents was astonishing: he could be acerbic, pugnacious, learned and compellingly logical, witty and profoundly moving by turns, and many of those who heard him never forgot his eloquence or his power.[72] He tended to cast into the shade those with whom he shared a platform, though on occasions they fought back.

Once, addressing a meeting with Paton, Rubin spoke first and at length, ending by modestly saying that he was just the hors d'oeuvre, and Paton the main dish. When Paton's turn came he thanked Rubin ironically for his remarks. 'But when one goes to dinner at eight and one finishes the hors d'oeuvre at ten to nine', he said, looking at his watch, 'one must expect to take the main dish in moderation.'[73] On another occasion Rubin, during a meeting at Sea Point, was being troubled by a heckler who made an amusing remark at Rubin's expense. 'That's not funny,' said Rubin severely to the audience, and then looked over his shoulder to see Paton laughing his head off on the platform behind him.[74]

In Natal Paton usually spoke in tandem with Peter Brown. One of the many young people who were attracted to the Liberal Party on hearing them was Tony Morphet, later to become a distinguished academic, who remembered his surprise at discovering that this famous writer was not only alive, but politically active. He first encountered Paton at a meeting in Pietermaritzburg in 1959, when Morphet was a naïve student of 19:

> I naturally assumed that authors lived in England or were dead, and I was amazed to see him appear, and absolutely astounded at the timbre of the speech, the clipped, very controlled, slightly Americanized speech it sounded, and the very powerful emotional push underneath the tightly controlled language. I thought God Almighty, this is something else. Peter [Brown] was by comparison diffident and large and helluva nice, but you thought, The Power is with Paton.[75]

Morphet was right: the power was with Paton, and he recognized the responsibility this placed on him. His time taken up with Party work and travel, he wrote relatively little in these years of the mid-1950s, and what he did produce was mostly journeyman work. In 1954, apart from the book he did with Dan Weiner, and in which his text very much played second fiddle to Weiner's splendid photographs, he also wrote for the American publisher Lippincott a book he called *The Land and People of South Africa*.[76] This was essentially a school text, part of a series of 'Portraits of the Nations', but Paton took it seriously, doing a great deal of research to get his facts on geography, history, and sociology right.

His most characteristic chapter was the one he entitled 'The Future'. In this he distinguished eight political groups in South Africa: those who stood for white domination, total apartheid, white leadership, the common society (this group naturally included the Liberals), Communism, the African National Congress, the Indian Congress, and the Coloured People. He reaffirmed his faith in the notion of a society common to all South Africans, and made a prediction, particularly bold in 1954:

> When *total apartheid* is seen to be impossible, what will the Afrikaner intellectuals and religious leaders do? Will they choose white domination or the common society? Surely, with their intellectual qualifications and their moral views, they must choose the *common society*.
>
> *What an important thing for South Africa this would be! I cannot think of any more important or more exciting thing that could happen. The danger of revolution would recede. The danger of conflicting nationalism would recede. Millions of hopeless people in South Africa, both black and white, would begin to look to the future with hope.*[77]

What is striking about the passage is Paton's belief that change in South Africa would come about only when the Afrikaner leaders came to the realization that it must. He did not imagine that they would be forced from power by blacks, or that Liberals by some electoral ju-jitsu might take their place. The initiative, he believed, would remain with the Afrikaner until the Afrikaner chose to give it up. The startling accuracy of his prophecies was the result, not just of deep knowledge of his native land, but of his long and intense study of the Afrikaner mind. He was to die before President de Klerk's speech in Parliament of 2 February 1990 or the release of Nelson Mandela, but he would not have been surprised by these developments as so many other observers were surprised.

Meanwhile the apartheid juggernaut rolled on, crushing all opposition. In November 1954 the 80-year-old Malan retired, being replaced as Prime Minister by J. G. Strijdom, who cared even less than Malan about constitutional propriety. Under Strijdom the Nationalists altered the constitution so as to make it much more difficult for the courts to challenge apartheid legislation, and they subsequently moved to disenfranchise Coloured voters in the Cape. A Women's Defence of the Constitution League, popularly known as the Black Sash, was formed in May 1955 to protest at these changes. Dorrie Paton was a founding member, and she and the other members (all white, since blacks could not have protested with impunity) would stand silently, wearing black sashes as a sign of mourning, at any public gathering where a government minister was likely to be present. Individual ministers were embarrassed by this attention, but the course of government policy did not swerve by a hair.

For his part Paton was trying to enlist the support of some of his powerful foreign contacts, particularly in America, to bring pressure to bear on the South African government. When the government began to prevent black students from taking scholarships abroad, for instance, he and the energetic priest Trevor Huddleston wrote in 1955 to such American figures as Mrs Eleanor Roosevelt asking for their intervention.[78]

Huddleston and Paton were increasingly friendly at this time. Always an activist, Huddleston was now losing patience with the slowness of the Anglican church to take action on behalf of blacks. His work among Africans in the sprawling Johannesburg slum, Sophiatown, gave him a good view of the increasingly difficult conditions in which blacks were forced to live, and he had many contacts in the ANC. One of these was Oliver Tambo, later to serve as President of the organization, who in 1954 was the victim of a banning order under the Suppression of Communism Act. Angered by the inaction of the church, Huddleston wrote and published in the *Observer* of 10 October 1954 an article entitled 'The Church Sleeps

On', in which he called on Christians in South Africa to resist the apartheid laws, and for Christians all over the world to isolate South Africa through a cultural boycott of the country. In later years Huddleston would boast, with some justification, that this article was the first move in the series of boycotts that did so much to depress the South African economy.[79]

The destruction of Sophiatown, in which stood Huddleston's large Church of Christ the King, affected Huddleston deeply. The place was a vast and festering slum that cried out to be rebuilt. The government, however, intended not just to rebuild it, but to move its black inhabitants out and replace them with whites, the area having been 'rezoned' for white use under apartheid legislation. The new white suburb, as a further token of the arrogance of the government, would be named Triomf, meaning 'Triumph'. Huddleston became the chief white spokesman against the removal: he addressed meetings, held press conferences, wrote newspaper articles, and appealed to prominent people everywhere. When the removal took place, with the precision of a military operation, on 10 February 1955, foreign journalists from all over the world reported on it, and the words many of them quoted, and the face many of them photographed, were the words and the face of Trevor Huddleston. From this time on he was seldom out of the limelight for long.

Huddleston went on to take an active part in the Congress of the People, called by the ANC at Kliptown, south of Johannesburg, on 26 and 27 June 1955. Paton and the Liberal Party had not attended this event, knowing that though it had been proposed by Professor Z. K. Matthews, the black Vice-Principal of the African University College of Fort Hare and a most respected figure in the ANC, it was modelled on a Communist Assembly of the People held in 1944–5. Paton also knew that the 'Freedom Charter' which was to be adopted had been drafted with the help of Congress of Democrat members, had not been distributed in advance to members of participating bodies, and had not been checked by either Matthews or Lutuli.[80]

During September 1955 Huddleston spent his annual holiday with the Patons, most of it sitting at a small card-table, writing a book about his recent political campaign. It was to be published as *Naught for Your Comfort*, and it would be a best seller: its title contained an ironic reference to the subtitle of *Cry, the Beloved Country: A Story of Comfort in Desolation*. As he sat at the card-table one morning, Huddleston was brought a letter: it was from the Superior of his Community, Father Raynes, and it recalled him to England.[81] Huddleston's methods had annoyed not just the South

African government, which longed to rid itself of this meddlesome priest, but the Bishop of Johannesburg, Geoffrey Clayton, and even the Archbishop of Canterbury, Geoffrey Fisher, who had warned Huddleston that for a Christian to use the weapon of force is to sink to the level of the enemy he fights. Huddleston remained unconvinced, and his Superior had now recalled him for his own good and that of the Community of the Resurrection.

To be withdrawn from South Africa in this way was the most profound shock to Huddleston. He told Paton subsequently that he could see no light, and dreaded going back to England.[82] Paton came to believe that in a sense Huddleston never got over it, and certainly his life thereafter suggests an obsession with the experiences he had had in Johannesburg, an obsession which at times was to lead him into terrible depression.[83] His life became a life of protest, and Paton was to think it pitifully loveless. 'To tell the truth', Paton wrote, 'sometimes my heart aches for him.'[84]

Paton considered it vitally important to maintain a balance in his own activities, believing that too complete an involvement in South African politics would, to use Yeats's phrase, 'make a stone of the heart'.[85] Accordingly he interspersed political activity with periods of travel and writing which helped him to distinguish the things he could change from those he could not. On 2 November 1955 he set off on another of his trips to the United States, which he loved more on each visit. On this occasion he had accepted the invitation of Father John Patterson, the headmaster of a remarkable church school, Kent School in Connecticut, to spend two months teaching there and taking part in a seminar on 'The Christian Idea of Education'. He had been approached the previous year by Louis Stone, an ex-pupil of the school, to make this visit, and had with difficulty convinced the forceful Stone that he could not possibly stay for a whole year.

Kent School had been founded by Frederick Sill, a member of the Episcopal Order of the Holy Cross, in 1906. Sill chose a magnificent wooded site in the valley of the Housatonic River and started from scratch, his first building being an old farm house. By the 1950s Kent was richly endowed with magnificent Georgian-style brick and stone buildings, and enough money to fly Paton and various other luminaries across the world in great style to its seminars. Paton, who had seen the place briefly in 1954 and would return two or three times in later years,[86] lived in the Visitor's Lodge, walked with delight in the flaming autumnal woods, taught English poetry, and talked to the boys about South Africa. The Kent system puts the boys in charge of a great part of their own discipline and of the running of the school. They struck Paton as mature, well informed, and inquiring, and the quality and nature of the teaching seemed to him greatly superior to anything he had come across before. The library exceeded anything he had ever seen outside a university. Kent, he rapidly concluded, was one of the great schools of the world.

The seminar itself was a high-powered meeting, drawing together experts in many fields. Two of the participants made an impact on Paton. One was Cyrus Vance, an alumnus of the school who would become Secretary of State under President Carter, and still later would act as the United Nations mediator in the Bosnian conflict during 1992–3. The other was the theologian Reinhold Niebuhr, whom Paton had first met at the British conference of Christians and Jews in 1946. Niebuhr had recently suffered a stroke, but delivered from his chair a paper which Paton thought the equal of the masterly address Niebuhr had delivered in 1946. He was to correspond with Niebuhr and his wife Ursula until the end of their lives.

Not all the papers were fascinating. During an address by a Dr Florovsky, a Kent staff member who was sitting next to Paton became aware of a steady, but gradually increasing sound. Paton had dropped off to sleep and was snoring. His neighbour prodded him awake, and Paton was amused rather than embarrassed.[87]

He enjoyed himself thoroughly at Kent, avoiding loneliness there by socializing with the staff, receiving visits from American friends, and striking up new friendships. One of these was with Edmund Fuller, the admirer who had met him briefly in Annie Laurie Williams's office in 1950, and who was now teaching at Kent.[88] Some of the other staff members of the school, notably the headmaster, Father John Patterson, and his wife Betty, became and remained Paton's friends for the rest of their lives.[89] After Patterson started a school in Rome in 1962, the Patons visited him and his wife Elizabeth there in 1971, and spent some strenuous days with them. The man who had been instrumental in getting Paton invited to the school, Louis Stone, a huge man with tumultuous energy, Paton was never so close to. He found Stone on occasions slightly menacing, and was not surprised when he in later years had a severe breakdown and committed suicide.[90]

After finishing his stint at Kent, Paton gave addresses to the English Speaking Union and to Princeton University on 4 December 1955, 'by which time,' Dorrie had predicted memorably, 'I should imagine he will be feeling like the Queen of Sheba when she beheld the glory of Solomon.'[91]

It was true that he found the wealth of the Ivy League universities slightly overwhelming. At a party in New York he was introduced to W. H. Auden. 'I must confess that we did not "click". My clearest recollection of him is that I had never seen such a ravaged face in my life.'[92]

He was taken to the Yale-Harvard football game in a snow-storm: 'I sat in the falling snow for two hours with 60,000 other halfwits,' he wrote caustically to Dorrie, 'but have suffered no ill effects whatever.'[93] His speaking engagement at Yale was called off because of a blizzard. He had then intended to go back to California, and to do again what he did at the end of 1949: hide himself in the redwood forests and write.[94] In the event it was not to California that he went, but to Long Island. *Too Late the Phalarope* had been adapted for the stage by the playwright Robert Yale Libott (unsuccessfully, as it turned out)[95] and was being funded by a wealthy American producer, Mary K. Frank, who had offered Paton the loan of her holiday house at Montauk Point. Accordingly he spent Christmas and New Year there, not enjoying himself much. For several weeks he tried keeping monastic Hours, reading the prayers aloud, but soon found he could never get up for Prime or work well between the fixed times of the programme.[96] Nor did the sight of the grey waves of the north Atlantic rolling in under a grey sky cheer him up. One sign of the stress he was under is that for the first time, in January 1955, his letters begin to complain of the migraines that were to plague him for many years,[97] at times becoming so bad that he would walk around his room in agony, stepping over the bed, onto the floor, around the foot of the bed, back onto the bed, on and on and on.[98]

He did not write, or even begin, another novel at Montauk Point, but he did make good progress with a play,[99] the first he had attempted since his failure to write one for Luther Greene. It was about the dying explorer David Livingstone being carried to the coast by his bearers, and it was deeply religious: Paton intended it for church performance, and he originally called it simply *David Livingstone*, in time retitling it *Last Journey*.[100] He also wrote at least one short story, of which he gives no details in his correspondence, and a poem for John Patterson, beginning,

> My plan be in Thy mind O God
> My work be in Thy hands
> My ears be ever swift to hear
> The words of Thy commands.[101]

But the writing he did on Long Island did not satisfy him: 'Sometimes I think I should be psycho-analysed & find what is the deep root of my lack of creative joy, & anxiety', he wrote to Dorrie, about whom he was dreaming a great deal, dreams in which she was associated with locked doors and windows.[102] 'I suppose you know what I'm talking about,' he told her after detailing his latest dream of sexual frustration; but if she did she made no sign.[103] At his repeated urging she joined him in America for the last weeks of his stay, her fare being paid by Kent School, but her presence seems to have done nothing for Paton's productivity.

Two other influences prevented him from writing really well on Long Island. The first was that, as had happened at Lane's Flat, he felt that no urgent novel was waiting to be written: 'I am aware of my own loss of creative power', he wrote mournfully to Dorrie.[104] The second was that South African politics were urgently summoning him back home: in his absence the South African Liberal Party had elected him National Chairman, early in 1956. What she saw as the leftward movement of the Party had persuaded Margaret Ballinger to resign as leader in December 1955. Paton accordingly took over the leadership of the Party, and Peter Brown became Vice-Chairman. Paton was to serve as National Chairman until June 1958.

The Patons flew home via Rome, arriving back in Johannesburg on 18 April 1956, as Paton was to put it, in 'the land of the Liberal Party and the Security Police'.[105] He must have known that he was going back to face the intensified attacks of a state which increasingly used its power in a tyrannical way, and he knew, too, that he would be hard put to guide the Liberal Party through the rapids that lay ahead. What he could not have foreseen was that one great danger would come from within the Party itself. Some younger Liberals were growing increasingly impatient with the Party's policy of remaining within the law. And Paton, for all his skills and all his wisdom, would not be able to contain them for ever.

18

STALKED BY THE STATE 1956–1960

> There is always another one walking beside you
> Gliding wrapt in a brown mantle, hooded
> I do not know whether a man or a woman
> —But who is that on the other side of you?
>
> T. S. ELIOT

Paton returned to South Africa to find the government enraged by Huddleston's *Naught For Your Comfort*, and the Liberal Party fighting hard against the Group Areas Act, which was dispossessing thousands of African and Indian families. 'Trevor [Huddleston]'s book has aroused great anger,' he wrote to John Patterson in America, '& we are now in the throes of the debate "can you change anything by angering your opponents?", a question that is, in certain important ways, quite irrelevant. It assumes that the only worthwhile aim is to *change*, whereas it is equally worthwhile to *tell the truth*.'[1] But a good many younger members of the Party could not find comfort in this reflection: they wanted to accomplish change. Among them was the son of a former Governor-General of South Africa, Patrick Duncan.

Duncan was an enthusiastic, warm-hearted idealist of a type to whom Paton felt deeply attracted, and his brilliant blue eyes and bright pink cheeks made it hard for Paton to refuse him anything. He was almost unendurably earnest. But he was also hot-headed, excitable and wholly unpredictable. Having decided on a course of action, he would take advice from no one. Once he had joined the Liberal Party he threw himself into the work with all his heart. He asked Paton and Brown to appoint him the National Organizer of the Liberal Party, and to the horror of some Cape Liberals, they acceded. Duncan and his wife Cynthia had plenty of money, and they founded a magazine, *Contact*, which Duncan controlled and for which Paton (in spite of suspecting that *Contact* was funded by the CIA)[2] began to write a series of articles which he called 'The Long View'. Over the years he was to put an extraordinary amount of time and energy into these carefully thought-out pieces, of which he produced three series; he never failed to produce his column, even when *Contact*, under government harrassment, had been reduced from a professional magazine to a few mimeographed sheets.[3] But Duncan was not satisfied with words alone. He wanted action, and felt increasingly frustrated by the apparent powerlessness of Liberalism.

Paton was frustrated by it too. In May 1956 he wrote to tell Leslie Rubin that he [Paton] leant towards the left, not the right, in the Liberal Party, that he wanted close co-operation with the ANC, and would not shrink from them because in certain matters they were tied to the Congress of Democrats. He also said he was sorry the Liberal manifesto had ever committed its members to employing only constitutional means in achieving its objects, and would like to take out that clause, 'if only to declare to the Government that we are adamantly opposed to them, & that we do not identify legalism with morality'. He added that some Liberals must be prepared to make sacrifices.[4]

This letter gives the impression that Paton was prepared to countenance illegal actions by members of the Liberal Party, and even to take such actions himself. If that was what he meant, it seems to have been a passing mood, for he not only continued to operate within the law (unjust though he thought many of the laws were) but also took pains to make it clear to fellow Liberals that he disapproved of illegal activity, and particularly disapproved of any use of violence against the state. In this he was being eminently practical, for the South African security police were efficient and ruthless, and the government had at its disposal draconian laws with which to deal with dissent. If Liberals were to go on holding up the lamp of non-racialism, they had no choice but to stay within the law. Such a course required great patience, maturity, and faith: fortunately, Paton had all three.

During the next decade he worked tremendously hard for the Party he now led, travelling, speaking, writing constantly. Each year there was the strain of the Party congress, at which he was the main speaker, rallying his troops against the series of repressive laws which continued to stream out of Parliament in Cape Town. The apartheid cords continued to tighten around the Liberal Party; even though it stayed within the law its members increasingly suffered harassment, visits from the police, threats of bannings, and arrests.

In the early years of his leadership of the Party Paton could still joke about the security police and about such ideologues as Verwoerd. It was about Verwoerd that he wrote the second and last of his Afrikaans poems, the only obscene ballad he is known to have produced. He wrote it on a particularly bibulous evening with Leslie Rubin, and it was roared out by Rubin and Paton to the tune of 'My Bonnie Lies Over the Ocean', with another Cape Liberal, Professor Tom Price, at the piano. Pearl Rubin, trying to sleep in a bedroom above, hammered on the floor in vain[5] as they sang Paton's five verses, which, drawing on a well-known wartime song, alleged that Verwoerd had only one testicle. The first verse ran:

Ons het 'n eerwaardige Minister
Sy naam is ou Hendrik Verwoerd.
Die arme ou het net een balletjie
En daaroor is vreeslik ontroerd.
　　Sny af! O sny af!
O Sny af sy enigste balletjie
　　Sny af! O sny af!
O sny af sy enigste bal.[6]

A literal translation would run, 'We have an honourable Minister / His name is old Hendrik Verwoerd. / The poor chap has just one small ball / And is terribly distressed about it. / Cut off! Oh cut off! / Oh cut off his only ball! / Cut off! Oh cut off! / Oh cut off his only ball.' In subsequent verses Verwoerd, gnashing his teeth in rage on his parliamentary bench, bites off his sole testicle by error. The Nationalist Party, realizing that a man unmanned cannot rule the world, gathers to beseech the Almighty to restore their leader's lost powers, but God refuses. The ballad ends with a robustly derisive chorus mocking the Nationalists for trying to 'play the game without a ball'.

But increasingly the Liberals realized that Verwoerd (who took over from Strijdom as Prime Minister in September 1958), his security police, and his laws were no joking matters. On 5 December 1956 the police, who had been investigating links between the Congress of Democrats and the ANC since the Congress of the People in 1955, swooped on COD, ANC and Indian Congress members[7] all over the country, arresting 156 of them and charging them with treason. Among the accused were Chief Lutuli, Z. K. Matthews of Fort Hare, Oliver Tambo, and three clergymen, one white and two black. The trials that followed were among the longest and largest ever held, and they produced great anti-South African feeling in Liberal circles around the world, particularly as it became clear the government's case was flimsy in the extreme. By August 1958 (when the trials actually began in earnest) the number of the accused had been whittled down to 91; by November 1960 it was down to 30, and on 29 March 1961 all without exception were acquitted of the charge of treason.[8]

The Treason Trials served to mobilize opposition to the government both inside and outside South Africa. Paton himself was arrested on 6 December 1956, the day after the mass arrests, for addressing a protest meeting of all races in Durban called to appeal for funds to defend the accused.[9] The meeting had been called without the permission of Durban's mayor, and Paton was tried on the charge of addressing an illegal gathering, and fined on 24 May 1957. He did not enjoy the experience of sitting in the dock, but when foreign supporters, particularly those from the United States and Sweden, wrote asking what action they could take on his behalf, he played the whole matter down.[10]

Through the early months of 1957, though, he was furiously active in organizing a Defence Fund to gather money to support the accused in the Treason Trials, and their families. This fund, which was soon to be known as the Defence & Aid Fund, gathered some money in South Africa, but was much more successful in getting funds abroad, chiefly from Britain, but also from the Scandinavian and other governments. Among its chief organizers abroad were Trevor Huddleston and Canon John Collins.

Huddleston was now writing wildly excited letters to Paton asking his help in returning to South Africa (if necessary against the will of his Superior) so that he, Huddleston, could stand trial too.[11] Paton consulted Bishop Ambrose Reeves of Johannesburg, and did his best to calm Huddleston down, reminding him of his vows: 'on no account should you contemplate disobedience'.[12] Huddleston listened to this advice, throwing his frustrated energies into public protests and fund-raising instead.

John Collins, it will be recalled, had invited Paton to speak in St Paul's Cathedral in March 1950, and he had visited the Patons during a brief trip to South Africa subsequently. His existing views confirmed by what he had seen, he welcomed the chance to begin collecting funds for Defence & Aid, and in his appeals and advertising he routinely invoked Paton's name, with Paton's blessing. Paton himself did a great deal to get the Fund launched in the United States, where his contact from Kent School, Louis Stone, was temporary Chairman of the Fund during 1957.[13] Paton asked Stone to raise one million dollars, a third of which he intended to use for the defence of the accused, and two thirds for the Liberal Party.[14] In the event, the Party seems to have got little or nothing, though it, and Paton in particular, became vital conduits through which money flowed from Defence & Aid sources abroad to ANC and COD members in South Africa.

Both Huddleston and Collins were considerably more sympathetic to the aims of South African Communists than to Liberal moderation. Collins, indeed, openly stated his conviction that his ministry was often 'best exercised in co-operating with non-Christians, including communists, than with church people'.[15] Paton found it increasingly galling to see the large sums flowing from Christians abroad to political extremists in South Africa, while the Liberal Party, which sought to avoid violence, was starved of funds. He endured this stoically for years, but in time he reached flashpoint.

Meanwhile apartheid legislation continued in an apparently ceaseless stream of ever more repressive acts. The Native Laws Amendment Bill of 1957 particularly disturbed Paton. It was aimed at banning any multiracial meeting in a white area. This would make impossible Liberal Party meetings in white areas, but more strikingly still, it was intended to apply the resultant Act to church services too. Public protest was particularly sharp, and the Anglican Archbishop, Geoffrey Clayton, for whom Paton had great respect, led his episcopal synod in telling the Prime Minister that he could not advise his clergy to obey a law which excluded people from church on racial grounds. Though Clayton died suddenly at this point, the government did in fact modify the Bill, but only so as to expose the offending African rather than the church concerned to prosecution. Verwoerd and his associates had no intention of admitting that there were things that belonged to God rather than to Caesar. But the Liberal Party announced that it intended to ignore the Bill, and it did so. So did the churches, and even Verwoerd had the sense to take no action. The 'church clause' of the Native Laws Amendment Act became a dead letter. It was a tiny victory for Paton and those who thought like him, but a real one.

There were not to be many such victories. There was, throughout the years that followed, a very real chance that Paton and his friends in the Liberal Party would be arrested at any moment, or banned. He accepted this risk, though Dorrie was terrified for a time. 'I went through a very unhappy period', she wrote to his American agent, Annie Laurie Williams, 'knowing what Alan was prepared to do, but I'm glad to tell you that I'm no longer afraid and know we shall be given an inner strength to bear what sufferings might be inflicted on us.'[16] Many Liberals were preparing themselves for suffering in this way, and they were right to prepare.

But life was not just a matter of waiting haplessly for blows to fall. Paton did not travel as often to the United States in these years, but he began to see more of Africa. In June and July 1956 he joined a harebrained expedition to the Kalahari Desert, travelling in the back of a five-ton truck, armed with a revolver in case dangerous tribesmen were encountered, and looking for a mythical lost city said to lie in the Aha mountains. The leader was a Major D. C. Flower, and his expedition consisted of five adventurers, listed in the expedition's preparatory papers as 'driver', 'navigator', 'photographer', and so on, and Paton, listed as 'scribe and bottlewasher'.[17] Paton wrote a lengthy and in parts hilarious account of this crazy journey, in the course of which they broke down often, got bogged in soft sand with regularity, were roasted by day and frozen by night, and travelled over 3,000 miles without achieving their object. They found neither the city nor the Aha mountains, but Paton seems to have enjoyed himself.

There were other African trips in these years too. In January 1958 he accepted an invitation to the first All-African Church Conference in Ibadan, Nigeria, as an Anglican delegate, and was photographed with about sixty other delegates on the lawn of the university in that city. From Nigeria he wrote one of his most hard-hitting *Contact* articles, on the need for a non-racial democracy in his homeland.[18] He had never visited west or central Africa before, and was sufficiently impressed by the brief taste he had of it to plan a lengthy trip by road from South Africa to the Belgian Congo (now Zaire) a few months later, in July 1958. On this trip he was accompanied not only by Dorrie, but by both his sons and their wives.

David, now 28, had taken a Master of Arts degree and then decided to do medicine, after which he specialised in radiology. His father supported him through part of this lengthy programme of study. David had met his wife, Nancy, in 1951, while they were students together at Natal University in Pietermaritzburg, and they married in 1954. Nancy thought the Patons a very loving and friendly family; among other things, she liked the family games of 'reformatory bridge', and the way in which everyone cheated Dorrie Paton in order to enjoy the rages she fell into.

Jonathan, now 21, had met his wife, Margaret, in July 1956, when they were student delegates together at a NUSAS conference in Pietermaritzburg. He was studying to be a teacher, she training to be a librarian. Their marriage was at first opposed by the Patons, who thought Jonathan too young to marry, but on meeting Margaret they gave in and the marriage took place on 4 January 1958 (not 1954 as Paton has it in his autobiography).[19]

The party of six travelled to central Africa in Paton's latest car, a very large red Pontiac. This car had attracted unfavourable comment from some of the left-wing members of the Liberal Party, who considered that Paton was, in his younger son's words, 'yielding to capitalist decadence'. Jordan Ngubane, on the other hand, thought its colour hinted at communism. One of the younger members of the Party, Peter Rodda, made up a song to the tune of the 'Internationale', of which the first lines ran,

> Alan Paton's car is painted red.
> 'I like it not,' Ngubane said.[20]

In this red monster the Paton party drove from Durban, through the Transvaal and the Rhodesias, to the Belgian Congo, where they saw the mountains of the Moon, the pygmies of the rain forest, and the great doomed estates of the white farmers. Paton paid all the bills and provided a good deal of the entertainment along the way, inventing endless word games, which he invariably won. These usually involved atrocious puns of the sort in which The Three Scots Villains had excelled so many years ago at Natal University College. At the Oribi Gorge, for instance, during an earlier trip to the Natal south coast, he had asked, 'Why is this gorge so named?' and following a baffled silence from the others, gave the answer, 'Because a man cried out here, I've been stung by a wasp—*or a bee*.'[21] Groans from his audience.

He also loved games involving tests of memory. One of his favourite exhibition pieces was to play Kim's game, in which the others would produce a list of perhaps twenty objects, ranging from household items to trees seen or cars passed. This would be read out to him and he would then recite the list from memory, in the right order. He was very seldom caught out, and his daughter-in-law Nancy found this a particularly impressive demonstration of his intellect.[22] However she also found him stern and exacting on occasions, as when he gave her and Jonathan's wife Margaret a tongue-lashing for daring to come in to dinner at their first hotel wearing slacks. They had little choice, for they had been given strict instructions to bring only the tiniest suitcase each.[23]

In addition to these trips and his political work, he managed to keep up his writing during these difficult years, though not much of it was directly imaginative. In 1957 the Natal Indian Congress asked him to write a booklet on the Group Areas Act of 1950, and in particular on the terrible effect it was having on Indian communities forced from their homes and businesses. The result was a powerful pamphlet which Paton called *The People Wept*, which pulled no punches in laying bare the nature of what Paton rightly called 'a callous and cruel piece of legislation',[24] and which did not endear him to the South African government. A rather similar pamphlet, *The Charlestown Story*, which he wrote in 1959, gave a moving and angry account of the destruction of a town in which Africans had lived and owned their houses since 1911.

He was commissioned by the Pall Mall Press in Britain to write a booklet, *Hope for South Africa*. He wrote it in the first half of 1958—the typescripts of some chapters actually being sent to him to correct in the Congo[25]—and it was published at the end of the year. As might have been expected, it was a detailed examination of the state of the country, with an analysis of what could be done about it. Paton concluded, as he was doing in many of his articles at this time, that the only alternative to black or white dictatorships in South Africa was for all her people to live and work together, and the only democratic, multi-racial party arguing for this solution was the Liberals.

He continued to produce occasional poems, but even these were political in content: a good example is 'My Great Discovery', published in August 1958. In this sardonic little narrative a South African scientist invents a means of changing the race of any individual through a course of injections:

> Five straight injections
> Position, lumbar
> In colour, umber
> Taste, very like cucumber
> Effect, inducing slumber
> And if I may remind you
> Five in number—
> These five injections could erase
> In just as many days
> The pigmentation
> From any nation.

The scientist reports his discovery to the Cabinet, which is aghast. Then Verwoerd, described as 'another Minister / Looking quite sinister / Just like the papers say', asks if it would be possible to turn a white man into a black by these injections. Told that it would, he orders that the inventor should be changed into an African forthwith, and the terrified man, to avoid this fate, destroys his laboratory,

> And all work exploratory
> Plus my assistants
> Whom at this distance

I spared the degradation
The gross humiliation
Of working for a caitiff
Who had gone naitiff.[26]

He also finished, in early December 1958, his play, *David Livingstone*, on which he had been working at intervals since 1956. His focus was on the admirable behaviour of Livingstone's two faithful servants, Chuma and Susi, who carried the body from Ilala through the territory of hostile tribes to Zanzibar, resisted the attempts of various whites along the way to persuade them to bury the body in Africa, but who then, when they reached the coast after vast labour and many privations, were simply 'frowned out of notice'[27] by those who received the body and transported it to England for burial in Westminster Abbey.

Paton's focus was on blacks who had absorbed the Christian message better than the whites who brought it to them, but he was concerned not to make his play crudely satirical:

> One of my problems has been to make this a true play and not to stress too much the heroic quality of the blacks and the insensitive quality of the whites. For this purpose I have imagined Lt. Murphy to be a warm-hearted young man who did his best to make reparations to Susi and Chuma for the thoughtless reception they received at Bagamoya.[28]

Paton travelled to Lusaka, Northern Rhodesia, to attend the first performance by the Waddington Players of *David Livingstone* (now retitled *Last Journey*) in the cathedral there. The play itself is deeply religious, its funereal close having the nature of a religious service, and it could only be performed in a church. Though Paton had hopes of giving it wider circulation, his agent Annie Laurie Williams thought otherwise, and there is little doubt that she was right.[29]

He also made several of his regular attempts to start another novel, trying, as he had tried before, to duplicate the conditions under which he had produced *Cry, the Beloved Country*. In June 1959, for instance, he travelled to the beautiful town of Knysna on the Cape coast, and stayed in a guest-house, Lagoon House, run by a widowed German woman, Marta, and used mainly by holidaying clergymen. He loved walking and driving into the forests surrounding the town, or going up into the nearby mountains, but he wrote little and gradually slid into depression. Disagreements with his hostess, a domineering woman of the kind he particularly disliked, added to his irritation.

It may have been here that he began a novel which we could call *John X. Merriman Bhengu* after its protagonist. John X. Merriman was a nineteenth-century Cape politician, while the name Bhengu was borrowed from one of Paton's black Liberal friends, Hyacinth (Bill) Bhengu, a Durban attorney who was banned in the 1960s. Paton's fragmentary story tells of a brilliant black writer whose talent is discovered by an Afrikaans officer while Bhengu is serving in the South African armed forces in north Africa. Paton seems to have intended to have Bhengu return home and there use his talents for the good of his people, before being stifled by apartheid legislation, but in the event he abandoned the work after completing only two chapters.[30] Though the plot was promising, there is nothing about *John X. Merriman Bhengu* to suggest it had fired Paton's imagination.

In the past, when failing to write a novel, he had produced an occasional short story, which he thought very much a second-best result. It was characteristic of him to downplay both his stories and his poems, though he shows remarkable power in both genres. He now began polishing a story he had already substantially written, 'Sponono', about the Diepkloof boy he had employed as a gardener at Anerley. Early on in his stay at Knysna, he also produced one of his most characteristic short stories, 'A Drink in the Passage', about a white man who invites a black friend up to his flat for a drink, but then is too embarrassed to sit down with him in the living room, so that they stand and drink in the passage. This story was almost entirely factual, based on an anecdote told to Paton by a black musician he had befriended at this time, Todd Matshikiza.[31]

Matshikiza was not just a performer, but a composer of great talent, and Paton had been drawn to him in the hope of collaborating with him. In July 1957 Paton had written something in a form entirely new to him, the libretto of a musical. There is little doubt that he had been inspired to attempt this by *Lost in the Stars*, which, despite Paton's dislike of it, had been a great financial success. In 1957, having been requested by the South African Institute of Race Relations to write for them something depicting 'a day in the life of Cato Manor' (an African suburb of Durban),[32] Paton got the idea of writing a story set in Emkhubane, described by him as 'a squalid suburb of Durban, poor but full of life'.[33] He finished his script in July 1957. He then set about looking for a composer to produce the music, and Todd Matshikiza was suggested to him by a Transvaal Liberal, Neil Herman, since Matshikiza (then working for the Gillette company as a razor-blade salesman) had written the music for a recent and very successful Johannesburg musical, *King Kong*.[34] He proved to have just the combination of talents, though his habitual dilatoriness meant that getting the music out of him took a long time. Malcolm Woolfson was the director, a Natal Liberal, Violaine Junod, was the theatrical assistant, and the black cast was recruited by word of mouth.

The completed musical, *Mkhumbane*, was scheduled to open in the Durban City Hall on 28 March 1960, before a mixed-race audience, for Paton refused to have any play of his performed before a segregated audience: 'Better no theatre at all than colour-bar theatre,' he believed.[35] Eight days earlier, on 20 March 1960, a demonstration organized by the Pan-Africanist Congress (an anti-communist offshoot of the ANC led by Robert Sobukwe) against the mandatory carrying of passes by blacks in white areas had got out of control at a police station in the black township of Sharpeville, 20 miles from Johannesburg, and the police fired into the crowd. Sixty-nine people were killed and 180 injured, many being shot in the back as they tried to flee. There was worldwide condemnation of the government for this act, and Chief Lutuli called for a National Day of Mourning on 28 March. The opening night of *Mkhumbane* was accordingly postponed until 29 March, and Paton prepared for trouble. There was none. 'In the whole of South Africa the Durban City Hall was of all places the most untroubled,' he wrote with relief. 'I remember it to this day as a kind of miracle.'[36]

Paton donated the proceeds of this first night to the South African Institute of Race Relations, and the musical went on to run trouble-free for the rest of the week for which the hall had been booked. Even at *Mkhumbane*'s opening, though, there were signs of the times: a group photograph was planned, but the photographer objected to having Matshikiza's wife, Esmé, stand with the whites, and the photograph was abandoned. And plans to take the performance to Johannesburg were given up as impractical under the circumstances.[37]

South Africa was in ferment, with violence in many places, alarmed whites queuing to obtain gun licences,[38] black marches through frightened white suburbs, and widespread intimidation of those who tried to work during the strikes called by the ANC. A large armed party of Zulus marched into Durban through the plush suburb of Berea; 30,000 Africans marched into Cape Town. The government's response to protest was to proclaim a state of emergency, on 30 March 1960, followed by more repression: 18,000 people were detained during the weeks following Sharpeville. Among them were many Liberal Party members.

Peter Brown was arrested on 31 March 1960 in front of his three small children, who wanted to know what Daddy had done. Brown had in June 1958 persuaded Paton to give up the post of National Chairman of the Liberal Party to give him more time to write, and had himself assumed the Chairmanship.[39] In addition to him, other Liberals to be imprisoned in Pietermaritzburg alone included Hans Meidner (Natal Chairman), Elliot Mngadi (party organizer for northern Natal), Frank Bhengu (National Committee), Derick Marsh, Zephaniah Zuma, Robert Zondi, Jay Gangai, Roy Coventry, Albert Cebekulu, Jacob Mbongwe, and Peter Kumalo,[40] and this Liberal honour-roll had its parallel in other parts of the country.

Yet even in prison white Liberals could not escape the privileges the apartheid system offered them. Gertrude Cohn, a Johannesburg Liberal arrested for her links with the Communist Party (which she had in fact left thirteen years before) recorded the politeness of the police sent to seize her, the way they stopped their search of her papers on hearing it was her birthday, and came solemnly to shake her hand and wish her many happy returns; the way white prisoners, in contrast to the punitive conditions under which blacks were held, were given sheets and blankets, were allowed to spend hours sunbathing, and were even supplied with black servants to clean their cells. 'Only in South Africa would white prisoners have had black servants!' she wrote shamefacedly later.[41]

It was not just individuals who were acted against. On 8 April 1960 the African National Congress was banned, and with it Robert Sobukwe's Pan-Africanist Congress. Paton considered these moves utter folly: 'Thus black people were in effect denied the right to organize politically, and this has had terrible consequences for us all,' he was to write.[42] When he, Mrs Albert Lutuli, Fatima Meer, and Manilal Gandhi's widow led night-long fasting and prayers to call attention to the numbers imprisoned, on 31 May 1960 (the day on which fifty years of Union were celebrated in South Africa), they were joined by hundreds of Indians and Africans.[43] In addition to his duties as President of the Liberal Party, Paton became acting National Chairman in Brown's place.

Why was Paton himself not arrested? The answer is probably a simple one: he had written *Cry, the Beloved Country*, and though the government had not yet learned to care much about its image abroad, it shrank from the publicity that would have surrounded the arrest of the country's best-known writer. He felt oddly guilty at being free while Peter Brown and so many of his other friends were in prison, and at the Liberal congress in 1960 told an apocryphal story according to which Brown and the others, irked that they should be behind bars while he remained free, wrote a message, tied it to a stone and threw it brazenly over a wall. According to Paton it read, 'Paton, for God's sake hide the revolvers!'[44] After the first year of Paton's 'The Long View' column in *Contact*, from February 1958 to February 1959, Brown had taken on the task. Paton now resumed writing 'The Long View'; in *Contact* on 16 April 1960 he paid tribute to Brown's courage and identified himself with it: 'I say to my friend Brown, wherever he may be, "We shall try to prove ourselves men and women who have honour and courage also".'[45] Brown, offered his freedom by the government, had refused to leave prison unless his black colleagues were also released, a stand which won many more black members for the party.[46]

Paton was angered and discouraged by the mass of white South Africans who not only did not protest at what was happening, but supported it. He was particularly irritated by those who were public figures and who in private would express sympathy with people like himself, but in public chose to remain silent. About men like the famous golfer Gary Player, who refused to make any political comment, he would make biting little jokes: Player, he said, lived in a world of eighteen holes and could not see out of one of them.[47]

Paton himself continued to play a high-profile role in publicizing abroad what was happening in South Africa, as if daring the government to act against him. He became acting National Chairman of the Party while Peter Brown was in prison. And he went on trying to live as if the apartheid laws did not exist, filling his house with people of all races, so that his sister Dorrie,[48] and even such old friends as Railton Dent and his wife,[49] ceased to visit him, not wanting to be 'implicated' in his protests against the government.

His relations with Africans were increasingly relaxed. Neil Herman was to recall inviting Paton to dinner at his home in Johannesburg. Paton got there before him, and Herman found him in the kitchen, talking animatedly to the black cook. She later asked Herman who he was, borrowed and read Herman's copy of *Cry, the Beloved Country*, and told him, 'If we have visitors like that in this house, then only great blessings can come upon it.'[50]

Herman was among many Liberals who lost heart after Sharpeville, and in the early 1960s began leaving South Africa for England, the United States, or Australia. It was certainly not a time for those of weak nerves. On Friday 1 April 1960 Paton's phone rang at midnight, but when he answered it the caller at once rang off. An hour later there was a knock at his bedroom window, and he saw with fear that there in the dark was his daughter-in-law Nancy. She had been dispatched by the Liberal Party in Johannesburg to warn him that he was about to be arrested, that Bishop Ambrose Reeves of Johannesburg was under the same threat, and that Reeves, terrified of imprisonment, had decided to flee to Swaziland which was then a British Protectorate.

Paton at once realized that Reeves had made a dreadful mistake, and determined not to make it himself. He would not run from the police, and if arrested he would go to prison as Peter Brown and many other Liberals had done. As it turned out, he was not arrested, and he used his liberty to fly secretly to Swaziland in a small plane piloted by a Pietermaritzburg friend, early in April 1960, intending to persuade Reeves to return to his flock while it was still possible to do so. He also wanted to find out what Reeves had done with the books and records of the Defence & Aid Fund. He found that Reeves had no intention of returning. He wanted nothing more than to be told that he had done the right thing. Having assured himself that the Defence & Aid documents were safe, Paton returned perilously to Pietermaritzburg, low cloud forcing his small aircraft down in a field; luckily no one was injured. Reeves was to go on to England on 12 April 1960, reviled by many members of the diocese he had abandoned, and coldly received by his brethren in Britain. There he refused the jobs found for him by the Archbishop of Canterbury (whom Paton urged by letter to do something for Reeves)[51] and lived out the sad remainder of his life as a parish priest with a dwindling congregation, dying in December 1980 at the age of 81, his life, Paton thought, having been ruined by that panic-stricken flight to Swaziland.

The despairing flight of many with liberal views began in these years, and was to accelerate through the 1960s. Paton's close friend Leslie Rubin, still a Vice-President of the Liberal Party, left South Africa in 1960, initially for Ghana, but in the longer run for the United States, where he would spend the rest of his life. This was a great blow to Paton, who had reached the stage of joking with Rubin even about his Jewishness, jokes which Rubin enjoyed. 'Not a single Rabbi caught yet with his pants down,' Paton would write; 'this is undoubtedly the effect of a strict Yeshiva Bogger education' ['Yeshiva Bucher' is Hebrew for a Talmudic College student].[52] He could even joke about more sensitive points, referring in one letter to 'The Jewish Women's Anti-Circumcision Guild', and then adding a footnote: 'Sorry, it is the Anti-Vivisection Guild. But greater men than I have made such mistakes. Disraeli supported Penal Reform because he thought it meant circumcision.'[53]

Rubin had been one of very few people to know something about Paton's sex life, though in fact he did not care to enquire too deeply. There are odd little signs that on his visits to Cape Town, where he stayed with the Rubins, Paton permitted himself a sexual licence he did not risk nearer home.[54] He could write to Rubin in November 1959, for instance, in planning a visit to Cape Town, 'It is my intention to confine the Cape Town visit to the biography [*Hofmeyr*, which he was then writing], drinking, and sex, and I thought you could help me with the first two.'[55] On another occasion, when Rubin had collected him from the airport and was driving him in to Cape Town, Paton suddenly looked up at him and said, 'No semen in the sacs.' Rubin did not respond, and Paton never returned to the subject.[56] Now this friend, one of the very few people to whom he could talk about details of his private life, was apparently lost to South Africa.

Patrick Duncan, with his mercurial enthusiasms, also slipped from view, for in the aftermath of Sharpeville he was banned from all political activity. All through the 1960s the dispiriting trickle of those saying goodbye continued, and many of them were the most active and committed Liberals. The Party seemed to be bleeding to death.

Paton felt terribly under siege at this time, not least because of his constant fear of arrest, and because the constant surveillance of his house by the security police intensified. His telephone was tapped (and would continue to be tapped, not constantly but randomly, for years) and his mail was intercepted at intervals from 1960 onwards. When writing to Liberal Party friends, or those connected with the Defence & Aid Fund, he would adopt a *nom de plume* when putting his address on the outside of the envelope ('Hymie Kantelovitch, Majiga Court', for instance)[57] and in the body of the letter he would use a most elaborate system of allusions and initials to conceal references to individuals. These often make for difficult reading: 'I don't know if Dr You-know-tiddle ee um dadoo dah doo [Dr Yusuf Dadoo, militant leader of the Transvaal Indian Congress, who had fled abroad in 1960] is about, but a message from his pals JN [probably J. N. Singh, a radical member of the Natal Indian Congress] & GMN [Dr Monty Naicker of the Natal Indian Congress] is to this effect, that they feel his big contribution is to be made in the land of the temple bells [India] & not in the lands of tranquillisers, john collinses, or pavlovas [the United States, Britain or Australia].'[58]

If this sounds paranoid, Paton had reason for paranoia from 1960 onwards, with many of his political allies imprisoned or on the run, and the forces of repression apparently triumphant. The liberal wing of the United Party had split to form the Progressive Party in 1959, but there was every sign that its members would be rejected by the voters at the next election as Liberals had been. In fact only one, Helen Suzman of Houghton, would survive their first election. Liberal candidates like Guinevere Ventress, wife of the mayor of Kloof, continued to be resoundingly and humiliatingly rejected.[59] Yet other prominent Liberals, including the parliamentarian Walter Stanford, Dr Oscar Wollheim (the party's first national chairman) and a leading lawyer, Gerard Gordon QC, defected to the Progressives, adding to the pressure on Paton.

Repression was given added edge by the racial violence and anarchy now affecting the Congo, and by an assassination attempt on Prime Minister Verwoerd, who was shot in the head on 9 April 1960, but survived. Continued mass demonstrations on 31 May 1960 were contained by the army, and Paton expected another round-up of political opponents of the government. 'No one can say whether the net will be thrown more widely or more narrowly in another emergency,' he wrote philosophically to Canon John Collins in England. 'If more narrowly, I shall escape I think. If more widely, I shall not.'[60] At the same time Dorrie Paton was writing to Betty Patterson in America, 'I hold on to my faith with every ounce of strength I've got. It is the one thing that keeps me from despair.'[61]

Canon Collins was continuing to channel money through Paton to deserving causes in South Africa, and it was probably this contact, rather than his work for the Liberal Party, that drew government attention to him. The South African government was intensely suspicious of individuals or organizations receiving large sums from abroad, perhaps believing that the funding had Communist origins. The harassing effect of continual police surveillance on Paton was such that his migraines returned with renewed force, and by August 1960 he was suffering attacks for periods of days at a time.

To make matters worse, Dorrie's health also deteriorated from 1960 onwards. She had a serious chest infection in May 1960. In July 1960 Paton took her, his son David, and daughter-in-law Nancy on a holiday to South West Africa (now Namibia) where Dorrie came down with another chest infection which rapidly developed into pneumonia. She had to be flown from Windhoek to Johannesburg, where she was hospitalized for weeks.[62] Even when she was discharged she seemed to have breathing difficulties. Though neither she nor Paton knew it, this was the first sign of the lung ailment that was to kill her in 1967. She was diagnosed as having chronic obstructive pulmonary disease, a term that meant little to her. She had always been a heavy smoker, seldom without a cigarette between her fingers or her lips in spite of Paton's disapproval of the habit in women, and she was now in the first stages of emphysema. Nor even now did she give up smoking.[63] Her gradual, terrible slide towards death would be the steadily darkening background to the next seven years of Paton's life. His hard time had begun in earnest.

But though his world seemed to be falling apart, he would not give up hope. In September 1960 he made a determined effort to persuade Chief Lutuli to launch a new, black-led United Democratic Party with Professor Z. K. Matthews. He seems to have suggested it himself to Lutuli and to have been rebuffed.[64] He then asked Mary Benson, who had worked for the Defence & Aid Fund in London, to persuade John Collins or Michael Scott to write to Lutuli urging him to 'shed the Left' and start a new party:

> With L[utuli] & Prof M[atthews] at the head, we might get more white support, & LP [the Liberal Party] will probably be willing to merge, & accept African leadership. It is an important thing . . . Money should go into this now.[65]

There is no evidence that Collins or Scott did contact Lutuli, and certainly Collins was much more interested in supporting the Left with money than in 'shedding' it in favour of the Liberal Party. Nor, to be honest, was there much chance that a political party with Lutuli and Matthews at the head would have got wider white support than the Liberal Party did.

Paton is likely to have continued to urge this course on Collins in person when he travelled to London in September 1960, partly to talk to his publishers about a book of short stories he was planning, but chiefly to tell the world what was going on in his country. Here he spent some time with his son Jonathan, who was now at Cambridge preparing for an academic career: his wife Margaret was pregnant with their first child, Paton's first grandchild, Nicholas, who would be born on 11 February 1961.[66]

Paton attended the annual conference of the Liberal Party of Great Britain in Eastbourne and made a well-received speech. He stayed with John Collins and his wife Diana at 2 Amen Court, London, and observed with fascination the money that poured in for Defence & Aid work with every mail delivery; but Collins showed a continued and deeply hurtful reluctance to channel any but small sums to the Liberal Party of South Africa.

On 2 October 1960 Paton flew on to the United States to receive the annual Freedom Award for his Liberal Party work. American luminaries from the literary and political worlds arrived in such numbers that many had to be turned away from Freedom House in New York. This was a great honour which cheered many Liberals; the Freedom Award had been won in the past by such figures as Winston Churchill, Franklin Roosevelt, Dwight Eisenhower, and George Marshall. President Eisenhower sent a message of congratulation, and the poet Archibald MacLeish gave the main welcoming address.[67] In his reply Paton said he asked only one thing of his audience, and that was that they should continue to concern themselves with the problems of his country. He revisited his friends at Kent School, and at the Divinity School at Yale, carrying the same message. And then he embarked on a speaking tour, giving twenty addresses in Virginia, Atlanta, New York, Boston, Chicago, and Toronto, in each place delivering the same speech, in which he sketched South Africa's history, described her present miserable circumstances, and asked the world not to forget what was happening there.

Meanwhile he was being watched. The South African Department of Foreign Affairs had at the beginning of 1960 sent out a directive, stamped 'Most Secret', to all heads of South African missions abroad, to send back to the Department all press clippings concerning Paton.[68] He was carefully monitored all the way by the South African embassies in the United States and Canada, and in New York became aware of this.[69] He was, in fact, very careful about what he said, though in many places journalists tempted him to condemn his country's policies in language that would sell newspapers.

He slipped only once, but it was to prove enough. During a recorded interview in Toronto with the Canadian Broadcasting Corporation, to be broadcast on 31 October 1960, he was asked whether South Africa was now a Nazi state. He replied that it was not—and then added, 'but it is a good imitation of one'. He was to regret that throwaway line, though at the time he thought little about it. 'When I speak in public I suppose these consular fellows must get absolutely furious,' he remarked unconcernedly.[70]

By the beginning of November 1960 he was back in London, seeing Mary Benson again: she introduced him to Arthur Koestler, who seems to have made little impression on him. Benson, still not entirely recovered from the pain of breaking away from Michael Scott, was struggling to make her way as a writer. Years later she recalled the encouragement Paton gave her at this time. Perhaps the warmth of their friendship and his empathy for Benson's problems lay behind the letter he now wrote from London to Dorrie in Kloof:

You think you are no longer physically attractive to me. I think you have always separated sex from love, but if you can't change, you can't. But you must never doubt my love for you. I think you are a lovely woman, & have been a loving mother & wife. And I wouldn't change you for anybody . . . Don't think you are old. We shall love each other until one goes, but if you are the one to go, I shall never forget you. One of the most wonderful things about our marriage has been that we think alike about all the things that are most important. Shall kiss you from head to foot, *if you want to.*

Your loving husband
Alan.[71]

This letter well illustrates Paton's ability to triumph over the most difficult circumstances. Though he went on longing to the end of Dorrie's life for a changed relationship with her, he accepted that he would not get that change, and affirmed his love for her in moving terms in spite of all. That element in him—he would have called it faith—that made him refuse to despair of his country and walk away from it, also made him persist with his marriage in spite of its pain. He seized on the smallest sign of affection from his wife. 'I am glad you wrote "I love you so"; quite a thing from you,' he told her.[72]

For all that, when they were apart he missed her gaiety, her flow of jokes and leg-pulls which stopped him taking himself too seriously, her relish for life. After her death her sons were to remember her doggerel poems, the way she seemed to lighten and animate parties without dominating them, and her sheer enjoyment of almost everything she did. Her sense of fun emerges vividly in the diaries she occasionally kept, for instance during her trip to Italy in 1954. In this she records, hilariously, her confusion when a porter had gone to get a ticket, or 'chit', for her. She, not knowing the word, had leaped to quite another conclusion: 'The porter said "Wait here—I go get chit." Well, well . . . I was prepared for most things in Italy, but hardly this.'[73] Paton, depressed by his own travelling, could have done with her to keep him on an even keel.

In November 1960 he spent three weeks in Geneva for a meeting of the World Council of Churches, his third, and helped in the making of a WCC film about South Africa, produced by John Taylor. He then returned to London to address a group called Christian Action; then to Brussels on 31 November 1960 to talk to a producer considering a revival of *Mkhumbane* (a proposal that eventually came to nothing because Todd Matshikiza could not be induced to co-operate, being sunk in apathy and depression),[74] and finally he flew back to South Africa from Frankfurt.

When he stepped off the plane at Jan Smuts airport on 4 December 1960 his usual honour guard of hard-eyed security policemen was waiting for him, and when he presented his passport to the immigration officer it was not returned to him. Instead he was handed a letter from the Secretary of the Interior, dated 2 December 1960, informing him that his passport had been withdrawn, and giving no reasons.[75] Paton made a defiant public statement that he hoped to have his passport returned 'by a government representative of the peoples of South Africa.'[76]

British papers headlined this news: *The Times*, the *Daily Mail,* the *Daily Telegraph*, the *Yorkshire Post*, and the *Guardian* all carried reports and protests in the next few days. The South African government was not unprepared for this reaction, and it answered in *Die Transvaler*, once edited by Verwoerd and now his government's mouthpiece, on 7 December 1960. 'British Papers Frenzied over Alan Paton', ran the contemptuous headline, and there followed a long leading article which accused Paton of seldom criticizing South Africa while he was in the country, but of using what it called his 'poisonous tongue' abroad to try to provoke the interference of foreigners in South Africa's affairs. This, it said, could not be tolerated, and accordingly Paton's passport had been withdrawn 'indefinitely'. In the event he was to be without a passport for a full decade.

Paton had been expecting this blow at intervals since August 1955, when a vitriolic attack in the Afrikaans publication *Dagbreek* had hinted that he should be prevented from travelling and 'traducing his country abroad'. All the same, now that the blow had fallen he felt it deeply. He loved travelling, and had been planning to go with Dorrie towards the end of December 1960 to Tanganyika (now Tanzania) where Trevor Huddleston had just been installed as bishop of Masasi; this was now clearly impossible.[77] Kent School had been inviting him to return for a full year; his friends in England were dangling similar enticing offers. But from now on he would be confined to South Africa, and would be lucky to avoid banning or imprisonment. The next decade, though he did not foresee it yet, would be the hardest of his life.

19

HARD TIMES
1960–1968

Worm, be with me.
This is my hard time.
THEODORE ROETHKE

Paton had arrived back in South Africa in time to take part in a conference of all the main Christian churches in South Africa other than the Catholic church. The Cottesloe Convention, as it became known, took place on 8–14 December 1960. It was intended to bridge the widening gulf between the Afrikaner and English churches and to help South Africa in the aftermath of Sharpeville. For Paton, a particular pleasure of this Convention was his meeting with Nelson Mandela, on bail from his trial for treason. He congratulated Paton on the vigour of his public refutation of the charges of disloyalty levelled against him by the government media when his passport was withdrawn, and he told Paton, 'You're a fighter'. Paton took this as a great compliment, coming from such a source.[1]

But he fought best, as he well knew, with his pen, and his books needed no passport to make their voice heard. In these circumstances he decided to settle down and write hard, rather than brood on the apparently endless series of government triumphs. 'The Nats [Nationalist Party] are going from strength to strength—there seems to be no stopping them,' Dorrie Paton wrote to her American friend Betty Patterson, but she added, 'And yet even at my most depressed—deep down, I know they cannot triumph.'[2] This was Paton's faith too, and if one circle of activity were closed to him, he would busy himself in another while waiting for change. He would not leave South Africa on an exit visa as some of his Liberal colleagues were doing: 'It's a strange thing that the more repressive they [the government] get', Dorrie wrote, 'the more people like ourselves realize the utter impossibility of clearing out.'[3]

Paton busied himself with two writing projects in the next few years. The first was a collection of the short stories he had written over the past decade,[4] and which he now proposed to entitle *Debbie Go Home*. For American publication Scribner would retitle it *Tales From a Troubled Land*. This had been accepted for publication by both Jonathan Cape and Scribner during his last trip abroad, and he now set about polishing up the proofs. For this task he needed complete peace, and that was unavailable in the house at 23 Lynton Road, where the stream of casual visitors was unending. He therefore rented a cottage on the beach at the Natal coastal resort of Ramsgate, and he and Dorrie would retire to this beautiful house each Monday, returning only on Friday.[5] There was no telephone in the beach house, and he was able to set politics aside during the week and devote himself to literary work.

Tales from a Troubled Land was published in America in April 1961 and in Britain and South Africa, as *Debbie Go Home*, the following month. It got respectful reviews in Britain and America, though the reviewers tended to have more to say about Paton himself than about his book. Malcolm Muggeridge, in the *Daily Herald*, picked out 'A Drink in the Passage' as the best story, and remarked that in the darkness that seemed to be growing in South Africa, Paton was a shining light.[6] Duff Hart-Davis, in the *Sunday Telegraph*, compared the stories favourably with *Cry, the Beloved Country*, and remarked that 'this collection of ten stories is in Alan Paton's most effective vein'.[7] The *Times Literary Supplement* said the volume 'will help to confirm Mr Paton's reputation as a writer who never has to raise his voice in any effort to convince us of his creative power and his sincerity'.[8] American reviewers were similarly welcoming of *Tales from a Troubled Land*. The *New York Herald Tribune* called him that rare thing, 'the writer with an unfaltering social conscience who is also an artist';[9] the *New York Times* commented on his wisdom as an observer and his skill as a writer;[10] the *Chicago Sun-Times* gushingly spoke of his 'liveliest love' and 'large creative pity'.[11] Most of the critics took the opportunity of reminding their readers that Paton had recently been deprived of his passport, the *New York Herald Tribune* marvelling at the 'massive stupidity' of a government that could do such a thing to such a writer.[12]

Paton himself thought of these stories as rather slight: they were the kind of work he did when he could not write a novel. It is true that several of them more closely resemble anecdotes than carefully worked stories, and almost all of them were, in fact, based on real incidents. In 'A Drink in the Passage', Paton makes only superficial changes to Todd Matshikiza's experience, changing the African from a musician into a sculptor, for instance. The stories based on his Diepkloof days are closer to fragments of autobiography than to imaginative story-telling; even the names of such characters as Sponono are retained without alteration. Dorrie wrote to Betty Patterson at this time,

> I agree with your criticisms of them [the short stories]—actually all the stories were drawn from life & are basically true. Sponono was certainly a strange person. Once when he was working for us, we refused some request of his—I've forgotten exactly what—& in a fit of pique he took the notes we had paid him for wages & burned them.[13]

Slight though the stories may appear, though, several of them penetrate to the very heart of South African society. An example is 'Debbie Go Home', the story in which a girl of a mixed-race family wishes to go to a debutante's ball at which she will shake the hand of a white administrator, and faces the opposition of her father and brother to what they think a demeaning ceremony. The family, divided in heart and blood, becomes South Africa in microcosm, and with the apparent effortlessness that comes to only the best artists Paton conveys the pain of apartheid in a way that is unforgettable. Just as powerful is 'A Life for a Life', an utterly convincing tale which in a few pages conjures up a world of pagan cruelty and injustice. And 'The Waste Land' is an understated masterpiece of violence and terror, in which a man hunted by a criminal gang kills one of its members and then discovers he has killed his own son.

What few of the critics seem to have noticed is the change that had come about in Paton himself. There is a seeming hopelessness or resignation in *Debbie Go Home* that is utterly unlike the quiet but profound optimism of *Cry, the Beloved Country*. Even more than *Too Late the Phalarope*, *Debbie Go Home* has a stoical acceptance of the fact that there are wrongs that cannot be righted. The Diepkloof-based stories, in particular, stress the chasm that lies between Principal and pupil, black and white: though the narrator reaches out his hand repeatedly to his charges, he is ultimately powerless to change the course of their lives. The same melancholy pervades 'A Drink in the Passage', in which the white man and black long to touch each other, but are unable to do so. In 'Sponono', relations between Sponono and the narrator reach an impasse, desired by neither but incapable of resolution. The iron had entered Paton's soul, and *Debbie Go Home* shows it plainly.

The dramatic quality of the stories attracted several playwrights and producers, as Paton's earlier work had done.[14] One of these was a young

Indian-born director, Krishna Shah, whose work Paton had first seen in October 1961. Shah had had considerable success in New York with a Rabindranath Tagore play, *King of the Dark Chamber*, and he had brought this to Durban. He was eager to work with Paton, and during June and July 1962 the two of them collaborated on turning three of the Diepkloof stories from *Debbie Go Home* ('Sponono', 'Ha'penny', and 'Death of a Tsotsi'), into a three-act play, which they called *Sponono*. Shah provided the stagecraft and theatrical expertise; Paton did the writing, adapting his own work. The play opened in Durban on 12 December 1962, and was such a success[15] that Shah decided to take it to Broadway. Paton invested $5,000 in this venture,[16] and lost it when the play failed to make an appeal to sophisticated black audiences in New York. They seem to have disliked the fact that the Principal, the figure of authority, was the one white actor in a cast otherwise wholly black, and Paton came to realize, with disquiet, that his play was interpreted as a relic of the colonial past.[17] 'For Krishna and me, it is fundamentally a story of two people who reached out to each other but could not touch except briefly', he wrote to Mary Frank, the producer,[18] but this was not the way the American audience saw it. The play was to be published in 1965 by Scribner, and republished by David Philip in 1983.

The other book that occupied his time during the 1960s was his long-deferred biography of his friend J. H. Hofmeyr. The obstacle posed by his sharp incompatibility with Hofmeyr's mother had been removed when the implacable old woman died on 27 July 1959. Urged on by Peter Brown, Paton resumed work in that year, and spent every moment he could spare from his political work in writing the biography during 1960 and 1961. A first draft was finished by the end of 1962, but there was much editing work to be done on the sprawling 300,000-word manuscript.[19] The editing was done by Paton's friend Leo Marquard, and during 1963 Paton drove from Natal to Cape Town and back several times, each time enjoying a different route.[20] *Hofmeyr* did not appear until 1964, when it was published by Oxford University Press in Cape Town.

Paton had put an enormous amount of work into it, often under very difficult circumstances, and he was justifiably proud of the result. In later years, if asked what he considered to be his best book, he would nominate *Hofmeyr*. Certainly he performed a herculean labour in assembling a vast archive, and extracting from it a flowing and generally interesting narrative. But he was always in danger of getting bogged down in details, and there are lengthy passages when the character of Hofmeyr disappears beneath waves of South African history. This is hard to avoid when the subject played such a prominent part in that history: the meaning of the man is likely to be lost without the context, yet the context is so complex that it threatens to overwhelm the man.

In a letter to Trevor Huddleston Paton gave an analysis of his aim in writing *Hofmeyr*: 'It is the story of a Christian, and a very loyal one too, with great flaws in his character (not of the flesh of course)—a white South African who was moving steadily towards emancipation, although he did not quite achieve it.'[21] In this one sees clearly the two elements of Hofmeyr that Paton most admired, and the motives that persuaded him he should write this book at this crucial time: Hofmeyr was a Christian, and he was a political Liberal of a rather wavering and inchoate kind. Quite plainly Paton saw his work on Hofmeyr as a contribution to the political debate in South Africa: far from squirrelling away in his study while others were fighting the good fight, he saw this writing as the most important and immediately relevant contribution he could make to his country's guidance.

Yet it is arguable that his commitment to the values he believed Hofmeyr represented weakened the biography. There is a piety in his treatment of his friend that leads him to gloss over Hofmeyr's flaws and failures, though he is too honest to omit them entirely. He mentions Hofmeyr's rudeness, his gross feeding habits, his lifelong immaturity, and other personal failings, but having mentioned them he does not revert to them, though they were a major and constant part of his subject's personality. Hofmeyr's errors in judgement, notably his cruel persecution of Stibbe at the University of the Witwatersrand, his failure to take a stand on vital issues such as the removal of African voters from the common roll, and his subsequent resignation from the Ministry over a comparatively trivial matter, Paton mentions but makes no judgement of. At times like these the reader expects more from a biographer than a mere report of events, no matter how detailed. And finally, it must be said, the character of Hofmeyr himself remains elusive throughout Paton's long book. We get a sense that Paton himself did not understand what went on behind those heavy-lidded eyes under the thick pebble lenses, a sense that after the mass of facts and documents have been sifted, Hofmeyr, with his giggle, his brilliance, his unctuousness and immaturity, remains a mystery to his biographer and to us.

By October 1962 Paton was under renewed pressure from the security police. Afrikaner Nationalism might have appeared to be triumphant, for it had in May 1961 achieved the long-sought goal of making South Africa a republic, and, incidentally, taking it out of the Commonwealth; it had declared illegal its black opposition, the ANC and the PAC, and it had banned and imprisoned individuals with impunity. Its majority in parliament continued to grow, election by election, and it could pass any laws it chose. By rapid stages the police were given powers to detain suspects without charge, and in solitary confinement, for 12 days (1962), 90 days (1963), 180 days (1965), for an unlimited period if authorized by a judge (1966) or even without such authorization (1976).[22] Armed with these frightening powers, it continued to harass clearly law-abiding and pacific individuals like Paton, who in October 1962 wrote to Canon John Collins, 'The Security Police have been to see me, and have asked all kinds of questions about my hours of work, about the places I must necessarily visit in order to write, etc. This may not mean with certainty that I shall be put under house arrest, but it certainly points in that direction.'[23]

Paton showed stoicism in talking so calmly of house arrest, for the danger was real and immediate. House arrest meant confinement to the victim's home, usually during the hours of darkness, though occasionally for as long as 23 hours a day. It was in general worse than banning, which involved confinement to a magisterial district or part of one, with other restrictions. House arrests and bannings were first imposed for five years at a time in October 1962, and they soon began to be applied to Liberals.

By July 1964, in addition to many ordinary members of the Party, those banned included Peter Hjul, the Cape chairman; Randolph Vigne, the National Deputy Chairman; Jordan Ngubane, the National Vice-President; Joe Nkatlo and Terence Beard, the Cape Vice-Chairmen; Adelain Hain, the Pretoria Secretary; Hammington Majija, a Cape executive member; E. V. Mahomed, the Stanger Secretary; David Evans, the Natal Coast Region Secretary; H. J. Bhengu, the National Vice-President; Eddie Daniels, a member of the Cape Executive; John Harris, a National Committee Member; Elliott Mngadi, the National Treasurer; and, on 31 July 1964, Peter Brown, the National Chairman. Nor did the persecution stop there. Between July 1964 and September 1966 another 26 Liberals were to be banned.[24]

From the photographs of this period, it is clear that the strain Paton was under aged him rapidly during the sixties. He was thinner about the face, the grim lines around his mouth more deeply incised, and his hair was snowy white by 1963 when he was in his sixty-first year. He had originally believed that apartheid would consume itself quite quickly; now he was not so sure. He told the American interviewer Studs Terkel, in May 1963,

'During the past few years . . . I have wondered if there is not greater stability in our present regime than I had wanted to see. Originally, I had hoped for an early end. By studying South African history one sees that a great number of contemporary realities have been present for many years.'[25] Fighting apartheid, he was realizing, was going to be a matter of the long haul.

Despite this realization he would not contemplate the use of violence, as other Liberals were beginning to do. When the *Cape Times* misquoted him as supporting violence on 29 November 1960, he took legal action against the paper in March 1961 and got a printed apology and full retraction on 26 June 1961.[26] In his address to the Liberals' sixth annual congress in 1960 he had reaffirmed,

> This party is openly and publicly committed to a policy of non-violence. . . . It will not consent to the use of violence by others, not encourage it, nor connive at it. A Liberal Party can never aid, or itself become, a terrorist organization, nor do I think it is in any danger of doing so.[27]

He was not just addressing his remarks to the younger and more impatient members of the Party, but speaking also to the government and the banned organizations, Congress of Democrats, ANC, and Pan-Africanist Congress. Liberalism would remain true to its faith, increasingly marginalized though it might seem as the clash of nationalisms grew fiercer.

But others in the Party lacked both Paton's patience and his pacifism. On 10 March 1963 Patrick Duncan, who had shortly before resigned from the Liberal Party, wrote to tell Paton that he could no longer accept that non-violent methods could do much to advance equality and non-racialism in South Africa. 'A fight is, in my view, coming,' he wrote, 'and I intend to play my part in that fight.'[28] The military wing of the ANC, Umkhonto we Sizwe, had on 16 December 1961 begun its campaign of bombings, at first concentrating on targets such as power lines and post-offices, and a violent organization named Poqo, sponsored by the PAC, had approached Duncan to help them procure guns.[29] Poqo did not shrink from taking lives, nor was Umkhonto's campaign bloodless. Duncan's letter of resignation, sent to Peter Brown through the mail, spoke of these matters and contained grave allegations against the Liberal Party, Paton, and Brown, and angered Paton because of the danger that it might have been intercepted by the police. 'Were you being ruthless and dishonourable, or just unbelievably careless?' Paton asked him bitterly.[30]

But worse was to come, for Duncan had at least resigned from the Party before taking the road of violence. On 11 July 1963 many of the leaders of Umkhonto were arrested on a farm at Rivonia near Johannesburg; Nelson Mandela, who had gone underground in May 1961, had been captured in August 1962, and was put on trial with the Rivonia prisoners. The Rivonia Trial, which took place in 1964, was to result in the life imprisonment of Mandela and the others. Meanwhile, inspired by the Umkhonto bombing campaign, a white student group calling itself the African Resistance Movement (ARM) had been formed to carry out sabotage, and though Paton did not know it, several of its prominent members were also members of the Liberal Party.

One of these was the leader of the National Union of South African Students, Adrian Leftwich. Other Cape members of the Party who joined the ARM included Randolph Vigne, the National Deputy Chairman, and Neville Rubin, son of Paton's friend Leslie Rubin. Paton had suspected that Vigne might be engaged in illegal activity when he stood for the Liberal Party in the general election of October 1961; he continued to entertain these suspicions when Vigne was banned in February 1963. In the Cape Town gardens he asked Vigne if he was engaged in illegal activity, and Vigne, without actually lying, gave Paton to understand that he was not. Adrian Leftwich, younger, more energetic, but also much more unstable than Vigne, had been questioned by Paton as early as 1961, and had deceived Paton in the same way. Paton had actually told Leftwich, 'If you are contemplating violence, you must not stay in the party,'[31] but Leftwich had reassured him.

Vigne and Leftwich recruited a planning committee which included Robert Watson, a man with a knowledge of explosives, and gathered a group of members including Leftwich's girlfriend, Lynette van der Riet, a Coloured member, Eddie Daniels, an economist, Norman Bromberger, and Michael Schneider, another young man. According to Leftwich, the ARM had about 32 members in all, mostly in the Cape and Johannesburg.[32] On 18 August 1963 they initiated a campaign of blowing up power pylons, radio masts, and railway signal cables. They had a number of failures, and at least four successes, the last being their destruction of a power pylon on 20 June 1964. On 4 July 1964 Leftwich's flat was raided by the police and incriminating documents discovered; Leftwich and Van der Riet were arrested, and within 96 hours Leftwich agreed to talk.[33] On 24 July 1964 another member of the ARM, John Harris, apparently acting alone, placed a bomb in Johannesburg Central Station, and it killed an elderly woman. The police burst into Leftwich's cell to tell him this news, and said to him, 'You'll hang.' The prospect so terrified him, he was to say, that he immediately agreed to give evidence for the State.[34] Twenty members of the Liberal Party were arrested; four fled the country, including Vigne, who simply got a friend to book a passage on a liner, and at the last moment took his place on board.[35]

At the trial that followed, in November 1964, Leftwich faced in the court, and betrayed, the members he had himself recruited to the ARM. They were given sentences as long as fifteen years, and Leftwich was released on 24 November 1964 to leave the country, telling the newspapers that he intended to marry Van der Riet, the daughter of an Afrikaans railway clerk.[36] In future years he was to wander from Israel to the United States before settling in Britain. 'Did he not know', Paton was to write later, 'that there can never be freedom for one who betrays his friends?'[37] Before he left he wrote Paton a rambling letter in which he said he had acted out of 'madness and irrationality', but rejected the right of anyone to criticize him.[38]

John Harris, meanwhile, was tried in October 1964 and sentenced to death by hanging for his station bomb. After the sentence was passed he broke down and clutched the knees of his lawyer, Ruth Hayman, begging her to save his life. 'This I find almost nauseating,' Paton was to write later, with unusual sharpness. 'If you are ready to take other people's lives, then you should not howl about anyone taking your own.'[39] Harris begged in vain, and was executed on 1 April 1965. In later years Paton would find it in him to forgive Leftwich, but he never forgave Harris.[40] Of Vigne he was to say in 1965, 'He used the force of his personality to persuade younger people to undertake sabotage and then left them to face the consequences.'[41] But in due course he forgave Vigne too.[42]

All the same, Paton at the time felt bitterly wounded by what the Liberal members of the ARM had done to him and to the Party. He did not dispute their right as individuals to turn to violence, though he continued to condemn it. But he considered that they had had a duty to leave the Liberal Party, as Patrick Duncan had done, before taking up the bomb. That they should have deceived him was bad enough, but the damage they had done to the Liberal cause was what really hurt him. 'It was a shocking experience for the National Chairman [Peter Brown] and the National President [Paton],' he wrote, adding, 'It justified the Minister [B. J. Vorster, Minister of Justice]'s famous remark that the communists killed people, but the liberal led people into ambush so that they might be killed.' He also said of the members of ARM, 'The damage they have done to the cause of liberalism is incomputable.'[43] Tony Morphet, a young Liberal who had got to know Paton in the 1950s, met him at this time and found him obsessed with the ARM:

> I found him in deep crisis about the African Resistance Movement. He talked obsessively, on and on, about the ARM people in the Liberal party, and completely obsessively every time I saw him he talked about John Harris, and Randolph Vigne, and Adrian Leftwich, and how they had betrayed the Liberal party. And the Leftwich case in particular had a special meaning for him, and how he could have turned State's Evidence against John Harris and his former comrades.[44]

Paton had reason to feel deeply wounded, for the Liberal cause, to which he had given eleven years of his life and so much effort and idealism, had been deeply wounded. The state's attitude towards Liberals hardened at once. On 30 July 1964, within weeks of Leftwich's arrest and Harris's bomb, Peter Brown was placed under a banning order for five years, being confined to an area of Pietermaritzburg and prevented from taking part in any gathering. Paton expected his own banning at any moment: it did not come. Dr Edgar Brookes became the National Chairman of the Liberal Party, in Brown's place.[45] Brown's banning order was after five years to be extended for another five, so that he suffered ten years of this misery, and though he never complained, his growing gravity showed Paton the price his friend was paying.

Other Liberal members, such as the architect Walter Hain, who, with his wife Adelain, was known to have shown sympathy for Harris, and had supported his wife and parents in their ordeal, were hounded out of the country: any firm Hain worked for was visited by the police and warned that it would get no government contracts if it continued to employ him. The Hains eventually had no choice but to leave. 'So we lost two of the strongest and bravest members of our party,' Paton wrote bitterly.[46] This kind of harassment happened to Liberals all over the country.

Paton had got a clear view of the way Liberalism was officially regarded when he agreed to give evidence at the Rivonia Trials in the hope of mitigating the sentences of Mandela and his co-accused. Here he was closely questioned by the prosecuting counsel, Dr Percy Yutar, who did his best to link Paton to the Communist party, and, failing in that, implied that he was a fellow-traveller of the Communists. Yutar tried to get Paton to admit support for the actions of the ANC, quoting from his Canadian interview with the CBC in 1960, in which Paton had predicted that violence would follow in South Africa. Yutar repeatedly implied that Paton had known the ANC planned violence, and had supported it. Paton as repeatedly refused to express support for what the ANC had done: 'I have not really come here to defend any actions of any persons—I have come here to appeal to the Court for clemency.'[47] But the clear implication was that Liberals and men of violence had been tarred with the same brush.

By August 1964 Liberal Party meetings were being harassed and turned out of halls booked for them, so that in Mooi River, for instance, they were obliged to use a cattle auction ring, with Paton speaking from the auctioneer's box, his speech drowned out by a rainstorm on the iron roof.[48] The police, who had always lavished attention on him, now stationed themselves permanently outside his house, photographing his visitors; they also tapped his phone with particular assiduity and intercepted his mail, making little effort to hide what they were doing. They pursued him relentlessly as he travelled around the country trying to rally the dispirited Party.

On 21 September 1964 he was followed on a trip to the town of Alice, and when he parked his car outside a hotel in the nearby Hogsback, a beautiful forested area, he came out in the morning to find its windows smashed by large stones. He made a report to the police,[49] who solemnly promised to seek the culprit, but he heard nothing further from them. In 1976, however, he was given information, through Donald Woods,[50] editor of the East London *Daily Dispatch*, that a former security policeman, Donald Card, could testify that the windows had been smashed by the head of the Alice security police, Warrant Officer G. A. Hattingh. Hattingh had followed Paton around Alice, taking hundreds of photographs of him, and when Paton protested, Hattingh told him, 'Mind your own bloody business.'[51] Paton tried to bring suit for damages against Hattingh, but the case had to be withdrawn in 1978 when Card became too frightened of his former colleagues to testify and the case collapsed. Hattingh then claimed R10,000 from the *Daily Dispatch*, but his case was dismissed in April 1983 after Paton had given evidence against him.

This was not the only damage to his car: in January 1966 his parked vehicle was tampered with again, unknown persons pouring grinding paste into the oil in his sump. The result was the complete loss of the engine. 'I traded in the car but lost R420', he told Leslie Rubin.[52] This incident was to go into his third novel, *Ah, But Your Land is Beautiful*. Again he was certain the culprits were the security police. Other members of the Liberal Party were suffering night-time shots through the windows of their homes.[53]

In the mid-1960s, facing such harassment and the siege of the Liberal Party inside South Africa, Paton felt bitterly angry when his friends in Britain appeared to criticise the South African Liberals for renouncing violence, for, in fact, sticking to their principles. He was particularly angered that those running the Defence & Aid Fund in Britain, notably Canon John Collins, should make repeated appeals in Paton's name for money, and then send it to the men of violence in South Africa rather than to Liberals who were trying to avoid bloodshed. In a challenging letter to Mary Benson, whom he had begun to suspect took this line, he wrote,

> For years now my name has been used to collect money for people whose views are not the same as mine. Yet we have not been able to get anything for the non-violent Liberal Party . . . One of the aims of Defence & Aid Fund is to support the establishment of a non-racial society, but in fact the Fund has given hardly a penny to Liberals who work to this end. One does not establish a non-racial society by blowing up a railway train.[54]

But his real anger was directed against John Collins, who had not only refused money for the Liberal Party, but had asked Paton not to approach him directly with any such request in the future.[55] Paton was well aware that the bannings of many Liberals were directly linked to their work for Defence & Aid: David Craighead, a Johannesburg actuary and a most committed Catholic, was one of these.[56] Banned in 1965 for his work as the Johannesburg Chairman of Defence & Aid, he would be forced to leave South Africa in May 1966.

In June 1965 John Collins made a speech to the United Nations in New York, in which he said that the aim of British Defence & Aid Fund was not just to provide legal defence, but also to buttress the morale of the underground movements in South Africa so that they could do their work easily and happily.[57] Paton thought this speech almost incredibly foolish, and likely to bring about the banning in South Africa of Defence & Aid and the banning or imprisonment of more of those who, like himself, worked for the organization.[58]

After carefully considering his position Paton wrote to Collins on 3 August 1965, telling him he thought Collins's speech 'incomprehensible' in its folly, that he had already been feeling disquiet over the distortions and omissions in Defence & Aid appeals in Britain, and that Defence & Aid in South Africa would now be known as the 'South African Defence & Aid Fund', in the hope of distancing itself from the parent body.

And to this powerful official letter he appended a personal note, even stronger:

> There is another thing I must in all honesty say to you. I do not think your D & A Fund has done much to carry out its third aim [the achievement of a non-racial society]. If it had, it certainly would have given greater help to the L[iberal] P[arty]. I doubt if we have received 1% of the funds you made available. You could at least have helped us to meet some of our more material difficulties, but you chose not to do so. We fully approve of the legal help D & A has given to those charged with political offences, even when serious crimes such as murder have been involved. Yet were saboteurs in your eyes the only kind of resisters worth sustaining? Did you think persons like Peter Brown of no consequence? Did you know that the overwhelming burden of the work of D & A was being done by people whose own causes you did not, or perhaps would not, support? I urge you to answer these questions, for your answers are vital to our friendship. And if you answer that you did not think that saboteurs were the only persons worth assisting, will you explain why you gave such negligible help to others?[59]

Nothing shows Paton's passionate nature better than the incandescent rage of this letter; nothing shows his intelligence better than the power of this unanswerable logic.

Collins did not attempt to answer directly. He could not have done so. Instead he dispatched his wife Diana, whom Paton liked and admired, to South Africa to talk Paton round in October 1965.[60] She succeeded in persuading Paton to go on with his work for Defence & Aid: as he remarked stoically, 'I could not [resign], because who then would have done our work?'[61] And regular sums now came to the Liberal Party from Defence & Aid, most of them reaching Paton through Laurens van der Post,[62] and most of them being spent on the Liberal journal *Contact*.

But Paton forgot neither Collins's pro-Communist bias, nor the ruthless folly of his United Nations speech. In later years, when Diana Collins was lobbying hard for her husband to be awarded the Nobel Peace Prize,[63] Paton declined to write in support of him, and was only with difficulty talked into doing so.[64] He considered that after such a speech, and after such support for violent organizations, the Prize for Peace was singularly inappropriate.[65]

All the same, he tried hard to preserve his friendship with the Collinses, whom he liked personally, and some of his most interesting observations about morality and religion were written to Diana Collins. 'Love for others', he was to write to her on one occasion, 'is not a substitute for religion, but is in fact of the very essence. It was St Theresa who said that one could not be sure of loving God, but one could be sure whether one loved one's neighbour or not, and it was through loving one's neighbour that one learnt to know and love God. This is my own view of religion.'[66]

He showed the same rigorous and practical regard for truth and justice when refusing the request of Lewis Nkosi, the exiled black writer, to contribute to Nkosi's paper *New African*. Nkosi had published a list of persons supposedly executed in South Africa for political reasons, including twelve executed not for political reasons but for murder. He had also implied that political executions took place in South Africa for reasons other than sabotage and murder. 'I cannot lend my support to half-truths of this kind,' Paton wrote firmly. 'Our case is strong enough, and distortion is not needed . . . I cannot contribute to your paper'.[67] On a subsequent occasion he also rejected Nkosi's paper *South African Bulletin*, because of an attack on Nadine Gordimer and the snide attitude of Nkosi and another black writer, Ezekiel Mphahlele, to liberals: 'In these days I do not choose to support any paper which tries to sow confusion amongst those outside the laager.'[68]

These external pressures were compounded by the increasing distress of Paton's home life, for Dorrie was by now clearly dying. The chest infection which had first manifested itself in May 1960 had steadily worsened and had been diagnosed as emphysema. This terrible disease, the gradual dying of the lungs and the consequent slow suffocation of the patient, had probably been brought on by her lifelong chain smoking. By the end of 1966 she was having regular episodes of such breathlessness that she had to be hospitalized and given oxygen at nights; presently she was obliged to take the oxygen cylinder with her everywhere.

In the midst of this gathering darkness, Paton also decided to have his mother, who was now 87 years old, come to live with him. Although she had had little to do with Paton since his decision to leave the Christadelphians, her mind was now going and she had become too much for her daughter Eunice ('Dorrie') to cope with. Paton brought her to his home on 11 March 1965, and she suffered a stroke a fortnight later, dying on 30 March 1965.[69] 'Alan's change from Christadelphianism broke our mother's heart,' his sister Dorrie was to say years later,[70] and though this was an exaggeration, her death must have filled him with gloomy thoughts.

But it was his wife's decline that filled him with grief. Neither she nor Paton could be in any doubt as to the future course of her illness. In September 1965, when Dorrie Paton was struggling constantly for breath, her sister Rad, another heavy smoker, died of emphysema. It is characteristic of this condition that sufferers have great trouble in exhaling. The long slow compression of the chest as the patient fought to breathe out, the sudden gasp for breath like a box opening, then the agonizing pause before the exhausting downward struggle began again, Paton and Dorrie had watched in Rad: they knew what lay in the future for Dorrie.

Yet, and for Paton this was the most painful part of the illness, they could not talk about it. Lack of communication about the most intimate matters in their marriage had always been a barrier between them; now Dorrie simply refused to acknowledge that there was any serious problem with her health, and Paton had not the heart to force her to acknowledge that she was dying. As a result he had to go through the terrible cheerful pretence that she was suffering a series of minor ailments, when he badly needed to talk about what was happening to them both.

By March 1966 he was writing to friends, 'There is no medical possibility of her regaining the use of parts of her lungs';[71] while Dorrie, in her diary, was opining that she was run down, and recording, 'Dr O'K[eefe] gave me second B12 injection. Feel they are already pepping me up.'[72] After recording a week of having gasped constantly for breath she could write, '[Dr] Caney very pleased with me—said I'm looking picture of health.'[73] She knew that she had emphysema, of course—'emphysema giving me hell', she could write[74]—but something in her denied it, and as the condition worsened she mentioned the disease less and less. On 25 October 1966 she had a long talk with her doctor, and recorded that he had told her that if she continued the treatment there was no need for anxiety or alarm. 'Cheered me up immensely,' she wrote happily in her diary that night. It was a kind of courage, but it did not make things easier for those around her. The heat of Natal worried her at night, and Paton had air-conditioning installed in their bedroom during one of her stints in Mariannhill hospital. This gave her comfort for a time, but heat was not her real problem.

Police harassment and Dorrie's decline came together on 18 May 1966, when the security police launched the swoop on Defence & Aid offices which Paton had foreseen at the time of John Collins's United Nations speech. Two senior security policemen knocked on Paton's door that morning, informed him that the Defence & Aid Fund of South Africa had been banned at midnight, and produced a search warrant. He showed them into his study, and the younger, looking at the shelves that lined each wall, said in Afrikaans, 'God, man, the books!' They removed all the Defence & Aid papers Paton showed them, but took Paton's word that there were none in other drawers or in the rondavel where he did his writing. The older of the two, seeing Dorrie straining for breath, commiserated with her about her asthma, saying his wife had asthma too.[75]

The Johannesburg *Star*, reporting the raid on Paton's home under the headline 'Who Cares?', commented that such incidents showed that 'South Africa has moved much nearer to traditional Fascism than is generally realized . . . the invasion of the esteemed brings home how exposed we all are'.[76] But the raid got little press attention, for other Liberals were suffering even more: it was at this time that house arrest was added to the mathematician Ken Hill's banning, the lawyer Ruth Hayman was placed under house arrest, and another Liberal, Heather Morkill, was banned in Pietermaritzburg. Paton's anger against Collins was richly justified.

His work for Defence & Aid came to a halt after this event, but he did not cut down on his own extensive donations to such causes as African education. Nor did he cease to call for Defence & Aid money to help others, such as the writer Bessie Head, for whom Paton appealed to Diana Collins in April 1966.[77] He was even more circumspect from now on about the source of the money that came to him in envelopes from Laurens van der Post, and he blamed Collins for what had happened: 'He more than anyone else caused this calamity because of the irresponsible statements he has made', he told Professor Z. K. Matthews.[78]

In July 1966 Dorrie spent nearly a month in hospital, her lungs very bad; her dead sister Rad occupied her mind a good deal. 'A year ago since Rad died,' she recorded in her diary on 5 September 1966: 'Been thinking of her a lot today.' Paton himself was unwell, his migraines returning in force, and the need for a haemorrhoids operation in September 1966[79] combining with the continuing political troubles to distress him.

On 6 September 1966 Prime Minister Verwoerd was stabbed on the floor of Parliament by a parliamentary messenger and died. 'A[lan] very restless and disturbed,' records Dorrie's diary that night. In December 1966 Paton published 'Dr Hendrik Verwoerd: A Liberal Assessment', in which he described the architect of apartheid as 'ruthless to a degree not necessary even by his own standards'. And he added, 'He permitted the banning of people whose only offence was that they had shown a courage and tenacity equal to his own.'[80] Verwoerd was succeeded by an even more inflexible figure, B. J. Vorster, who during the war had been interned for his Nazi sympathies.

At this time of strain inside his family and without, Paton was supported by his Liberal and Christian friends. He who had strengthened so many others over so many years was now grateful for the regular visits of a retired Archbishop of Central Africa, Edward Paget, and his local parish priest, Murray Dell; of Indian friends such as the Bughwans, Dennis a photographer and Devi professor of Speech and Drama at the University of Durban, Westville; the lawyer Pat Poovalingam, who, with Paton, served on the board of trustees of the Phoenix Trust;[81] Pondi Morel, a great friend of Dorrie's and a most faithful Liberal; and many others. Paton loved to play croquet with Paget and the Bughwans, and would take out all his frustrations in this aggressive game. 'He was a devil on the croquet lawn,' another croquet partner, Anthony Barker, was to say of him. 'The glee—you don't need to have glee in your eyes when you sky someone off into the rhododendrons [imitates Paton chuckling horribly]. He transmogrified into a demon of the worst sort.'[82]

His birthday parties in January each year had become too much for Dorrie, and the task was taken over by a group of Liberal women, including Devi Bughwan, Pondi Morel, Guinevere Ventress (wife of the mayor of Kloof, and a onetime Liberal candidate), Susan Francis (a vivacious Canadian whose husband Goondie was a local insurance broker), Janie Malherbe, and others. They would prepare the food and bring it ready made, set out the drinks, and run the elaborate festivities with their formula of a long toast to Paton, a doggerel poem read out to great laughter and applause, and then Paton's reply, which amounted to a state of the nation (or at least state of the Liberal Party) address. The number of guests increased annually until they numbered a hundred or more. Some guests, such as Chief Gatsha Buthelezi of the Zulus, whom Paton had met in the late 1950s, arrived with sizeable parties of retainers and bodyguards who had to be accommodated at short notice. All this was clearly beyond Dorrie's failing strength.

Paton himself was finding his work increasingly too much for him under the twin assaults of political strain and Dorrie's decline. He found he could no longer keep up with his correspondence, which poured into Lynton Road in a ceaseless stream; answering his more urgent letters took hours each day, as it had done since he published *Cry, the Beloved Country*. He longed to retire from public life and public writing, but felt that to do so would be to let down friends such as Peter Brown. 'I am in a position . . . of not having the courage to run away,' he wrote wittily.[83]

The Collins affair, the bannings, the house-arrests and searches, the hostility of the government, the mutual ill-will of groups opposed to the government: he was sick of it all. The American Consul General in Durban, Bill Toomey, whom he was meeting in preparation for a visit by Senator Robert Kennedy in June 1966, urged upon him the need for the opposition groups to 'give up their selfishness and pigheadedness and come together',[84] but Paton knew it could not be done.

He did, in fact, dine with Kennedy and his wife in Durban on 7 June 1966, and after the meal they drove through the streets to a meeting at Natal University with a police escort wailing in front of them. Mrs Kennedy remarked that it must be a novelty for Paton to have the police clear the traffic for him like this. Paton snorted. 'That's true,' he said. 'They're usually following me.' The Kennedys thought this a good joke.[85]

When Kennedy was criticized in the government-controlled media for daring to comment on South Africa's problems when he had only just arrived in the country, Paton responded with a brilliant little parable, comparing South Africa to 'a room full of people with all the doors and windows closed, and all the people smoking and drinking and talking. And a stranger from outside opens the door and exclaims, "Phew! What a fug in here". And they shout at him: "How can you know? You've only just come in".'[86]

He continued his activities as well as he was able. He began collaborating with Edward Callan, by now a professor of English at the University of Western Michigan, on a book of Paton's political writings, which was to be called *The Long View* after the title Paton had given his regular column in *Contact*. Callan had recently finished a critical book, *Alan Paton*, published by Twayne, the first full-length book to appear on Paton's work.[87] In the course of his research Callan had collected the political articles from *Contact*. Since Paton had told him he considered these articles an important part of his literary work, Callan offered to approach a publisher on his behalf, and Paton agreed.[88] It was an agreement he never regretted.

Paton also began making plans to write a second biography, that of Archbishop Geoffrey Clayton, who had had such an impact on Paton while he was bishop of Johannesburg. Paton had doubts about this project, chiefly because he could not understand Clayton. On the other hand, this problem had not prevented him from writing about Hofmeyr.

He continued to travel around, addressing Liberal Party gatherings, and visiting and encouraging anyone working against racism. One of these visits was to Nqutu in Zululand, where a pair of medical missionaries with the Society for the Propagation of the Gospel, Dr Anthony Barker and his wife

Maggie, were running a mission hospital in very difficult circumstances. Barker had had the idea of a prize-giving to encourage his nurses, and Paton agreed to present the prizes and attend a performance of *Sponono* put on by the black staff. This was an evening he greatly enjoyed.[89] He was particularly impressed by the natural way in which the young white medical students at the hospital, many of them Afrikaans, mixed easily on terms of complete equality with the black nurses. 'Watch the Afrikaner,' Paton told the Barkers. 'The English speaker often has his bolt-hole in Birmingham, but the Afrikaner's here for good. He's the one to watch.'[90] His conviction that the English speaker was ultimately an onlooker at the great clash of Afrikaner and black nationalisms did not stop this English-speaking onlooker from trying to avert it.

But, even in public events, death seemed to be all around him. On 4 June 1967 Patrick Duncan, that erratic idealist, died of anaemia in London, and Paton wrote a warm tribute to him for *The Times*.[91] And in July 1967 Chief Lutuli died after having been struck by a train near his home. Paton, with the new American Consul Red Duggan, attended the funeral at Groutville, north of Durban, and Edgar Brookes and Paton, President and Chairman of the Liberal Party, gave the orations. The government of South Africa, Paton said of Lutuli, 'took away his freedom, but he never ceased to be free. Indeed, he was more free than those who banned him.'[92]

Dorrie's coming death grew to fill his life. By February 1967 Paton was writing, 'She has little lung to breathe with. She is very courageous, but I do not wish her to go on if life is to be such a burden.'[93] In summer the heat tormented her, in winter the danger of chest infection grew. 'I shall be glad when the summer is over, but then, of course we will be coming into winter,' Paton wrote wearily to Leo Marquard.[94] The dying woman's diary makes increasingly terrible reading. 'Depressed. Gave way. So ashamed of my weakness. Full of self-pity. Disgusting,' she wrote on 8 January 1967. 'Gasped the whole day,' she recorded on 27 January. 'Do hope I sleep tonight.' Then follows, in handwriting that is broken and wandering, 'Had another lousy night.'

Her refusal to acknowledge what was happening to her adds to the pity and the terror. A month later, 'Seems as if I must be on oxygen a good time still.'[95] In April 1967 she was recording feeling sorry for her nurses. Many of her thoughts were for others, even when she was being slowly choked. 'Wish I could be brighter,' she recorded on 1 April 1967. 'Feel I must depress everyone.' She would wake in the morning after a drugged sleep, quivering with terror at having to face another day of this garrotting. But what terrified her above all was death, and this meant the subject could not be mentioned.[96]

On 6 April 1967 Aubrey Burns arrived in South Africa for his first visit to the country,[97] his trip paid for by the percentage of the royalties he was still receiving from *Cry, the Beloved Country*.[98] On 12 April Paton drove him to see Ixopo and Carisbrooke, and they called in at the church of St John in Ixopo, where the Patons had been married in 1928. Paton knelt and gave thanks for their married life, now coming to its end, and that evening he told Dorrie, in hospital, what he had done. 'Her whole face lit up with joy,' he was to write movingly.[99] There had been too few such moments of communication between them.

Running through her diary, like a counterpoint of life fighting death, are brief notes from Paton, all of them about the garden at Lynton Road. 'Kept oxygen on all night,' Dorrie wrote on 3 April 1967, and Paton added, 'Started sowing seeds.' 'Coughing a lot today,' she noted two days later: 'Calendulas up!' he responded. 'Night staff late in fixing me up,' she wrote the next day: 'Stocks up!' added Paton.

He was fighting to keep his mental and spiritual balance in these terrible months, and by July he was working hard on a new book, which he was to call *Instrument of Thy Peace*. It consisted of a series of twenty-one meditations based on the prayer which he believed to be by St Francis of Assisi, a saint to whom he had had a particular feeling of affinity at least since December 1955, when he had recorded reading all he could about him.[100] He quoted the prayer at the beginning of his book:

> Lord, make me an instrument of Thy peace. Where there is hatred, let me sow love; where there is injury, pardon; where there is doubt, faith; where there is despair, hope; where there is sadness, joy; where there is darkness, light.

This prayer is in fact not by St Francis. At its first known appearance, in an Italian prayer-book of the first half of the twentieth century, it was attributed to 'William the Norman', but it is almost certainly modern.[101] That fact would not have disturbed Paton, who considered that the prayer summed up the faith of St Francis, no matter who its author had been.[102]

Instrument of Thy Peace, with its meditations on aspects of this prayer, is a book of the deepest spirituality, written without affectation and in the simplest possible language, deeply and continuously moving. It is not ground-breaking theology. But it is a most practical and profound series of homilies on such problems as the nature of pain, or how to cope with sin, despair, fear, and weakness. This little volume, finished on 15 December

1967[103] and first published in 1968, was to be reprinted more often than any other of Paton's books except *Cry, the Beloved Country*, and no one reading it could wonder why. It is a book of extraordinary practical power, and the fruit of much reading and long thought. Many of the meditations end with words which Paton prayed many of the days of his life, and lived too: 'Help me this coming day to do some work of peace for thee'. He was to describe *Instrument of Thy Peace*, in later years, as an 'attempt to explain what is worth holding on to in the Christian faith at a time when so much dogma and theology is not only being challenged, but is in some cases being invalidated.'[104]

Paton would have laughed if anyone had told him that this is a book showing deep sanctity. But it is true that he, who had once been so filled with ambition, pride, and a sense of the righteousness of his own causes, had showed extraordinary spiritual growth in the last decade. The struggle of the past few years, which had so marked his face, had marked his soul also. Roy Campbell, Paton's favourite South African poet, had written of Mazeppa,

> Out of his pain, perhaps, some god-like thing
> Is born. A god has touched him, though with whips[105]

The words could have applied to Paton too. Certainly he was much closer to sanctity than he had been in the days when he had written to Hofmeyr of his desire for it, while he plied the whip himself at Diepkloof.

He needed the strength his book offered others. Dorrie's struggle for breath was now unceasing, and she could only sleep with more and more powerful sedatives. The time came when she could no longer walk on her own, and Paton would wheel her around the garden. 'I'm going to plant stocks there, and calendulas there, and take out all the balsams—we have enough anyway,' he told her on one of these walks. 'Yes, you told me all that yesterday,' she replied with asperity, and he rejoiced to see something of her old fire.[106] At the end of September 1967 she was diagnosed as having stomach cancer, which because of her emphysema was inoperable.

He himself was under terrible pressure from worry, and on 11 September 1967 he was diagnosed as suffering from a duodenal ulcer, and advised to move to a separate bedroom in order to sleep better. Though they had not shared a bed since very early in their marriage, Dorrie was pained by the thought that they would no longer share a room. 'A[lan] told me he must get away for some nights from our bedroom, as I'm getting him down. This has been the unkindest cut of all whether or not I deserved it,' she wrote despairingly. 'My only peace is in oblivion.' Paton added beneath, 'Do not think I told D. that she was getting me down.'

At the end of September she went into hospital in Durban for intensive care, but Paton soon brought her home to die, hiring extra nurses to supplement the two he had employed since April. He returned to sharing a room with her from 6 October 1967, which was the day Jonathan and his wife Margaret arrived, having been told that Dorrie was now sinking fast. The dying woman said to Margaret, when the two of them were alone, 'You know, if Mr P wants to get married again you must let him get married because he hasn't had an easy time with me.' And Margaret understood that Dorrie was speaking of their sex life.[107]

By 12 October 1967 she was continuously sedated, her diary by now a pitiful scrawl, but she was aware of her grieving family gathering round her: 'so many people but under sedation' she managed to write. Her last diary entry is 17 October 1967, 'Sedated [illegible] unpleasant evening.' On 22 October she was still alive, and David and Jonathan went home to Johannesburg until they should be summoned again. That night Paton woke several times to look at the oxygen, but he was sleeping when Dorrie died at 6.55 a.m. on 23 October 1967. The black nurse Queenie woke him with the sorrowing words, 'Our mother has gone.'[108]

At the huge funeral which followed, on 25 October 1967, the church was packed with all races: 'Just such a crowd as D. would have loved,' Paton noted in her diary, which he had taken over. That same day the telegrams began to pour in, and by the next day he was noting, 'Many many letters.' By 29 October the crowd had dispersed. 'First evening alone,' he wrote in her diary; 'Missed you very much.' On 31 October he was reading the hundreds of letters and grieving over them: 'Each fresh batch sets me aching again, perhaps weeping—So full of praise for Dorrie's virtues, so full of concern for me. They write of her tolerance, her brightness, her courage, her loyalty under all circumstances—They bring back the woman who so enjoyed life before illness overtook her.'[109] His marriage had begun in pain and ended in friendship, and now that Dorrie was gone he was devastated. 'Wish you were here tonight,' he wrote on 24 November 1967, and some days later, 'Answered 100 letters. Wept much.'[110] On 1 December 1967 he noted, almost with detachment, 'I wanted to die today.' He needed a new beginning, and he was about to find it.

20

FALL AND RISE 1968–1969

Sin is the writer's element.
FRANCOIS MAURIAC

'Alan is in a bad way,' Paton's Liberal friend Pondi Morel wrote to Leslie Rubin at the end of 1967, '& what he really needs now is to get away from SA even if only temporarily but I am afraid he won't. You would be distressed to see him.'[1] Many of his friends were distressed to see him at this time. After Dorrie's death he lived alone in the big house at Kloof, cared for by devoted servants but not caring for himself. He had never been much interested in clothes, but now he became decidedly shabby; he ate very little and drank far too much. One or two whiskies had been his usual evening allowance, but now he drank alone and to excess. He would arrive unexpectedly at the doorstep of Liberal friends such as Guinevere and Bob Ventress, saying he had 'come for a couple of whiskies', and it would be evident that he had had more than a couple even before arriving.[2]

Worst of all was his sense that he had lost spiritual direction, and that his faith was wavering, that it had at least in part been derived from Dorrie. 'I realised that my spiritual strength . . . was to a phenomenal degree dependent on the presence of another person', he was to admit.[3] Spiritual strength and direction were vitally important to him, and he was badly demoralized at this time. This demoralization showed even in his handwriting: this was the period when his shapely, sloping and beautifully legible script began its alteration to the crabbed, jagged hand of his later years, at its worst resembling barbed wire.

His sons, filled with pity, invited him to come and live with them in Johannesburg, but he declined. His friends invited him out to dinner as often as they could, and he would frequently dine with Goondie and Susan Francis, or Dennis and Devi Bughwan, or the Poovalingams, or with Tony

Mathews, professor of Law at Natal University, and his wife, Pam, who was a Liberal Party member, or Pondi Morel, or the Ventresses in Kloof. But he was sad company at this time, and he knew it. 'Thanks & grief are in a kind of equilibrium', he wrote to Elizabeth Patterson, 'but sometimes grief turns the scales. We had a long life together, with perfections & imperfections. Laus Deo.'[4]

His correspondence was now in great confusion, letters remaining unanswered for month after month. Even Edward Callan, patiently waiting for the introduction which Paton had promised for *The Long View*, wrote in vain until he at last heard from Paton, after a silence of nine months, in January 1968.[5] Paton then wrote a warm tribute to Callan's painstaking scholarship and sureness of touch. 'I opened the parcel [containing *The Long View*] with the greatest trepidation because I find it extremely difficult to read anything which is written about myself. It is a tribute to your work that not once did you embarrass me.'[6] He was frankly amazed that anyone as far away as North America could collect as much detailed and accurate information on South African politics and letters as Callan had, and in the future he would mine Callan's detailed, accurate notes in writing his own books, particularly his autobiographies. *The Long View* cheered him at a time when he badly needed cheering, and so did the very good reception of *Instrument of Thy Peace*. In later years Paton was to hear with pride that Mother Teresa of Calcutta liked the book.[7]

Best tonic of all, though, was the fact that he was writing again. On 1 December 1967 he had noted his grief at Dorrie's death, and written, 'I cannot trade for ever on having looked after you. That job is done, & one must find another. Shall I write our life?'[8] This idea rapidly took hold of him, and on 16 December 1967 he noted, 'Started writing "Letters to You"'. At what stage he got the idea of interleaving these contemporary 'letters' to Dorrie (the letters he so wished he had written her during her life) with a series of autobiographical chapters reaching back to the start of their marriage is not clear, but it was a master-stroke. The resulting book, which he was to call *Kontakion For You Departed* (published in the United States as *For You Departed*) is a remarkable and original piece of autobiography, at once judicious and deeply moving.

A kontakion is an Eastern Orthodox hymn for the dead, and Paton had heard the beautiful example he quotes in his book, the Kiev Kontakion, at the annual Memorial Day services of St John's College, his sons' school,[9] and at the funeral of Archbishop Clayton of Cape Town in 1957. He owned a phonograph record of the St John's choir singing this moving piece, and played it after his wife's death; with its references to the place where there will be neither sorrow nor sighing, it seemed to him to speak of that holy mountain mentioned by Isaiah, and towards which he hoped he himself was moving.

Friends who had known Dorrie well were to criticize *Kontakion For You Departed* as too intimate for publication, but writing it gave Paton deep satisfaction. There is no doubt that it did more than almost anything else to get him through the grieving process. One of the signs of this was his growing ability to joke again in letters to friends like Leslie Rubin and his wife Pearl. 'Give my love to Pearl', he wrote in February 1968, 'and tell her that she is a wonderful woman for sticking to you these many years.'[10]

This mood of slowly returning buoyancy was almost destroyed by two events during 1968. The first, long anticipated, was the dissolution of the Liberal Party. In 1966 the Nationalist government had introduced the Prohibition of Improper Interference Bill, which made it a criminal offence for a person to belong to any non-racial political organization. Since the Liberal and Progressive Parties were the only legal organizations to fit this description, they were clearly the targets. The Bill was debated and amended during 1967, and in 1968 it became law as the Prevention of Political Interference Act. The Progressive Party, with utmost reluctance, shed its non-white members. The Liberal Party was faced with the choice of abandoning its principles and following suit, or dissolving itself. It chose dissolution, and Paton travelled around the country in April and May 1968 speaking at closing meetings in Johannesburg, Hambrook (where the members were almost entirely black), and Durban.

The forced closure of the Party was a blow to him, but he did not take it as hard as some of his friends expected he would. For one thing, he considered that the aim of the Party was not to come to power, but to hold up before South Africans the possibility of a non-racial, non-violent course, an alternative to the cruel hypocrisies of apartheid and the brutal fraud of communism. This educative function would not cease just because the formal structures of the Party had disappeared. 'The lessons taught by the Liberal Party,' he would write nineteen years later, 'are only now beginning to be learned by many South Africans.'[11]

There was also a measure of quiet relief in his reaction to the dissolution of the Party, for he was deadly tired of political life. In March 1968 he had told Mary Benson, 'It is very difficult to write about my revulsion from public life,' adding, 'I have always had a concealed resentment against public life because it prevented me from writing.'[12] Now the demise of the formal machinery of the Party freed him from the constant public speaking, the annual conventions, the regular and inescapable journalism, that being Chairman and then President of the Party had involved. One side of him was glad to be rid of it. For another side of him, however, this was an important link with the past snapped, another kind of bereavement. Not for a moment, though, did he consider that his years of thankless, ceaseless work had been in vain.[13]

Another even more traumatic, and much more personal, event took place on 6 March 1968. Paton had been invited that evening to have dinner with Goondie and Sue Francis on their farm, Roches Point, at Hillcrest, not far from the Toc H TB Settlement where Paton had lived for a year in 1953. After the visit, during which he had drunk at least two whiskies and a good deal of wine, Paton left the Francises to drive back to Kloof alone. Some hours later a retired black detective of the security police, Mr Alfred Khumalo, heard a car horn sounding outside his house in the African township of Clermont. When he went to investigate, two people ran away from the vehicle into the darkness, and in the car he found a white man apparently dead.

The unconscious figure was Paton: Khumalo recognized him, for he had once been assigned to watch Paton's house and note all his movements. Paton had been throttled, and remained insensible for over an hour.[14] Khumalo summoned both an ambulance and the police. When Paton recovered consciousness he told the police that after leaving the Francis home he had picked up two male African hitch-hikers. He often picked up black hitch-hikers, a rare and dangerous habit in a white South African; it was one of the ways he stayed in touch with blacks, and in doing it he showed the sympathy of the white man in *Cry, the Beloved Country* who picks up the bus boycotters. His passengers, Paton told the police, asked him to drive them to Clermont, and on arrival in the township they assaulted him and robbed him of his watch and wallet containing about R20. He declined to be hospitalized; the police took him back to his home in Kloof, and he spent the next day in bed. He subsequently made a good recovery, suffering only a sore throat.

That might have been the end of the matter, had it not been for the police investigation that followed. Paton's alleged assailants were arrested on 15 May 1968 and brought to trial. They were a known criminal, Derek Ndhlovu, and a 29-year-old woman, Alice Ngcobo, and some of Paton's stolen things were found in their possession. The case came up for hearing on 16 May and again on 30 May 1968 before being adjourned until July.

Before the magistrate in Pinetown, Mr P. H. Castell, on 16 May, Ndhlovu alleged that Paton had given him a lift and asked him to procure the sexual services of a black woman, and Ngcobo, in a separate statement, corroborated this.[15]

Paton left the hearing in the greatest distress, with good reason: a charge under the Immorality Act mandated a six-month prison term. More than that, as *Too Late the Phalarope* makes abundantly clear, it involved terrible disgrace. Writing in 1980, Paton would say about the Immorality Act,

> If a white man of any substance, a minister of religion, a lawyer, a schoolmaster, is found guilty of breaking this law, his life is ruined, even if the court suspends the punishment. At the time I write this, three white men have committed suicide in the last few weeks rather than face trial.[16]

Newspapers featured the story of the Chairman of the Liberal Party being accused of trying to procure the services of a black woman. None of them seems to have raised the parallels with *Too Late the Phalarope*, but they were bad enough. He tried to conceal from those around him what was happening, telling his new secretary, who found him bedridden on the day after the assault, only that he had a sore throat. On 24 July he gave her a strange little note, telling her she was not to open it until she got home. She feared she was being sacked, but the note merely said, mysteriously, 'There may be something tomorrow that you won't like, and I'll understand if you don't want to come and work for me again.'[17] The 'something' was the trial in the Durban Regional Court.

Before the hearings Paton had sought the legal advice of a Liberal Party friend. The friend was subsequently to tell Jonathan Paton that his father had only on a second visit mentioned the presence of the black woman: 'I should have told you. There was a woman too.' The friend had then, he told Jonathan, advised Paton to deny all knowledge of her.[18] And he had assured Paton that all would be well. Paton pressed him irritably for reasons. 'How the hell do you mean? What evidence have you got?'[19] But the friend could do no more than express an optimism which Paton badly wanted to share.

At the trial on 25 July 1968 Paton gave evidence that he had picked up two men and been throttled and robbed by them. No woman had been present. Ndhlovu was then allowed to cross-question Paton, and again alleged that Paton had picked him, Ndhlovu, up alone and asked him to obtain the sexual services of a woman for Paton. Ndhlovu claimed he had agreed to do this, warning Paton that women in Clermont would not accept less than R10. Paton, Ndhlovu alleged, had agreed to this sum, also offering to pay Ndhlovu for his help.

They had driven to Clermont where Ndhlovu said he had without difficulty obtained Alice Ngcobo's co-operation. She had wanted to 'have an affair' with Paton in the car or on the grass, but he had objected, saying he wanted to take her home for the night. Ndhlovu claimed that as a result of this disagreement he had realized that he was not going to get his money, so he had seized Paton by the neck while Alice Ngcobo searched and robbed him. In the struggle that followed Paton had managed to press the car horn.[20] His assailants had taken R31.60 and Paton's watch and coat, and on the arrival of the retired policeman, they had run away, Ngcobo losing her shoes in flight; and in some nearby bushes they had divided the spoils, Ndhlovu taking most of the money and the watch. Alice Ngcobo was dissatisfied with her share, according to Ndhlovu, and when they met the next day he threatened to hit her, as a result of which she reported him to the police. It was this information that had put them on Ndhlovu's trail.

Ngcobo's statement to magistrate Castell agreed with Ndhlovu's on the fact that Paton had sought to secure her sexual services, but differed in claiming that she had run away as soon as the assault began.[21] She also was allowed to question Paton at the trial. She asked if one of the people he gave a lift to was a woman, and Paton answered 'No'. 'Then how do I come into the picture?' asked Ngcobo. 'I don't know,' replied Paton.[22]

He need not have worried about the outcome of the trial, for his legal friend's optimism proved justified. The judge, Mr F. J. Dietzsch, sentenced Ndhlovu, who had previous convictions for robbery, gambling, and being in possession of a dangerous weapon, to two years corrective training after finding him guilty of robbery. He discharged Ngcobo on the grounds that though she had been present at the scene of the crime, she had not assaulted Paton, and her story that she had run away before the robbery took place 'might reasonably be true'. He commended Paton for being a most impartial witness. As for the vital matter of Ndhlovu's allegations that Paton had tried to procure Ngcobo's services for immoral purposes, Dietzsch ruled that this 'may be true or may be a fabrication', but that the court did not deem it necessary to elaborate further on the matter.[23]

Paton was hugely relieved. He phoned his sisters to tell them in person what the outcome of the case had been. He arranged for a special service of thanksgiving to be held in his local church, and asked David and Jonathan to attend it with him, flying both his sons down to Natal so that they could appear at his side as a public expression of family rejoicing.[24]

David, who was not a Christian, was much put out by this peremptory summons to worship. Paton also took them to the American consulate in Durban, having previously been invited for cocktails that evening. He was soon surrounded by admirers. David took Jonathan aside and said, 'There he is, centre of attraction now he's out of trouble. Cock of the walk again.'[25]

Paton clearly considered himself to have been vindicated. His sons were not so sure. Although Ndhlovu and Ngcobo had fallen out with each other, their stories agreed on the central fact that Paton had tried to procure Ngcobo's services. Nor could Paton explain what the woman had been doing there at all. And there was one other detail the two sons found unsettling: according to Jonathan, Paton had asked a doctor to examine him, and he had been found to be suffering from an infestation of pubic lice. Paton told the doctor he had 'picked them up in the veld', but the physician knew they were spread chiefly through sexual contact.[26] Perhaps most conclusive was the lawyer's account, given to Jonathan, of his discussions with Paton.

Exactly what did happen that night of 6 March 1968 will probably never be known with certainty. In my view the evidence (and not just that adduced at the trial) leads toward the conclusion—though it does not prove—that Paton did try to procure the services of a woman. Moreover, the lice suggest that this was not the first time he had done so. Even while his wife was alive, as we have seen, he sought sexual solace outside the marriage, hints of this being made to friends like Leslie Rubin;[27] they would be confirmed in his autobiographies. Now that he was a widower he must have felt freer. Nor would the prospect of breaking the Immorality Act have troubled him, since he considered it an unjust and hateful law. His long and frustrating marriage had now come to an end, and he was lonely and demoralized; such an adventure as that alleged by Ndhlovu and Ngcobo is not improbable.

As for the policeman, Khumalo, the only independent witness who could have seen Alice Ngcobo running from the scene, when contacted by reporters as soon as the news broke he said he would say nothing without Paton's permission, and at the trial he gave evidence that he had seen Paton being dragged out of his car by two men and assaulted, evidence that contradicted Paton's, Ngcobo's, and Ndhlovu's accounts of the assault having taken place inside the car where Paton was found. Paton subsequently paid for the education of Khumalo's son Zasi, a benefaction lasting many years.[28] It was quite in character for Paton to show his gratitude in this way for his rescue, for over the years he paid for the education of many young Africans, but even his admitted generosity does not wholly allay the suspicion that Paton was also rewarding Khumalo for keeping quiet.

Certainly the whole matter was deeply embarrassing to Paton, not least because of the effect produced by the throttling of the leader of the Liberal Party by an African, and his rescue by a retired member of the security police. 'I presume that the whole business was cunningly organized [by the government]', wrote Professor Prestwich of the University of Natal to Paton, expressing a widely held suspicion, but Paton played the whole matter down and wanted nothing more than to have it forgotten.[29] He had had a bad fright, and the parallels with his own father's strangling may not have escaped him.

Certainly the event gave him a taste of mortality, for in March 1968 he made a new will, leaving 20 per cent of his estate to each of his sons, and the remainder to various institutions. The institutions themselves are interesting: he proposed to leave 10 per cent to the black seminary at Alice, 10 per cent to the Indian Phoenix Settlement on whose board he served, 10 per cent to the Anglican church, 10 per cent to the University of Natal, and 10 per cent to the Institute of Race Relations, with many smaller bequests to his faithful servant Anna Makhaye, his cook Theresa Sibisi, the gardeners Sikali Ngcongo, Zodwa, and Constance, and so on. Royalty and other continuing income he left to his sons, writing to inform them what he had done.[30] In future years, however, he would alter his will several times.

Once he was through this crisis he went rapidly on with *Kontakion For You Departed*, being helped and encouraged by his new secretary. She was Anne Hopkins, and she was to play a most important role in Paton's life. Paton had agreed to interview her in January 1968 at the urging of their mutual friend Goondie Francis. Francis was deeply worried by Paton's decline after Dorrie's death, and he was equally worried about Anne Hopkins, who was going through an extremely painful marital breakup.

She had been born Anne Margaret Tindale Davis in London on 21 October 1927, and was therefore twenty-four years younger than Paton. Her parents were divorced when she was 6, and she was sent off to a boarding school 'to get her out of the way'. She hated the school, and in later life was to say that these traumatic childhood experiences, combined with the stiffly formal training of her mother, made her almost incapable of showing affection, and contributed to the failure of her own first marriage. Her education had been badly disrupted by tuberculosis, and her mother had insisted that she take a secretarial course, arguing that she would 'always be able to get some sort of a job'.[31] She worked for a while for the director of the Royal Ballet School, Arnold Haskell, before wanting something more exciting from life. In 1950 she joined the Foreign Office as a secretary, was posted to Greece, and there met, and in 1951 married, a divorced wing commander in the RAF, Paddy Hopkins. Although serving in the British forces, Hopkins had been born in South Africa,[32] and when he retired as a Group Captain in 1962, he and Anne, and their two children Athene and Andrew, sailed to Durban.

They set about building a house for themselves, and meanwhile rented a house in Kloof. This rented house happened to be next door to the Paton home at 23 Lynton Road. The Hopkinses knew about Paton: that is, they knew that he was a dangerous revolutionary, probably an anarchist, intent on overthrowing the government. Why else would he have the security police virtually camping at his front gate? They intended to have nothing to do with him, and he made no approaches to them.[33] Paton's gardener, Sikali Ngcongo, irritated the Hopkinses with frequent noisy parties and confirmed their view of their neighbour. In a conversation years later, Anne Hopkins recalled her first meeting with Paton.

Then one day I was standing at my window looking out, and over from Alan's garden came five little birds, they flew in the air and then dropped dead. And I thought This is the end, I've had enough, this man poisons birds.

So I looked in the telephone directory for Paton. That's the thing about him, Alan was a humble man, it never occurred to him to have an unlisted number. Mr Paton, I said, You are poisoning the birds. And I gave him a terrible dressing down on the phone. And I buried the birds and thought nothing about it.

About half an hour later shuffling up the long drive came a chap, a little old man in a filthy old raincoat and a grotty hat. And I thought Who on earth's this? He came to the door, explained who he was, and he said Where are the birds? I said Well I buried them. And he told me some cock and bull story about how they must have flown into the white wall, and got a fright. And I let him have it. And he shuffled away again, you see.

And he went and told Dorrie about this harridan who lived next door. And that's how I met him.[34]

Anne Hopkins was capable of 'letting him have it' in a big way. She was plain-spoken to a marked degree, she had a precise English accent which to many South African ears sounded intimidating, and when she was annoyed it acquired real bite. She told Paton about his gardener, his behaviour, his poisoning of the birds, and his general undesirability.[35] It was not a promising beginning.

When Anne's marriage broke up in 1967, Goondie Francis, who in Anne's view 'fancied himself as a match-maker',[36] suggested she should act as Paton's secretary, and brought the two of them together at his house on 21 December 1967, two months after Dorrie's death. Paton, although he had not forgotten their first encounter, found he liked her. She was attractive and she did not defer to him: if he pulled her leg in conversation, she could give as good as she got.

On 5 February 1968 he asked Anne around for an interview. Once more she was impressed by his shabbiness and carelessness in dress. He wore a cotton shirt, old cotton shorts, shoes and socks, and a straw hat. He glared at her over his spectacles, that pale blue glare that unsettled so many of his visitors. After a few questions he gave her his foreword for Edward Callan's volume *The Long View*, and asked her to type it and return it to him. When she did, he looked at it and asked her, 'Mrs Hopkins, how do you spell foreword?' 'F-O-R-W-A-R-D', spelled Anne promptly.

Now Paton hated careless spellers. In 1953 he had composed a witty poem for a young woman, Margaret Snell, who had written him a mis-spelled letter:

Confused no doubt by words like 'choose'
You go on writing 'loose' for 'lose',
Soon you'll be writing 'ruse' for 'rues'
And 'Bruce' for 'bruise' & other blues [errors].

If 'loose' is 'lose' then Heavens knows
What you will make of 'noose' and 'nose',
You'll squeeze your oranges for Jews,
And look for friends in old Hoose-Hoose.

How bright in church when pews are puce,
How shocked the ear when moos the moose,
How sad the sheep when fleas are fleece,
How glad the world if peas were peace.

Come Margaret dear, attention police,
Go down upon your bended niece,
And ask for even half the nous
Which Providence allows a louse.[37]

Now he briskly explained to Anne Hopkins the error of her ways, and she was sure she had lost the job. Trying to recover the situation, she said, 'I really can spell quite well usually. I can spell exorbitant, E-X-H-O-R-B-I-T-A-N-T.' Paton looked stunned, and then suddenly he laughed, a laugh Anne never forgot. 'His laugh completely transformed his face, and I realized that the ferocious glare was really a twinkle.'[38] And so she became his secretary.

Anne Hopkins rapidly transformed Paton's life by bringing organization and order into it. On days when he had to go out she would sort through his chaotic files, and after much labour reduced the piles of dusty papers, unanswered letters, unpaid bills, unread proofs, and unreconciled bank statements to chronological order, safely put away in a steel filing cabinet. Presently she began organizing his finances, asking him if he knew he was R5,000 overdrawn at the bank, and getting his accounts in order. In a short time she had made herself indispensable.

As she got to know his working methods, Anne Hopkins was amazed by Paton's ability to predict the length of time it would take him to complete an article or chapter, and his complete professionalism in keeping to a timetable. 'I've just been asked for a 5,000-word article', he would announce. 'I'll start it tomorrow, and it'll be finished for you to type on Friday afternoon.' And it was.

On a personal level they were getting to know each other better too. She began bringing him his tea each morning and having it with him. 'Shall I stick some more water in the pot?' she asked him once. 'Certainly, Mrs Hopkins. I presume you brought the glue?' said he, chortling at his own joke.[39] On hearing her say that she intended giving someone 'a piece of my mind' he gravely advised her not to: 'You haven't much to spare, you know.' She typed *Kontakion For You Departed* as he wrote it, and gave it her approval, which he recorded in chapter 40 of that volume: 'I never knew your wife, but she springs to life in these pages.'[40] He was delighted by this comment: it was as if Anne were meeting Dorrie, and liking her.

When the assault case came up, Paton believed that Anne might not want to work with him again, but her loyalty never wavered. She had absolutely no doubts that he was innocent.[41] Years later she was to say,

> Of course I stood by him. But it was unpleasant for him. But he was very dignified about it all, because it was horrible to have this slur cast upon him, that he'd gone out to look for a woman. It was nasty. He was exonerated of course, but it was not nice. But it was stupid, he did silly things like picking up people and giving them lifts, he was asking for trouble. He was lucky not to have been killed.[42]

And this loyalty also must have warmed Paton towards her.

Meanwhile he was receiving letters from some of the women he had known well while Dorrie was alive, making tentative, tactful, but unmistakable offers of marriage.[43] Some of them were former Liberals in Natal, each of whom told him that she was the right woman for him to marry. Four or five of them actually arrived from Canada, the United States, or Europe, to visit him with the same object in mind. It was Anne who typed Paton's politely negative responses to the letters, and she watched as he dealt tactfully with the visitors.[44] He was naturally flattered by these offers, and later told Leslie Rubin about them.[45]

Mary Benson, in South Africa because her father was dying, invited him to meet her in Pretoria, to which she was restricted by the terms of her visa. He went, with a reluctance he did not hide from Benson, in August 1968, but instead of taking her to a smart restaurant as she had expected, he took her to Fountains, a picnic area, where they walked in the winter sunshine and talked. Or rather Paton talked, chiefly about *Kontakion For You Departed*. He could not resist quoting all the praise he had received from friends who had read it—'The best thing you've done since *Cry, the Beloved Country*!'[46] Mary Benson was making sounds of assent, 'Hmm. Uhuh. Mmm,' and was startled when Paton suddenly stopped in mid-sentence and gave her one of his glares: 'Do you always make those extraordinary noises?' he asked. After they had had lunch they sat in the car, and Paton said, 'Now tell me about you.' But after Mary Benson had spoken for a little while she noticed that he had fallen fast asleep.[47] She was not so much hurt as amused by this conflict between her expectations and his ego.

It was soon after this meeting in Pretoria that Paton conceived what he called the 'fantastic idea' of marrying Anne Hopkins. The two of them began going to films together, and Paton took Anne to his favourite restaurant, Saltori's in Durban. She invited him in to her house for coffee after the dinner, and in the course of the evening remarked to him, 'I think you should marry again.' He replied, 'If I marry again, it will be to someone like yourself.'[48] Anne took this as a proposal, and after giving it careful thought over a period of some weeks, told him she agreed to accept it.

Paton consulted some of his friends, including Peter Brown and the headmaster of King's School, Nottingham Road, John Carlyle Mitchell, and found them hostile to the idea of his marrying Anne. She was utterly unlike Dorrie; she was politically conservative, cool but outspoken, not South African, too young, ill-educated, strong-willed. In short, she was quite wrong for Paton. Peter Brown, in particular, told Paton gently but firmly, 'She won't do.'[49] Paton was downcast for days by this opinion, which he communicated to Anne: but it is a testament to the strength of her attraction for him that he ultimately ignored the advice of

Brown, whose judgement he respected more than anyone else's. In later years, it is fair to add, Brown came to show great loyalty to Anne, and to like her.

Her divorce had come through in July 1968, and in October Paton did something rather daring: he took her away for a week to a cottage which Peter Brown owned in the Drakensberg and where, until Brown's banning in 1964, Brown, Paton, and a few selected men friends had gone for a weekend of drinking each year on or about December 16. It was an isolated farmhouse near Giant's Castle, innocent of such conveniences as running water; there was an earth privy in the garden, and a river to wash in. The view of the mountains, however, was superb. In this idyllic but primitive retreat Paton and Anne spent a week getting to know one another, to see if they were, as Anne put it, 'entirely happy in one another's company'.[50] They were. They bathed in the Bushman's river, they walked and read. Anne found Paton a most loving companion, and by the time they came back from the Drakensberg they were sure of each other.

She had already met his sons briefly, when they came down on 23 March 1967 to see Paton receive an honorary doctorate from his old university in Pietermaritzburg. This was the third such degree he had received, for in addition to the honour from Yale in 1954, he had been given an honorary D.Litt. (in absentia) by Kenyon College, Ohio, in 1962. Now Paton invited Jonathan down to Natal, before Christmas 1968, and there told him and Margaret he had a surprise for them. Jonathan was to remember the conversation:

> Then you could see something welling up. He suddenly said, gruffly, 'I want to tell you two something.' 'Yes?' 'I'm thinking of getting married.' Well! 'Who're you going to marry?' 'Thinking of marrying my secretary.'
>
> Once he'd come out with it, he was a different person. Full of beans and bounce, couldn't wait for us to meet Anne, we all went on a family picnic with Anne, and I remember how Anne at once spotted that the sons were not overjoyed, particularly David. And there were lots of whisperings between her and my father, almost like teenagers. Anne was 25 years younger than him, and he was in his element. Occasionally you'd see their shoes rubbing together. That sort of thing, you know. It was almost as though they were being a bit naughty.[51]

The truth seems to have been that Paton rather enjoyed surprising and even shocking his sons: not many men of 65 get the chance to. They naturally resented the speed with which Anne had replaced their mother, but they could scarcely resent Paton's evident happiness. And though neither of them came to like Anne, both of them could see that she was devoted to Paton, and also that she was very good for him. Pondi Morel, Pat

Poovalingam, the Francises, and others could see that without Anne, Paton would probably have written nothing further, and perhaps would have drunk himself into the grave within five years. Anne saved him, organized him, and gave him nearly twenty more extraordinarily productive years.

There is no doubt that Paton recognized this potential in her when he agreed to marry her. 'A most sensible thing',[52] he was to call his second marriage, though it was his heart rather than his head that ruled at the time. As for Anne, she was of course marrying a world-famous author, but she was also taking on alarming responsibility. Why did she take it on? I asked her two years after Paton's death. 'I loved him,' she said simply. Then after a pause, 'I actually loved him. I really loved him. Although he was 65 and I was 40, I loved him very much. And he loved me. Oh yes, very much. Although I think that later on he thought he'd made a mistake in many ways, because I'm not easy. But no, we were very happy, actually.'[53] And it was true, every word.

The engagement was announced in the newspapers on 24 November 1968, the Johannesburg *Sunday Times* carrying a large photograph and banner headline, 'Paton to Marry Divorcée'. The news evoked intense interest. Anne's daughter, Athene, then 14 and at her boarding school, heard the matron remarking, 'Look, there's that woman on the back page of the *Sunday Times*', and years later could remember saying very loudly, very aggressively, 'That woman happens to be my mother.'[54] She had been badly hurt by her parents' divorce, and was as little pleased by the prospective marriage as Paton's sons were.

The fact that Anne was a divorcée posed particular problems, for the Anglican church could not marry them. In 1962 Paton had been invited to become a member of the Anglican order of St Simon of Cyrene, an honour extended to laymen who had given particularly distinguished service to the church; he now felt obliged to offer his resignation from the Order, and was very hurt when the Archbishop of Cape Town, Selby Taylor, accepted it. He then felt he should resign from the council of St Peter's College of the black seminary at Fort Hare, and this resignation also was accepted. Bishop Inman of Natal, a more sympathetic man than Selby Taylor, now offered a way by which Paton could marry in the Anglican communion: since Anne's husband had himself been a divorced man when they married, Inman suggested that her first marriage could be the subject of a decree of nullity. And would that mean, Paton icily enquired, that her children would be considered illegitimate?[55]

His back was up: this was the greatest crisis in his relations with his church. In the event it was not to last long; Bishop Inman readmitted him and Anne to communion, and Paton was soon reinstated to his former positions of distinction. All his life, however, he was to be troubled at intervals by his church's disapproval of his marriage to a divorced woman, and there exist letters of protest he wrote to his bishop about the matter, getting no satisfaction from the replies.[56] In spite of this he never wavered in his devotion to Anglicanism.

Given his feelings at the time, though, he chose to be married in an old Methodist church in central Durban, by the Reverend Robert Irvine, on 30 January 1969. It was a pressure-cooker day of the type at which Durban excels, and as he placed the ring on Anne's finger Paton noticed a drop of perspiration trickling down her nose.[57] Very few people were present: his children and their wives, and a few friends including the Francises. After the ceremony the party went to Saltori's restaurant for lunch, and presently Goondie Francis went out into the street and returned triumphantly waving a newspaper poster: 'Alan Paton Married'.[58] The newly-weds spent their first night in the Imperial Hotel in Pietermaritzburg, and then went on to Peter Brown's Drakensberg cottage to resume their honeymoon. Neither of them seems to have thought it odd that a good deal of their time in the Drakensberg was spent in correcting the proofs of *Kontakion For You Departed*, with its intimate picture of Paton's life with Dorrie.

He told himself that from this point on he was going to retire from active life, and merely be an observer of South African politics and a family man. He rather enjoyed the role of grandfather: it is from this period that his first letters to his grandchildren date, the earliest to survive being a letter to Jonathan's son Nicholas, then aged seven:

Dear Nick,
Thank you for the drawing of a man, a chair, two lamps, and a grandfather clock with a small clock on top.

Why does the man need two clocks?

I do not think he is very good looking, not as nice looking as your Dad.

Much love from Grandad.[59]

To Nicholas's sister Pamela he wrote,

Dear Pamela,
How is my mischievous granddaughter? When are you coming to see me? Have you swallowed any more beads? I have never swallowed a bead, I am too sensible. Are you getting bigger? I hope to see you again soon, and I hope you are being a good girl.

Love from Grandad.[60]

As they aged, though, he sometimes found his grandchildren trying, particularly Jonathan's son Anthony, who was a difficult and noisy child. 'I suppose you could bring the hamster,' he once replied to a request about bringing pets for the holidays, 'I am sure he will not be any worse than the children.'[61] This letter was dictated to Anne, and typed and signed by her in Paton's absence. The impersonality of such letters gradually became a source of friction between Paton and his sons. They felt they had lost the direct contact with their father which they had once enjoyed.

Anne came between Paton and some of his friends too, for she guarded him fiercely against those who would draw him away from his writing. Many people had got into the habit of phoning him whenever they pleased, or simply dropping in to see him. Now the phone was invariably answered by Anne, and she could be very sharp in defence of Paton. 'What do you want?' she would ask in her direct way, and callers who declined to explain themselves would not be put through to Paton. More than one of his old friends were so upset by this questioning that they stopped phoning him.[62]

Nor could they any longer drop in to see him and simply walk round the house to his writing-rondavel, because Anne had persuaded him to move house after their marriage. She associated the Lynton Road house with Dorrie, and wanted a new house to make her own. She and Paton did not move far; they found a big dilapidated house for sale in Botha's Hill, on the very crest of the hill that gives that leafy dormitory settlement its name: it was number 14 Botha's Hill Road, and it cost them R22,000.[63] In moving to it from Kloof, Paton and Anne went through his and Dorrie's papers, and burnt a great many of them. It was possibly at this time that he destroyed Dorrie's love letters to himself, and also destroyed those letters from him to her that were written at times of stress between them.[64]

The Patons named the Botha's Hill house 'Lintrose'. They altered the place, enlarging one bedroom and the kitchen, adding a bathroom, and building a fine study for Paton. The house as it eventually developed consisted of two long, low wings forming a splayed V; it was light, airy, and its big windows opened everywhere onto the garden. To reach Paton's study the visitor had to pass from the front door at the point of the V through a passage, then through a small dressing room off the Patons' double bedroom, and so into the new study at the end. From this study French windows opened on to the acre of garden, where terraces bounded by beautiful stone retaining walls would soon lead down to a magnificent croquet lawn, and there was a fine view over the hills down towards Durban.

This arrangement differed in one important respect from the writing room Paton had had in the garden at Lynton Road: a caller wanting to see him had to knock at the front door and pass Anne's inspection, and she did not hesitate to turn away those who arrived without an appointment. Paton's time was no longer frittered away by casual callers. His friends protested bitterly to him and he would sometimes sympathize with them, but the truth was that he valued Anne's defence of his writing hours. In the past he had had to hide himself in sea-side cottages to do his writing, as he did for years when producing *Hofmeyr*; Anne solved this problem for him at a stroke. His marriage had changed the whole complexion of his life. 'I did a very wise thing when I married again, and am very happy,' he wrote to Leslie Rubin.[65]

21

BETTER THAN RUBIES
1969–1973

At least I have loved you;
Though much went wrong,
This was good,
This was strong.

SARAH TEASDALE

Paton rejoiced in his second marriage, but there were inevitable strains. These came partly from the difficulties of adjustments each had to make, but partly, too, from the need for their children to adjust. Paton's sons found Anne unsympathetic, and David came actively to dislike her and ceased to stay in his father's house as he had been in the habit of doing. She was not sorry; she considered him uncouth, objecting (for instance) to his coming in to breakfast wearing nothing but a pair of brief underpants.[1] David denied ever doing this.[2]

Jonathan continued to come for holidays with his family, for he had remained very dependent on Paton, who lent him money to buy a house, paid his school and doctor's bills, and generally supported him in any emergencies.[3] Paton was also continually on the lookout for ways of furthering his son's career, though he considered that Jonathan would have gone further as a schoolteacher than as an academic.[4] But Jonathan, though he continued to see his father often, disliked the way Anne reproved his children for boisterousness and no longer felt at home in Paton's house: it had in fact become Anne's house too, naturally enough.

The greatest difficulties, though, were the result of the jealousy Anne's daughter Athene and Paton felt about each other. Even before their marriage Anne had warned Paton that Athene would always come first with her, and he was aghast, as well he might be: he cannot but have remembered Dorrie's words about not loving him as she had loved her first husband. He experienced flashes of jealousy on such occasions as when they were motoring near Durban, and Anne said, 'There's our house'. He was astonished, for Botha's Hill was nowhere in sight, and then realized with a great pang that she was referring to a house in Winston Park where she had once lived with her first husband.[5]

Athene, an active and highly intelligent teenager, came home each weekend, and she resented the amount of time her mother spent with Paton. Paton, for his part, resented Athene's lack of respect for him, her demands on Anne's time, her increasing use of the telephone, and in due course, when she began to drive, her borrowing of Anne's car so that Anne had to borrow his. He thought Athene a wastrel, and would go round behind her turning off lights. Presently, after they had been married for four or five years, Anne bought Athene a car of her own, and Paton was enraged, wrongly believing she had used his money for this purchase. He accused her of deceit, there was a bitter quarrel, with threats of divorce, and Athene had to be told she could not come home for the weekends. She expressed hatred of Paton, and he reciprocated and then would repent and feel ashamed of his behaviour. Gradually, as Athene matured, she and Paton came to declare a truce, and in the long run were to grow to like one another, but this festering situation was the background of his marriage for nearly a decade, and it put great strain on Anne and himself.[6] With Anne's son Andrew he got on better, for Andrew was younger, came home less often from his boarding school, and was temperamentally much less inclined to clash with Paton.

Anne was determined to continue to live a life of her own, and she did. She played tennis regularly and helped to organize the tennis club, she drove an ambulance as a volunteer for many years, she helped with a blood transfusion service, she took woodworking classes and in time became a highly competent carpenter, and she had a large circle of friends whom Paton did not share. Although her political stance was well to the right of Paton's, she had a profound concern for the poor, and helped to organize a feeding scheme for malnourished Zulu children in the Valley of a Thousand Hills. She became the President of this, the Emolweni Schools Feeding Scheme, which began by giving regular meals to perhaps 50 children, and rapidly expanded until it was feeding nearly 8,000 and running a clinic for those needing medical attention.[7] Paton admired this activity, but had very little to do with it himself.

He recognized Anne's strength and leaned on it from early in their relationship. This sometimes led her to think him uncaring of her. On one occasion, early in their marriage, they had gone out in two cars, and left a dinner very late. Paton got into his car and roared off, leaving Anne to follow as best she might, though she was nervous of driving alone in the dark. He always drove very fast, and by the time she got home he was sitting comfortably in an easy chair sipping a whisky. 'Well, you might have waited for me,' she told him angrily, and he looked up in surprise.[8] He expected her to cope, and in spite of protests like this, she did.

He continued to write, producing a great deal of journalism. The Liberal Party could no longer function as a party, but he intended that its ideas should continue to be disseminated through the press, and he contributed to, and helped to fund, journals through which it could make its voice heard. One of these was a new magazine, *Reality*, whose first issue appeared in March 1969, carrying an editorial written by him, in which he argued forcefully for the common society, for universal franchise, and for the need to defend human rights against white and black nationalism. His aim, and that of the new magazine, would be what he fought for most of his life: 'to proclaim and pursue our ideal of a new South Africa, in which a citizen's worth and place and future will not be determined by his race and colour, but by his willingness to serve the country that belongs to us all.'[9]

Reality was the journal of the disbanded Liberal Party, and for those in the know its full title made this plain. The former Liberal Party journal had been called *Liberal Opinion*: the new journal was called *Reality: A Journal of Liberal Opinion*. Paton served on the editorial board, with Colin Gardner and others, though after the magazine was established he did not spend a great deal of time working for it.[10] What he did instead, in the early years of the journal's existence, was fund it, partly out of his own pocket, partly with Defence & Aid money which reached him from Diana Collins via Laurens van der Post, always in letters suggesting that the money was part of Van der Post's American royalties.[11] The sums were R2,000 or R3,000 a year, and when they proved inadequate, Peter Brown stepped into the breach.[12]

Paton continued to fight for Liberal ideals in his personal relations too, and one of the few things that could make him break with a friend was the suspicion that the other was showing signs of racism. When a nephew of Pondi Morel asked him to his wedding, but failed to invite the Indian friends they had in common, Paton wrote him a sharp and sad letter:

> Some of our mutual friends—who have always been welcome in our homes, and in whose homes we have always been welcome, have not been invited to either wedding or reception. It is your business whom you invite or whom you do not.

But if these friends are to be excluded, then we have no wish to be included. It is an important principle for me and I cannot set it aside for any person . . . What our future relations will be after this, I do not know. But they will certainly not be restored until you have restored relations with those others, who have been deeply hurt by their exclusion.[13]

He was particularly sensitive to Indian concerns in 1970, for he was playing a big role in the Gandhi Centenary celebrations, opening a Gandhi Memorial Clinic, Library, and Museum, presiding over a meeting of 6,000 people at the Phoenix Settlement which he continued to serve as a board member, and in June 1970 giving 11 addresses to Gandhi memorial gatherings in a single week.[14]

With Anne's encouragement he had begun writing poetry again, the first he had written since 1956. The first of these new poems was a bitter commentary on the factional fighting which had begun to afflict the black population of Natal as the ANC recruited the young, and traditional chiefs used violence to maintain their ascendancy over their people. Untitled, it begins with the words 'I am the Law', and Paton wrote it in February 1970. The last two stanzas run,

Did you hear of Zuma's wife
Seven months with child?
They went through her house like lords
Her brothers of the homeland.
Zuma came hurrying home
In time for lamentation.
I cannot look at his eyes
They have seen a new thing
Quite unspeakable,
The new freedom of the homeland hills,
And the new winds bringing death.

I am the Law and the Power and the Glory
I come down from the throne
And say to the people, you put me there
And when you wish, remove me.
I shout at them, God damn you, smile
For if you do not smile
Your brothers of the homeland
Shall come like lords to your house.[15]

Another poem he wrote early in 1970 was inspired by the death in police detention of a Muslim cleric, Imam Haron, in Cape Town in September 1969. Many prisoners died in police detention as a result of brutal interrogations, the police commonly giving such explanations as that they had slipped and fallen. Paton called his protest over this example 'Death of a Priest', and addressed it to the Minister of Justice:

> Most Honourable I knock at your door
> I knock there by day and by night
> My knuckles are raw with blood
> I hope it does not offend you
> To have these marks on your door . . .
>
> Most Honourable the sorrow is not my own
> It is of a man who has no hands to knock
> No voice to cry. A sorrow so deep
> That if you had it for your own
> You would cry out in unbelieving anguish
> That such a thing should be.
>
> Most Honourable do not bestir yourself
> The man is dead
> He fell down the stairs and died
> And all his wounds can be explained
> Except the holes in his hands and feet
> And the long deep thrust in his side.[16]

The South African darkness seemed to be deepening, and in the general election of 1970 the Nationalists showed they were as entrenched as ever, while the Progressives, who carried the torch of moderation after the disbanding of the Liberals, were unable to get more than one member into Parliament. That member continued to be Helen Suzman, whose plush constituency of Houghton was a demonstration of the curious fact that in white South African politics it was the wealthy who were on the left of what remained of the white political spectrum. Part of the explanation, no doubt, was that the white working class felt menaced by the possible rise of the black working class, in a way that the wealthy did not. Paton wrote to Suzman congratulating her on her victory, but expressing his sober opinion that there would not be much change in South African politics during the next five years.[17]

He did not confine his attention, or his writing, to South African politics alone. He also continued to take a great interest in the politics of the United States, which was now deeply mired in Vietnam. The agony of that war showed itself in the protests which erupted all over the country, and Paton hated to see America apparently tearing itself apart. When four students were shot dead by National Guardsmen during a protest at Kent

State University in Ohio in May 1970, he was deeply shocked. 'The news of Kent Ohio is terrible,' he wrote to Jonathan, then a visiting lecturer in Kalamazoo. 'Their photos were in the D[aily] N[ews] last night—four nice-looking kids.'[18] His distress was plain in the poem he wrote about Kent State, 'Flowers for the Departed'. The dominant image is of the flowers that flame each spring in the Kalahari, the desert which Paton had visited in July 1960 with Dorrie, and where her final illness first showed itself. In the poem he devotes a stanza to each of the Kent State dead:

> Allison Krause, for you this flower
> Desert-born in a distant land
> Suddenly, in rain miraculous
> Flamed into life and lit with orange fire
> The arid plain. So may your seed,
> Returned untimely to the earth
> Bring back the beauty to your desert land.

And the poem, which draws on Whitman's famous elegy for Lincoln, 'When Lilacs Last in the Dooryard Bloom'd',[19] ends with a rebuke to the country he loved second only to South Africa:

> America, for you these flowers
> Would we could reach out hands to comfort you
> But we dare not
> We dare not touch those fingers dripping
> With children's blood.

This poem had a great impact in the United States when it was published in the *New York Times* on 5 May 1971. It was set to music by the American composer Daniel Jahn,[20] and it played a part in the passionate debates in Congress which followed the killings, when Congressman Ogden R. Reid read it into the Congressional Record.[21]

Paton's reputation as a man of letters continued to grow, partly through the efforts of Edward Callan, who had now established himself as the foremost Paton scholar. In 1969 Callan collaborated with Professor Rolf Italiaander of the Free Academy of Art in Hamburg (whose 'Plakette' [plaque] had been awarded to Paton *in absentia* in 1961), on a bibliography based on that in Callan's *Alan Paton*; Paton was pleased with the result when it appeared in 1970.[22] Callan and Paton met in July 1969, the first time for more than twenty years, when they both attended a conference at Rhodes University in Grahamstown, and years later Callan would remember how Paton had walked with him round an exhibition of paintings by nineteenth-century English settlers, and pointed out that all the flowers in the paintings were English, not South African. The newcomers were not yet South Africans, but 'transplanted Englishmen'.[23]

In Grahamstown Paton saw the world première of Athol Fugard's play *Boesman and Lena*, which he thought a most powerful work. He heard through a mutual friend, the writer Jack Cope, that Fugard regarded *Boesman and Lena* as an attempt to make amends to his wife, Sheila, for all she had been through in their marriage. Paton had met Fugard at rare intervals from the early 1960s on, and was to have more contact with him in later years, when Mary Benson began working on Fugard's notebooks. He heard that Fugard had expressed a low opinion of Paton as a dramatist. 'If he did so then I think he was foolish,' Paton observed tartly. 'I have never had any pretensions to being a dramatist. However it does not pain me at all if he did make these remarks.'[24]

At this conference Paton and Nadine Gordimer read from their work from the same stage, being introduced by Edward Callan. Paton's reading of part of *Kontakion For You Departed* struck Guy Butler, professor of English at Rhodes University, as deeply moving; he called it 'one of the best one-man shows I have ever attended.'[25] Paton and Gordimer had an uneasy relationship. Paton retained his reputation as South Africa's premier novelist, despite all Gordimer's talent and energy, so that to some extent she felt the effect of his shadow; for his part, he could not but be aware that by comparison with his own, her stature as a novelist was growing steadily.

Gordimer had the habit of undercutting Liberals in her writing. Again and again one sees in her stories scathingly vivid portraits of wealthy and privileged characters who, from a range of motives, get a kick out of revolution, and who (like Flora Donaldson, the Liberal who occurs in both the story 'Something for the Time Being' and in *Burger's Daughter*) 'fellow-travel beside suffering as a sport enthusiast in a car keeps pace alongside a marathon runner.'[26] This kind of sneer made Paton very uneasy about her. He found her writing very cold: 'She has almost a clinical power of observation, and I cannot remember experiencing any emotion while reading anything she has written,' he told Jack Cope, and coming from him this was a severe judgement.

He was too honest to conceal it from Gordimer, letting her know that he felt her treatment of her characters lacked warmth. She replied that her character Mehring (in *The Conservationist*) was of the species she despised more than Vorster or Treurnicht, Afrikaner politicians, while Bray (in *A Guest of Honour*) was a character she felt close to; but this had nothing to do with 'warmth'. 'I can't', she confessed, 'see warmth/lack of warmth as a literary quality.'[27] Paton responded that *Burger's Daughter*, in which Gordimer was paying a debt of gratitude to the Communist leader, Bram Fischer, showed this warmth precisely because she was repaying a debt. 'She did not reply to this unanswerable argument,' he told Edward Callan.[28]

But while he found her novels sometimes tedious, and considered that she 'put in the sexual pieces for effect',[29] Paton admired her short stories (and novellas such as *July's People*)[30] and liked her personally. For her part Gordimer considered that too many of his characters were 'quite simply Alan Paton speaking',[31] though she thought this less true of *Too Late the Phalarope* than of *Cry, the Beloved Country*. In spite of their reservations about each other's work the two often exchanged friendly and admiring letters.

There were increasing marks of recognition for Paton in these years: in July 1968 he was first elected the Honorary President of the National Union of South African Students (NUSAS). NUSAS had been founded by Paton's old Liberal friend, Leo Marquard, in 1924, and Paton took on the Honorary Presidency with great pleasure, not least because NUSAS was at this time one of the very few remaining channels of vigorous political protest. It was a significant tribute to Paton's stature that he was elected Honorary President by an organization whose student leaders were more radically inclined than Paton. He regarded the post as one giving him the duty to advise students, and the right to be consulted by them. When they failed to consult him, he would be annoyed and would write to tell them so: he refused to be a mere figurehead.

He was also receiving offers of honours from abroad, particularly from America, where his readership continued to be huge and where he had many friends. These offers he had to turn down, for he still had no passport, but he was very much tempted to apply for one again. The order confiscating his passport had been indefinite as to time, the implication being that he could apply for one when he chose, and might or might not be given one at the discretion of the government. He had decided that he would not apply for a passport while other Liberals, particularly Peter Brown, could not travel more than a few miles from their homes. Brown had been banned in July 1964 for five years, and Paton hoped very much that the ban would not be renewed. In July 1969 it was renewed for another five years, the Minister of Justice as usual providing no explanation of this flagrant injustice.

'We have condemned before today this supra-legal process of banning, and we do it again today,' Paton told a meeting called to protest the renewal, in Pietermaritzburg on 8 August 1969. 'Another five years of a kind of imprisonment have been imposed on Mr Peter Brown. Yet his offence is unknown. He has not been charged with any offence. He has not been brought before any court and proved to be guilty . . . It is monstrous is it not, that the Minister of Justice should ban Peter Brown on the grounds that he is furthering the aims of Communism?'[32] But the protests as usual were in vain.

Brown, however, encouraged Paton to travel, if only so that his voice should continue to be heard in the world. There was a strong incentive to take this advice: in January 1970 Paton was offered an Honorary Doctorate of Divinity by Edinburgh University, and he was very keen to accept this honour. In addition, he had now begun to think that he would like to write the biography of South Africa's greatest poet, Roy Campbell. The work had been begun by a distinguished scholar, Professor W. H. Gardner of Natal University, but Gardner had died in January 1969 before writing a word, and his wife and son, Colin, who had been a very active Liberal Party member, were happy to accept Paton's suggestion that he take on the task. Paton had written a well-researched lecture on Campbell, which he delivered at the University of Natal in 1970, and though he had met Campbell only once, he wrote to Campbell's old friend Laurens van der Post, 'I feel that I could do justice to his life'.[33] To do this work he would need to consult the poet's widow, Mary Campbell, who lived in Portugal. Accordingly he applied for a passport in February 1970, not very optimistically: 'Have applied for a passport but doubt extremely,' he told Jonathan.[34] To his considerable surprise he was granted one, though it was valid for one year only, and restricted his travel to Western Europe, the United States, Canada, and Brazil.[35]

He now planned to go overseas for the first time in a decade in 1971, but before leaving he intended to finish his research on a biography of Archbishop Geoffrey Clayton. He had first agreed to do this book in November 1966, and had done a good deal of preliminary research during 1967, but the illness and death of Dorrie had driven it out of his mind. During 1970 he began researching again with a will, making repeated trips to Cape Town to work on the Clayton papers there. He stayed at Bishopscourt, the beautiful residence of the Archbishop of Cape Town, where his host was the incumbent, Selby Taylor, a rigid and ascetic figure. Paton soon found himself missing his evening whisky, and turned to his friends for help.

One of these was David Welsh, a former Liberal Party member whom Paton had met in Natal and liked. In a conversation years later, Welsh, who in 1980 would be appointed professor of Politics at the University of Cape Town, recalled the circumstances:

> Selby Taylor is austere. The first night Paton was given a sherry before dinner, and that was it. So the phone rang the next morning, early. Alan never said who he was. 'David? Would you get me two bottles of scotch, and bring them to Bishopscourt on your way back from the university. And David? Would you be very discreet about this.'
>
> I got the whisky, and I got one of those box files, and put the bottles in them. Went to Bishopscourt and rang the bell. Flunkies and chaplains came round and I asked for Alan. In due course he came down the stairs. The flunkies were still standing round, and I said 'Here are the research notes you wanted.' And I passed him the file. He took it, and said 'Thank you for these notes.' Then he leaned close to me with that very simian little smile. He murmured, 'And David—they gurgle.'[36]

On a subsequent visit, a few months later, Paton again called on Welsh when life at Bishopscourt had got him down. 'David? I want you to invite me for dinner tonight.' 'Certainly, Mr P. What time shall I come for you?' asked Welsh, thinking 7.30 p.m. might be appropriate. 'About 4 p.m.'[37]

Anne regretted that he had taken on the job of writing Clayton's life: 'He is working terribly hard and for what?' she told Margaret Paton. 'To write a book for a very limited public about someone who sounds (to my jaundiced mind) to be less than interesting!'[38] She was not the only one who felt this: Laurens van der Post, who did his best to encourage Paton to write another novel, in later years was to regret the time Paton had spent on 'that big book about his bishop'.[39]

Yet for Paton the writing of *Apartheid and the Archbishop: The Life and Times of ✠ Geoffrey Clayton* was a labour of love akin to the writing of *Hofmeyr*. As Hofmeyr had been Paton's political mentor, Clayton had been his spiritual guide for many years. And the Clayton book, like *Hofmeyr*, was more than a biography: it was a history of the rise of apartheid, and a vivid account of the battle that had been waged against it by people like Paton. He summarized the book's themes for Edward Callan:

> It has five main themes: the church, the conflict between church and state, the incompatibility between the Dutch Reformed Churches and the others, the politics of the times, and the strange personality of Archbishop Clayton, with many anecdotes. My job was to keep the first theme within bounds.[40]

And in a later letter to Callan, Paton added that Clayton's life was 'an allegory of the Christian way'.[41]

Like *Reality*, the Clayton biography was a way of keeping the ideals of Liberalism before the South African public. Paton told Tony Morphet that this book was being written because it was his duty. Morphet remembered Paton's defence of the task:

> 'The real duty is, this is the next book after *Hofmeyr*. And how am I going to tell the story of apartheid if it's not through Clayton?' I [Morphet] said, 'It's a very strange way to tell the story of apartheid.' He said, 'No it's not, not at all, not at all. The key thing in the book is Clayton's refusal to accept the church clause, and his letter to Verwoerd that he will command the church to disobey. And then he goes to bed not knowing what the outcome will be. Defiance.'
>
> And I thought, 'Well even so, it's a lot to write to get to that.'[42]

It was quite plain to others besides Anne that the Clayton book was not going to be a big seller. Paton had a great deal of trouble finding a publisher. The book had been commissioned by the religious publisher SPCK, who had envisaged a short monograph and were taken aback when presented with the life of Clayton in 180,000 words. Reluctantly they had to inform Paton that they lacked the resources to publish it. Paton, annoyed by this, tried a number of publishers without success, before having the luck to find David Philip, a cultured, Oxford-educated Capetonian who was about to set up his own publishing business, leaving Oxford University Press after seventeen years of fruitful association. Philip naturally hoped to take some authors with him, and Paton was to be the most important of these. It was in fact Paton's agreement to let Philip have *Apartheid and the Archbishop: The Life and Times of ✠ Geoffrey Clayton* that gave Philip the confidence to take the plunge on his own. His aim was to provide an outlet for writers in opposition to the government, which was having marked success at stifling its opponents through censorships and bannings.

David Philip had pronounced Liberal views and a good sense of humour, and Paton took a personal liking to him. He strongly supported what Philip hoped to do for oppositional publishing. Henceforward he would often stay with David and Marie Philip on his trips to Cape Town, and he came to love their comfortable house in Claremont. *Apartheid and the Archbishop: The Life and Times of ✠ Geoffrey Clayton*, which Paton finished writing in November 1972, did not prove a big seller when it was published on 28 September 1973, but it was a steady one, and it satisfied both Paton's and Philip's hopes of it, winning the CNA Literary Award in 1973.

Meanwhile Paton had rediscovered his pleasure in travelling. He and Anne left South Africa on 21 May 1971, Paton's first trip abroad since the loss of his passport in 1960. Once it had become known that he was able to travel again, other institutions had realized that he might accept their offers of honorary degrees, and on this trip he was to receive an honorary doctorate from Trent University in Ontario, Canada, another from Harvard, and the DD from Edinburgh, thus doubling the total of his honorary degrees at a stroke.

The Patons flew via Rio de Janeiro to New York, spending three days in each of those cities before flying on to Canada. They arrived in Toronto on May 26, and stayed with one of the members of the Liberal diaspora, Richard Robinow, whom Paton had become very friendly with while he was an active member of the party in Durban. Robinow came of an old and distinguished Jewish shipowning family in Hamburg, but was himself a most committed and faithful Christian. The author of a number of books,[43] he had worked as an advertising executive after coming to South Africa in 1935. In 1962 he and his Afrikaans wife, Beatrice, had moved to North America, where Beatrice built a most successful career as a medical librarian.

The Robinows showed the Patons Toronto, and drove them to Trent University in Peterborough, Ontario, for the degree ceremony on 28 May 1971. By way of recompense for Robinow's kindness, Paton preached on 30 May in the Bloor Street United church where Robinow was a regular worshipper. The great challenge of the church, he said, was to heal the wounds of the world without being corrupted by the world.

> This means, and it cannot mean anything else, that there is a continual tension between the Church and the world. The Church, being human as well as divine, cannot help being corrupted by the kind of society in which it is placed. We have that very bitter story in South Africa of a black man who is cleaning the church and another man comes in and says, 'What are you doing in here?' And he says, 'I'm cleaning the church.' And the white man says, 'That's all right, but God help you if you pray.'[44]

On 31 May 1971 the Patons flew to California to stay with Paton's old friend, Aubrey Burns, in San Francisco, and on 6 June 1971 Burns took them off by car through the High Sierras, to Yosemite, which Paton loved more each time he visited it, and then to Las Vegas. From there they drove to Bryce Canyon and so to the Grand Canyon, which Paton had last seen in 1947. This time he did not risk his ankle by climbing down into it; instead Anne photographed him standing solitary on the edge contemplating the majestic scene, a stumpy, comical figure under a huge floral sunhat.

On 15 June 1971 the Patons flew back to the east coast, where on 17 June Paton received an honorary D.Litt. from Harvard, after which he had the further honour of making the 320th Commencement address in Harvard Yard. In this he urged his young audience to involve themselves in the world, and to live for a purpose. He was urging on them, in fact, the conclusion he had come to under the influence of Railton Dent while he was himself a student:

> The only way in which one can make endurable man's inhumanity to man, and man's destruction of his own environment, is to exemplify in your own lives man's humanity to man and man's reverence for the place in which he lives. It is a hard thing to do, but when was it ever easy to take upon one's shoulders the responsibility for man and his world?[45]

Paton made a profound impact on many of his audience. The economist Barbara Ward at the time said to him simply, 'It sounded like the Bible.'[46] Years after his death Anne Paton would recall 'the almost mesmeric power he had as a speaker. As long as I live I will remember with clarity the way he dominated the proceedings when giving the Harvard lecture.'[47]

After another brief stay in New York, where Scribner's put on an elaborate reception for him, at which he met the writer Edmund Fuller again, he and Anne flew to Britain, where on 15 July 1971 Paton was made an honorary Doctor of Divinity in Edinburgh, admitted to the degree by the touch of a cap sewn, bizarrely, from John Knox's breeches.[48] He was particularly proud of this degree, and boasted of it to Trevor Huddleston, who was now Bishop of Stepney, and who replied tongue-in-cheek that he had some time ago received a doctorate from Glasgow, a much better place.[49]

Following their Edinburgh visit, Paton spent a bibulous weekend with Leslie Rubin in a Sussex village, Forest Row, and he and Anne then hired a car and drove through France and Spain to Portugal, where they spent some time in the beautiful hill-town of Sintra. Here, in an isolated and rather primitive country house, lived Mary Campbell, the widow of the poet Roy Campbell. Although he had yet to finish the biography of Clayton, Paton began interviewing Mary Campbell in preparation for the book about her husband.

He found her a difficult subject to interview, and as the work progressed he became less and less happy about doing it. Roy Campbell had been a great teller of tall stories, boasting in his autobiography, *Broken Record*, 'I am not the one to wish to bore you with a list of facts.'[50] Mary Campbell took the same attitude to the truth, particularly where it affected her marriage, which had been a difficult one, and her husband's political views, which had been confusedly right-wing. Campbell's younger daughter, Anna, was not much help either: deeply emotional and poorly educated, she alternated between truculent boasts about her father's greatness, and sudden outbursts of sobbing. Anna's wealthy companion, Rob Lyle, who had befriended Campbell in his last years and supported him financially, was a man of icy reserve and sudden suspicions, of whom Paton was to remark that he would rather interview an iceberg in a typhoon.[51] 'To put it bluntly,' Paton was to write, 'I found Mary Campbell devious and the other two not much better.'[52] Campbell's elder daughter, Teresa, was Paton's main source of information, but there was much she did not know. It soon became clear to Paton that the Campbell book was not going to be easy to write. He knew the danger signs: he had suffered eleven years with Mrs Hofmeyr.

He was to return to Sintra in 1973 in an attempt to get more information from Mary Campbell about the extra-marital affairs both she and Roy Campbell had had, and which had soured their lives and marked Campbell's poetry. In particular, Paton had now begun to realize that Mary Campbell had had lesbian lovers during her marriage. But he was too embarrassed to ask questions about such matters with the degree of directness that would have elicited direct answers, and Mary Campbell evaded him. Harold Nicolson once memorably remarked that a good biographer needs to be 'a snouty little man', but Paton was unable to snuffle through the Campbells' private lives. Even when dealing with Clayton, he had had to be prodded, by Tony Morphet, into saying that Clayton was a homosexual, and having stated the bald fact, he had said no more. 'I don't want to go into that,' he had growled at Morphet.[53]

He grew more unhappy about the Campbell project, partly as a result of these difficulties, partly because of the cost of writing it and the likelihood that he would not recover these costs, partly because a life of Campbell did not fit the pattern into which all his writing fell. His novels, his poems, and his political journalism all carried the same message, a religious and political message about living justly and dealing rightly by others. His biographies of Hofmeyr and Clayton had carried the same message, with their analysis of the rise of apartheid and the fight of just and deeply moral people against it.

But Campbell was a much more ambiguous figure than Hofmeyr or Clayton. After a period of youthful rebellion against the 'colour bar' of the 1920s he had left South Africa, lost interest in the country's racial problems, and become a right-winger who supported General Franco and Mussolini before the war, and who all his life made occasional anti-Semitic comments. Paton found him an increasingly unsympathetic figure to work on. 'Am now contemplating that fascist anti-Semite, Campbell,' he wrote gloomily to Leslie Rubin.[54]

When I visited him at Botha's Hill in 1974, he quickly recognized me as the salvation he had begun to seek for, and asked me if I would like to write the Campbell book instead of him. Would I! It was the kind of thing every young researcher dreams of. 'Alexander was a sober and serious young man,' he was to write later, 'but he was obviously pleased and excited by the prospect.'[55] The truth is that I was so overwhelmed I could hardly speak, though neither he nor I could foresee that he had changed the course of my career decisively.

Like Gardner before him, Paton had not written a word of the biography, but he had collected a mass of papers and manuscripts, and he passed these on to me together with the Gardner papers. During the next eight years he supported me in all my struggles to get at the truth about Campbell's extraordinary life, and sympathized with me when my attempts to tell the whole truth led to threats of legal action from Rob Lyle, as a result of which part of the Campbell biography had to be cut, in particular that portion in which I told the extraordinary story of Campbell's self-castration while tormented by sexual temptation in 1952. From the start Paton had thought of me as his 'rescuer';[56] as he watched me struggling in the toils he increasingly congratulated himself on his escape.

He himself was fighting a feeling of sterility and lack of inspiration at this time. After his death there was found among his papers a 'Daily prayer for one's work', dated 17 January 1972, of which he notes, 'Written at what I hope is the end of a period of sterility'. It reads,

> O God my Maker, from whom all my gifts proceed, I pray earnestly that You will help me to use them. Remove from me all sloth, uncertainty, vanity, self-satisfaction, and fear, that might hinder me in my work.
>
> Help me to be industrious, & help me to continue working even when the work seems poor. I do not ask to be saved from melancholy, but I ask the strength to get up & do something, for You & others and myself, when I am fallen into it.
>
> And help me to struggle continually to think more of others and less of myself. Lord listen to the prayer. Amen.[57]

He did suffer from occasional fits of melancholy in these years, partly because of his continual failure to produce imaginative writing worthy of publication, partly because of the difficulty of adjusting to his new marriage.

Ironically, as his output of creative writing seemed to be falling, he achieved wider public recognition, for universities now seemed to be queuing to give him honorary degrees. On 8 April 1972 he received one from Rhodes University in Grahamstown, and gave an address in which he urged the students to help bring about the changes that would be necessary to save South Africa from even worse racial violence. He cited one of his friends, Chief Gatsha Buthelezi, now chief executive councillor of the KwaZulu legislative assembly, as a 'wise man' who was working in the right direction; he praised the students' union, NUSAS, of which he continued to be Honorary President, as a force for change; and he urged students not to lose faith or hope.[58] But he felt that the officials at Rhodes were not sympathetic to his message, whatever the students thought. He wrote sardonically to his friend Leo Marquard, founder of NUSAS,

> Both the Vice Chancellor [Dr J. M. Hyslop] and the Chairman [Professor J. A. Gledhill] are clots of the first magnitude and presiding over it all is the Buddha-like figure of W. G. Busschau [the Chancellor], who smiles and says nothing. Whether this conceals profound learning or profound indifference I could not say. He and the Vice and the Chairman all congratulated me on my speech, but it all meant nothing.[59]

The cutting edge of this letter shows the extent to which Paton felt himself to be isolated at this time, and it was true that with increasing honours had come increasing hostility directed towards himself in particular, and liberal thinkers in general. The early 1970s had seen the beginning of what was to become known as the Black Consciousness movement, which drew on the American Black Power movement and on 'liberation theology', and whose best known leader was to be Steve Biko. Biko and his friends rejected white liberals, of whom Paton was a leading representative, regarding co-operation with them as part of the psychological conditioning which made blacks see themselves as underdogs. Blacks were to find ways of working out their own salvation, and they would do it alone.

From 1971 on, then, Paton increasingly found himself under attack from young black radicals, who saw his writing (particularly *Cry, the Beloved Country*) as depicting compliant Uncle Tom-style blacks as the ideal, and who saw his political stance as palliating apartheid and thereby ensuring its survival. He was hurt by these attacks, feeling that the sufferings of Liberals like Peter Brown entitled them to some respect, but they did not come as a great surprise to him. 'Black Power is the inevitable consequence of White Power and White Arrogance,' he told the Rhodes students in April 1972. 'If you regard yourself as a white liberal, or even as a white radical, you would be foolish to be angered or hurt by it.'[60] Foolish it might be, but he was hurt at this rejection by those for whom he had fought for so long.

This split between Liberals and radicals was reflected in the editorial board of *Reality* too. Paton was among those who resisted altering the name of the magazine, and its editorial content, to reflect more radical thinking; he continued to resist any actions that might harm others, and found himself increasingly out of step with those liberals who felt Paton was 'the liberal who makes it more difficult for the oppressed to conceive of the situation in revolutionary terms,' as Colin Gardner (a friend of Paton's) was to express it.[61] When Paton was overruled, in November 1972, and the journal's title became *Reality: A Journal of Liberal and Radical Opinion*, Paton submitted with good grace, sending Gardner a little poem about Paton's conversion from 'trad Lib' to 'rad Lib'. All the same he did not change his mind about the use of violence or any action that hurt others: he continued to believe that the end, no matter how noble, could not justify ignoble means. For this stance, which he had maintained all his life and would uphold to the end, he was to be increasingly reviled by many of those he had given so much to help.

But he was playing little active role in politics now, not least because of his age. On 11 January 1973 he turned 70, and the usual enormous birthday party was even bigger than usual. Anne found Paton's annual party ('The Party', as she called it)[62] one of the chief crosses she had to bear, and she did her best to reduce its size. She began by relieving the various devoted Liberals of their annual job of bringing along a huge dish each: Dennis and Devi Bughwan, for instance, brought a great pot of breyani each year, and others including Sue Francis and Guinevere Ventress both catered and organized before Anne took charge. For some years Anne did all the catering herself, and then handed everything over to a caterer. She coped with the mess caused by dozens of people trampling through the house by having a big marquee tent erected in the garden, for it always seemed to rain on the night of the party.

What she never succeeded in doing was cutting the guest list down to manageable proportions. When she did persuade Paton to strike off people who never invited them back, she got phone calls from affronted ex-guests demanding to know why they had not been invited. The truth was that many Liberals regarded Paton's party as a sort of official function of the Party, and so in part it was. The long political speech proposing a toast to Paton, the doggerel poem by Edgar Brookes, Paton's 'state of the nation' reply, these were the remnants of his active political life as much as a celebration of his seventieth birthday.

The crowd at these parties was often not far short of a hundred, and it was always multi-racial, Catholic Archbishop Denis Hurley rubbing shoulders with Zulu leader Gatsha Buthelezi, and Pat Poovalingam chatting to Pondi Morel. Brookes's poems were a standing joke, and Paton's doggerel response often not much better:

> It has been said that Edgar's verse
> Has year by year got worse & worse
> Then why this year this sudden change,
> A deeper depth, a stronger strength,
> And oh alas a longer length,
> A wider-ranging kind of range?[63]

And the poem usually went on to mention by name such friends as Selby Msimang, Archbishop Paget, or Peter Brown. Paton thoroughly enjoyed his party, and to Anne's complaints about the heat would reply rather plaintively that he could not help having been born in January.[64] But the party grew smaller year by year as the older Liberals died, and as more of them continued to leave the country. The truth was that Paton's political role was in decline. His reputation as a man of letters, however, was growing, for like W. B. Yeats he was about to experience a flowering in old age.

22

STORMY TWILIGHT
1973–1974

I said to myself, be still, and wait without hope
For hope would be hope for the wrong things; wait without love
For love would be love of the wrong things; there is yet faith
But the faith and the love and the hope are all in the waiting.

T. S. ELIOT

On 3 March 1973 the Patons were on the move again, this time flying to the United States in great style on the Concorde, for Paton had been elected the Chubb Fellow at Timothy and Dwight College, Yale. This lucrative fellowship involved only the giving of a lecture, and Paton much enjoyed his brief time at Yale, though he found that the black students there avoided him. Presumably few of them heard his public lecture, in which he defined liberalism in terms that might have won them over:

> By liberalism I don't mean the creed of any party or any century. I mean a generosity of spirit, a tolerance of others, an attempt to comprehend otherness, a commitment to the rule of law, a high ideal of the worth and dignity of man, a repugnance for authoritarianism and a love of freedom.[1]

After the days at Yale Paton gave another address at Princeton, where he met his old friend Liston Pope for the last time. Pope had remarried and spoke of going to live in Norway, and the next year, 1974, he would travel to Scandinavia and die in Trondheim, the city in which Paton had begun *Cry, the Beloved Country*. 'I don't suppose anyone exemplified better St Paul's words that the spirit is willing but the flesh is weak,' Paton was to remark compassionately of him.[2] After the Princeton visit the Patons flew to California to see Aubrey Burns once more, and then flew home via Honolulu, Japan, Hong Kong, Singapore, Tehran, Israel, Istanbul, and Greece, a zig-zagging journey devised by Anne and much enjoyed by both of them.

At one stage on this trip Paton became annoyed by Anne's bossiness, told her that from now on he would deal with their travel documents, and demanded them all from her. She took him at his word, and in Tokyo left him to deal with the immigration officials while she sat at a distance to observe what happened. Paton reached the head of the long queue, and was asked for passports. He dug hastily into his briefcase, and after a long scrabbling search found them. The official then asked for the immigration forms, and found that Paton had not filled these in. The queue then had to wait, with growing signs of impatience, while Paton painstakingly filled in the forms, at intervals casting appealing looks in Anne's direction. She pretended not to see him, though she was convulsed with laughter. 'Why that queue of people didn't assassinate him I don't know,' she was to say. When at last he rejoined her, he thrust the passports at her with a growled 'Here, you take them back', and thereafter was content to let her deal with all such formalities and arrangements.[3]

They were back in South Africa on 28 March 1973, to find Liberalism of the style Paton still believed in increasingly under attack from the so-called radicals, both black and white. With radical blacks, who dismissed all liberals with contempt, he did not attempt to argue. But when the attacks came from those he had considered his friends and allies, he was often stung into defending himself and the Liberal position. In June 1973 he was distressed to read a privately circulated document written by an old friend, Fatima Meer. A member of the Indian Congress, she had long been a supporter of the Liberal position, but now turned on Paton and Buthelezi, accusing them of collaborating with the government. Paton was terribly hurt, as he told Buthelezi:

> I will not trust her any more. I think she has ambitions to become the black power leader, but she has no hope . . . I do not feel at all sentimental towards people who are willing to stab you in the back when you are not looking. She has spent her life flirting with liberals on the one side and black power on the other . . . She has probably found the liberal too cold a lover and prefers the more passionate embraces of black power. She will end up by having no lover at all.[4]

He had had earlier disappointments of this kind, as we saw when Mary Benson had seemed to criticise Liberals in 1964, and when it was reported to him that Ela Ramgobin, one of those with whom he had worked for years on the Phoenix Trust, told a group of visiting Americans that Paton had become a supporter of apartheid. The resulting correspondence between him and Ramgobin ended in a vituperative letter from her. Such words, Paton told her, 'are easy to write, but their damage is hard to undo. When you write such words as these, you close and lock the door against the person to whom you are writing. No one can open the door again but you.'[5]

Several such doors were shutting against him. In May 1974 Eddie Webster, later to be professor of Sociology at the University of the Witwatersrand, wrote an article on black consciousness for the journal *Dissent*, which was published by the students' union, NUSAS, of which Paton was still an honorary Vice-President. In the article Webster distinguished between traditional liberals and despairing liberals, and contrasted them both very unfavourably with radicals. Paton wrote to Charles Nupen, then leader of NUSAS, objecting strongly to this attack and reminding him that the founder of NUSAS (Leo Marquard, who had died in March 1974) was 'a liberal of a splendid kind'.[6] Nupen returned an evasive answer[7] with which Paton refused to content himself: later that year he asked not to be re-elected to the honorary Vice-Presidency of the organization.[8]

There were many opponents of apartheid who after the mid-1970s began to suspect, and increasingly to say, that Paton was moving to the right in his old age. Among the elements involved in reaching this false conclusion were three that stood out: they were Paton's attitude to boycotts against South Africa, his apparent abandonment of the notion of the common society, and (from about 1989) his friendship with Chief Mangosuthu 'Gatsha' Buthelezi.

Paton's attitude to boycotts against his country was complex and apparently contradictory. As early as September 1958 he had strongly supported the notion of a sports boycott when a visit by a West Indian cricket team had been mooted.[9] In December 1961 he had, in his capacity of Vice-President of the South African Sports Association, written a strong letter to *The Times* of London protesting against a planned visit by the British Lions.[10] On 16 April 1964 he wrote to two Maori members of a New Zealand rugby team about to visit the country, urging them not to come.[11] His belief was that a sport boycott would have a much greater effect on white South Africans than on blacks, that, in fact, it was precisely targeted at those whom the boycotters hoped to influence. On similar grounds he supported a cultural boycott by actors and playwrights.[12] Rather inconsistently, he opposed an academic boycott: 'People like myself strive to keep alive liberal and civilized ideas in this country, and we need all the help we can get.'[13] He recognized his own inconsistency: 'Strangely enough, I approve of sports boycotts, I am a "don't know" about arts boycotts, but

I am totally against an educational boycott. Not logical, but one can't be too logical here.'[14]

He strenuously opposed a trade boycott, on the grounds that it would hurt the very people it was designed to help, black South Africans. It was Paton's view that as the South African economy was damaged and unemployment grew, it would be black workers who would find themselves unemployed. He first made public this opposition when he was giving evidence in mitigation of sentence at the Rivonia trials, in June 1964. On the stand he said that though he had for some months, during 1961, advocated trade sanctions, he had changed his mind because of the damage such sanctions would do to the mass of the South African population.[15] In his view, foreign corporations should instead try to improve the salaries and other benefits of their non-white employees. He told his audience at Harvard University in June 1971, 'I stand not for the withdrawal of American investment but for this dramatic improvement in salaries and benefits.'[16]

This position exposed him to the criticism, increasingly severe, of old friends such as Trevor Huddleston, who was proud of the role he had played in encouraging trade boycotts of South Africa, and who did not like to hear of the resulting misery among black workers. Huddleston was to attack Paton publicly, on American television on 20 September 1985, to Paton's distress. Paton in fact had much to lose, and nothing to gain, from his stance on economic sanctions: it is a testament to his courage and his honesty that he took the stand he did, and maintained it till the end of his life in face of increasingly harsh criticism.

The second point of contention between Paton and some of his former allies was his apparent relinquishment of the notion of the common society, which had been central to Liberal Party policy. Paton did not of course abandon the central tenet that all South Africans should be treated equally by society; but increasingly he came to believe that the common society might be more quickly achieved as a federation than as a unitary state. The first open signs of this shift in his thinking came in August 1970, when he wrote a paper for a new organization, the Study Project on Christianity in Apartheid Society (SPROCAS), which his friend Tony Mathews, professor of Law at Natal University, had persuaded him to join, rather against his better judgement.[17] In this SPROCAS paper, Paton argued that the new organization, while setting forth the goal of the common society, should also adapt itself to what he called 'the realities of the situation'. As he told Leo Marquard at the time,

It seems to me that for better or worse separate development is the pattern, and that this will create new centres of power which will, I hope, shake the monolithic authority of the Nationalist party. There are signs that Gatsha Buthelezi and Matanzima [Kaiser Matanzima, the Transkei leader] will make greater demands of the government and that this will create a further cleavage in the Nationalist party. I also said in the course of my paper that one might proceed by means of a federal constitution towards a common society, but that I did not think that SPROCAS could, or should, provide any blueprints for any kind of society.[18]

By November 1973 he was aware of the hostility his move away from the idea of a unitary state was evoking among radical opponents of apartheid, who saw his pronouncements as tending towards an acceptance of the government's bantustan policy. Paton wrote to Mangosuthu Buthelezi about these critics,

They still hold the view of a unitary society. I myself held this view very strongly for fifteen years, but my reason tells me that history is not going that way and we are more likely to achieve a federation or confederation, though even that is not a certainty, especially if the other homelands fear domination by the Transkei or KwaZulu.[19]

He put forward these views with increasing vigour as the 1970s advanced. He also did his best to persuade statesmen such as Cyrus Vance, then American Secretary of State under Jimmy Carter, to give South Africa more time to work out the federal state. One of his articles in *Reality*, 'The Americans and Us', put the point with particular clarity in 1977,[20] shortly after Paton had travelled to meet Vance in Washington,[21] and drew a protest from his old friend Edgar Brookes. 'From you more than from any other person I learned the gospel of universal suffrage leading to majority rule,' Brookes wrote. 'Now you seem to have abandoned it to some extent.'[22]

In his reply Paton said that he had no doubt that South Africa was destined eventually to become a unitary state, with universal franchise and majority rule, but that he saw a federal state as a stepping stone which the Afrikaner rulers of the country might possibly accept. If they were not offered such a compromise, he foresaw that military and economic pressures would be applied to South Africa from the outside world, and the result would be the laying waste of the whole country for a generation. And he told Brookes, who had accused him of seeking a way of evading the franchise issue,

This [federal proposal] may seem to you to be 'a way out of the franchise issue', and I cannot prevent you from thinking so. But in fact it is an attempt to find a way

> out of desolation and the destruction of Afrikanerdom. It may be that the destruction of Afrikanerdom is inevitable because of its inability to adapt to a changing world. But for the time being I feel it is my duty to try and prevent it, and to encourage the West not to lose its patience with our rulers. If they do lose patience, they will I think inevitably abandon us to our fate, and perhaps even hasten it by the employment of economic sanctions. I have no doubt that our fate would be to be ruled by a Marxist-authoritarian state, and this I would try to prevent.[23]

Brookes remained sceptical that the outside world would accept any solution based even remotely on apartheid, or that the Vorster government would accept any solution not based on apartheid. There were many like Brookes who thought Paton wrong on this issue.

The third element that brought Paton into disrepute with the radicals was his friendship with Mangosuthu Buthelezi. Paton had known the Zulu leader, whose family had the hereditary right to advise the Zulu Royal house, since the late 1950s, when Buthelezi had been expelled from Fort Hare university for involvement in a demonstration organized by the ANC Youth League. The two men liked and admired each other personally. Buthelezi was among those who regularly attended Paton's birthday parties from the time that they started in the 1950s. On at least one occasion he proposed the toast at these parties, and Paton replied with a poem to him:

> You know from Cape Town to the Congo
> Each Chieftain has his pet imbongo, [praise-singer]
> Who cries out praises on the spot
> Whether they are deserved or not.
> Alas imbongo have I none
> So I have got a chief for one
> And from his distant northern fief
> He comes to Botha's Hill this chief
> And having heard imbongos many
> He does not yield the palm to any . . .
> And thank you Irene [Buthelezi's wife] for coming as well
> And telling Gatsha what to tell.[24]

Buthelezi was a Christian and a believer in evolutionary change, like Paton, and he early turned to Paton for advice on such occasions as when, in December 1963, he was arrested and menaced by the security police.[25] Paton was delighted when Buthelezi became chief executive officer of the KwaZulu national assembly in June 1970, feeling sure that his friend would pressure the government, in ways it would find hard to resist, to allow black South Africans greater freedom. 'Naturally', Paton wrote encouragingly to Buthelezi, 'your duty is to get whatever you can for KwaZulu and its people, even in a framework of which you do not approve'.[26] But radical opponents of apartheid, black and white, considered that Buthelezi risked becoming the creature of Vorster's government by appearing to co-operate with the bantustans policy, and they condemned Paton for supporting him. This reaction grew stronger as Buthelezi's Inkatha movement challenged the claim of the ANC to speak for all black South Africans.

For his part Paton saw Buthelezi as a bulwark against the black nationalism which he feared as much as he feared its white equivalent. He also saw Buthelezi as being tugged in too many directions. 'Gatsha is under great pressure, too great,' he wrote to Red Duggan, who had been the American consul in Durban. 'He has too many advisers. He has too many calls. He is up against the big boys, & if they break him, they bend KwaZulu to their will & greatly strengthen the forces of Black Power.'[27] He said much the same thing in a laudatory article on Buthelezi which he published in June 1974.[28]

For all Paton's admiration of Buthelezi, and his characteristically Natalian admiration of the Zulus as a people, however, he did not find it easy to maintain good relations with him. Buthelezi often ignored his advice, at least in the sense of not responding to it in his letters, and increasingly as the 1970s wore on he would bridle at what he took to be criticism of himself. In April 1977, for instance, he responded with irritation to Paton's recommendation that he should read more and think more. He read as much as he could, he told Paton with asperity, and added, 'As for time for thinking, I do not think for someone leading an oppressed people like me I cannot do any more than think on my feet.'[29]

In August 1980 Buthelezi rejected Paton's advice more publicly, by criticizing him and Helen Suzman in the KwaZulu Legislative Assembly as 'the kind of white liberals who think they know what is best for the blacks'.[30] Paton was saddened by this, but did not respond directly to his friend. In a letter to an American scholar, though, he referred to Buthelezi's 'extreme sensitivity to criticism', and said that 'many of his friends, both black and white, are very troubled by his proneness to anger . . . There are undoubtedly paranoid elements in his make-up'.[31]

Yet Paton and Buthelezi maintained their friendship, and in July 1982, when Paton wrote to Buthelezi apologizing for what the government was doing to blacks, and to South Africa, Buthelezi replied, 'I want you to know how humbled we were by your apology to us for what the government is doing to us and to South Africa. You have fought what they stand for all your life.'[32] Perhaps encouraged by this, Paton in January 1985 advised Buthelezi to shorten his speeches and his articles:[33] the response was an extraordinarily intemperate blast in which Buthelezi repeatedly and icily addressed Paton as 'Sir' and told him he 'could not care hoots in Hell' how other people wrote.[34] Paton, deeply shaken, replied, 'I should have known better than to presume to offer you advice on the length of your speeches. When you call me "Sir" then I know that I am in trouble. I shall never presume to offer you advice again.'[35] Nor did he, though he and Buthelezi continued to exchange occasional letters until Paton's death. Buthelezi had lost the best and most sympathetic adviser South Africa had to offer him.

For Paton this was one more instance of those he was most keen to help turning upon him, but he was neither embittered nor tempted to change course. In these circumstances he liked to quote words which he attributed to William the Silent, and which he had first heard quoted by the Master of Timothy Dwight college at Yale: 'It is not necessary to hope in order to undertake, and it is not necessary to succeed in order to persevere.'[36] He did what he could for all the people of South Africa, not because he thought they would thank him for it (on the contrary, many reviled him), or even because he was confident that his efforts would improve their lot: he did what he did because he thought it right, and he was so constituted that he would go on doing right, no matter what the consequences.

The attacks on him and on liberalism continued and intensified through the 1970s. This was the period when public figures denounced liberalism as quietism, and proclaimed themselves radicals. One such was Nadine Gordimer. In an interview she gave in London in December 1974, Gordimer was quoted as saying, 'Liberal is a dirty word. Liberals are people who make promises they have no power to keep.' Paton was stung by this, and in an interview he gave the Johannesburg *Sunday Times*, he asked whether she included in her condemnation the many liberal organizations which were working actively to improve the racial situation: the Christian Institute, the Institute of Race Relations, and the TEACH and LEARN educational funds run by daily newspapers. He pointed to the sufferings of Liberals like Peter Brown, the lawyer Ruth Hayman, and the architect Walter Hain, who had had their careers ruined by banning. 'Many of them made greater sacrifices for their liberalism than Miss Gordimer has ever made for her radicalism,' he said pointedly.[37] Subsequently he sent her what he called 'a nice Valentine in the *Natal Witness*', by way of making peace again,[38] and she wrote to assure him that 'we have an enduring friendship that has not changed, so far as I'm concerned'.[39] But the sneering attacks from other born-again radicals did not stop, and he had to endure them stoically to the end of his life.

The radicals had already been answered by Albert Lutuli, who in his autobiography, *Let My People Go*, had written of the Liberal party before its dissolution, that:

> their effectiveness is not to be measured in votes, but in the appraisal they have forced on whites. The Liberal Party has been able to speak with a far greater moral authority than other parties with white members because of the quality of the people at its head—such as Alan Paton, Senator Rubin, Margaret Ballinger, Peter Brown, Patrick Duncan and others. Moreover it has tried to take its stand on principles and not on expediency—a new thing in South African politics.[40]

The truth is that the radicals attacked Paton, not for any perceived change in his political stance, but precisely because he had not changed. He had always been a moderate, a believer in non-violence, one who hoped by reason and persuasion to turn Afrikaner nationalists and black nationalists alike from their collision course. He hung on his study wall a framed cartoon by Jock Leyden, published in a Natal paper in 1976 and entitled 'The Gentleman of the Lamps'. In it two powerful locomotives, labelled 'Black Power' and 'Afrikaner Nationalism' are rushing towards each other on the same set of tracks, while Paton, waving red warning lamps and shouting at the top of his voice, tries to avert disaster.[41]

At the time of the Liberal Party's foundation, Paton had occupied a centrist position in South African politics. In 1960, following the banning of the Communist Party and the ANC, he was on the far left of the legal political spectrum. By the end of the 1970s, after the rise of black consciousness, he was back in the centre again, and by the beginning of the 1980s he had begun to seem inclined towards the right: yet his views and his statements had not changed. It was his continued belief in reason and suasion, rather than violence, that earned him the contempt of the radicals after 1970.

Colin Gardner was to say, after he had become a member of the ANC, 'The banning of the ANC and PAC made Alan the far left of a truncated spectrum. But once the screen widened again, he and the Liberal party were left more or less in the middle. If Paton moved to the right, he moved by the simple process of staying where he had been.'[42] The views he had put forward in the late 1950s and 1960s had seemed radical to most white South Africans then; by the 1980s they seemed less extreme to many whites, and insufficient to blacks. As for his continued belief that peace would only come through talking with the government, many radicals came to think this a kind of treachery to the cause.

Paton defended himself when he was attacked by former white friends, and endured in silence when he was attacked by blacks. He lived in hope even when he was laughed at for hoping, and he did whatever he could to help others. The journalists who frequently rang him for interviews began asking him why he had wasted his life. And he would reply, 'At the time I stood up for blacks, it was the right thing for me to do. Whether I was going to succeed in helping them or not was not the main question. If I had it to do again, I'd do it again.'[43]

One liberal he was able to help at this time was the South African he admired above all others, Peter Brown. Brown, it will be recalled, had been banned on 1 August 1964 for five years, and his ban had been renewed for another five years in 1969. As this second banning order drew to an end in 1974, Paton was very anxious lest it be renewed again without explanation, and he appealed to his friend Helen Suzman, the Progressive Party's single Member of Parliament, to do what she could to prevent this from happening. 'I believe you have most unexpected friends in high places,' he wrote.[44] Suzman promised to try: 'My "unexpected friends in high places" sometimes, just sometimes, toss me a little bone to keep me quiet.'[45] But she warned Paton to keep the matter under his hat, since Peter Brown had in the past been very resentful of the notion that anyone might intervene on his behalf. Paton did keep the matter secret from Brown, but he rejoiced openly and sincerely when the ban was not renewed and Brown, after ten weary years of utterly unjust forced withdrawal from society, was able to rejoin the world. Liberalism, despised and marginalized, retained its quiet power.

23

THE PUBLIC MAN
1974–1979

Whatever prevents you from doing your work has become your work.

CAMUS

Paton's trips abroad continued to be the release he needed from South Africa's intractable problems. His friend Red Duggan, now retired from the American consular service and living in Oregon, had arranged for him to be given another honorary degree, this time from Willamette University. The Patons flew to the United States on 29 October 1974, and saw Aubrey Burns in San Francisco before driving up to Oregon for the degree ceremony on 20 November. They loved the north-west of the United States, and drove up through the Rockies to see British Columbia too before returning to South Africa via Britain. They agreed that they could happily live in Oregon.

The fact that he was now in his seventies made Paton feel he was approaching the end of his journey: in an address he gave at the opening of the Johannesburg College of Education on 23 November 1974, immediately after his return from the United States, he quoted to the students about to go out into the arena of the world the words of the Roman gladiators, *morituri te salutant*, 'We who are about to die salute you'. He then changed them to *moriturus vos saluto*, 'I who am about to die salute you'. And he added, 'I at the end of my journey am glad to wish you godspeed at the beginning of yours.'[1]

This intimation of mortality persuaded him to begin a review of his life, and two books resulted. The first was a collection of Paton's shorter writings, published as *Knocking on the Door*, including many of his previously uncollected poems, selected and edited by his Liberal friend Colin Gardner, who had in 1972 become professor of English at Natal University, a post his father had held before him. Colin Gardner was a gentle, bearded scholar whose intelligence and learning Paton greatly admired, though he treated him with faintly ironical respect: 'This is Pro*fes*sor Gardner,' he would say, stressing the second syllable of 'professor' as if holding it between forceps. He took to phoning Gardner with literary queries: 'Colin, how do you use avatar in a sentence?' he would ask, having rung Gardner in the middle of a class. 'Oh I know what the dictionary says: I want to know what *you* think.' He had a whole range of experts he regularly called on for information in various fields, academics like Gardner, Tony Morphet, Ben Marais, Braam de Vries, and Michael Chapman, journalists like Ian Wyllie, politicians like Helen Suzman, Ray Swart, or Alex Boraine, all of whom served on the 'Paton brains trust' for years. Paton much enjoyed digging through his files for long-lost poems to include in *Knocking on the Door.* 'You would be surprised to know what odd things we have dredged up,' he told Guy Butler, another academic friend. 'Some quite good I may say.'[2] *Knocking on the Door*, begun in 1974, was published in August 1975 by David Philip, the second of Paton's books to carry Philip's imprint.

The other task of revaluation he had embarked on was the first volume of his autobiography, which he called *Towards the Mountain.* The title indicates the central myth that informed Paton's view of all life, and certainly his view of his own. He saw his career, as he had seen Clayton's, as an allegory of the Christian way: 'this is the way I think about these things', he had told Edward Callan.[3] Life was a Bunyanesque pilgrimage, and the goal was that holy mountain spoken of by Isaiah, where none would hurt or destroy.[4] 'The pursuit of this "holy mountain", though very imperfectly done,' Paton said in a sermon he delivered in England, 'is the story of my poor life as a Christian.'[5]

The goal could never be reached in this life, but to cease to strive for it was a failure of faith. Paton might turn aside from the path; he might fall into what he thought of as sin; but he would never forget what he was struggling towards, or pretend that it could be reached by ignoble or violent means. His was a world-view, Tony Morphet was to say, which could have been shared by Christians at any time for the last two thousand years. 'He came to see himself, quite early, as not exactly Christian on pilgrimage, but as sustained by his significance in Christian cosmology.'[6] It was this philosophy of life which allowed Paton to see all of his existence as integrated and significant, and to sustain hope even when hope seemed dead. 'The question of hope or despair is one which Christianity has answers to,' Colin Gardner was to say. 'The need to maintain hope even in hopeless circumstances is one of the things he got from Christianity.'[7]

He began *Towards the Mountain* late in 1974, and enjoyed writing it after the struggle he had had with the lives of Hofmeyr and Clayton. 'The amount of effort and concentration required in writing a biography is many times greater than that required for an autobiography,' he told Laurens van der Post, to whom he was explaining why he had abandoned the Roy Campbell book. He added that he could not understand Campbell, nor admire him.[8]

Even autobiographical writing has its pains, though, and he took a long time over *Towards the Mountain*. By March 1976 he had still not finished, and was complaining that he was bored with his subject, a comment that was only half a joke.[9] He told Leslie Rubin, 'It's the kind of story that once you put it down you can't pick it up again.'[10] But by 13 September 1976 he was nearing the end, and he seems to have finished a first draft, perhaps only taking the story up to 1939, by the end of the year. He sent this opening section to Scribner's for their approval. Then the editing process began, and he found it more than usually annoying to have a young and inexperienced editor at Scribner's picking his carefully crafted prose to pieces. Through much of 1977 he struggled to keep his syntax and his temper, eventually sending a cable of furious protest to Charles Scribner, who defended his editor.[11]

By the beginning of 1979 Scribner persuaded Paton to publish only the first part of the book, stopping with the publication and reception of *Cry, the Beloved Country*, because, Paton admitted, 'he has despaired of waiting for the whole thing'.[12] The long process was to drag on into 1980, and *Towards the Mountain* was eventually published in October of that year by Scribner in New York, and David Philip in Cape Town.

He was still writing the book when he travelled to England alone in April 1975 to take up a Visiting Fellowship at Clare Hall, a postgraduate college at Cambridge. He travelled first to the University of York, where he gave a paper on Roy Campbell, and where, to the astonishment of some of his Liberal friends, he stayed with Adrian Leftwich, who was now an academic there. He travelled up to York by train with two Liberals by then living in Britain, Marion Friedmann, with whom he was often to stay in London, and Peter Rodda, a friend of Jonathan's who had known Adrian Leftwich as a student. Rodda was helping Paton out with his luggage at York when someone approached and held out a gloved hand: Rodda thought it was a porter, and was shocked, when he looked up, to find that it was Leftwich. 'I shook his hand rather than make a scene,' said Rodda later. 'He hadn't put *me* in prison. He drove Marian and me to the residence where the other guests were staying, and Alan went home with him.'[13]

Rodda believed Leftwich had asked Paton for forgiveness for his treachery. In fact he had not, but Paton forgave him amply, believing he had suffered enough from guilt. Hereafter he would visit Leftwich several times, and other Liberals joked that the hardest thing Leftwich had to bear was 'having Paton visit him every couple of years and turn over his conscience!'[14] Leftwich himself showed little gratitude for Paton's forgiveness: in later years he was to show that he had no understanding of Paton's motivation when he remarked, 'I always wondered whether he got a charge out of dealing with the ostracized. I felt about his approaches to me what one feels in people who like to work among lepers: there's something called leprophilia.'[15]

Leftwich understood much better the continuing hostility of others whose trust he had betrayed. When he drove Paton from York to Liverpool, he had to drop Paton in the city and drive away, because the person Paton was seeing next was David Evans, a Liberal and ex-member of the ARM who had served five years in prison as a result of Leftwich's testimony. Evans, a sad exile living in a tiny terraced house 'in a street which is one of a thousand streets',[16] as Paton put it, had no desire to set eyes on Leftwich ever again.

Paton worked very hard during his three months in Cambridge, giving seventeen talks or addresses, making himself accessible to any student who wanted to see him, and dining in Clare Hall on most nights. I went to hear him preach in the University church, Great St Mary's, his silver-haired figure seeming very small in the pulpit. I was struck by the way he combined profound spirituality with a series of very funny anecdotes from his autobiography, then much on his mind. He told a story of stealing a toffee from the counter of a baker's shop, and had to stop because the laughter of the congregation made him completely inaudible. His delivery was made the more effective by his slow, grating voice, with the unique accent, raw Natal with a dash of American, and his perfect sense of timing: the result was a performance of great power.

He did not live in Clare Hall, but in a large boarding-house found for him by the College at 3 Wordsworth Grove, a quiet back street not far from the Cam. The house and garden were large, and Paton, who had never been in England in spring before, was astonished by the volume of bird-song. On one occasion I asked him if he was managing to write in Cambridge. 'How could anyone write with that?' he said, gesturing to a blackbird pouring its song into the sparkling morning. His landlady was Mrs Violet Swann, and in later years she remembered him vividly:

> When he came he looked so fierce, the way he glared over his glasses, that I was afraid of him. Gradually I became used to him and saw he was kind, but I never quite got over it: I was always careful that everything I did for him had to be right. He was very particular about time: if his breakfast was due at 8.30, it had to be there at 8.30. Lunch and dinner he had at the College, and any guests he had he took there.
>
> He always had breakfast in bed, wouldn't mix with the other guests who had breakfast in my communal dining room. They were very curious about him: 'Who's having breakfast in bed?' He had a firm manner: no nonsense with him. If there was something he didn't like on his tray, 'You take that away' he would say, just like that. He was used to being served, oh yes, his little wife must have run round him.[17]

In spite of this initial impression, Vi Swann and Paton became good friends, and on his subsequent trips to England, he came to see her twice, travelling up from London to do so, and in 1976 he stayed with her again, for two weeks. She treasured a signed copy of *Cry, the Beloved Country* which he gave her: 'He was a wonderful man, you don't meet many like him', she was to say towards the end of her life.[18]

At Clare Hall he made little impression, for he spent much of his time in the University Library, and at the College meals did not speak much unless spoken to. One of the Fellows, Dr Audrey Glauert, remembered him as 'rather arrogant and used to being made a fuss of. He sat as though expecting people to lionize him, and when they didn't he sulked.' She also thought him used to being served, and a 'typical South African'.[19] How Paton would have liked this last remark to have been true!

The University of Cambridge had considered giving Paton an honorary doctorate during his visit, and in this regard had written to the former bishop of Johannesburg, Ambrose Reeves, and to Trevor Huddleston asking them for their opinions. Reeves, for all that Paton had stood by him nobly during and after his flight from South Africa, did not return the compliment: he suggested that Cambridge consider Israel Maisels QC, a South African lawyer, or Nadine Gordimer, and the University quietly let the matter of a degree for Paton drop.[20]

Anne joined him during his last few days in Cambridge, and the two of them then travelled by car to the Hebrides on 16 May 1975. Paton had long wanted to see the islands, and he enjoyed the experience. 'I cannot say that the weather was good,' he wrote, 'but it was the kind of weather one expects in the Hebrides.'[21] They returned to South Africa at the beginning of June, and Paton resumed his home life, writing steadily at his autobiography, handing over the Campbell papers to me on 14 June 1975, travelling to Cape Town for a poetry festival, writing an article for a Toc H commemorative booklet (and being very annoyed when they cut it without his permission),[22] helping prepare for publication the poems of Aubrey Burns, who had now had a stroke,[23] writing political commentaries for papers such as the *Sunday Tribune*. One of his journalist friends, Ian Wyllie, was now editor of the *Sunday Tribune*, and Paton over many months wrote the unsigned leading article for him.[24]

He had begun to see signs of improvement in the political situation in South Africa, at a time when most other opponents of apartheid considered they were going from bad to worse. Prime Minister Vorster was talking of important changes, but making few or none. But Paton was heartened by the speeches of the Cape chairman of the National party, P. W. Botha, who was to succeed Vorster as Prime Minister in September 1977. Paton wrote to Red Duggan,

> Mr P. W. Botha said at the Cape Nationalist Congress that if there were any people there who did not believe that all men were equal in the sight of God they must not regard him as their leader. It is really fantastic. If I had said that a couple of years ago, the Nats would have rushed round to explain that heavenly equality was all very well but surely I was not such a fool as to think men were equal in gifts etc. Times change.[25]

Many liberal observers discounted such speeches by P. W. Botha as hypocrisy; Paton, with his characteristic faith, continued to hope that they portended real change. Real change was coming, but not in the way he hoped.

Paton had begun the year 1976 by writing to Gatsha Buthelezi that he had 'forebodings about the year ahead of us'.[26] His forebodings were amply justified. On 16 June 1976 the student riots in Soweto broke out, ostensibly over the issue of compulsory instruction in Afrikaans, and spread rapidly through townships all over the Reef. By 21 June violence had also engulfed the townships of Pretoria, and by September had spread to the Cape. The crowds burned government offices, schools, buses, beerhalls, libraries, anything remotely connected with government, and they also burned to death policemen, local councillors, and 'collaborators' with the government. The unrest started in this way continued to flare for years, the level of violence rising to terrible peaks in 1985 and 1986. By then the aim of black radicals was to make the country ungovernable, and in black townships they succeeded for a time, though the government's response was an iron-fisted clampdown and the declaration of a state of emergency.

In a series of newspaper articles, Paton pressed home the message now being borne in on white South Africans by the violence all around them: South Africa was in crisis, and it was the immoral system of apartheid that was the cause. He called on white South Africans to 'repent of our wickedness, of our arrogance, of our complacency, of our blindness'. The only hope for the country was rapid change by the government. It must at once begin serious talks with the leaders of the black community.[27]

The South African government, through its Minister of Justice, James T. Kruger, replied scoffingly that Paton's articles represented nothing more than one man's opinion, that not even a hundred thousand opinions make a fact, and that Paton appeared to be guilt-ridden.[28] Paton riposted with a letter filled with facts, pointing out, for instance, that the South African government outlayed nearly R500 per annum on a white child's education but only R28 on that of a black child, that black pensions were less than a third of those of whites, and so on. And he warned that Kruger's blindness to these things 'causes many of us to doubt whether the Afrikaner Nationalist is psychologically capable of saving himself, and therefore I believe of saving us all'.[29] He continued to believe that salvation for South Africa would only come when the Afrikaner had decided to change course.

On his own local level he was active too, pressing his bishop to amalgamate the white Chapelry of Hillcrest with the overlapping black parish of the Holy Spirit. Paton felt shamed by the fact that the great majority of white and black Christians worshipped separately, and he urged bishop Phillip Russell to begin creating combined black and white parishes which would worship together. In due course he was to achieve this aim in Hillcrest.

He continued to keep up old literary friendships, like those with Laurens van der Post and Uys Krige, and to make new literary friends. Among these was André Brink, a professor at Rhodes University whom Paton had met on one of his occasional visits to Grahamstown. Paton solicited from Brink an article for *Reality* on Afrikaans writing; Brink produced it in January 1970, arguing that the Afrikaans writer was culpable for not using the freedom he had within the apartheid system to oppose that system. Brink was going to use this freedom himself in the next few years, and to discover its strict limits, when his *Kennis van die Aand* [subsequently published in English as *Looking on Darkness*] became the first Afrikaans novel to be banned, in 1974.[30] Paton read it with admiration in January of that year. Subsequent Brink novels he liked less, writing to Brink that *Dry White Season* was a novel of unmitigated pain without catharsis;[31] and when *Chain of Voices* appeared in 1982, Paton found he could not bring himself to read it.[32] He found Brink's later books lacking in enjoyment.[33]

Another new literary friend was the Afrikaner novelist Karel Schoeman, whose novel *Na Die Geliefde Land* ('To [or 'After'] the Beloved Country') Paton thought the best novel to appear in Afrikaans for many years, and not just because of its echo of his own first book. With its predictions of what South Africa might be like after a black takeover, *Na Die Geliefde Land* seemed to Paton both courageous and alarming, and he reviewed it generously. The result was a correspondence, first with Schoeman's energetic mother, and then with Schoeman himself. Schoeman was to ask Paton's advice about novels he was planning, telling him in 1977 that he hoped to write a *jeugroman* [literally, 'youth-novel'] about Gandhi's stay in South Africa, but admitting that he knew little about the Indians in his own country.[34] Recently returned to South Africa from a period in Britain, he told Paton he was feeling 'lost and unhappy', troubled by the 'sense of strange, aimless turmoil and chaos just under the surface' in South Africa.[35] Paton was troubled by them too: in an article he wrote for the *New York Times* in 1976, he spoke of black violence as akin to death knocking on white South Africa's door, and said, 'After 73 years of life in South Africa I, for the first time, ask myself the question: Do I want to be here when there comes that last imperious knocking on the door?'[36]

He had no intention of leaving the country permanently, but he continued to take advantage of generous offers from American institutions to fly him to the United States. On 29 April 1977 he flew via New York and Detroit to Flint, Michigan, where he and Anne lived on the campus of the University of Michigan for a fortnight. Paton gave occasional classes and many interviews, and on 12 May 1977 received an honorary doctorate, his ninth.

The Patons were then driven to Kalamazoo, Michigan, where they stayed with Paton's old friend Edward Callan, now a distinguished professor at the University of Western Michigan. Paton gave a lecture to an enormous audience in the university auditorium in Kalamazoo, a hall which seated 3,000 people, and he had numerous other engagements. Anne made sure that they flew first class, and that Paton was given time for a nap each afternoon, for he was now 74: but his energy was remarkable, and he seemed to enjoy himself thoroughly. He particularly appreciated the recuperation afforded him in his time with the Callans, where he met Callan's wife Claire for the first time. He liked the Callans' large quiet house, and carefully studied the birds that came to feed in the garden.

The talks he gave at Flint and Kalamazoo were not literary, but historical and political. At such a time of crisis for his country, he felt he had a duty to interpret for foreigners what was happening in South Africa, and his talks had such titles as 'A Total View of South Africa' and 'South Africa—Its System of Criminal Justice and Internal Security'. But he also, in his interviews, urged Americans to overcome the demoralization they suffered from after the Vietnam war, and to take pride in their Constitution and system of government. America, he told them, was still highly regarded as a democratic example.[37]

But it was not just university audiences he wanted to influence: he hoped to make contact with the highest levels of government in the United States, and before he left South Africa had attempted, through the journalist Walter Cronkite (who had interviewed him earlier in the year), to set up a meeting with President Jimmy Carter in Washington. This Cronkite was unable to do, but he arranged for Paton to meet Cyrus Vance, Carter's Secretary of State.[38] This change was probably not entirely a disappointment to Paton, who disliked what he considered Carter's sanctimoniousness. 'Why the devil he thought it was his duty to put the world to rights I don't know,' he was to write after Carter's defeat by Reagan.[39]

The meeting took place on 2 June 1977 at the State Department. The tall, patrician Vance listened courteously as Paton urged on him the psychological inability of the Afrikaner nationalist to accept what South Africa needed and would inevitably get, in Paton's view: 'a common society, and a unitary state, a universal franchise, and majority rule'.[40] If attempts were made to force them on the Afrikaner, Paton warned, 'he would rather be destroyed than yield'. The result would be war, 'and we would enter a period, lasting how long I do not predict, of grief and desolation, during which our cities, harbours, railways, industries, medical services, and worst of all, our agriculture would collapse'. Instead, Paton argued, the Americans should press the South African government to plan for 'a federal constitution, and for a radical sacrifice on the part of white South Africa, to consolidate the black states that would be members of the federation'.[41]

Whether or not it was the result of Paton's message, President Carter made a statement on 25 July 1977 to the effect that he did not expect South Africa to change overnight, nor indeed even within a period of two years.[42] The United States government continued to engage the South African leaders in dialogue during the Carter years, rather than risk driving them into a corner by punitive sanctions. The South African government, however, was not grateful for Paton's intervention: as he wrote to Cronkite after his return home, 'I hear Mr Pik Botha [the Minister of Foreign Affairs] and others are not very pleased with my meddling in their affairs.'[43] And he wrote to Cyrus Vance, 'The truth is that these Nationalists think the country belongs to them.'[44]

He was back in South Africa for only a few months before returning to the United States, on 12 October 1977, for a lecture tour built around a stay at Harvard, where he delivered three formal lectures in November. Before going to Harvard Paton gave lectures at Princeton, and at several campuses in the South, where he, a Mark Twain fan, was delighted to see the Mississippi. He and Anne also travelled to Seattle to speak at a packed service in St Mark's cathedral on 3 November 1977. They then visited their friend Bunny Duggan[45] in Oregon, and drove down to San Francisco to see Aubrey Burns once more. Though Paton did not know it at the time, this was to be his last meeting with Burns, whose health was now failing slowly. By 1978 he could not write or remember much as a result of a series of strokes, and he was to die on 15 February 1983. Paton continued to write to him to the end.

The Patons flew to the East Coast on 10 November 1977, Paton lecturing at St Paul's School in Concord, Massachusetts, and again at Kent School where he stayed with his old friend, Edmund Fuller. Fuller found him depressed about the situation in South Africa, where the violence, the school-burnings, and the murders were continuing unabated, and where the inquest on the young black consciousness leader, Steve Biko, who had been beaten to death by the South African security police,[46] was about to report its findings: it would exonerate the police.

On 14, 15, and 16 November 1977 he gave the William Belden Noble lectures at Harvard. He entitled his lectures, 'Help Me Where Faith Falls Short', 'South Africa in Dark Times', and 'The Wolf and the Lamb'. Both he and his hosts were expecting anti-South African protests from his audiences. In the event his lectures were not interrupted, but pamphlets were circulated before the lectures attacking Paton as a member of 'the privileged, white group', and claiming that he stood for a 'paternalistic, white liberal position that has been discredited by the black youth and workers of South Africa'.[47] These views were also put forcefully to him during the question period that followed each lecture. The university itself came under fire for inviting a white South African to give the lectures.[48] The anti-apartheid movement was now in full cry on American campuses, demanding disinvestment, and though Paton stoutly defended his own position, he remarked to a friend, 'I would certainly not undertake any such tour again.'[49]

On 5 December 1977 Anne left Paton in America, and returned to England and then South Africa to be with her children during the school holidays. Paton was discomfited to be left in America to cope on his own, but cope he did. He went on to Washington for more lectures, including one in the Library of Congress, and then travelled to England to stay with his old Liberal friend Marion Friedmann and her husband Allan in London.

The Friedmanns put on a big party for Paton, to which they invited many of the Liberal diaspora: David Craighead, Gertrude and Gerhard Cohn, Peter Rodda, Randolph Vigne, David Evans, Peter Hjul, Walter and Adelain Hain, Mary Benson, and many others. At this party Paton made a speech in which he reviewed what was happening in South Africa, and gave his views on the future, a 'state of the nation' speech of the type he made each year at his birthday party. Hereafter whenever he returned to London, he would stay with the Friedmanns, and the gathering would be repeated. These parties became and remained a focus for the South African Liberals in Britain, even though many of them had little patience with Paton's calls for the world to give Vorster's government more time. He remained the Liberal leader, respected and attended to, even though the Party had long gone, and events seemed to be making its programme increasingly irrelevant. Paton's continued hope that change was coming seemed mocked by the results of the general election in December 1977, in which the Nationalists took many seats from the demoralized United Party in Natal and elsewhere. The polarization of South African politics, which had so marginalized moderates like Paton, seemed irreversible.

He made a brief trip to Germany from England, a pilgrimage, he was to call it,[50] to see the University of Munich, and to visit Dachau. The purpose of this pilgrimage was to honour the League of the White Rose, a group of young students who, with their professor, Kurt Huber, threw copies of their leaflet, *Pamphlet of the White Rose*, into a crowded hall at Munich

University on 18 February 1943, calling for the overthrow of Hitler. They were all arrested and shortly thereafter executed in Dachau concentration camp.[51] To Paton, who had first heard of them at the conference of Christians and Jews which he had attended in London in 1946, they were heroic figures, for they had done what he tried to do: in the worst of times they had stood up for what was right, undertaking without hope simply because undertaking was the right thing to do. He was grateful that he had not been required to pay for his beliefs the price they had paid for theirs. In Munich he laid a white rose on the memorial to the members of the League: 'this somehow makes my fears seem unimportant', he told Mary Benson.[52]

Back in South Africa in time for a sweltering Christmas 1977, he found the political situation so depressing that he could hardly bear to write anything except the constant stream of articles of political commentary for which journalists and editors never ceased to press him. In these and his huge correspondence he continued to fight against the growing pressure for trade sanctions against South Africa. 'I have no doubt whatever that it is the black people who would suffer as a result of disinvestment, and what is more, they may go on suffering for ten or twenty years, because the future is at the moment unpredictable,' he wrote. 'I live in a comfortable home, and certainly never go hungry, and I cannot believe that I have any right to be the cause of others going hungry.'[53]

Although political matters took up so much of his time, in 1978 he did plan a volume of essays on Natal, about which a South African publisher, Tafelberg, had approached him. It was to consist of ten 4,000-word contributions on various aspects of the province, and he approached various friends for an essay each: Colin Gardner on Natal literature, his old friend Reg Pearse on the Drakensberg, an Indian friend, Dr Bhana, on the Indian community, and so on. This book, which he was actively planning in May 1978, ran into major problems when some of the contributors did not produce the promised essays, or delivered something wholly unsuitable; Paton was still struggling with the project in November 1981, and trying other publishers, including Shuter & Shooter and Macmillan. In the end it was quietly abandoned.[54]

In June 1978 he also took on the job of writing 60 brief biographical paragraphs to accompany 60 cartoons of South African public figures, a task which involved much more work than he had expected. He and Anne were glad when it was done. Both of them were aware that in frittering his energies away on such tasks, Paton was marking time. These were retirement jobs, and why not? Paton was well beyond retirement age. But he would not mark time much longer. For by the end of 1979, for the first time in over twenty years, a new novel was germinating in his head. He was nearing eighty, but he was about to gather a late harvest.

24

UPHILL ALL THE WAY
1979–1984

Does the road wind up-hill all the way?
  Yes, to the very end.
Will the day's journey take the whole long day?
  From morn till night, my friend.

CHRISTINA ROSSETTI

In September 1979 Paton found himself most unexpectedly beginning to think of a novel again.[1] The inspiration was his autobiography: having ended *Towards the Mountain* at the point where he left Diepkloof, he was naturally considering how to write the continuation. This faced him with a major problem: he would have to mention many people still alive, and to say some very hard things about some of them: about Verwoerd and Vorster, of course, but also about Leftwich, Harris, and the others he felt had betrayed the ideals of the Liberal Party. Paton did not like saying hard things about anyone. Fiction offered him a way out, particularly fiction of an historical kind in which he could mention those whom he had liked and admired by their real names, and those whom he disliked or despised could appear in fictional guise. He found a model ready to hand, Paul Scott's *Raj Quartet*.[2] Paton much admired Scott's use of a complex chronological sequence which gradually reveals the political, personal, and racial conflicts in India in the period leading up to partition. He came to believe he could do something of the same sort for South Africa, depicting the period when the Nationalist government's grip on the country tightened into something approaching tyranny, the period which he knew so thoroughly.

Excited by this prospect, he began at once, and after a difficult start wrote fast. By 8 April 1980 he was enthusiastically telling friends that he had completed 40,000 words;[3] by 29 July he had done 90,000, the end was in sight, and he was telling Edward Callan (whose edition of *The Long View* he had found helpful during the writing of what was a complex historical novel of the years 1952 to 1958), that he was planning a trilogy, of which the second volume would cover the years of Verwoerd's premiership, 1958 to 1966.[4] The third would take the story up to the explosion of Soweto in 1976.[5]

He finished the writing in January 1981[6] and called this, his third novel and the first for twenty-eight years, *Ah, But Your Land is Beautiful.* He had little gift for punchy titles, though he told Edward Callan he intended this one to be 'both ironic and wistful'.[7] The book combined historical characters such as Huddleston, Patrick Duncan, and Albert Lutuli with others thinly disguised: Hendrik Verwoerd, for instance, appears as 'Dr Hendrik'. There were yet others who combined Paton's own experience with pure fiction: chief among these was the white protagonist Robert Mansfield, who Paton told interviewers was based upon himself: 'But I didn't like him, so I sent him off to Australia'.[8] His chief theme was the rise of apartheid, and he dealt with it by following its effects on a high Afrikaner civil servant, Van Onselen, an Indian family, a black man, Elliot Mngadi, who had played such a courageous role in the Liberal party, and Robert Mansfield.

The novel was published in September 1981 by David Philip in South Africa, and in March 1982 by Scribner in the United States, and Cape in Britain. It got mixed reviews. Many of them stressed the fact that this novel was the first of a trilogy, arguing that it could not be finally judged until the other two novels appeared.[9] They commented respectfully on Paton's energy at the age of 78.

But there was one long and thoughtful review which appears to have had particular impact on Paton, partly because it was by Martin Rubin, the son of his old friend Leslie Rubin. Writing in the *Washington Quarterly*, Martin Rubin remarked on the fact that Paton's projected trilogy was modelled on *The Raj Quartet*, and commented, 'But there is a far higher proportion of fact to fiction in Paton's work than in Scott's and, alas, a corresponding diminution of the artistry and technical skill needed to carry off so ambitious an enterprise.' He highlighted the extent to which *Ah, But Your Land is Beautiful* echoed Paton's earlier novels (in such incidents as that in which the Afrikaans civil servant is arrested for an offence under the Immorality Act, a strong echo of *Too Late the Phalarope*) and remarked, 'To a sad extent, whatever literary qualities *Ah, But Your Land is Beautiful* possesses represent a kind of feeding upon the author's own flesh, an act less surprising than poignant in a writer approaching 80.'

There was some unfair criticism in this review, as when Rubin argues that the powerful scene in which the judge washes and kisses the feet of a black woman during a Holy Thursday service in Bloemfontein does not succeed: in fact it is easily the most moving and powerful scene in the novel, and the writing takes fire at this point. But when Rubin ended with the observation that *Ah, But Your Land is Beautiful* is neither a work of true fictive art nor history, but falls between those two stools, Paton must have recognized the truth of the judgement. 'Is it too much to hope,' Rubin concluded, 'that, given the failure of this fictional form, Paton will resume his interrupted autobiography instead of continuing the novel-sequence and give us the truth, the whole truth, and nothing but the truth that he is the most qualified of all to provide?'[10]

Paton seems to have been stimulated by the final question. There had been other similar responses, among them a long letter to Paton from André Brink, praising the novel in general terms, but criticising the inconsistency of treatment of such major characters as Van Onselen, and asking why other characters, notably 'Dr Hendrik', should have been disguised at all. The implication of such remarks was that Verwoerd could have been more plainly set out in a history or memoir of the period. Reluctantly, Paton began considering abandoning the trilogy, and instead writing the second volume of his autobiography.

This decision was given impetus by the fact that he had begun the second volume of the trilogy in July 1981, before the reviews of *Ah, But Your Land is Beautiful* appeared, and had had great trouble with it.[11] 'If you detect a bitter note in this letter,' he wrote to me in September 1981, 'it is because there is no worse experience than to be trying to write a second novel while the first is not yet published.'[12] And once *Ah, But Your Land is Beautiful* had been published, and the reviews came in, his difficulties grew rather than diminished. By November 1981 he was telling Mary Benson that he was experiencing a writing drought.[13] By January 1982 things were still worse, as he explained to Nadine Gordimer:

> I have been in the slough of despond (and am emerging, I hope). It is extraordinary that one should experience this at age 78 (soon 79!), a total disinclination to do anything, a distaste for what one has written so far of the second book of the trilogy, a deep doubt of oneself, and of course, inevitably, a contempt for one's behaviour.[14]

This depression and sloth, unusual for Paton, continued and deepened through the first months of 1982, and he suffered persistent pain in his lower-left side. He had a range of tests and scans, and lost 22 pounds in weight. The doctors were baffled, suspecting kidney stones, until another scan revealed an aneurism of the aorta, and in April 1982 he was successfully operated on to give him another aorta of dacron. The surgeon, he thought, showed great unconscious wit in saying to him afterwards, 'That aorta will last you forever'.[15] He himself was convinced that he had not long to live.[16]

Although his health had up to this point been excellent, he had become a hypochondriac in old age, constantly taking himself to the doctor for ailments real or imagined, and on one occasion having to be restrained by Anne from taking strong tranquillizers straight after breakfast because his pharmacist had misread his doctor's handwriting.[17] He liked tablets and pills, and had a huge collection of bottles. This real illness brought him up with a jerk. He gave up any thought of continuing with the second novel of the trilogy, and seems to have destroyed what he had written.

There followed a long period of accidie in which he was content to recuperate, to enlarge his house, doubling the size of Anne's office and adding a sunny verandah at the end of 1982, and to enjoy the marks of distinction which continued to come to him. On 8 October 1982 the University of Natal unveiled a bust of him which it had commissioned from the sculptor Naomi Jacobsen; Paton, pleased and proud, made a speech, and Chief Buthelezi made another, in his honour.

On 11 January 1983 he turned 80, and his birthday party, always large, was particularly big and joyful that year. Anne took the unusual step of telling Paton firmly that this year he was not to ask any of the usual friends to propose the toast to him: she would do it herself. There were tributes from each of his sons, who had, as usual, come down for the party. Edgar Brookes read his atrocious verse,[18] and then Anne made a speech. 'Who is that?' she heard someone ask as she got to her feet.[19] They soon knew. She paid tribute to Alan in moving terms, but she also paused to give some of his friends a rap over the knuckles, so forthright that many of the guests had never heard anything like it. She told them that many of them had been coming to these parties for years, and she, Anne, had provided food and drink for them for years even though she was aware that several of them, when asked by Alan if he should marry her, had advised him not to.

Peter Brown and John Mitchell, who were the targets of this remark, looked distinctly uncomfortable, and Ian Wyllie, standing next to Paton, heard him snort in amusement. Then Paton said quietly to Brown and Mitchell, 'Now you know!'[20] The next Sunday Paton went to lunch with

Wyllie, as he often did, and Wyllie asked him, 'Did you know what she was going to say?' 'No,' said Paton, but again came that characteristic snort of amusement.[21] The truth was he rather admired Anne for being forthright enough to slap down his friends in this way, just as he valued her capacity to save him from unwanted intrusions by policing his phone calls and guarding his door.

Shortly after this memorable eightieth birthday party, Paton and Anne drove to Cape Town, where Paton's friend Tony Morphet, now head of the Department of Adult Education at the University of Cape Town, had organized a series of five lectures, entitled 'Alan Paton at 80—a celebration'. They were given by Denis Hurley (Catholic archbishop of Durban and one of Paton's friends and admirers, as well as his occasional croquet opponent), Colin Gardner, Richard Rive (a distinguished Coloured writer and head of a Cape Town training college), René de Villiers (Leo Marquard's nephew, and a distinguished journalist), and Paton himself, and they were a great success. Paton received a standing ovation after his lecture, and Anne, watching from the back of the packed hall, saw his eyes shining with pride and pleasure.[22]

He was living more quietly now, no longer flinging himself at new experiences, content to accept what came to him. His relationship with Anne, after some very rocky patches, had settled into a pattern that pleased them both. She took over the practical details of his life, organizing him, and he was content to be looked after. She dealt with almost all financial matters, she ensured that his massive correspondence did not get neglected, and if it did she would answer the less important letters for him. She typed his birthday messages to his children and grandchildren, and she signed the very small cheques they got as presents;[23] she organized the servants and saw to the smooth running of the household. Paton's one sphere of interest around the house was the garden, which he continued to oversee with great skill and knowledge. But in everything else he depended on Anne. She did the driving when they went out, and if they came back to find the hot-water boiler pouring water through the kitchen ceiling, Paton would disappear to his study saying tranquilly, 'I'm sure you'll be able to deal with it, dear': and she did. If a window stuck or a door squeaked, Anne, who was by now a skilful carpenter, would fix it.

Her carpentry had produced one quarrel between them. Paton liked to read while standing, and Anne made him a set of high bookcases with a lectern-like reading shelf. They cost her a good deal of work, and she was very proud when she presented them to him. Paton tried them out and rejected them at once: 'Don't like them', and he would not have them in his study. Anne, deeply hurt, eventually gave them to her daughter Athene.[24] There was a ruthless side to his honesty: a little judicious and tactful lying would have made things easier for them both.

The ill feeling between Athene and Paton, which had reached a climax during the mid-1970s when she began bringing a boyfriend to Paton's house, had eased when she married in 1978. It is likely there was an element of jealousy in Paton's resentment, not just of Athene's calls on Anne's time, but of Athene's boyfriend. His own strict upbringing made him very intolerant of intimacy between young men and women, and the mere presence of Athene and her friend made him irritable. Her marriage to a Cape Town archaeologist, Martin Hall, changed that, and the arrival of their daughter Nerissa produced an instant bond between them, for Paton loved Nerissa from the moment she began to speak.

His evident fondness for this grandchild was hard for his sons to accept, for Paton seemed to show a decreasing affection for their children. Part of the reason for this was simple: Jonathan's and David's children were now in their late teenage years, and were beginning to bring to the Botha's Hill house girlfriends and boyfriends of their own, with consequent strain and occasional explosions.

The most dramatic of these came on 17 January 1981, when Jonathan and his family were staying at Botha's Hill after the annual birthday party. Jonathan's son Anthony, now 19, had brought a beautiful girlfriend named Petie. Anthony annoyed Paton by holding hands with her in church,[25] and when Paton subsequently walked into Petie's room to find Anthony sitting on the bed with her holding her hand, he erupted. Jonathan, in the next room, heard him growl 'like a tiger', a sound remembered with dread from childhood, and then heard him shouting at the young people. When Jonathan tried to defend them, Paton roared at him, 'The rule in our house always was, girls don't go in boys' rooms, boys don't go in girls' rooms!' Since he had had no sisters, Jonathan was baffled by this.[26] Paton was, of course, quoting his own terrible father from more than seventy years before. The result was a furious row between father and son. Anne and Jonathan's wife, Margaret, returning from an outing, found Jonathan in the street outside, 'with a face like a beetroot', crying out that they must leave at once.[27]

With David there was a crisis of a more lasting sort. In 1976 he and his wife of many years, Nancy, were divorced as the result of a long association David had had with a radiographer, Maureen Finegan. He and Maureen then began living together, but chose not to marry for the time being.[28] Paton's sympathies were entirely with Nancy; such was his shock at hearing of his son's proposed divorce that he did not speak for two days,[29] and he later told his sister, Dorrie Arbuthnot, that David was 'a wicked man'.[30] When David proposed to come down to Botha's Hill some time later, bringing Maureen with him, Paton asked such friends as Ian Wyllie what he should do,[31] and then told David that he and Maureen could not share a room. The result of this confrontation was that David never slept under his father's roof again. Only in his last two years of life did Paton agree to stay with his elder son once more, but had then relaxed sufficiently to laugh at an obscene joke Maureen told him.[32]

Scenes like the one with Jonathan, of which Paton was ashamed once he had cooled down, grew less and less frequent as he moved into old age. His reduced competitiveness showed itself, too, in his declining love of croquet, surely the most aggressive of all games. His grandchildren began to find they could beat him. 'Well done, Anty, well done,' he would say to Anthony, but he did not look pleased.[33] Presently he stopped playing altogether, and the splendid croquet lawn stood disused except when Jonathan and his family came down for the Christmas holidays.

There was another sign of increased calm: the 'train-dreams' from which he had suffered for many years gradually came to an end. These were dreams in which he was rushing for a train, late, and had to give a lecture which he had not yet prepared, and they were often so bad that Anne would wake him up to stop him from thrashing about. Now, in old age, he slept so quietly that she would sometimes lean over him to see if he were still breathing.[34]

Though his physical activity slowed, he retained his love of life and his sense of humour. He was tremendous fun to be with. Anne was to recall many examples of his hilarious clowning, as when, one Christmas night, she had presented him with a new briefcase:

> Later, after he had gone to bed—or so we thought—there suddenly appeared in the sitting-room an amazing apparition. It was Alan wearing a straw hat, one of my silk scarves, a tennis ball in a net round his neck, my tennis peak under the hat, and a kitchen towel hanging from his waistband like a waiter. In his right hand he carried the new briefcase. He stood there clowning away, while we fell about in hysterics. This was typical Alan; one never knew when he was going to do something quite dotty.[35]

By the early 1980s, though, he had reached the age when the obituary columns of his daily paper held increasing personal interest for him. In

January 1983 Canon John Collins died, and though Paton had never forgotten the folly of Collins's New York speech, he wrote a letter to Diana Collins of sincere tribute and condolence. In February 1983 Aubrey Burns, whose mental decline in the last years had reduced him to silence, died at last, and Paton felt the snapping of this link with the first reader of *Cry, the Beloved Country*. Some time before his death a friend, Ruth, had asked Burns about Jesus, and he replied, 'Jesus? Jesus? No, I don't know him.' Ruth said, 'Do you know Alan Paton?' 'Oh yes,' he said, 'of course I know *him*.'[36]

That same month Vincent McGrath died, a dedicated American fan of Paton and collector of all his publications, to whom Paton had written so often that he came to have a considerable affection for him, and had taken the trouble to visit him on one of his trips to America. And in June 1983 Neville Nuttall, with whom Paton had shared that first, disappointing bottle of port as a student, died after a long illness.

Paton himself was not working at his old pace, but he was far from retiring. When asked to give a reading at an Arts Festival in Grahamstown in 1983 he declined, and when Pat Poovalingam, who was forming a new all-Indian Party, asked him to join, he replied, 'My Party days are over, and I don't want them back again.'[37] But he continued to accept invitations to lecture closer to home, and always prepared carefully for these occasions. His power as a speaker was undiminished. Commonly his audiences found these talks both very funny and deeply moving. Douglas Livingstone, an old friend and South Africa's finest living poet, wrote to him after a talk he gave at the University of Natal in April 1983, 'It has taken me a couple of days to recover properly from your talk on Tuesday evening. It was almost unbearably moving. You came over with maximum integrity, & you were at once puckish/modern/iconoclastic (—if you can sort out that mixture!), & above all with great courage.'[38]

Many different audiences in many places agreed with Livingstone's assessment of Paton as a public speaker. From the platform he exercised an extraordinary and almost mesmeric power, and he could hold any audience, from schoolchildren to the Harvard Convocation, entranced. His bilingualism helped in this with South African audiences: one admirer wrote to him, 'You tell a story in Afrikaans as though to the Marico born',[39] the Marico being a northern Transvaal area of hardscrabble Afrikaner farmers, immortalized in the stories of Herman Charles Bosman. Paton seemed to find the level of his audience effortlessly, whether he was addressing Anne's tennis club or the students of Kearsney College, near his home. The effortlessness, however, was an illusion, and his success as an orator cost him a good deal of work and thought. For all that, he gave himself unsparingly, particularly in the hundreds of addresses he delivered to school students. The first of these was in December 1948, at the school of which Reg Pearse was headmaster, and he went on delivering them until the year of his death.

Nor, even now, was he dormant as a writer. He decided in June 1983 that he would abandon any pretence of working on the second novel of the trilogy, and would instead begin on the second volume of his autobiography, hiring a student to do the preliminary historical research for him. The decision was a great relief to him, for his struggle with the projected second volume of the trilogy had become painful. On a good day he wrote a hundred words, on a bad one nothing, and there were many more bad days then good ones.

He went on publishing newspaper articles calling for the government to talk to black leaders: in May 1983, in an important article entitled 'The National Interest: Is There Such a Thing?' he pleaded with South African politicians, white and black, to work towards the definition of a new South African nationalism, based not on race, but on a sense of the over-riding interests of the country as a whole. He was calling, in fact, for a conference for a democratic South Africa, nearly a decade before it came to pass. At the time he seemed a mere dreamer, for his thinking was still far in advance of that of most of his countrymen.[40]

This article was written against the background of increasing talk of liberalization from the Prime Minister, P. W. Botha,[41] which, vague though it was, had the effect of alienating the conservatives within the Nationalist Party. Their leader, Andries Treurnicht, was forced out of the party in 1982, and formed the Conservative Party. Having expelled the right-wing threat in this way, P. W. Botha began with greater confidence to talk of 'healthy power sharing', by which he seems to have meant that there would be consultation with Coloured people, but not with blacks. In November 1982, the government announced, there would be a referendum among whites on the issue of a new constitution, allowing for limited sharing of power along these lines. An impassioned debate preceded the referendum, Treurnicht's group opposing it for going too far and the left of centre Progressive Federal Party, led by F. Van Zyl Slabbert, contending that it did not go far enough.

In one of his addresses to schoolchildren, at Woodmead school in Johannesburg on 23 July 1983, Paton disagreed with those on the political left who considered the new constitution a giant swindle, designed to keep blacks out of power for ever. 'I don't believe it,' he said, comparing the problem facing the National party with that of a man who has been riding a tiger for years, and now very much wants to get off. It was unrealistic to ask them to dismount in a single leap and embrace the tiger. Those who asked for this 'are asking for the moon', he said.[42]

Immediately after giving this address, Paton left for a brief holiday in the Okavango Swamps of Botswana. As a result he did not see the newspaper headlines implying that he had turned his back on his former political friends, and had expressed full support for P. W. Botha. 'Paton Supports PW's Plan', cried the *Rand Daily Mail*, and ' "The Constitution Plan is no Swindle" Declares Paton' said the *Star*. The Afrikaans press picked up this story, and, for the first time in Paton's life, praised him for his wisdom and penetration.[43] His friends, including Peter Brown and Colin Gardner, were horrified and rushed to distance themselves from him, and the Natal Indian Congress condemned him. When Paton returned from Okavango on 28 July 1983 and learned how his words had been misinterpreted, he was, as he put it, 'sick to the stomach',[44] and gave interviews to make it plain that though he did not think the new constitutional plan a swindle, he did not intend to support it either.[45] He also stressed that his attitude to Afrikaner nationalism had not changed: he detested it.[46]

This incident, and particularly the readiness of some of his close friends to believe ill of him, left a bad taste in his mouth, and there was more to come. In August 1983 Paton was nominated to fill the position of Chancellor at the University of Natal, where for some years he had served as President of Convocation. Just when it seemed certain that he would be elected Chancellor, the students nominated Nelson Mandela. Although Mandela had been in prison for many years and there was at this time no sign that he would soon be released, Paton felt he could not stand against such a man, and he withdrew his nomination with a good deal of regret. He had an experience of Winnie Mandela's ethics when he learned she was spreading the story that Paton had tried to pressure her husband to withdraw. Patiently he wrote to tell her that this was not true, and to express his sympathy for the cruel way the government had treated her, thereby implying that she had some excuse for her erratic behaviour.[47]

This letter was the more generous and forgiving because Paton had once before suffered at her hands. A year or so before the Chancellorship incident, Winnie Mandela had asked Paton to help her daughter Zinzi prepare for publication a collection of Zinzi's verse. Paton had generously invited the girl down to Botha's Hill, and had spent many days painstakingly going through her writing trying to put it into publishable form. He then sought out a publisher for her, and she left him with many expressions of gratitude. After some months he learned from the newspapers that Zinzi's volume had been published, but from the young woman herself he heard nothing. Saddened by this he wrote her a letter of mild reproof, telling her that it would have been courteous had she at least let him know the book was about to appear. Zinzi did not reply: instead Paton got a ferocious letter from Winnie Mandela, calling him a typical white racist and a phony liberal, and asking who he thought he was to expect Zinzi to thank him. When he attempted to explain his position the reply was more abuse of the same kind. 'Unbelievably vicious', Anne Paton called Winnie Mandela's letters; and Paton himself was so upset by them that he destroyed them.[48]

After incidents like these he was glad to get out of South Africa for another whirlwind trip to North America and Britain, on 2 September 1983. This was chiefly a pleasure trip, and the Patons did a good deal of sightseeing and visited a few of Paton's many American fans. This tour Paton thought of as his last, and his reason for seeing his oldest and most faithful fans in this way was to say goodbye to them.

He and Anne flew to California, hired a car, and drove to the Yellowstone National Park and Grand Teton Park, before heading across the desert to Salt Lake City and then to Denver. From there they flew to Texas, where a long-time fan, Rebecca Canning,[49] met them at Fredericksburg and drove them to Houston, a trip which he and Anne much enjoyed. The Patons then travelled to New Orleans, and on to Hilton Head, South Carolina, to see Bill Toomey, who had been American consul in Durban for some years, and they then travelled up the east coast to spend some days driving round New England.

Here they saw several other admirers. One of these was Father John Pesce of Hartford, Connecticut, who guided them through his city and impressed them as a demon driver. Another was Ursula Niebuhr, widow of Reinhold Niebuhr and herself a woman of great distinction, whom Paton much admired. Yet another was Holly Rood, a young woman who had been writing to Paton since 1967, who had first met him at Harvard in 1971, had visited him in South Africa in 1980, and with whom he had flirted agreeably.[50] Her mother, Janet Rood, welcomed the Patons with open arms and gave them advice on the best places to see the Fall in Vermont without being trampled by what she called 'leaf-peepers'.

The Patons flew to Britain on Concorde, for which they had by now developed a great liking, on 7 October 1983. Here there was Paton's usual stay with the Friedmanns in London, the big party for the Liberal expatriates, where Mary Benson saw him again, and a visit to Cambridge to hear Trevor Huddleston preach. In the course of his sermon Huddleston told the congregation that he had wasted his life.[51] Paton was not surprised: 'I do not think that a man of his stature can get much joy out of trying to stop Zola Budd [a South African athlete] running in Britain,' he remarked to Diana Collins.[52]

Paton had kept up his links with Huddleston, for whom he continued to feel great sympathy. After his period in Tanzania and a spell as bishop of Stepney, Huddleston had become Archbishop of the Indian Ocean, a position which sounded grander than it was. He had spent some years based in Mauritius, a lovely but dull island where he found few minds as keen as his own, and suffered terribly from boredom and loneliness. The Patons visited him there in 1980. 'I am sorry to say that he is in a very depressed mood,' Paton told Gonville ffrench-Beytagh. 'The truth is that he does not like Mauritius. I don't think there is any doubt that his heart is still in Sophiatown.'[53] The community of exiles in England generally he thought a little sad, and increasingly out of touch with the South African situation. Some of them thought the same of him.[54]

Paton was back in South Africa in time for the first of the engagements that made up his regular end-of-year festivities: the weekend party at Peter Brown's cottage in the Drakensberg. This males-only occasion was always held on the weekend closest to 16 December, which in South Africa at that time was a public holiday commemorating the Boer victory over the Zulus at Blood River in 1838. Paton entirely rejected the notion of celebrating the victory of white South Africans over their black fellow-countrymen, and by way of signalling their disrespect for this day, he, Peter Brown, the headmaster John Carlyle Mitchell, and other liberals such as Sam Chetty and Bill Hoffenberg had begun what became a tradition of spending the weekend at Brown's cottage, drinking, walking, and talking.

David Welsh[55] was present at just one of these weekends, in 1976, and recalled that a great deal of alcohol was drunk. The others drank beer, but Paton was supplied with whisky and did heroic deeds with it. During these weekends he, who very seldom swore in normal conversation, became completely uninhibited—'real locker-room stuff', Welsh was to call it. The drinking would start with lunch, and continue steadily until the early hours of the morning. On a fine afternoon they would sit on the verandah of the cottage, looking out at the mountains, and throw their empty beer cans onto the lawn in a sort of competition. On the first day of the weekend the idea was to hit another can. On the second it was to avoid hitting another can.[55]

It was not only Paton who was uninhibited during these weekends: the others hid their respect for him, and would chaff him mercilessly. He did not always like it, and on one memorable occasion grew so angry that he slammed out of the cottage, got into his car and drove noisily away. He travelled a few miles, realized he was making a fool of himself, and drove back. When he entered rather sheepishly, the others, who had been contemplating the ruin of the weekend, greeted him joyfully. 'Paters, you're a great man, you've come back,' said John Carlyle Mitchell.[56] After that they treated him with more caution.

Welsh had not learned this lesson, and when he woke from the first night's drinking he and Paton made their way together down to the river to wash, since, it will be remembered, Brown's cottage had no running water. Paton was looking very much the worse for wear after the night before, but Welsh tactfully said nothing about this until, just as they were climbing through a wire fence, Paton spotted a bird singing quite near him. 'David you know, I think I'm like St Francis of Assisi,' he told the astonished Welsh, 'because I can talk to these birds and they respond.' Welsh said, 'I don't think that's the reason. I think they find you interesting because you look so seedy this morning.' Paton did not laugh. 'He didn't like jokes being made against himself,' Welsh was to say. 'He would certainly put you down, but he didn't like you doing the same.'[57] For all that, once Welsh got to know him better he often pulled Paton's leg in this way, and the truth was that part of Paton enjoyed not being treated as the grand old man he had become.

25

JOURNEY COMPLETED 1984–1988

I must lie down where all the ladders start
In the foul rag-and-bone shop of the heart.
W. B. YEATS

By 1984 Paton was beginning to slow down. In that year he started writing the second volume of his autobiography, *Journey Continued*, but he made very slow progress with it. 'Do I think I shall live for ever, or do I not care?' he asked Mary Benson.[1] He continued to agree to ceremonial engagements, such as addressing a reunion of old boys of Maritzburg College in December 1983, or, on 23 February 1984, opening the splendid new Brenthurst Library on the Johannesburg estate of the magnate Harry Oppenheimer, whom the Patons saw socially from time to time.[2]

He also continued to take the closest interest in the politics of the United States (where Ronald Reagan was about to be re-elected President for a second term) and of his own country. In May 1984, when the government was talking about destroying Crossroads, the huge squatter shanty town that had sprung up outside Cape Town, Paton wrote to the Prime Minister, P. W. Botha, appealing to him to desist, 'not only out of my concern for the people of Crossroads, but also out of my concern for our country and yourself'.[3] Botha responded to this with clichés and platitudes. 'The Government', he wrote, 'is doing everything to the best of its ability to persuade the squatters to discontinue squatting.'[4]

Pressure on the South African government during these years was steadily increasing, as a result of the anti-apartheid movement's growing success in persuading the outside world of the justice of its cause, a victory in which Paton's writings had played a major part since 1948, and as a result of the economic sanctions which Paton had steadfastly opposed. As a military threat the ANC had proved largely impotent, but the country had been kept on the boil as a result of the steady cycle of violence in the townships. In September 1984 these took the form of an upsurge of attacks on local government buildings and the murder of black council chairmen and mayors. On 21 March 1985, the anniversary of the Sharpeville shootings, the police shot dead twenty mourners at a black funeral at Uitenhage, and the spreading violence that resulted reached new heights in 1985 and 1986. Another state of emergency was declared on 20 July 1985, and it was to be renewed annually until 1990.

Against this background of violence and political polarization the humane values of Liberals were ever more contemptuously rejected by the antagonists on both sides. The ingratitude and contempt of a Winnie Mandela Paton could shrug off stoically, but when journalists whom he had respected, or academics he would have liked to respect, showed that they had written off what Liberals like himself had devoted years of their lives to, he was angered. Newspaper articles purporting to list the main liberal organizations, but omitting to mention the Liberal Party, drew letters of protest from him, and he was particularly stung when an apparently authoritative book did much the same thing.

White Power and the Liberal Conscience,[5] by a young scholar named Paul B. Rich, spent only six and a half pages on the Liberal Party, in spite of its title, and, in Paton's view, dismissed it contemptuously as a party of 'crooks, fools, opportunists, tools of capitalism, and manipulators'. Inside the cover of his copy of Rich's book Paton scribbled wittily, 'Author: Rich. Book: POOR', and he used this as the headline of a review of the volume, in the course of which he made an eloquent defence of the party to which he had given many of the best years of his life. 'The Liberal Party, whatever its merits, or demerits, was the most concrete and articulate expression of the liberal conscience in this century, and amongst its members were some of the best human beings that I have encountered in my long life.'[6]

Other Liberals, including the eminent political commentator David Welsh, were depressed by the inaccuracies and omissions of Rich's book,[7] and determined that something should be done to correct the impression it left. The outcome was a conference on the heritage of the Liberal Party, organized by the Institute of Social and Economic Research, and held in Grahamstown in July 1985. This drew sixty or so former Liberals from all over the country, and it turned into a tribute to what Paton and his Liberal associates had done over many years: they had kept alive the ideals of liberalism over many dark years in the face of illiberal nationalism, both white and black.

Paton himself opened proceedings with a talk in which he tried to set straight the history and achievements of the Liberal Party, and in which he paid an elaborate tribute to Peter Brown, who had not even been mentioned in Rich's book. He stressed the extent to which Brown, Ruth Hayman, and other Liberals had been motivated by conscience: 'people do certain things because they think the doing of them is right'. Having opened the conference in this way, he enjoyed himself, meeting former Liberals he had not seen for years, holding court, and staging the public jokes at which he had excelled as a student more than sixty years before. One of the scheduled speakers was Tony Mathews, professor of Law at the University of Natal, but at the last moment he was taken ill and David Welsh had to read his paper for him. As Welsh finished and the applause died, Paton was on his feet: 'Mr Chairman, I just want Professor Welsh to know that that is the best lecture I ever heard him deliver.'[8]

His love of travel persisted, and he and Anne continued to undertake trips, some of them by car inside South Africa (to the Drakensberg, at least once a year, to the Kruger Park, or the Cape), some of them to places more remote. In November 1984 they flew to South America and did an extraordinary trip by boat down the Amazon. 'My last journey,' he wrote to Mary Benson, and added '(I hope? I fear? I lie?)'[9] He also saw the Iguazu Falls, which impressed him mightily. They reminded him of the Victoria Falls, which he had seen on his unhappy first honeymoon, and this, and the background of terrible violence in his country, lay behind the fragmentary poem the experience inspired, in which the falls are Victoria and Iguazu, and the 'destiny' human unhappiness, political violence, and his coming death:

The tributary widens, I sense an urgency in the water
The trees and the sky seem to say to me
Some greatness is imminent.
The stream quickens, it is aware of its destiny
And I am fearful & exalted,
For I shall see the great falls and the smoke that thunders
Of which I have heard since childhood.
Now I yield to the current, now I struggle against it
I am uplifted, I am cast down by this solemn journey.[10]

He was still pressing forward with his autobiography, but with glacial slowness and many struggles. Though his mind was as clear and sharp as ever, he increasingly lacked physical energy and the will to write.[11] He was now eighty-two, and for all the terror and violence of his country, he was growing increasingly serene and philosophical. 'I must admit that the state of the world is somewhat depressing,' he wrote to Ursula Niebuhr, 'but I have a shrewd suspicion that it always has been, and that the world does not go through happy phases and unhappy ones. We do, and therefore we are inclined to believe that the world is doing the same thing. I think this struggle between good and evil is perennial. I think the really bad thing is when you begin to think that good is losing the battle.'[12]

Paton himself never lost his faith that good would win. In May 1985 he gave the Hoernlé lecture to the South African Institute of Race Relations. His title was 'Federation or Desolation', and he put forward the view that P. W. Botha's 'New Dispensation' was not going to work because it continued to exclude blacks, while the United Democratic Front's[13] demands for universal suffrage in a unitary state could only be achieved by violence and revolution. The alternative, he argued after consultation with political leaders from Buthelezi to Alex Boraine, was a federal system.

As a result of this lecture he was interviewed on South African television, in October 1985, and made a great impression: fan mail came in for days. 'The TV interview was very good, partly because I no longer wanted to impress anybody!' he wrote to Edward Callan. 'So I impressed a great number, & enjoyed an Indian summer of fame.'[14] After this success he made more frequent appearances on the government-controlled radio and television. He was enjoyably interviewed on television by his son Jonathan in January 1986, and by the academic Michael Chapman in April 1986. These South African interviews were a new phenomenon, though he had long been interviewed by such international luminaries as Malcolm Muggeridge (in June 1977), and Bernard Levin (in March 1981), with both of whom he struck up friendships.

He was trying to slow down, and write fewer articles and speeches. Among his last active participations in politics were his discussions with Peter Brown, in 1985 and 1986, about the possibility of founding a new Liberal Democratic Party, discussions which came to nothing; and his participation in the discussions between leaders of Natal and KwaZulu, known as the Indaba, in October 1986, where he enjoyed meeting David Steel, the leader of the British Liberal Party.

Though his public role was circumscribed, he continued to devote a part of his income to private acts of charity. It is impossible to estimate the number of black children whose education he paid for, both at school and at university, for his correspondence with them was too voluminous to preserve and Anne weeded the files steadily. This is what Wordsworth called the best portion of a good man's life: 'His little, nameless, unremembered, acts / Of kindness and of love.'[15] Nor did he confine these acts of kindness to his countrymen; young people would write to him from Zimbabwe or Zambia asking for his help, and they would get it.

A young Zimbabwean, Arthur Mamvura, ended a long correspondence with his benefactor by writing, 'I find it impossible to forget you after what you did for me . . . I also need to thank you about what you have done for the whole world. Indeed you will go down in history as one of the men with noble acts.'[16] An elderly black woman, Edith Obose, whom he had befriended and helped over many years, wrote to him in 1986, 'Thanks God for sparing us such a man as yourself. A man who knows no line across colour. A man who goes out of his way to help the needy. May you be spared more useful years.'[17] For Paton this kind of private tribute hugely outweighed the public disparagement of Liberalism by his detractors on the left and the right. Referring to the violence racking South Africa, in an article he published at the end of 1985, he called that year the unhappiest of his life, and then said that it had been saved for him by his personal relations.[18] It was not just his family and friends, but contacts with people like Arthur Mamvura and Edith Obose that he had in mind.

In spite of describing each journey he took as his last, he continued to travel. In April 1986 he made a car journey to the Karoo with his son Jonathan, ostensibly to see Halley's Comet, which he could remember his stern father waking him to see in 1910. Then, he recalled, the comet seemed to stretch from horizon to horizon; now, he and Jonathan scanned the clear semi-desert skies with binoculars in vain. Halley's Comet in 1986 was a fizzer. At last, on a night in Bloemfontein when Paton had had so many whiskies he scarcely cared whether they saw it or not, Jonathan found a small fuzzy blob through the binoculars and insisted his father should look. Paton peered blearily at the sky for an instant before handing back the glasses contemptuously. 'Halley's Comet?' he said. 'Poof!' And he stumbled off to bed.[19]

Jonathan, on this trip when they visited many of the beautiful towns in the Karoo, saw aspects of his father he had not known before. One night in Beaufort West they were sharing a motel room, and Paton, who had been drinking freely and reminding his son of previous trips they had made to the Karoo in Jonathan's childhood, seemed deeply troubled about something. He paced the room, apparently in anguish of mind, and when Jonathan asked him what was wrong, he said, 'I'm a sinner. I'm a terrible sinner.' Then coming up to Jonathan he embraced him and kissed his son full on the lips. Jonathan, who for years had regretted the fact that his father showed him so little overt affection, was astonished. 'Something was troubling him a great deal,' he was to say of this strange scene.[20]

In June 1986 Paton travelled to Germany as the guest of the German government and PEN International, a writer's organization, and he and Anne also took the opportunity to have a holiday in Austria. At the conference in Hamburg, on 'Contemporary History as Reflected in Contemporary Literature', he read 'To a Small Boy who Died at Diepkloof Reformatory', sharing the stage with the black South African poet Sipho Sepamla, who read his poem on Nelson Mandela. After the conference Paton was invited to meet the German President, Richard Von Weizsäcker, who asked him about disinvestment. When Paton said he was opposed to it, Von Weizsäcker opined 'that black people would suffer in the short run but in the long run they would win'. Paton replied, 'And how long would the long run be?' and Von Weizsäcker acknowledged the point at once.[21]

In October 1986 Paton paid a final visit to the United States, this time to receive another honorary degree, from La Salle University, his twelfth and last.[22] It was on this trip, during the South African Airways flight across the Atlantic, that he wrote his last known poem.[23] His subject was the violence racking his country, and in particular the most hideous death invented in modern times, the 'necklace'. In this the victim is tied, often in a sitting position, a motor tyre hung round his or her neck, a bottle of petrol poured into it, and a match thrown. The result is a slow burning to death, the youthful killers, calling themselves 'the Comrades', often dancing the toyi-toyi and chanting or singing to drown their victim's screams.[24] Paton's poem, which refers to the fact that parents were sometimes burnt by their own children for suspected 'collaboration' with the authorities, is entitled 'Necklace of Fire':

I send you a present, my love, my love,
Though not of my own desire
The comrades say it's my duty, my love
This present of terrible beauty, my love
The necklace of fire.

They say I need not be present, my love
To stand at the funeral pyre
So long as you know that I sent it, my love
The necklace of fire.

It is the children, my love, my love
It is the children who now require
That I must send you this present, my love,
The necklace of fire.

They say they must die for the cause, my love
But you say obey the laws, my love
So the children say you must die, my love
With the necklace of fire.

It is the children, my love, my love,
Oh how can I tell you, my love, my love?
It is our children, my husband, my love
It is our children who now require
That it is your wife who must end your life
With the necklace of fire.

It will be tomorrow, my love, my love
It is tomorrow that the comrades require
And if I say that I cannot agree
Then the comrades may also order for me
The necklace of fire.[25]

It was on this trip that he saw Edward Callan for the last time, Callan having driven to see the degree ceremony in Philadelphia in mid-October 1986. A journalist who interviewed Paton in Philadelphia left a vivid portrait of him sitting stiffly, 'a small, white-haired figure in a dark, boxy suit, with round-toed black shoes and thick white socks'.[26] His dress sense had not much improved, in spite of all Anne's efforts, and with his intense stare and puckered mouth he looked more like a wise old baboon than ever.[27]

On this American trip he also paid a visit to someone he had last seen in 1948, the photographer Constance Stuart Larrabee, who had photographed him at Anerley. She had in 1985 published a catalogue of the photographs she took then, to accompany an exhibition of her work entitled 'Go Well, My Child', quoting from *Cry, the Beloved Country*. She was now living in the beautiful Maryland city of Chestertown, and the Patons thoroughly enjoyed their brief stay with her. Paton took particular pleasure in sitting on her huge porch, watching the broad Chester river slide by and the geese coming in from Canada.[28] The Patons finished this American visit with a trip to Key West in Florida.

The accession to power in the Soviet Union of Mikhael Gorbachev and his pronouncements about the need for rapid change there Paton regarded with great and growing hope, and he followed the negotiations between Gorbachev and Reagan with keen interest. 'Even if it is only a change in vocabulary it is something to be thankful for,' he wrote. 'I don't think we shall blow each other up after all.'[29]

By May 1986 he had completed the writing of what he considered the most difficult section of his second volume of autobiography, which was to be published as *Journey Continued*. This problematical section covered the years of the African Resistance Movement's acts of sabotage, Harris's station bomb, and Leftwich's betrayal of his comrades, and having finished it he sent it to Leftwich and Randolph Vigne for comment. Between them they suggested many changes and omissions, most of which Paton acceded to. But where they wanted him to omit the sneering comments of the judges at the ARM trials, or Paton's references to Harris's actions as a 'totally futile deed', Paton dug his heels in. 'No, I certainly do not intend to alter my words "totally futile deed". After all I am writing about my own opinions, not anyone else's.'[30]

By September 1986 he was on the last chapter, and planning the short epilogue in which he summed up the years after 1969. He wrote this in slow stages during 1987, the end becoming rather fragmentary as he tired. For all that, the volume, which was published in May 1988, is very precise as to dates and facts; like the first volume, it omitted details mainly to spare others like Adrian Leftwich from embarrassment.

One of those Paton spared from embarrassment was himself: he said nothing about his reputation as a flogger while he was a teacher and Principal of Diepkloof; he concealed much of the real difficulty of his marriage; he appeared quite unaware of his children's views of his deficiencies as a parent; and though he gave details of one extra-marital affair, that with Joan Montgomery, he drew a discreet veil over his friendship with Mary Benson. Of the events surrounding his assault by Derek Ndhlovu and Alice Ngcobo he said nothing at all.

Journey Continued also showed Paton's growing lethargy in its lack of sparkle. The writing is flat and dutiful, with little of the vibrant energy so typical of his letters. His mind was as clear as ever, but his spring was gone, and he knew it. He withdrew from the last formal position he held, resigning from the editorial board of *Reality* in June 1987 and explaining that his back and hearing were giving him trouble.

By the end of 1986 he was consciously preparing for death, reading the poems on the subject in Rabindranath Tagore's *Gitanjali*, and finding them very beautiful, as he told Mary Benson.[31]

On the day when death will knock at thy door what wilt thou offer to him?
Oh, I will set before my guest the full vessel of my life—I will never let him go with empty hands.

Throughout 1987 one finds uncharacteristic little remarks—they are not complaints—about his health, scattered through his still-huge correspondence. He had had a prostate operation in January 1986, and it was not a success; he was obliged to get up four or five times each night. 'You are quite right when you say "the burden of years" is more than an expression', he told an American fan, Father John Pesce.[32] Among his trials was increasing pain in his gums, which made it very difficult for him to wear his dentures.[33]

He continued to take great pleasure in his garden, though he did little or nothing in it himself now, being content to give directions to the three gardeners; he continued to rejoice in birdsong and the appearance of the first clivia. And he took great pleasure in watching the development of Athene's daughter, Nerissa, whose childish ways sweetened his last years. In his speech at his birthday party in January 1987 he mentioned seeing the miracle of language as the little girl learned to speak.

He continued to enjoy his whisky, and his word game, for which a special hour was set aside each evening. The word game was printed each day in the *Natal Mercury*, and it consisted of a jumbled nine-letter word. An example was:

IDI
USU
JOC

The object was to find the word,[34] and then to make as many as possible other words each with a minimum of four letters, and each containing the central letter ('s' in the example above). Paton was outstandingly good at this game: where lesser players might find thirty or forty words, he would commonly find sixty or seventy. He loved competing with anyone who would join him at it. I did, and was utterly humiliated. 'Sixty-seven so far', he announced, looking up. Then, sardonically, 'How's Cambridge doing?' Cambridge had twenty-three. One of the few people who could beat him was Anne's daughter, Athene. He did not enjoy losing to her.[35]

He continued to take a pride in his old school, Maritzburg College, to which he had returned many times over the years to present prizes and make the end-of-year address. At one of these, in December 1963, he had written in wet cement a poem of his own about the College, composed in the 1930s while he was teaching there; it can still be seen in the southern wall of the old gymnasium. In 1983 he had attended a reunion of his contemporaries at the school, giving the address, and had found it an amusing experience to see the schoolboys of seventy years before changed into wizened old men—himself not the least wizened of them. In 1988 the College celebrated its 125th anniversary, and Paton contributed his reminiscences of the school to a special supplement of the *Natal Witness* to mark the occasion. It was another rounding up of the past, another farewell.

At his last birthday party, on 11 January 1988, at which he turned 85, his speech was filled with literary quotations, mostly Yeats and Tagore, all of them dealing with death. The guests listened with increasing disquiet. 'What is going on here?' hissed Professor Colin Webb of Natal University to Ian Wyllie.[36] The answer was plain enough: Paton was signing off, and wanted his guests to know it. 'I am pining away, but remain very cheerful,' he wrote to Diana Collins a week or two later.[37] He was not just cheerful, but active: at his request, Anne began planning another trip for them, to China this time.

He published one last book in January 1988, quite painlessly: it was *Save the Beloved Country*, a volume of his speeches and articles on South Africa since 1965, edited by Hans Strydom and David Jones, inspired and seen through the presses by a literary agent, Frances Bond. Paton had nothing to do but write a brief introduction, and travel to Johannesburg for an enjoyable book-launching.[38]

And he made one last trip to the Drakensberg in March 1988 with Anne and Constance Stuart Larrabee, who was visiting South Africa; she drove with Paton to the scenes from *Cry, the Beloved Country* which she had first photographed in 1948, and also photographed him in front of the Pine Street house in Pietermaritzburg, in which he wrongly believed he had been born. They spent some pleasant days staying at Champagne Castle hotel in the mountains, where Larrabee took one of the last photographs of him.[39] Here it rained heavily, and Paton, making his way to the dining room from his luxurious rondavel, found his way blocked by a torrent across the path. Two Zulu waiters, very black in the darkness, appeared and, picking Paton and Larrabee up in their arms, carried them safely across.[40] Anne wished she had had her camera handy to capture this emblematic scene.

By February 1988 Paton was putting the finishing touches to the second volume of his autobiography, and was grateful to be nearing the end with it. 'My pace of working has become very slow and I do not think I shall write anything substantial any more,' he told Ursula Niebuhr.[41] He said as much in the moving final paragraph of the book: 'I shall not write anything more of any weight. I am grateful that life made it possible for me to pursue a writing career. I am now ready to go when I am called.'[42]

He did not have to wait long for the call. In March 1988 he found he could not swallow in the evenings: 'Too much whisky,' said Anne. After three days he could not swallow solids at any time of day, and had increasing difficulty ingesting liquids. Fifteen days after the first symptoms had manifested themselves he had a gastroscopy which revealed a tumour, closing the oesophagus and already too large to be removed surgically with any chance of success. An operation to force a tube through the tumour was undertaken, but failed.

Paton had always regretted not being able to tell his wife Dorrie that she was dying, and to talk through the matter with her. He and Anne, by contrast, had often discussed death, and agreed that they would tell each other what was coming if necessary. Anne now waited until he had begun to talk hopefully of eating his dinner after the failed operation, and gently told him that the operation had not succeeded, that he would never be able to swallow, and that he had a well-advanced cancer. She wept as she spoke, and he patted her hand comfortingly. He showed neither fear nor regret, but gratitude: 'Well, thank goodness I finished my autobiography,' he said.[43]

During the next few days, fed through a drip, he sank gradually towards death, his mind occasionally wandering. When Peter Brown came to see him he took fright: 'Brown, am I under arrest?' he asked. 'Not yet, old chap,' said Brown soothingly.[44] Anne tried to keep all but close friends and family away, for the news of his illness had been picked up by the papers, and the telephone at Botha's Hill never stopped ringing in these last days. Badgered for a statement on his health, she drew on a childhood memory of the death of George V in 1936. 'My husband's life is drawing peacefully to its close,' she said.

One of those who did manage to get through to see him in his private ward was Cathy Brubeck, who had got to know Paton when she was the Secretary of the Liberal Party in Pietermaritzburg in the late 1950s.[45] Jonathan Paton was staying with her in Durban, and, very much against Anne's orders, invited Cathy Brubeck to visit his father at St Augustine's hospital. Paton's eyes opened as she approached his bed, and he recognized her. 'Oh, Cathy,' he said, and his lips pursed for the kiss she gave him. Even on his deathbed he appreciated an attractive woman.

His sister, Dorrie Arbuthnot, came to see him, though she had never quite forgiven him for abandoning Christadelphianism. Even his autobiography, *Towards the Mountain*, had distressed her, with its claim that the holy mountain of Isaiah could not be attained in this life, when good Christadelphians knew that Christ would establish it in very fact at Jerusalem. She arrived to find two bishops at his bedside, the Anglican bishop Michael Nuttall (son of Paton's old friend Neville Nuttall), and the Catholic archbishop Denis Hurley. They were, she said, 'both talking what I thought was a lot of nonsense. Hurley told him that he was on his way to the mountain of the Lord. He said I can see you travelling there.'[46]

Pneumonia set in, and Anne, showing the medical staff the 'living will' Paton had signed some years before, refused to allow him to be given antibiotics. That night she returned to the hospital and found he had been sat bolt upright because of the fluid in his lungs, and he was distressed because no one answered when he called. Anne settled him down in his bed. 'I want death,' he said. 'It's coming soon,' she reassured him, and he relaxed and went to sleep. The next morning she found that he had been taken off the drip, his veins having collapsed. She decided to take him home. Jonathan Paton and his wife Margaret travelled with Paton in the ambulance, on 11 April 1988. 'Am I going home?' he asked. 'Yes, Dad,' said Jonathan, and Paton sighed with relief. It was just four weeks since his illness made itself felt.

He lost consciousness for the last time soon after arriving home, and sank steadily. Early in the morning of 12 April 1988, the day of his death, he seemed to be in pain, groaning, 'Anne, Anne',[47] and a doctor was called to administer a sedative. As Anne was to describe it, 'The hours passed, but that strong young heart went on beating. Alan lay peacefully on his side, one hand between his knees, the other up by his neck, as he had lain so often in sleep. Slowly his breathing grew quieter and quieter. Slowly the beating of his heart grew more gentle, and at 5.10 a.m. it stopped.'[48]

CONCLUSION

As soon as the news went out to the world, the phone at Botha's Hill began to ring, and it went on ringing apparently endlessly, day and night. Then the telegrams started, and piled up steadily on Anne's desk. After three days the letters began arriving, sacks full of them. There were tributes in the local paper and in papers across the world. Radio and television stations from Michigan to New South Wales rang anyone who had known Paton for tributes they could put on air. South African politicians, from the State President P. W. Botha to Chief Buthelezi, Colin Eglin, and Pat Poovalingam, paid tributes to Paton as a man 'who kept hope alive'.[1] The African National Congress was silent. Anne and Paton's sons were asked to write tributes; Jonathan completed Paton's last article, left unfinished at his death, and it was published in *Time*, as 'A Literary Remembrance', on 25 April 1988. In it Paton, anticipating his end, quotes several poems about death, including his favourite, Whitman's 'When Lilacs Last in the Dooryard Bloom'd'.

The body was cremated at the Stellawood crematorium on 16 April 1988, after a small funeral service in Highbury Chapel to which Anne invited few people other than Jonathan and Margaret, David, Maureen, and the Patons' five servants. The servants were very shocked to see that the coffin was just a plain wooden one with rope handles. Its simplicity was at Paton's request: he had once asked Anne to make him just such a one. Walter Atherstone officiated, as assistant parish priest and a great friend of Paton's, delivering a moving tribute. Paton's sisters were distressed at not being included,[2] but Anne, feeling the world pressing in from all sides, wanted to keep the ceremony small and intimate.

Following this private ceremony there were others, very much grander, befitting the passing of the public figure. There was a big memorial service in Pietermaritzburg, at which bishop Michael Nuttall spoke, and many of Paton's surviving school and university friends, including Reg Pearse and Vic Harrison, came together once more to remember him. There was a grand memorial service in St Mary's Cathedral, Johannesburg. And on 21 April 1988 there was a huge thanksgiving service in St George's Cathedral in Cape Town. There were joyful hymns, a reading of Paton's favourite passage from Isaiah 11, addresses by Helen Suzman, David Welsh, and Eddie Daniels, and the organist Barry Smith played the Kiev Kontakion which Paton had so loved. On 11 May 1988 there was an even larger service in the cathedral of St Peter and St Paul in Washington, DC, at which Bishop Cabell Tennis, who had been Paton's host in Seattle in 1976, gave the address. The congregation was sprinkled with ambassadors, and Paton's publisher, Charles Scribner III, rubbed shoulders with Chester Crocker of the State Department, both delivering tributes. And on 28 June 1988 there was a memorial service in St Paul's, London, where Paton had first preached in 1950. Here Peter Brown, Diana Collins, and the South African novelist Dan Jacobson delivered the tributes, David Steel read from Paton's writings, and Anne Paton, in England for her son's wedding, was present to meet the Liberal Party diaspora for the first and last time.

The last memorial ceremony was also the most intimate. For months after the cremation in April 1988 Paton's ashes had remained at the crematorium, for Anne could not decide how they should best be disposed of. After the cremation Bishop Nuttall had suggested a service of interment, but Anne felt she could not face another ceremony. Paton had once light-heartedly suggested that his ashes be scattered on his beloved croquet lawn. Anne considered for months before coming to a decision in October. On the first Sunday in October 1988, at the time when the Patons would normally have been in their local church, she took the little box of ashes down to the lawn which had been turned into an arboretum during the last few years. In the last months of his life Paton had had a path cut across it, a strange path leading towards the bed of clivias and azaleas under some large trees, but stopping short. Anne had asked him at the time why he was making this path which never reached its beautiful goal. 'It's not a path, it's a road,' he said, smiling, but she never got an answer to her question. Now she began to see that it had been for him an emblem of that road towards the holy mountain which we never reach in this life. She walked along the path scattering his ashes on the lawn on both sides, thinking as she did so of all the things she had loved him for.[3] It was a ceremony entirely befitting one side of Paton's nature. There is no memorial stone, plaque, or tablet to him in that garden today; he needs none, for the garden itself, beautifully preserved by its new owners,[4] is an eloquent memorial of its creator.

Under the terms of Paton's last will, which disposed of assets of R788,989, the royalties of *Cry, the Beloved Country*, much the greater part of his income, went into trust for his grandchildren. His sons got much less than they might have expected, an omission which was a perverse result of Paton's disapproval of David's home life. Having largely cut David out of his will he did much the same to Jonathan, apparently from a misguided sense of equity, an injustice which Jonathan felt bitterly. Anne was left the usufruct of the large house at Botha's Hill, but not enough money to keep it up. Accordingly she sold it, and she also sold Paton's papers and manuscripts, most of which, through the generosity of two of Paton's wealthy friends, went into the Brenthurst Library in Johannesburg and the archives of the Alan Paton Centre at Natal University Library in Pietermaritzburg.[5] This Centre, founded with a bequest from Paton to the university (though he did not tie the bequest to that or any other purpose), is a magnificent memorial to him, and easily the most comprehensive archive of the papers of any South African literary figure. Anne Paton presented the Centre with the entire contents of Paton's study, as well as his diplomas, hoods, and other personal items, and the *Sunday Tribune* newspaper made a large contribution of money. There are other memorials too: Kent School set up an Alan Paton Memorial Prize, and the Johannesburg *Sunday Times* established an annual R15,000 Alan Paton Prize for non-fiction writing on South Africa.

In 1993 Paton's family continued to thrive. David worked as a radiologist in Johannesburg. Jonathan was a much-loved teacher of literature at the University of the Witwatersrand. Anne Paton continued to live in the area where she and Paton learned to love each other, and though she kept alive his memory, she infused it with strong jets of realism. 'He wasn't a saint, you know,' she said after his death. 'He was difficult to live with. He depended on me totally. I ran his life.'[6]

It is true that Paton was no plaster saint, but he was a most faithful believer. Christianity was for him a religion in the fullest sense: it provided him with an all-embracing philosophy, a comprehensive guide to life and living. It was the linchpin of his existence. He tried, and sometimes failed, and always tried again, to live by it. He was, in fact, a rich and strange mixture of sinner and saint, of simplicity and complexity, pride and humility. It was the constant, lifelong struggle with himself that gave him the capacity to understand the weaknesses of others, and to forgive them when they wounded him. It was this struggle too that gave him his extraordinary capacity for growth and development. The internal tension is what infuses the best of his writing and what made him the outstanding novelist he was.

His was the full and fortunate life of a profoundly talented man, as his range of careers suggests: teacher, prison reformer, writer, and politician. As a writer he will be read as long as the English language is read, for *Cry, the Beloved Country* is a great novel. He produced no work of fiction to equal it, but he was not a one-book man. *Too Late the Phalarope* suffers only by comparison with its incomparable predecessor, and Paton's other writings, biographies, autobiographies, poems, short stories, volumes of political essays, are never less than good, and are often models of their kind even when they have fallen into the shadow of his other productions: *Instrument of Thy Peace*, for instance, is one of his least considered, but finest, achievements.

As a political leader his record is less clear. In a limited sense success, as political leaders tend to think of it, eluded him. The Liberal Party he helped to found and lead never came within reach of power, nor looked like doing so. It remained a minor player on a stage dominated by two great nationalist forces, and its efforts to mediate between them were greeted with suspicion and derision. When it was driven out of formal existence in 1968 it was not greatly lamented by those larger forces it had tried to divert from collision. Yet the achievement of Paton's Liberalism lay in the abiding value of the ideas and ideals it kept alive through a period of darkness and intolerance, and those ideals Paton continued to propagate patiently, faithfully, until the end of his life. His prescription for a sharing of power, his urging of constitutional talks between the opposed sides,[7] have proved since his death to have been uncannily prophetic. His analyses of the South African situation, often derided in his lifetime, have been proved remarkably accurate, and his abiding faith that his country would in time be able to work out its problems should remain an inspiration and a source of hope. He was a prophet, not just in the biblical sense, but in a sense that places him in the company of Thomas More, Locke, the framers of the American Constitution, and writers from Milton to Aleksandr Solzhenitzyn.

He maintained with all his power Judaeo-Christianity's unique affirmation of the worth and dignity of the individual. Running through all the richly varied fabric of his life is the unbroken thread of his commitment to what he saw as the moral course: he sought out (and believed he had found) the right way to live, which was to live for others. That way he pursued to the end.

His Christian creed was probably best summed up in practical essentials: to uphold human rights and dignity; to lift the downtrodden; and to promote a common society in opposition to the polarization of apartheid.

He glorified God in loving his fellows. He hated the power-hungry, exercised intelligence and independence, and had faith in the decency, tolerance, and humanity of the common man. He liked to quote the praise of Sir Robert Shirley, who 'did the best of things in the worst of times, and hoped them in the most calamitous',[8] and they were words that applied to himself. He showed immense courage, not least in never abandoning hope of a better world, not attainable in this life, but certain in the next for those who kept faith. His was a life lived to the full, and lived unstintingly. He fought the good fight to the end; he ran the race to the finish; he kept the faith.

Key to Abbreviations of Manuscript Collections

AL	In the possession of Mr Adrian Leftwich, York.
AP	In the possession of Mrs Anne Paton, Gillitts.
APC	Original or copy in the collection of The Alan Paton Centre, University of Natal, Pietermaritzburg.
B	In the Oppenheimer collection, Brenthurst Library, Johannesburg.
BD	In the possession of Mrs Bunny Duggan, Kirkland, Washington state.
BP	In the possession of Mrs Elizabeth A. Patterson, San Fransisco.
EC	In the possession of Professor Edward Callan, Kalamazoo, Michigan.
EF	In the possession of Mr Edmund Fuller, Chapel Hill, North Carolina.
FJP	In the possession of Father John Pesce, West Hartford, Connecticut.
GC	In the possession of Mrs Gertrude Cohn, London.
JP	In the possession of Mr Jonathan Paton, Johannesburg.
K	In the possession of Mrs Peg Kinnison, Washington.
Kent	In the archives of Kent School, Connecticut.
LC	In the manuscripts collection of the Library of Congress, Washington.
LR	In the possession of Mr Leslie Rubin, Santa Barbara, California.
MB	In the possession of Miss Mary Benson, London.
MD	In the possession of Mrs Mabel Dent, Pietermaritzburg.
MW	In the possession of Mrs Margaret Worthington, West Hartford, Connecticut.
PFA	In the possession of the writer.
ROP	In the possession of Mr Reginald Pearse, Winterton, Natal.
RR	In the possession of Mr Richard Robinow, Toronto, Canada.
RV	In the possession of Mr Roy C. Votaw, Santa Rosa, California.
SB	In the possession of Mrs Shirley Broad, Greenwich, UK.
Texas	In the Harry Ransom Humanities Research Center, University of Texas at Austin.
TH	In the possession of Bishop Trevor Huddleston, London.
TM	In the possession of Dr Anthony Morphet, Cape Town.
UCT	In the manuscripts collection of the University of Cape Town Library.
Wits	In the Hofmeyr Collection, University of the Witwatersrand Library.
WL	In the possession of the Weill-Lenya Foundation, New York.

Notes

1. BEGINNING

1. *Towards the Mountain*, 3.
2. Ibid. 6.
3. Ibid. 15.
4. To some this grace and dignity have seemed pretension: Tom Sharpe, in his satirical novels set in Pietermaritzburg, finds it an easy target.
5. *Towards the Mountain*, 4–5.
6. Constance Stewart Larrabee, for instance, photographed him there just before his death in 1988.
7. Mrs Eunice Arbuthnot, Paton's sister, believes her parents may not have moved out of the Ridleys' home until just after the birth of Alan: quoted in unpublished letter to the writer from Mrs Joicelyn Leslie-Smith, 17 March 1993.
8. She was actually christened Eunice, but was always known as Dorrie; to avoid confusion with her mother Eunice, after whom she was named, I shall call her Dorrie in the rest of this narrative.
9. Unpublished memoir entitled *Alan: Journey Begun* by Mrs Eunice ('Dorrie') Arbuthnot: PFA.
10. Unpublished and unfinished play, written about 1948: APC.
11. Unpublished letter, Paton to Michael Black, 10 February 1986: APC.
12. Unpublished letter, Alan Paton to Jonathan Paton, 19 March 1983: JP.
13. Scottish Record Office, Edinburgh, Register of Births and Deaths. Anderston was notorious for overcrowding and disease in the nineteenth century, and was the subject of several commissions aimed at improving conditions there. Dennistoun at this time was being 'gentrified', but was still a mixed area with a wide socio-economic range of population.
14. The name is so spelled in the records of James Paton's birth in Glasgow.
15. The daughters' given names were Grace Wilson and Elizabeth Stewart. There was also a fourth child, Robert, who seems to have died in childhood. Unpublished letter, Alan Paton to Jonathan Paton, 19 March 1983: JP.
16. This dating is based on the evidence of Mrs Eunice ('Dorrie') Arbuthnot (interview with her, Pietermaritzburg, 31 January 1990); Alan Paton believed his father came to South Africa only in 1901—*Towards the Mountain*, 7.
17. Interview with Eunice ('Dorrie') Arbuthnot, Pietermaritzburg, 31 January 1990.
18. Ibid.
19. Published in *The Eisteddfod Poetry Book* (Cape Town, Maskew Miller, December 1921), 51.

20. Interview with Eunice ('Dorrie') Arbuthnot, Pietermaritzburg, 31 January 1990.
21. *Towards the Mountain*, 16–17, and photographs of James Paton in the possession of Jonathan Paton.
22. *Towards the Mountain*, 7.
23. Ibid. 14.
24. Ibid.
25. Unpublished memoir, 'Alan: Journey Begun' by Mrs Eunice ('Dorrie') Arbuthnot: PFA.
26. Interview with Eunice ('Dorrie') Arbuthnot, Pietermaritzburg, 31 January 1990.
27. Interviews with Reg Pearse and Victor Harrison.
28. Interview with Eunice ('Dorrie') Arbuthnot, Pietermaritzburg, 31 January 1990.
29. Ibid.
30. Ibid.
31. *Towards the Mountain*, 38.
32. Interview with Eunice ('Dorrie') Arbuthnot, Pietermaritzburg, 31 January 1990.
33. Ibid.
34. Ibid.
35. *Towards the Mountain*, 9.
36. Interview with Eunice ('Dorrie') Arbuthnot, Pietermaritzburg, 31 January 1990.
37. *Towards the Mountain*, 10.
38. Interview with Eunice ('Dorrie') Arbuthnot, Pietermaritzburg, 31 January 1990.
39. *Towards the Mountain*, 19.
40. Interview with Eunice ('Dorrie') Arbuthnot, Pietermaritzburg, 31 January 1990.
41. *Towards the Mountain*, 14.
42. Ibid. 19.
43. Ibid. 14.

2. CHILDHOOD AND YOUTH

1. This is the height listed in his passport, now in APC, though it may have been an overestimate: the passport is wrong about other details, such as the colour of his eyes, which it gives as grey.
2. Unpublished memoir, 'Alan: Journey Begun' by Mrs Eunice ('Dorrie') Arbuthnot: PFA.
3. I am indebted for this information to Mrs Joicelyn Leslie-Smith, who interviewed Mrs Dorrie Arbuthnot in Pietermaritzburg on 18 June and 26 June 1992 for me.

4. Victor Cuthbert Harrison was born on 22 October 1901, the son of an architect and civil engineer. He attended the same schools as Paton, but read law at Natal University College. Subsequently, and for the rest of his long life, he practised law in Pietermaritzburg, and maintained his friendship with Paton until the latter's death. He was still thriving, and still practising law, in 1991.
5. Interview with Victor Harrison, Pietermaritzburg, 30 January 1990.
6. Untitled novel, referred to hereafter as *John Henry Dane*, ch. 2: APC.
7. *Towards the Mountain*, 22.
8. Ibid.
9. Pages 21–3.
10. *John Henry Dane*, ch. 4: APC.
11. *Towards the Mountain*, 23.
12. It is worth reminding northern-hemisphere readers at this point that in South Africa the academic year, both for schools and for universities, coincides with the calendar year, starting in February and ending in November.
13. *Towards the Mountain*, 24.
14. Ibid.
15. Unpublished memoir, 'Alan: Journey Begun' by Mrs Eunice ('Dorrie') Arbuthnot: PFA.
16. Ibid.
17. Ibid.
18. I am indebted to Professor John Honey of Osaka International University for this information.
19. Interview with Victor Harrison, Pietermaritzburg, 30 January 1990.
20. It is now 16 Echo Road, and surrounded on all sides by a housing estate, but it can still be glimpsed from Bulwer Street. I am indebted to Mrs Joicelyn Leslie-Smith and Professor Colin Gardner for this information.
21. *Towards the Mountain*, 28.
22. Ibid. 29.
23. Ibid. 30.
24. Unpublished memoir, 'Alan: Journey Begun' by Mrs Eunice ('Dorrie') Arbuthnot: PFA.
25. It is now in APC.
26. Unpublished memoir, 'Alan: Journey Begun' by Mrs Eunice ('Dorrie') Arbuthnot: PFA.
27. Ibid.
28. *Too Late the Phalarope*, ch. 14.
29. *Towards the Mountain*, 37.
30. *John Henry Dane*, ch. 3: APC.
31. He uses the term 'Home' for Britain in letters to R. O. Pearse well into the 1920s.
32. Fischer, Reid, and Morris.
33. From a common Latin tag, 'Pro aris et focis pugnare', meaning literally 'to

fight for altars and fires', or all that one holds dear. I am grateful to Professor Mary Chan of the University of New South Wales for this information.

34. *Towards the Mountain*, 31.
35. Simon Haw and Richard Frame, *For Hearth and Home: The Story of Maritzburg College*, 214.
36. Interview with Mrs Eunice ('Dorrie') Arbuthnot, Pietermaritzburg, 31 January 1990.
37. Ibid.
38. *Towards the Mountain*, 31.
39. Ibid. 32.
40. Ibid.
41. Ibid. 25.
42. Ibid. 32.
43. Ibid. 31.
44. Ibid. 38.
45. Interview with Victor Harrison, Pietermaritzburg, 30 January 1990.
46. *Towards the Mountain*, 40.
47. Ibid. 42.
48. PFA, presented to me by Mrs Eunice ('Dorrie') Arbuthnot.
49. *Towards the Mountain*, 42.
50. When one of his staff at Diepkloof proved to be homosexual, Paton was the one staff member who would have the man in the house.
51. *Towards the Mountain*, 42.
52. Interview with Mrs Eunice ('Dorrie') Arbuthnot, Pietermaritzburg, 31 January 1990.
53. *Towards the Mountain*, 14.
54. Interview with Mrs Eunice ('Dorrie') Arbuthnot, Pietermaritzburg, 31 January 1990.
55. Ibid.
56. *Towards the Mountain*, 33.
57. *Towards the Mountain*, 33–4.
58. Simon Haw and Richard Frame, *For Hearth and Home: The Story of Maritzburg College*, p. 192.
59. Ibid.
60. Interview with Victor Harrison, Pietermaritzburg, 30 January 1990.
61. *Towards the Mountain*, 44.
62. Interview with Victor Harrison, Pietermaritzburg, 30 January 1990.
63. I am indebted for this information to Professor John Honey of Osaka International University.
64. *Towards the Mountain*, 43.
65. Simon Haw and Richard Frame, *For Hearth and Home: The Story of Maritzburg College*, p. 187.
66. Interview with Victor Harrison, Pietermaritzburg, 30 January 1990.
67. Ibid.
68. Ibid.

69. Unpublished memoir, 'Alan: Journey Begun' by Mrs Eunice ('Dorrie') Arbuthnot: PFA.
70. *John Henry Dane*, ch. 4: APC.
71. Interview with Mrs Eunice ('Dorrie') Arbuthnot, Pietermaritzburg, 31 January 1990.
72. Interview with Victor Harrison, Pietermaritzburg, 30 January 1990.
73. *Towards the Mountain*, 45.
74. Ibid.

3. NATAL UNIVERSITY COLLEGE

1. Unpublished memoir entitled 'Alan: Journey Begun' by Mrs Eunice ('Dorrie') Arbuthnot: PFA.
2. *Towards the Mountain*, 57.
3. Ibid. 58.
4. Ibid.
5. Ibid. 57.
6. Simon Haw and Richard Frame, *For Hearth and Home: The Story of Maritzburg College*, 212–13.
7. Interview with Victor Harrison, Pietermaritzburg, 30 January 1990.
8. In 1914 Paton had been one of 78 pupils admitted to Maritzburg College, so that the total school numbers in 6 forms must have been close to 500.
9. Interview with Victor Harrison, Pietermaritzburg, 30 January 1990.
10. Interview with Mabel Dent, Pietermaritzburg, 21 June 1991.
11. *Towards the Mountain*, 60.
12. Ibid. 59.
13. Interview with Mabel Dent, Pietermaritzburg, 21 June 1991.
14. Unpublished letter, Reg Pearse to the writer, 21 May 1992.
15. *Towards the Mountain*, 63. He alludes here to Neil Armstrong's words on first stepping onto the surface of the Moon.
16. Paton describes him in this way in the obituary he wrote for the *Natal Witness* in October 1957: SB.
17. Interview with Mr Reg Pearse, Emkhizweni, Drakensberg, 7 February 1990.
18. Ibid.
19. Unpublished manuscript in the handwriting of C. J. Armitage: SB.
20. Interview with Victor Harrison, Pietermaritzburg, 30 January 1990.
21. *Towards the Mountain*, 64.
22. Ibid. 70.
23. This seems to have meant being a member of a reading group or circle.
24. Unpublished letter, Paton to Reginald Pearse, 20 March 1923: PFA.
25. In an unpublished novel written in 1922, one of the characters refers to having read that Eliot is the greatest modern poet, but another character replies, 'I find him rather dry'. *Ship of Truth*, verso of MS, 16: APC.
26. Unpublished letter, Paton to Pearse, 5 December 1923: PFA.
27. Ibid. His opinion of Masefield he was to repeat enthusiastically in his

fragmentary unpublished novel *Ship of Truth*, MS, 202–3: APC.

28. Unpublished manuscript, in the handwriting of C. J. Armitage: SB.
29. They are now in the possession of his daughter, Mrs Shirley Broad of London.
30. *Natal Witness* in the 'Topics' column published 'The Strange Case of Mr Methuselah Brown', 29 November 1919, and 'NUC Reduit' on 28 February 1920; also 'The Battle of Mussonville' on 6 March 1920, and 'Ye Brief Accounte of Ye Strange Visions seen on Ye Towne Hill', 24 April 1920. I am grateful to Mrs Shirley Broad for bringing these items to my attention.
31. The NUC motto is 'Stella Aurorae'.
32. Interview with Mrs Eunice ('Dorrie') Arbuthnot, Pietermaritzburg, 31 January 1990.
33. *Natal University College Magazine*, 7 (October 1922), 19–22.
34. Ibid. 9 (October 1923), 63–6.
35. The MS is dated 'Oct 1922' on p. 41, 'Dec 1922' on p. 63, 'March 1923' on p. 200, 'Nov 1923' on p. 210, 'Dec 1923' on p. 226, and 'Dec 25' on p. 263.
36. MS, 83–4, *Ship of Truth*: APC.
37. MS, 89, *Ship of Truth*: APC.
38. Unpublished letter, Dorothy Durose to the writer, 21 April 1992.

4. DISCOVERING THE WORLD

1. Years later, while Principal of Diepkloof, he would express envy of a tall subordinate, I. Z. Engelbrecht: *Towards the Mountain*, 178.
2. Unpublished memoir, 'Alan: Journey Begun' by Mrs Eunice ('Dorrie') Arbuthnot: PFA.
3. Ibid.
4. Interview with Leif Egeland, Johannesburg, 2 February 1990.
5. Interview with Victor Harrison, Pietermaritzburg, 30 January 1990.
6. Interview with Mrs Eunice ('Dorrie') Arbuthnot, Pietermaritzburg, 31 January 1990.
7. *Natal University College Magazine*, 7 (Winter term 1922), 41.
8. Unpublished letter, Mrs Dorothy Durose to the writer, 21 April 1992: PFA.
9. *Natal University College Magazine*, 7 (Winter term 1922), 41.
10. Unpublished letter, Mrs Dorothy Durose to the writer, 16 June 1992: PFA.
11. Unpublished memoir, 'Alan: Journey Begun' by Mrs Eunice ('Dorrie') Arbuthnot: PFA.
12. Unpublished letter, Mrs Dorothy Durose to the writer, 16 June 1992: PFA.
13. Ibid., 21 April 1992: PFA.
14. Ibid.
15. *Towards the Mountain*, 68.
16. Ibid.
17. Unpublished letter, 1 February 1923: PFA.
18. Included in unpublished letter to Reginald Pearse, 5 December 1923, and

first published in *Natal University College Magazine*, 10 (June 1924), 26.
19. Unpublished letter, R. O. Pearse to the writer, 21 May 1992: PFA.
20. *Towards the Mountain*, 67.
21. Interview with Leif Egeland, Johannesburg, 2 February 1990.
22. Interview with Victor Harrison, Pietermaritzburg, 30 January 1990.
23. Ibid.
24. Interview with Leif Egeland, Johannesburg, 2 February 1990.
25. *Towards the Mountain*, 67.
26. Ibid.
27. Ibid. 40.
28. Unpublished letter, 21 May 1992, from a correspondent who asks not to be named, to the writer. The incident seems to have taken place towards the end of 1919, since Walmsley House closed down at the end of that year.
29. This rule was affirmed by Mrs 'Snib' Sweeney, who boarded at Walmsley House in 1919: unpublished letter from a correspondent who asks not to be named to the writer, 2 April 1993: PFA.
30. Unpublished letter, 21 May 1992, from a correspondent who asks not to be named, to the writer.
31. *Towards the Mountain*, 43.
32. Interview with Victor Harrison, Pietermaritzburg, 30 January 1990.
33. *Towards the Mountain*, 67.
34. Unpublished letter, 3 January 1922: PFA.
35. Ibid.
36. Report on the Imperial Conference by Paton, *Natal University College Magazine*, 11 (October 1924), 10.
37. Unpublished letter, 9 January 1925.
38. *Towards the Mountain*, 79.
39. Ibid. 79–80.
40. Report on the Imperial Conference by Paton, *Natal University College Magazine*, 11 (October 1924), 12.
41. Ibid.
42. *Towards the Mountain*, 81.
43. Unpublished letter, 9 January 1925: PFA.
44. Report on the Imperial Conference by Paton, *Natal University College Magazine*, 11 (October 1924), 11.
45. *Towards the Mountain*, 81–2.
46. Publicity document prepared for Scribner's, undated, but probably 1948: 'Biographical Details of Alan Paton: News of Scribner Books and Authors': APC. The piece is unsigned, but can only have been written by Paton or Pearse, and various stylistic elements suggest Pearse rather than Paton.
47. Unpublished account of the walks by Reginald Pearse, entitled 'Three Men and a Handcart': PFA.
48. Ibid.
49. Ibid.

50. Ibid., and *Towards the Mountain*, 68.
51. Unpublished account of the walks by Reginald Pearse, entitled 'Three Men and a Handcart': PFA.
52. *Towards the Mountain*, 68.
53. Interview with Anne Paton, Gillitts, 29 January 1990.
54. Publicity document prepared for Scribner's, undated, but probably 1948: 'Biographical Details of Alan Paton: News of Scribner Books and Authors': APC. The piece is unsigned, but can only have been written by Paton or Pearse, and various stylistic elements suggest Pearse rather than Paton.
55. *Towards the Mountain*, 69.
56. Unpublished letter, 3 January 1922: PFA.
57. Unpublished letter, 16 March 1923: PFA. He repeats his gratitude to Pearse in another unpublished letter, 20 March 1923: 'If ever you look back on your life sorrowfully, & wonder what good you ever did, write down on the balance side what you did for me last July.'
58. Unpublished essay, 'God in Modern Thought', probably written in 1934: APC.
59. BBC interview between Bernard Levin and Alan Paton, 9 March 1981: BBC Written Archives Centre, Reading.
60. *Towards the Mountain*, 76.
61. Interview with Victor Harrison, Pietermaritzburg, 30 January 1990.
62. Unpublished letter, Paton to Pearse, 9 January 1925: 'I have been transferred to Ixopo, & am not too pleased about it, but am determined to make the best of it; Buss [the Ixopo headmaster] and I are rather friendly, which is something.' PFA.
63. Unpublished letter, 1 February 1923: PFA.

5. IXOPO

1. It was still called Stuartstown as late as 1927: D. E. Johanson, *Looking Down the Years: A History of the Ixopo School, 1895–1964*, 75. 'Ixopo', the name of the local river, is onomatopoeic, imitative of the sound of cattle plashing through marshy ground. I am indebted for this information to Professor Trevor Cope of Sydney.
2. Quoted in D. E. Johanson, *Looking Down the Years: A History of the Ixopo School, 1895–1964*, 24–5.
3. *Towards the Mountain*, 70.
4. Unpublished letter, 5 December 1923.
5. *Towards the Mountain*, 84.
6. D. E. Johanson, *Looking Down the Years: A History of the Ixopo School, 1895–1964*, 17.
7. Ibid. 78.
8. *Towards the Mountain*, 86–7.
9. Ibid. 87.
10. Ibid.

11. Interview with Harry Usher of Highflats, Natal, conducted for me by Andrew Campbell of Ixopo: letter to the writer from Andrew Campbell, 5 August 1990: PFA. Of the pupils whom I managed to contact, Usher had the most bitter memories of Paton.
12. Unpublished letter, Harry Usher to the writer, 20 September 1992: PFA.
13. Interview with Joe Kirk, Ixopo, 28 January 1990.
14. This ordinance of the Education Department limited even housemasters to two strokes only. But it was (one might say) honoured on the breech rather than in the observance.
15. Telephone interview with Mrs Doreen Hulley, quoting her deceased husband, Clifford Hulley, one of Paton's pupils at Ixopo: 21 January 1992.
16. Unpublished letter, Harry Usher to the writer, 20 September 1992: PFA.
17. D. E. Johanson, *Looking Down the Years: A History of the Ixopo School, 1895–1964*, 73.
18. Interview with Harry Usher of Highflats, Natal, conducted for me by Andrew Campbell of Ixopo: letter to the writer from Andrew Campbell, 5 August 1990: PFA. Also unpublished letter, Harry Usher to the writer, 20 September 1992: PFA.
19. Unpublished letter, first page missing and therefore impossible to date exactly (mid-1925?): PFA.
20. Unpublished letter, Harry Usher to the writer, 20 September 1992: PFA.
21. Interview with Joe Kirk, Ixopo, 28 January 1990.
22. *Towards the Mountain*, 87.
23. In the photographs of himself among the senior boys of Maritzburg College in 1930, he is dwarfed by several of his pupils.
24. D. E. Johanson, *Looking Down the Years: A History of the Ixopo School, 1895–1964*, 72–3: 'How happy we were down there [in the "Hospital"] and how splendidly the boys behaved! The prefects were in charge when I was prevented from taking duty and how well they maintained discipline!'
25. Unpublished MS, 46: APC.
26. Unpublished letter, Harry Usher to the writer, 20 September 1992: PFA.
27. Ibid.
28. Ibid.
29. I am indebted for this information to Professor Douglas Irvine of the University of Natal. Durban High School shared this reputation: Professor Irvine records getting 36 cuts there in a single period, and remembers the use by masters and prefects of canes, cricket stumps, and even cricket bats. Unpublished annotation to the MS of this book: PFA.
30. Unpublished letter, first page missing and therefore impossible to date exactly (mid-1925?): PFA.
31. D. E. Johanson, *Looking Down the Years: A History of the Ixopo School, 1895–1964*, 75.
32. Ibid. 137.
33. Ibid.
34. Ibid.

35. *Towards the Mountain*, 134.
36. In *Towards the Mountain*, 134, Paton mistakenly says the first camp was in 1925.
37. Interview with Reginald Pearse, Natal, 7 February 1990.
38. Interview with Shirley Broad, London, 1 May 1991.
39. Unpublished letter, first page missing and therefore impossible to date exactly (mid-1925?): PFA.
40. Ibid.
41. *Towards the Mountain*, 91.
42. Ibid. 88–9.
43. Her son Jonathan was to remember, as an example, her liking for jokes about urination: 'The Yangtse Kiang [Yellow River] is flooding,' she would say when Paton used the chamber-pot in their bedroom. Interview with Jonathan Paton, Sydney, 22 June 1993.
44. *Towards the Mountain*, 89.
45. Interview with Harry Usher of Highflats, Natal, conducted for me by Andrew Campbell of Ixopo: letter to the writer from Andrew Campbell, 5 August 1990: PFA.
46. *For You Departed* (also published as *Kontakion For You Departed*), 5.
47. *Towards the Mountain*, 89.
48. Ibid. 90.
49. The present owners of Morningview possess a contemporary photograph of the house, which has now (1992) been remodelled and reroofed with grey tiles that change its appearance considerably.
50. *For You Departed* (also published as *Kontakion For You Departed*), 9.
51. Ibid. 11.
52. *Too Late the Phalarope* ch. 12, 82.
53. *For You Departed* (also published as *Kontakion For You Departed*), 11–12.
54. 'You were chaste . . . I shall not easily forget the times when you showed your affection. They were mostly times when I had been away, perhaps for many months.' *For You Departed* (also published as *Kontakion For You Departed*), 81.
55. Interview with Anne Paton, Gillitts, 29 January 1990.
56. *For You Departed* (also published as *Kontakion For You Departed*), 21.
57. Ibid. 22.
58. *Towards the Mountain*, 94.
59. *For You Departed* (also published as *Kontakion For You Departed*), 34.
60. Ibid. 23.
61. Paton could not recall the date exactly in later years, and I have not been able to establish it from documentary sources.
62. *For You Departed* (also published as *Kontakion For You Departed*), 47.
63. Ibid.
64. Ibid. 17.
65. Ibid.

66. Ibid. 47.
67. Interview with Eunice ('Dorrie') Arbuthnot, Pietermaritzburg, 31 January 1990.
68. *For You Departed* (also published as *Kontakion For You Departed*), 51.
69. *Towards the Mountain*, 102.
70. Unpublished letter, Harry Usher to the writer, 20 September 1992: PFA.
71. *For You Departed* (also published as *Kontakion For You Departed*), 68.
72. Paton's second wife, Anne, considers it more likely that Dorrie destroyed the letters herself. Unpublished letter, Anne Paton to the writer, 16 March 1993: PFA.
73. *For You Departed* (also published as *Kontakion For You Departed*), 75.
74. Ibid. 68.
75. *Towards the Mountain*, 102.
76. *For You Departed* (also published as *Kontakion For You Departed*), 26.
77. *Towards the Mountain*, 95.
78. *For You Departed* (also published as *Kontakion For You Departed*), 70.
79. Paton was to claim that these were the first-ever candidates from Ixopo School; but the school history reveals that the first matriculant from Ixopo, M. Bailey, sat the examination in 1920, and that another eight had taken the examination before Paton arrived in 1924. All appear to have failed. D. E. Johanson, *Looking Down the Years: A History of the Ixopo School, 1895–1964*, 177.
80. *For You Departed* (also published as *Kontakion For You Departed*), 69.
81. Ibid.
82. Ibid.
83. *Towards the Mountain*, 102.
84. They signed an ante-nuptial contract, meaning that for legal purposes their possessions were not held in community of property. The priest was Alan Earp Jones or James.
85. These details are given in a letter from Pearse to his fiancée (later his wife), 2 July 1928: ROP.
86. Ibid.
87. *Towards the Mountain*, 108.
88. *For You Departed* (also published as *Kontakion For You Departed*), 77.
89. Interview with Jonathan Paton, Johannesburg, 17 January 1990.
90. Unpublished letter, 1 November 1960: APC.
91. *For You Departed* (also published as *Kontakion For You Departed*), 78.
92. 'The Tributary Widens', written in November 1984 after seeing another great waterfall, the Iguazu Falls in South America.

6. MARITZBURG COLLEGE

1. Unpublished letter, Paton to Hofmeyr, 21 April 1935: 'if I saw a chance of a big school of my own before I turn forty, I would stay [in the education system].' Wits.

2. Unpublished letter, Paton to J. H. Hofmeyr, 7 August 1931: Wits.
3. Unpublished letter, 27 May 1923: PFA.
4. Shirley Broad, interviewed in London, 1 May 1991.
5. Unpublished: APC.
6. Interview with Victor Harrison, Pietermaritzburg, 30 June 1990.
7. *Towards the Mountain*, 111.
8. Ibid. 90.
9. Interview with Deryck Franklin, Sydney, March 1990.
10. Letter from Creina Alcock, wife of Neil Alcock, to Paton, undated: APC.
11. S. Haw and R. Frame, *For Hearth and Home: The Story of Maritzburg College, 1863–1988*, 257.
12. *Towards the Mountain*, 114–15.
13. Unpublished letter, Hofmeyr to Pearse, 31 August 1932: 'Sometimes I am disposed to worry about the future of the camps. You and Alan have meant so much to them up to now—but we are all getting older, and one doesn't quite see who are going to carry on when the time comes that you feel you can no longer do as much.' Wits.
14. *Towards the Mountain*, 113.
15. *Natal University College Magazine*, 20 (1929), 97: signed 'ASP'.
16. Written in Pietermaritzburg, 22 March 1931: APC. At first entitled 'Retreat', it was published as 'The Hermit' in *Natal University College Magazine* (May 1931) and republished in *Knocking on the Door*, 8–9.
17. In *Towards the Mountain*, 106, he says he helped to launch Toc H in Ixopo in 1926, but this must be a misprint; other passages (103 and 111) make it clear that 1928 was the year intended.
18. *Towards the Mountain*, 104.
19. Ibid. 105–6.
20. This hostel work is referred to in his application letter for Wardenship of a reformatory: 21 April 1935: APC.
21. *For You Departed* (also published as *Kontakion For You Departed*), 81.
22. Ibid. 86–7.
23. *Natal University College Magazine*, 24 (1931), 22: signed 'ASP'.
24. Written on 16 February 1931: APC. Previously unpublished.
25. Written on 13 March 1931: APC.
26. This fragmentary essay, beginning 'Nog weer, om perde te neem . . .' is on the verso of the MS version of 'School' in APC, and can therefore be dated 1928 with confidence.
27. *Towards the Mountain*, 118.
28. Quoted in Leif Egeland, *Bridges of Understanding*, 106.
29. *Towards the Mountain*, 135.
30. *Journey Continued*, 13.
31. *Hofmeyr*, 136–7.
32. Unpublished letter, 7 August 1931: Wits.
33. Ibid.
34. Unpublished letter, Paton to Hofmeyr, 7 August 1931: Wits.

35. Interview with David Paton, Johannesburg, 3 February 1990.
36. *Maritzburg News*, 2 July 1930. 'Death of Mr J. Paton'.
37. Eunice Paton to Mrs Hutchinson, 29 June 1930, a letter Paton preserved in his copy of *The Churchman's Pocket Book and Diary, 1966*: APC.
38. Interview with Eunice ('Dorrie') Arbuthnot, Pietermaritzburg, 31 January 1990.
39. Victor Harrison, a member of the legal profession and therefore better informed on deaths that were the subject of inquest than most of the population, continued to believe all his life that James Paton's death had been a muti murder. Interview with him, Pietermaritzburg, 30 June 1990. Muti murders, of which children are usually the victims, are not uncommon in rural areas of Southern Africa even in the 1990s.
40. Interview with Eunice ('Dorrie') Arbuthnot, Pietermaritzburg, 31 January 1990.
41. *Maritzburg News*, 2 July 1930, 'Death of Mr J. Paton'.
42. Ibid.
43. *Towards the Mountain*, 13.
44. Ibid.
45. Ibid.
46. Ibid.
47. This by-law was enforced with varying degrees of strictness, and was simply abandoned on occasions when large numbers of Zulus came in to Durban for some special event, such as a funeral like that of the poet Roy Campbell's father in 1926.
48. Interview with Victor Harrison, Pietermaritzburg, 30 June 1990.
49. *Towards the Mountain*, 123.

7. REVALUATION

1. *Towards the Mountain*, 125.
2. *Natal University College Magazine*, 26 (May 1932), 34; reprinted in *Knocking on the Door*, 12. The MS is in APC.
3. Unpublished letter, Paton to Edward Callan, undated [18–21 March 1966]: EC.
4. One page bears the date 18 May 1930: APC.
5. Unpublished letter, Paton to Hofmeyr, 14 August 1932: Wits.
6. The publisher was Ernest Benn, London.
7. *General Botha, the Career and the Man*, (London: Constable, 1916).
8. Unpublished letter, Paton to Hofmeyr, 14 August 1932: Wits. Bowen's novel was published in 1931 by Collins.
9. Unpublished letter, Paton to Hofmeyr, 14 August 1932: Wits.
10. *Louis Botha*, Act 2: APC.
11. Ibid. Act 4: APC.
12. Ibid.
13. Unpublished letter, Paton to Hofmeyr, 14 August 1932: Wits.

14. Unpublished letter, Hofmeyr to Mrs Lewis Casson (later Dame Sybil Thorndike) 4 April 1933: Wits.
15. Report of the Repertory Play-Reading Society, Johannesburg, undated [12 December 1935], accompanying letter from O. V. Pay to Paton: APC.
16. Unpublished letter, Paton to Hofmeyr, 4 March 1934: Wits.
17. *Towards the Mountain*, 125.
18. Paton gave the date of his joining the Institute of Race Relations in *Apartheid and the Archbishop: The Life and Times of Geoffrey Clayton*, 117.
19. Unpublished letter, 4 March 1934: Wits.
20. Quoted in Leif Egeland, *Bridges of Understanding*, 77.
21. Unpublished poem, written 6 May 1931: APC.
22. Unpublished letter, 4 March 1934: Wits.
23. *Towards the Mountain*, 128.
24. Ibid. 129.
25. Unpublished letter, 14 June 1934: Wits.
26. *Towards the Mountain*, 129.
27. Unpublished letter, 14 June 1934: Wits.
28. D. E. Johanson, *Looking Down the Years: A History of the Ixopo School, 1895–1964*, 146–7.
29. 'Yet I do not like beer; I think it tastes foul.' Unpublished MS, 'Beer', APC.
30. TS at APC dated P[ark] R[ynie] 25 September 1934: published in *Natal University College Magazine*, (October 1934); reprinted in *Knocking on the Door*, 14.
31. Unpublished MS, 'Convalescence', APC.
32. 'Secret for Seven', unpublished MS 2: APC.
33. Ibid. 3.
34. Ibid. 2.
35. Unpublished letter, 14 June 1934: Wits.
36. *Towards the Mountain*, 130.
37. Quoted in 'Notes & Queries', *Natalia*, 19 (December 1989), 71.
38. According to Linda Chisholm, 'Education, Punishment and the Contradictions of Penal Reform', (*Journal of Southern African Studies*, 17/1 (March 1991), 25), the reformatories were Tokai, for white boys, Porter, for Coloured boys, Estcourt, for girls of all races, Houtpoort, for white boys, and Diepkloof, for Africans and some Indians. Of these Diepkloof was the largest in 1935.
39. Unpublished letter: Wits.
40. Unpublished letter, Paton to Hofmeyr, 21 April 1935: Wits.
41. Draft of letter of application, APC. I have not been able to find the actual application letter in the Education Department's files, but this draft is a clean copy, and was almost certainly preserved by Paton as an exact copy of what he had sent.
42. In *Towards the Mountain*, 138, he says he was paid £620 p. a. as a teacher, but this is one of the rare occasions on which his memory had let him down.

43. Draft of letter of application, APC.
44. Unpublished letter, 21 April 1935: Wits.
45. Unpublished letter, 24 April 1935: Wits.
46. *For You Departed* (also published as *Kontakion For You Departed*), 92–3.
47. 'Witty, even excoriating, yet tender and sincere: clever and eloquent, yet wise—Alan Paton perforce resigns his post as Natal Area Pilot upon becoming, on July 1st, the head of Diepkloof Reformatory, near Johannesburg—and it is safe to say that there is not a Natal member without (in the heart's corner) a sense of loneliness and loss.' *Compass*, June 1935.
48. *For You Departed* (also published as *Kontakion For You Departed*), 93.
49. Ibid. 104.

8 DIEPKLOOF

1. *For You Departed* (also published as *Kontakion For You Departed*), 93.
2. Interview with David Paton, Johannesburg, 3 February 1990.
3. *Towards the Mountain*, 167.
4. Ibid. 168.
5. Ibid. 168–9.
6. Ibid. 169.
7. Report on the Diepkloof Reformatory by C. N. Kempff, Chief Clerk, 6 February 1935: State Archives, Pretoria, file UOD, 1460, E/55/6/4.
8. Paton gives the figure as 400, but the official reports for 1936 make it plain that he exaggerated slightly. The average number of inmates during Paton's first year was 370. Inmate numbers rose sharply shortly after his arrival, however; there were 432 inmates in 1937, 531 in 1938, and 620 by August 1939. In the first years of the war they fell again, to 372 in March 1941, and Paton asked that magistrates be urged to send pupils to the reformatory. Annual reports, Diepkloof, 1936–41: State Archives, Pretoria, file UOD, Vol. 1700, E9/4/6.
9. Report on the Diepkloof Reformatory by C. N. Kempff, Chief Clerk, 6 February 1935: State Archives, Pretoria, file UOD, 1460, E/55/6/4.
10. According to the Inspector of Institutions report for February 1935, 'a heavy percentage of former inmates has run away from employers'. Ibid.
11. Report on the Diepkloof Reformatory by Miss Chattey, Inspectress of Domestic Science, 1 May 1935: ibid.
12. Inspection report by Miss Chattey, 1 May 1935: ibid.
13. Report on the Diepkloof Reformatory by Miss Chattey, Inspectress of Domestic Science, 1 May 1935: ibid.
14. Paton requested permission for the bread ration in a letter to the Secretary for Education on 13 July 1935: ibid.
15. The urban black population increased between 1921 and 1936 from 587,000 to 1,142,000: P. Kallaway (ed.), *Apartheid and Education: The Education of Black South Africans* (Johannesburg, 1984), 225.

16. *Towards the Mountain*, 145.
17. Report on the Diepkloof Reformatory by Miss Chattey, Inspectress of Domestic Science, 1 May 1935: State Archives, Pretoria, file UOD, 1460, E/55/6/4.
18. Report on the Diepkloof Reformatory by Z. Martins, Chief Clerk, April 1935: ibid.
19. Ibid.
20. Inspection report by Miss Chattey, 1 May 1935: State Archives, Pretoria, file UOD, 1460, E/55/6/4.
21. Report on the Diepkloof Reformatory by C. N. Kempff, Chief Clerk, 6 February 1935: ibid.
22. Report on the Diepkloof Reformatory by C. N. Kempff, Chief Clerk, 6 February 1935: ibid. Also see Paton's *Towards the Mountain*, 146.
23. Inspection report by Miss Chattey, 1 May 1935: State Archives, Pretoria, file UOD, 1460, E/55/6/4.
24. Report on the Diepkloof Reformatory by C. N. Kempff, Chief Clerk, 6 February 1935: ibid.
25. Annual Reports on Diepkloof, 1932–1948: State Archives, Pretoria, file UOD, Vol. 1700, E9/4/6.
26. Report on the Diepkloof Reformatory by C. N. Kempff, Chief Clerk, 6 February 1935: State Archives, Pretoria, file UOD, 1460, E/55/6/4.
27. He had inspected the plans on 3 July 1935: letter, J. E. Van Zyl to Paton, 10 July 1935: State Archives, Pretoria, UED, Vol. 14, E6/63.
28. Inspection report by Miss Chattey, 1 May 1935: State Archives, Pretoria, file UOD, 1460, E/55/6/4.
29. *Towards the Mountain*, 142.
30. Paton was to say that he had a staff of sixty, but he must have been thinking of a period nearer the end of his time at Diepkloof. The official report of Inspector Kempff on 6 February 1935 gives the figures I have quoted. State Archives, Pretoria, file UOD, 1460, E/55/6/4.
31. Unpublished letter, C. J. W. Kriel to the writer, 18 September 1992: PFA.
32. *Towards the Mountain*, 125 and 134.
33. Linda Chisholm, 'Education, Punishment and the Contradictions of Penal Reform: Alan Paton and Diepkloof Reformatory, 1934–1948', in *Journal of Southern African Studies*, 17/1 (March 1991), 24.
34. Cyril Burt, *The Young Delinquent*, 237.
35. Ibid. 238.
36. Annual Report on Diepkloof for 1936: State Archives, Pretoria, file UOD, Vol. 1700, file E9/4/6.
37. Paton to Secretary for Education: State Archives, Pretoria, file UOD, 1460, E/55/6/4.
38. Report of meeting between L. Van Schalkwyk and the Board of Diepkloof, 29 July 1935: State Archives, Pretoria, file UOD, 584, E14/103/9.
39. Inspector's report on Diepkloof, 13 August 1935, makes it clear that Paton

favoured a group of rondavels as the model for the free hostels, while the Inspector and the Committee of Visitors favoured a single building. But on the principle of the free hostel itself there was no dispute. Ibid.

40. Inspector's report on Diepkloof, 13 August 1935: ibid.
41. Paton to Secretary for Education, 19 August 1935: State Archives, Pretoria, file UOD, 1460, E/55/6/4.
42. Notes on a visit paid to Diepkloof on 22 August 1935 by J. D. and E. Rheinallt Jones: ibid.
43. Interview with Gonville ffrench-Beytagh, London, 2 May 1991.
44. Paton to Hofmeyr, 16 November 1935: Wits.
45. Interview with Gonville ffrench-Beytagh, London, 2 May 1991.
46. *Towards the Mountain*, 180.
47. Unpublished letter, Paton to Dr Gutsche, 16 September 1948: APC.
48. *Towards the Mountain*, 149.
49. 'The number of absconders for the last six months of 1935 was 48; for the first six months of 1936, 39. We hope, but dare not prophesy, that the tendency to decrease will be maintained.' A sceptical superior officer noted '?' in the margin. Annual Report on Diepkloof, 1936: State Archives, Pretoria, file UOD, Vol. 1700, E9/4/6.
50. Annual Report on Diepkloof, 1936: ibid.
51. *Towards the Mountain*, 146.
52. Annual Report on Diepkloof, 1936: State Archives, Pretoria, file UOD, Vol. 1700, E9/4/6.
53. *Towards the Mountain*, 174–5.
54. Ibid. 176.
55. Unpublished letter, C. J. W. Kriel to the writer, 18 September 1992: PFA.
56. *Towards the Mountain*, 164.
57. Ibid. 163.
58. Paton gives these figures in his first annual report, that of 1936: State Archives, Pretoria, file UOD, Vol. 1700, E9/4/6.
59. *Knocking on the Door*, 68–9. The poem was begun in March 1949.
60. Interview between Paton and Bill Hill, *The Island Packet* (South Carolina) 27 September 1983: 6-A. I am grateful to William Toomey for drawing my attention to this interview.
61. *Towards the Mountain*, 271.
62. Ibid. 181.
63. Unpublished letter, C. J. W. Kriel to the writer, 18 September 1992: PFA.
64. Paton gives these figures in his annual report of 1939: State Archives, Pretoria, file UOD, Vol. 1700, E9/4/6.
65. *Towards the Mountain*, 179.
66. According to an official visitor, these murals featured mainly rural scenes, including herds of cattle, but during the war tanks, aircraft, and heavy guns made their appearance. Report of J. A. Herholdt, 30 January 1943: State Archives, Pretoria, file UOD, Vol. 1460, file E/55/6/4.

9. PROVING

1. Unpublished letter, 16 November 1935: Wits.
2. Ibid.
3. The lists of pupils to be considered by the Diepkloof Board for release give only scanty details of home background, but make depressing reading. Cyril Burt's details of poverty among Londoners would have prepared him poorly for contact with Diepkloof pupils. State Archives, Pretoria: UOD, Vol. 1971, file REF E202/1/1.
4. Antonio Gramsci, *Prison Notebooks.*
5. 'This is my own, my native land', *Common Sense*, July 1946: reprinted in *Knocking on the Door*, 26.
6. 'Misadministration at Diepkloof', undated TS signed James L. Gubevu, in the Ballinger papers: Wits. I am indebted to Peter Kohler of Cape Town for bringing this document to my attention and supplying me with a copy of it.
7. Gubevu, now dead, gave this account to Mrs Joyce Dhlomo, a social worker: the account was published in the Johannesburg *Sunday Times*, 23 February 1992.
8. Jacobus Leodwyk [sic] to Sir Patrick Duncan, 16 November 1938: State Archives, Pretoria, file GG, Vol. 247, file 4/55/77.
9. Cecil Jafa Somdakakazi to Sir Patrick Duncan, 16 November 1938: ibid.
10. Cecil Jafa Somdadakazi to Sir Patrick Duncan, 27 November 1938, and Jacobus Leodwyk to Sir Patrick Duncan, ibid.
11. Unpublished letter, Paton to Hofmeyr, 24 September 1946: Wits.
12. Annual Report on Diepkloof, 1936: State Archives, Pretoria, file UOD, Vol. 1700, E9/4/6.
13. Annual Report on Diepkloof, 1937: ibid.
14. By 1937, for instance, it was down to 1,756 strokes although the number of pupils had climbed, and Paton was predicting that corporal punishment would die out altogether. He was too sanguine. Annual report, Diepkloof, 1938: State Archives, Pretoria, file UOD, Vol. 1700, E9/4/6. The figure did fall steadily, however; computed in terms of strokes per pupil per year, it was 5.1 in 1936, 3.7 in 1937, 3.1 in 1938, 2.5 in 1939, 2.4 in 1940; it then rose again to 3.1 in 1941. Annual Report, Diepkloof, 1939: ibid. Also Visitor's Report by Maj. H. S. Cooke, April 1942: State Archives, Pretoria, file UOD, Vol. 1460, file E/55/6/4.
15. Interview with Jonathan Paton, Johannesburg, 17 January 1990.
16. Unpublished letter, C. J. W. Kriel to the writer, 18 September 1992: PFA.
17. Ibid.
18. Annual Report on Diepkloof, 1937: State Archives, Pretoria, file UOD, Vol. 1700, E9/4/6.
19. *Towards the Mountain*, 194.
20. Annual report on Diepkloof, 1938: State Archives, Pretoria, file UOD, Vol. 1700, E9/4/6.

21. *Towards the Mountain*, 191.
22. Ibid.
23. Annual Reports on Diepkloof, 1938: State Archives, Pretoria, file UOD, Vol. 1700, E9/4/6.
24. Ibid.
25. *Towards the Mountain*, 191.
26. Annual Report on Diepkloof, 1938: State Archives, Pretoria, file UOD, Vol. 1700, E9/4/6.
27. Minutes of Diepkloof Board meeting, March 1942: State Archives, Pretoria, UOD, Vol. 584, E14/103/9.
28. *Towards the Mountain*, 192.
29. Official visitor's report, March 1939: State Archives, Pretoria, UOD, Vol. 1460, file E/55/6/4.
30. In his annual report for 1939, Paton refers to this opinion as having been expressed and expressed often. State Archives, Pretoria, file UOD, Vol. 1700, E9/4/6.
31. Annual Report on Diepkloof for 1939: ibid.
32. Unpublished letter, Paton to Hofmeyr, 23 August 1936: Wits.
33. Ibid. 7 April 1936: Wits.
34. T. R. H. Davenport, *South Africa: A Modern History* (4th edn.), 282.
35. Ibid. 284.
36. Unpublished letter, Paton to Hofmeyr, 2 March 1938: Wits.
37. Ibid.
38. Unpublished letter, Paton to Hofmeyr, 4 September 1938: Wits.
39. *Towards the Mountain*, 215.
40. Unpublished letter, Paton to Hofmeyr, 4 June 1939: Wits.
41. Ibid.
42. Ibid.
43. Unpublished letter, Paton to Hofmeyr, 4 September 1938: Wits.
44. Unpublished, untitled, unfinished play written about 1938: APC.
45. Unpublished letter, C. J. W. Kriel to the writer, 18 September 1992: PFA.
46. A photograph of him so dressed appears in *Towards the Mountain* following 161.
47. Unpublished letter, C. J. W. Kriel to the writer, 18 September 1992: PFA.
48. *Towards the Mountain*, 209.
49. Ibid. The verb 'opdonder' literally means to 'thunder up' the English, but it was close to an obscenity. There are many obscene English equivalents to what I have translated as 'knock the stuffing out of'; these would be closer to the spirit of the Afrikaans and the reader may supply them. One of Paton's staff who was there remembered a more bellicose expression still from the Afrikaner: 'Een van die dae gaan ons elke bleddie Engelsman uit die land uit binne in die see injaag' (One of these days we'll chase every bloody Englishman out of the country into the sea). Unpublished letter, C. J. W. Kriel to the writer, 18 September 1992: PFA.

50. *Towards the Mountain*, 210.
51. Ibid.
52. Ibid. 211.
53. 'Verworp' can also mean depraved, reprobate or castaway, but Paton showed that he intended to refer to rejection when he earlier entitled this poem 'Engelse Liedjie', (English Song).
54. He wrote this poem on 19 December 1938 (APC) and sent the final version, entitled 'Engelse Liedjie' [English song] to Hofmeyr on 22 December 1938: unpublished letter: Wits. The poem was published in *Contact* in July 1957.
55. The other is also the only known obscene ballad he produced, in 1956, beginning 'Verwoerd's in 'n ernstige knopie'.

10. WAR

1. T. R. H. Davenport, *South Africa: A Modern History* (4th edn.), 292.
2. *Towards the Mountain*, 216.
3. T. R. H. Davenport, *South Africa: A Modern History* (4th edn.), 300.
4. The Greyshirts, openly Nazi in inspiration, were led by L. T. Weichardt; the Stormjaers (storm chasers) were linked with the Ossewa Brandwag. T. R. H. Davenport, *South Africa: A Modern History* (4th edn.), 302.
5. *Towards the Mountain*, 219.
6. *Natal Witness*, 22 December 1940: 'Maritzburg's First War Casualty'.
7. 21 December 1940: APC.
8. Both dated 23 December 1940: APC.
9. *Towards the Mountain*, 219.
10. Ibid. 221; also Minutes of Diepkloof Board meeting, 3 November 1942: State Archives, Pretoria, file UOD, Vol. 1971, file REF E202/1/1.
11. *Towards the Mountain*, 233.
12. Unpublished letter, Paton to Mary Benson, 15 August 1948: MB.
13. Interview with Gonville ffrench-Beytagh, London, 2 May 1991.
14. *Towards the Mountain*, 225.
15. Interview with Gonville ffrench-Beytagh, London, 2 May 1991.
16. Ibid.
17. His involvement is listed in the publicity document prepared in late 1947 by Scribner's in preparation for the publication of *Cry, the Beloved Country*: AP.
18. *Towards the Mountain*, 229.
19. Unpublished letter, Paton to Edward Callan, 11 August 1981: 'I wrote the life of Clayton who was undoubtedly homosexual': EC.
20. Unpublished letter, Paton to Trevor Huddleston, 12 November 1966: TH.
21. *Apartheid and the Archbishop: The Life and Times of Geoffrey Clayton*, 182.
22. Ibid. 182–3.
23. *Towards the Mountain*, 239.
24. *Apartheid and the Archbishop: The Life and Times of Geoffrey Clayton*, 103.
25. Ibid.

26. Ibid. 117, fn.
27. *Towards the Mountain*, 240.
28. Ibid. 244.
29. Quoted in *Towards the Mountain*, 244.
30. Quoted in *Apartheid and the Archbishop: The Life and Times of Geoffrey Clayton*, 118.
31. 'Michael Scott: What Kind of Man Was He?', *Contact* 1/23 (13 December 1958), 9.
32. *Apartheid and the Archbishop: The Life and Times of Geoffrey Clayton*, 117.
33. Edward Callan (ed.), *The Long View*, 57.
34. Her death is referred to in the minutes of the Diepkloof Board, 2 May 1944: State Archives, Pretoria, file UOD, Vol. 1971, file REF E202/1/1.
35. *Towards the Mountain*, 253.
36. *Knocking on the Door*, 99.

11. PEACE

1. *Towards the Mountain*, 234.
2. *For You Departed* (also published as *Kontakion For You Departed*), 119.
3. Ibid.
4. Interview with David Paton, Johannesburg, 3 February 1990.
5. The joke was still being referred to at David's second wedding, on 15 May 1993.
6. *For You Departed* (also published as *Kontakion For You Departed*), 87.
7. *Towards the Mountain*, 235.
8. Ibid.
9. Interview with Jonathan Paton, Johannesburg, 17 January 1990.
10. Unpublished letter, Paton to Edward Callan, 1970: EC.
11. *Towards the Mountain*, 235.
12. Interview with Jonathan Paton, Johannesburg, 17 January 1990.
13. Ibid.
14. Unpublished letter, Paton to Dorrie, 1 November 1960: APC.
15. Unpublished letter, Paton to Dorrie, 23 December 1946: APC.
16. Interview with Jonathan Paton, Johannesburg, 17 January 1990.
17. Interview with Jonathan Paton, Sydney, 22 June 1993.
18. *For You Departed* (also published as *Kontakion For You Departed*), 65.
19. Ibid. 65–7.
20. Board of Management Minutes, Diepkloof, 5 August 1941: State Archives, Pretoria, UOD 584, file E14/103/9.
21. Board of Management Minutes, Diepkloof, 2 December 1941: ibid.
22. Unpublished letter, Paton to Hofmeyr, 6 August 1941: Wits.
23. Unpublished letter, Paton to Hofmeyr, 5 November 1941: Wits.
24. Ibid.
25. Ibid.

26. *Towards the Mountain*, 58.
27. *Forum* 6(44), 29 January 1944, 24.
28. *The Mentor* (June 1942), 5–7.
29. *Forum* 6(29), 16 October 1943, 25–7.
30. *Forum* 6(44), 29 January 1944, 24–6.
31. Ibid. 25.
32. Ibid. 26.
33. *Forum* 8(37), 15 December 1945, 7–8.
34. *Race Relations Journal* 12 (1945), 69–77.
35. *Nongqai* 37 (July 1946), 805–6.
36. Penal Reform League Pretoria, *Pamphlet No. 2*, 1948.
37. *Forum* 6(44), 29 January 1944, 25.
38. *Die Transvaler*, 27 January 1945.
39. *Towards the Mountain*, 255.
40. Ibid. 256.
41. Unpublished letter, Paton to Hofmeyr, 8 October 1944: Wits.
42. Unpublished letter, Paton to Hofmeyr, 19 July 1945: Wits.
43. Foch (1851–1929), French military hero of the First World War, longed for command over the British and Belgian forces as well as his own, so that he could co-ordinate the Allied attack on German lines.
44. Unpublished letter, Paton to Hofmeyr, 10 December 1944: Wits.
45. Quoted in Paton's unpublished letter to Hofmeyr, 19 July 1945: Wits.
46. *Hofmeyr*, 125.
47. Unpublished letter, Paton to Hofmeyr, 19 July 1945: Wits.
48. Ibid.

12. *CRY, THE BELOVED COUNTRY*

1. Unpublished diary notebook in Paton's hand, for July to mid-October 1946: he records the Rhodesian trip in this, and notes that he made an official report to the Education Department on what he had seen on 1 April 1946: APC.
2. Paton told his second wife this detail: letter from Anne Paton to the writer, 16 March 1993: PFA.
3. Unpublished letter, Paton to Hofmeyr, 31 March 1946: Wits.
4. Ibid.
5. It was in fact the second of these flights, so that this was almost as much a venture into the unknown for the aircrew as it was for Paton.
6. Unpublished letter, 15 July 1946: APC. This was only the second letter from him that she kept. After this date all his letters to her seem to have been preserved. The first was written in July 1938.
7. *Towards the Mountain*, 257.
8. Ibid., and unpublished letter to Dorrie, 15 July 1946: APC.

9. Unpublished letter, Paton to Dorrie, 15 July 1946: APC.
10. Ibid.
11. Unpublished letter, Paton to Dorrie, 11 August 1946: APC.
12. Ibid.
13. Unpublished letter, Paton to Dorrie, 31 July 1946: APC.
14. *Towards the Mountain*, 259.
15. Unpublished letter, Paton to Dorrie, 31 July 1946: APC.
16. Article by Paton, 'Great Conference of Christians and Jews at Oxford', *Outspan* 40 (1020), 13 September 1946, 34–5 and 103, 105.
17. Unpublished letter, Paton to Dorrie, 7 August 1946: APC.
18. Father Schneider was sent to Buchenwald in 1938, according to another note Paton made at the time.
19. Unpublished letter, Paton to Dorrie, 7 August 1946: APC.
20. 'Great Conference of Christians and Jews at Oxford', *Outspan* 40 (1020), 13 September 1946, 103.
21. Report to the Secretary of Education, 3 April 1947: I have not been able to find this in the Education Department files in the Pretoria National Archives, but a copy exists in APC.
22. Unpublished diary notebook for July–mid October 1946: APC.
23. *Towards the Mountain*, 264.
24. Unpublished letter, Paton to Dorrie, 1 September 1946: APC.
25. *Towards the Mountain*, 265.
26. Unpublished letter, Paton to Dorrie, 11 September 1946: APC.
27. Ibid.
28. Ibid.
29. Unpublished letter, Paton to Dorrie, 15 September 1946: APC.
30. The unique copy of this poem is in ibid.
31. Ibid.
32. *Towards the Mountain*, 266.
33. Unpublished letter, Paton to Hofmeyr, 19 September 1946: APC.
34. 'Absconding seems reasonable, somewhat higher than Diepkloof.' Unpublished letter, Paton to Dorrie, 15 September 1946: APC.
35. Except for a week he had spent in the Hebrides late in August 1946: letter, Paton to Hofmeyr, 28 August 1946: Wits.
36. *Towards the Mountain*, 267.
37. Ibid. 268.
38. Interview in Island Packet (South Carolina), 27 September 1983, p. 6-A. I am grateful to William Toomey for drawing my attention to this interview.
39. Paton's diary for 1946: APC.
40. Unpublished letter, Paton to Dorrie, 29 December 1946: APC.
41. Interview with Ian Wyllie, Gillitts, 19 June 1991.
42. Isaac Bashevis Singer remarked, 'The whole modern school of fiction in the twentieth century stems from Hamsun'. ('Knut Hamsun, Artist of Skepticism', 1967)

43. *Towards the Mountain*, 256.
44. Hamsun, K., *Growth of the Soil* (originally entitled *Markens Grøde*) (London, 1920: Gyldendal).
45. As an examination of the manuscript shows: B.
46. 1908–1960. Wright was born in Memphis Tennessee, was largely self-educated, and felt bitterly that blacks were victims of white capitalist oppression. He published *Black Boy* in 1945 (NY: Harper).
47. Unpublished letter, Paton to Dorrie, 22 September 1946.
48. Edward Callan, *Cry, the Beloved Country: A Novel of South Africa*, 11.
49. *Cry, the Beloved Country*, ch. 1.
50. He was to identify Ndotsheni as Nokweja in *Journey Continued*, 43.
51. Unpublished letter, Paton to Dorrie, 5 October 1946: APC.
52. *Towards the Mountain*, 271–2 and ibid.
53. Unpublished letter, Paton to Michael Black, 6 May 1986: APC.
54. Paton told Huddleston's flock in Mauritius that their bishop was the model for Father Vincent: interview with Huddleston, London, 11 May 1991.
55. Unpublished letter, Paton to Michael Black, 17 July 1986: APC.
56. Unpublished letter, Paton to Michael Black, 6 May 1986: APC.
57. Unpublished letter, Paton to Dorrie, 20 October 1946: APC.
58. Notably in 1920.
59. Unpublished letter, Paton to Dorrie, 14 October 1946: APC.
60. Unpublished letter, Paton to Dorrie, 27 October 1946: APC.
61. *Towards the Mountain*, 275.
62. Paton gives these scanty details of the novel's development in his letter to Hofmeyr of 9 April 1947: Wits. He told of Jacky in a powerful story in *Debbie Go Home*, 'The Divided House'.

13. *CRY, THE BELOVED COUNTRY* (CONTINUED)

1. Unpublished letter, Paton to Dorrie, 12 November 1946: APC.
2. *Towards the Mountain*, 276.
3. Unpublished letter, Paton to Dorrie, undated (18 November 1946?): APC.
4. Ibid.
5. *Towards the Mountain*, 276–7.
6. Unpublished letter, Paton to Dorrie, undated (18 November 1946?): APC.
7. Ibid.
8. He subsequently tried to obtain a print of it: unpublished letter to Edmund Fuller, 20 March 1948: EF.
9. *Towards the Mountain*, 277.
10. Ibid.
11. Unpublished letter, Paton to Dorrie, 25 November 1946: APC.
12. Unpublished letter, Paton to Michael Black, 6 May 1986: APC.
13. Unpublished letter, Paton to Dorrie, 3 December 1946: APC.

14. Unpublished letter, Paton to Dorrie, 10 December 1946: APC.
15. *Towards the Mountain*, 283.
16. Unpublished letter, Paton to Dorrie, 23 December 1946: APC.
17. Ibid.
18. Ibid.
19. Unpublished letter, Paton to Mary Benson, 15 August 1948: MB.
20. Unpublished letter, Paton to Dorrie, 23 December 1946: APC.
21. Unpublished letter, Paton to Dorrie, 6 January 1947: APC.
22. Unpublished letter, Paton to Dorrie, 29 December 1946: APC.
23. Ibid.
24. Unpublished letter, Roy Votaw to the writer, 5 March 1992: PFA.
25. Ibid.
26. Unpublished letter, Paton to Dorrie, 29 December 1946: APC.
27. Unpublished letter, Paton to Dorrie, 6 January 1947: APC.
28. *Towards the Mountain*, 291.
29. Unpublished letter, Paton to Dorrie, 13 January 1947: APC.
30. Unpublished letter, Paton to Dorrie, 6 January 1947: APC.
31. Quoted in A. Scott Berg, *Max Perkins, Editor of Genius*, 556.
32. *Towards the Mountain*, 292.
33. Ibid.
34. Unpublished letter, Paton to Dorrie, 13 January 1947: APC.
35. Unpublished letter, Paton to Hofmeyr, 9 April 1947: Wits.
36. Unpublished letter, Paton to the Burnses, 23 January 1947: B.
37. *Towards the Mountain*, 294–5.
38. Ibid. 295.
39. Unpublished letter, Paton to Dorrie, 22 January 1947: APC.
40. Not even Aubrey knew how she did it: Aubrey Burns to Paton, unpublished letter, undated [February? 1979]: APC.
41. Unpublished letter, Paton to Dorrie, 12 February 1947: APC.
42. A. Scott Berg, *Max Perkins, Editor of Genius*, 558.
43. Unpublished letter, Paton to Dorrie, 9 February 1947: APC.
44. Burns reminded Paton of these details in an unpublished letter, 14 April 1979: APC.
45. Unpublished letter, Paton to Dorrie, 9 February 1947: APC.
46. A. Scott Berg, *Max Perkins, Editor of Genius*, 558.
47. Unpublished letter, Paton to Hofmeyr, 5 February 1947: APC.
48. Unpublished letter, Paton to Dorrie, 9 February 1947: APC.
49. Ibid.
50. Unpublished letter, Paton to Dorrie, 12 February 1947: APC.
51. He gives his monthly salary (£60) in his letter to Dorrie of 14 February 1947: APC.
52. Unpublished letter, Paton to Dorrie, 12 February 1947: APC.
53. *Towards the Mountain*, 297. *Cry, the Beloved Country* sold more than a million copies in the Penguin edition alone before 1988.

54. *Towards the Mountain*, 298.
55. It is now in APC.
56. Unpublished letter, Paton to Dorrie, 13 January 1947: APC.
57. *For You Departed* (also published as *Kontakion For You Departed*), 86.
58. Ibid.

14. LOTUS EATING

1. This support is referred to in the publicity document 'Biographical Details of Alan Paton, News of Scribner Books and Authors', prepared by Scribner after the publication of *Cry, the Beloved Country*: APC.
2. A. Scott Berg, *Max Perkins, Editor of Genius*, 559.
3. Unpublished letter, Paton to Hofmeyr, 9 April 1947: Wits.
4. He makes his excuses in an unpublished letter, Paton to Hofmeyr, 20 March 1947: Wits.
5. The original, American edition was dedicated to Aubrey and Marigold Burns.
6. Unpublished letter, Paton to Hofmeyr, 9 April 1947: Wits.
7. Part I appeared in *Nongqai* (July 1947), 876–8 and 875; Part I in August 1947, 1013–15 and 1040.
8. Published by the Penal Reform League: Penal Reform Pamphlets No. 2: 1948.
9. Linda Chisholm, 'Education, Punishment and the Contradictions of Penal Reform: Alan Paton and the Diepkloof Reformatory, 1934–1938', *Journal of Southern African Studies*, 17/1 (March 1991), 39.
10. Unpublished letter, Paton to Mary Benson, 7 June 1948: MB.
11. *Nongqai* (August 1947), 1013.
12. Paton gives details of this school in his unpublished letter to Mary Benson, 7 June 1948: MB.
13. T. R. H. Davenport, *South Africa: A Modern History* (4th edn.), 320.
14. Unpublished letter, C. J. W. Kriel to the writer, 18 September 1992: PFA.
15. Unpublished letter, Paton to Mary Benson, 24 June 1955: MB.
16. They were not the first two published, as he mistakenly claimed in *Towards the Mountain*, 299; several papers jumped the gun, reviewing the book even before publication, among these being the Cincinnati *Enquirer* of 31 January 1948.
17. As Edward Callan has pointed out: *Alan Paton* (rev. edn., 1982), 34.
18. Frederick Stix in the Cincinnati *Enquirer* of 31 January 1948.
19. Meadville *Tribune-Republican*, 31 January 1948.
20. *New York Herald Tribune*, 2 February 1948.
21. *Christian Science Monitor*, 7 February 1948.
22. *New York Times*, 2 February 1948.
23. Unpublished letter, Paton to Hofmeyr, 22 February 1948: Wits.
24. Harry Hansen in the *Chicago Tribune*, 15 February 1948.

25. Quoted in an unpublished letter, Paton to Hofmeyr, 20 March 1948: Wits.
26. Paton quotes these words in his unpublished letter to Hofmeyr of 14 March 1948: Wits.
27. Maxwell Anderson (1888–1959) reached the height of his reputation in the 1930s, when he rivalled Eugene O'Neill. Among his best known plays were *What Price Glory?* (1924), *Elizabeth the Queen* (1930), and *Winterset* (1935), which has been called one of the most audacious experiments in world drama. In 1946 he had produced *Joan of Lorraine*, which had not matched his previous successes. *Winterset* was Paton's own favourite among Anderson's works.
28. Paton kept a copy of the latter cable in his diary for 1948: APC. The first cable appears to have been lost.
29. *Towards the Mountain*, 302.
30. Edward Callan, *Cry, the Beloved Country: A Novel of South Africa*, 17.
31. These figures are taken from the notebook headed '"Cry the Beloved Country" Receipts', in APC.
32. I have computed these figures from Dorrie Paton's account book for 1949: APC.
33. *Towards the Mountain*, 306.
34. The figures are given in Dorrie Paton's account book for 1949: APC.
35. Peter F. Alexander, *William Plomer*, 292.
36. This is made clear in the draft of a letter to Scribner, undated [May 1948?]: APC.
37. Unpublished letter, Paton to Mary Benson, 9 December 1948, lists these writers as having written to him. MB.
38. Unpublished letter, Paton to Hofmeyr, 22 February 1948: Wits.
39. Ibid.
40. Unpublished letter, Paton to Hofmeyr, 14 March 1948: Wits.
41. *Towards the Mountain*, 309. In fact, as Profesor Douglas Irvine of Natal University kindly pointed out to me, the Nationalists had gained their majority on a minority of the votes cast, due to a delimitation of constituencies favouring rural voters: not until 1958 would they win a majority of white votes. This explains why it was still possible in the 1950s for Liberals and others to think in terms of parliamentary politics as a potentially effective opposition to apartheid.
42. *Towards the Mountain*, 304.
43. Unpublished letter, Paton to Hofmeyr, 24 May 1948: Wits.
44. Ibid.
45. There is an interesting article on his association with Toc H to this point, 'An Author's Pilgrimage', *Compass* (May 1948), 72.
46. *Towards the Mountain*, 304.
47. Ibid. 305.
48. APC.
49. Unpublished letter, Railton Dent to Paton, 4 April 1948: MD.

50. Ibid.
51. Unpublished letter, Paton to Pearse, 26 October 1948: PFA.
52. Unpublished letter, Paton to Dent, 2 May 1948: MD.
53. Ibid.
54. Ibid.
55. Unpublished letter, Paton to Mary Benson, 15 November 1948: MB.
56. *Towards the Mountain*, 306.
57. Unpublished letter, Paton to Dent, 2 May 1948: MD.
58. Unpublished letter, Paton to Mary Benson, 15 August 1948: MB.
59. Dorrie Paton's account book for 1949: APC.
60. *Journey Continued*, 10.
61. David had in fact been boarding at St John's College for some years, at his own request.
62. Interview with Jonathan Paton, Johannesburg, 17 January 1990.
63. *Journey Continued*, 18: 'All these rages and glooms aside, the fact was that he enjoyed both holiday and school.'
64. Interview with Jonathan Paton, Johannesburg, 17 January 1990.
65. 'An Author's Pilgrimage', *Compass* (May 1948), 73.
66. Ibid. 75.
67. Unpublished letter, Paton to Maxwell Anderson, 16 May 1948: Texas.
68. Unpublished letter, Paton to Mary Benson, 20 May 1948: MB.
69. Paton gives these publication figures in 'The Story of *Cry, the Beloved Country*', in *The S.A. Publisher and Bookseller* (1950), 2–3.
70. Paton lists these editions in his interview with Horton Davies, published in *Outspan* (12 October 1951), 45. Further translations followed in later years.
71. Untitled transcript of broadcast, undated but almost certainly late 1948: APC.
72. Unpublished letter, Paton to Anderson, 22 June 1948: Texas.
73. Unpublished letter, Paton to Anderson, 16 May 1948: Texas.
74. Ibid.
75. *Journey Continued*, 10.
76. *Milady* (Aug. 1951), 'A Day with Mrs Alan Paton' by Rebecca Reyher.
77. Dorrie Paton's accounts book for 1949: APC.
78. Interview with Constance Stuart Larrabee, Chestertown, Maryland, 8 April 1992.
79. Unpublished letter, Paton to Mary Benson, 15 August 1948: MB.
80. *Knocking on the Door*, 64. The MS is in APC.
81. Ibid. 63.
82. Quoted in *Knocking on the Door*, 114–15, where it is wrongly dated '1952'. The MS (in APC) bears the date 'Anerley 11/8/48'.
83. *Knocking on the Door*, 98.
84. Ibid. 82. Dent's criticism is here exaggerated for dramatic effect.
85. Ibid. 99. Paton sent a copy of this poem to Mary Benson in a letter of 14 September 1948: MB.
86. Ibid. 72.

87. Unpublished letter, Paton to Mrs M. Y. Irwin, 27 May 1986: APC.
88. *Knocking on the Door*, 100.
89. Ibid. 124.
90. Ibid. 65.
91. Ibid. 73.
92. Ibid. 112.
93. Ibid. 111.
94. Ibid. 68.
95. Ibid. 70.
96. Unpublished letter, Paton to Pearse, 26 October 1948: PFA.
97. Unpublished letter, Paton to Mary Benson, 15 August 1948: MB.
98. Ibid.
99. Unpublished letter, Paton to Mary Benson, 14 September 1948: MB.
100. Unpublished letter, Paton to Mary Benson, 1 January 1949: MB.
101. *Journey Continued*, 12.
102. The article was reprinted in *Forum* (13 November 1948), 12–13. I have not been able to discover which British paper it was originally written for.
103. Whitman, 'When Lilacs Last in the Dooryard Bloom'd'. I am indebted to Professor Edward Callan for pointing out this link between Paton's lines and Whitman's.
104. Unpublished letter, Paton to Mary Benson, 1 January 1949: MB.
105. Untitled poem, undated, but on same sheet as a draft of 'Anxiety Song of an Englishman', which was written in 1948: APC. It is possible that this poem was intended as a second stanza of 'Anxiety Song of an Englishman'.
106. *Journey Continued*, 15.
107. Unpublished: APC.
108. Previously unpublished: the MS is in APC.

15. BLEEDING ONTO THE PAGE

1. *Journey Continued*, 13.
2. *Knocking on the Door*, 74.
3. It is in APC.
4. Unpublished letter, Paton to Luther Greene, 31 December 1948: APC.
5. The Levi novel is referred to in his letter to Mary Benson, 3 April 1949: MB.
6. Unpublished letter, Paton to Mary Benson, 3 April 1949: MB.
7. Unpublished letter, Paton to Luther Greene, 31 December 1948: APC.
8. Unpublished letter, Paton to Mary Benson, 11 March 1949: MB.
9. Unpublished letter, Paton to Mary Benson, 3 April 1949: MB.
10. Unpublished cablegram, signed 'Max and Kurt' to Paton, 17 March 1949: APC.
11. He was to say as much to the students of Michigan University at Flint, on 16 May 1977, in an address entitled 'A Writer in Residence': K.
12. *Journey Continued*, 17.

13. Unpublished letter, Paton to Mary Benson, 11 March 1949: MB.
14. *Journey Continued*, 18.
15. Ibid.
16. Ibid. 19.
17. Interview with Mary Benson, London, 29 April 1991.
18. Unpublished letter, Paton to Mary Benson, 31 May 1949: MB.
19. Ibid.
20. Mary Benson, *A Far Cry*, 53.
21. Unpublished letter, Paton to Mary Benson, 11 March 1949: MB.
22. Ibid.
23. Unpublished letter, Paton to Mary Benson, 31 May 1949: MB.
24. Unpublished letter, Paton to Mary Benson, 10 August 1949: MB.
25. Mary Benson, *A Far Cry*, 54.
26. Ibid.
27. Unpublished letter, Mary Benson to the writer, 21 May 1993: PFA.
28. Mary Benson, *A Far Cry*, 54.
29. *Sic*, in both the MS and the version in *Knocking on the Door*, 29.
30. Ibid., where it is mistakenly dated 1946–7. In his copy of the book Paton corrected this to '1949', and in the index, p. vii, wrote opposite this poem's title, 'London 1949'. APC.
31. *Journey Continued*, 20.
32. Ibid. 21.
33. Ibid.
34. Ibid. 22.
35. Quoted in the *New York Herald Tribune*, 20 October 1949.
36. Letter from the SA Ambassador to Washington to the Secretary for External Affairs, Pretoria, 7 November 1949: State Archives, Pretoria, SAB, BVV, 98, Ref 79.
37. *Journey Continued*, 22.
38. The diary is in APC.
39. *Journey Continued*, 24.
40. Ibid.
41. In the possession of the Weill-Lenya Research Center, New York, to whom I am grateful for permission to quote it: WL.
42. *Journey Continued*, 23.
43. *New York Times*, 6 November 1949.
44. *Life*, 7/12 (5 December 1949), 22.
45. She mentions the first night in her *A Far Cry*, 54.
46. A revival in 1992 by the Boston Lyric Opera under director Bill T. Jones proved particularly successful: *Christian Science Monitor* (15 January 1992), 14.
47. *Journey Continued*, 25.
48. Unpublished letter, Paton to Mary Benson, 22 November 1949: MB.
49. He mentions the Retief idea in a letter to Mary Benson, 22 November 1949: MB.
50. Unpublished letter, Paton to Mary Benson, 12 December 1949: MB.

51. Ibid.
52. I am indebted to Professor Edward Callan for pointing this out, in a lecture entitled 'Walt Whitman, Alan Paton and the WMU', delivered on 17 November 1992: EC.
53. Published as 'No Place for Adoration' in *Knocking on the Door*, 84.
54. Unpublished: the MSS and TS are in APC.
55. First published as 'A Psalm' in *South African Outlook* (1 July 1953), 111, reprinted in *Knocking on the Door*, 85.
56. Published in *The Christian Century* (13 October 1954), 1237–9, reprinted as an SPCK booklet, London, 1959, and again reprinted in *Knocking on the Door*, 86–93. The MS, with cancelled stanzas and many variants, in APC.
57. Paton's unpublished diary for 1949 and 1950: APC.
58. 8, 10, and 11 January 1950: APC.
59. Unpublished diary for 1949 and 1950: APC.
60. Not 'Eel river at my door', as he records the entry in *Journey Continued*, 26.
61. Unpublished letter, Paton to Hofmeyr, 9 April 1947: Wits.
62. Paton's unpublished diary for 9 January 1950: APC.
63. He made this suggestion in *Journey Continued*, 27.
64. Ibid. 26.
65. Unpublished letter, Paton to Mary Benson, 14 April 1950: MB.

16. RECOVERY

1. Unpublished letter, Paton to Leslie Rubin, 13 November 1968: LR.
2. *Journey Continued*, 30.
3. *Instrument of Thy Peace*, 52–3.
4. Paton's diary for 1 February 1950: APC.
5. *Journey Continued*, 32.
6. Paton told this story in an interview with the *Los Angeles Times*, 12 December 1967: 'S. African Novelist Not Appreciated at Home'.
7. Fuller, born in 1914, is a novelist, critic, biographer, and editor. In addition to *A Star Pointed North*, his novels include *Brothers Divided* (1961), *The Corridor* (1963), and *Flight* (1970).
8. This meeting is recalled in Fuller's letter to Paton of 21 March 1979: APC.
9. Unpublished letter, Fuller to the writer, 9 June 1992: PFA.
10. *Journey Continued*, 32.
11. Paton's diary for 1950: APC.
12. Interview with Mary Benson, London, 29 April 1991.
13. Ibid.
14. This unique copy of Paton's address was preserved by Mary Benson: MB.
15. APC.
16. *For You Departed* (also published as *Kontakion For You Departed*), 86–7. This is one of the rare occasions when his memory let him down: he wrongly attributes this meeting to 1948, though in that year he did not leave South Africa.

17. *Journey Continued*, 42.
18. Interview with David Johanson, 31 January 1990. The hall (memorial to the dead of the Second World War) is now a white elephant.
19. Unpublished letter, Guy Butler to Paton, 11 February 1981: APC.
20. Unpublished letter, Paton to Butler, 17 February 1981: APC.
21. This photograph is in the collection at APC.
22. The correspondence with Moloi has been preserved at APC: letters passed between him and Paton on the subject of the lobola (bride-price) in October 1952.
23. Unpublished letter, Paton to Mary Benson, 4 May 1950: MB.
24. *Journey Continued*, 43.
25. Unpublished letter, Paton to Mary Benson, 5 February 1951: MB.
26. *Journey Continued*, 46.
27. He expresses his bewilderment in ibid. 35.
28. Ibid. 47.
29. The original of this cable has been lost, but Paton transcribed it in his diary for 1948: APC.
30. Notably *The African Patriots* (1963), republished by Penguin as *South Africa: The Struggle for a Birthright* (1966), and *Nelson Mandela* (1986).
31. Unpublished letter, Paton to Irita Van Doren, 18 December 1951: LC.
32. Unpublished letter, Paton to Mary Benson 19 December 1951: MB.
33. *Journey Continued*, 49.
34. Ibid. 47.
35. There is a copy of this photograph in APC.
36. Unpublished letter, Paton to Professor W. H. Gardner, 7 October 1959: ROP.
37. Unpublished letter, Paton to Railton Dent, 1 July 1952: MD.
38. Interview with Mrs Mabel Dent, Pietermaritzburg, 21 June 1991. Paton's reasons for claiming that it was entirely written in England may have been financial: under double-taxing arrangements, the royalties earned on a book written and published abroad were not subject to South African taxes if fully taxed abroad, and sale of the script to an overseas buyer was permitted.
39. *Commonweal*, 28 August 1953.
40. *Los Angeles Times*, 23 August 1953.
41. *Christian Science Monitor*, 27 August 1953.
42. Washington *Star*, 23 August 1953.
43. *New York Times* Book Review, 29 August 1953.
44. Unpublished letter, Paton to Dorrie, 28 August 1953: APC.
45. *Natal Witness*, 31 August 1953.
46. Unpublished letter, Paton to Mary Benson: MB.
47. *Forum*, 6(44) (29 January 1944), 24–6.
48. *Journey Continued*, 52.
49. Ibid.
50. Ibid.

51. Unpublished letter, Paton to Irita Van Doren, 30 September 1952: LC.
52. Ibid.

17. THE POLITICS OF INNOCENCE

1. *Journey Continued*, 39.
2. Interview with Pauline Morel, Durban, 29 January 1990.
3. Letter to Kingsley Martin, 7 May 1963: *Letters of Leonard Woolf* (ed. F. Spotts), 451.
4. Edward Callan was born in Ireland in 1917 and lived in South Africa from 1938 to 1950, first teaching in Catholic private schools, and after the war with the Transvaal Education Department. He moved to the USA in 1952 and from 1957 taught at the University of Western Michigan where he became Professor of English, in 1983 being honoured with the title of Distinguished University Professor. His major publications include *Alan Paton* (1969, revised 1982), *Yeats on Yeats: The Last Introductions and the 'Dublin' Edition* (1981) and *Auden: A Carnival of Intellect* (1983).
5. Interview with Edward Callan, Kalamazoo, 5 April 1992.
6. *Knocking on the Door*, 113. The MS is in APC.
7. He gave this account in his Peter Ainslie Memorial Lecture of 29 August 1951, published in *South African Outlook*, 81 (2 October 1951), 148–52.
8. Unpublished letter, Paton to Mary Benson, 19 December 1951: MB.
9. *Journey Continued*, 53.
10. Unpublished letter, Paton to Mary Benson, 4 May 1950: MB.
11. *Journey Continued*, 67.
12. Unpublished letter, Paton to Mary Benson, 19 December 1951: MB.
13. *Rand Daily Mail*, letter to the editor, 14 October 1950.
14. Unpublished letter, Paton to Dorrie, 4 September 1954: APC.
15. Fragmentary MS dated 19 October 1952: APC.
16. Unpublished, untitled, and undated: APC. Paton sent a copy of this poem to Leslie Rubin on 13 September 1957: LR. However, it may well have been written considerably earlier.
17. Letter to the *Natal Mercury*, 25 November 1952.
18. At the first election of 'Natives' Representatives' in December 1953 four Liberals had been elected: Margaret Ballinger (Cape Eastern) and Walter Stanford (Transkei) in the House of Assembly, and William Ballinger (Transvaal and Orange Free State) and Leslie Rubin (Cape Province) in the Senate. Rheinallt-Jones, in spite of his liberal views, never joined the Liberal Party.
19. Paton published a tribute to her in 1960: 'Margaret Ballinger', *Contact* 3/10 (21 May 1960), 5.
20. One of Paton's best, according to Rubin, involved a diner annoyed by a sloppy waiter.

    Diner (*angrily*): Waiter, you've got your thumb in my soup.

Waiter (*unconcerned*): Yes sir. You see I have a boil on my thumb, and my doctor says I've got to keep it warm.

Diner (*enraged*): Then stick it up your backside!

Waiter: I do, between courses.

One of Rubin's stories which Paton never forgot, and which he was to tell to both his wives, involved a very drunk pianist at a party, sitting on a traditional stool with a seat made of interlaced thongs, known as riempies. Presently he goes out to the toilet, and comes back with his fly still open. As he is about to go on playing a woman approaches him and says confidentially, 'Do you know your balls are hanging through the riempies?' 'No', says the pianist, 'but if you hum it I'll soon pick up the tune'. Letters from Leslie Rubin to the writer, November 1992: PFA.

21. Letters to the writer from Leslie Rubin, 1992: PFA.
22. Letter to the writer from Leslie Rubin, 3 November 1992: PFA.
23. Robert St John, *Through Malan's Africa*, 112.
24. These details are drawn partly from *Journey Continued*, 68, partly from a detailed letter Paton wrote to Maggie Rodger, then doing research on Margaret Ballinger, on 25 March 1980: APC.
25. 'The Liberal Approach', in *South African Outlook*, 83 (1 October 1953), 156.
26. *Journey Continued*, 116.
27. Ibid. 74.
28. Ibid. 60.
29. These details are given in an interview with Paton published in *Spotlight*, August 1953: 'Alan Paton now works on the "Beloved Land"', 36–7.
30. Unpublished memorial speech delivered by Paton at the Settlement in October 1979 after McKenzie's death in September 1978: APC.
31. *Spotlight*, August 1953: 'Alan Paton now works on the "Beloved Land"', 36–7.
32. Unpublished letter, Paton to Hofmeyr, 24 September 1936: Wits.
33. Unpublished letter, Paton to Mary Benson, 24 June 1955: MB.
34. Unpublished letter, Paton to Dorrie, 16 August 1953: APC.
35. Unpublished letter, Douglas V. Steere to Paton, 27 October 1953: APC.
36. *Journey Continued*, 121.
37. Ibid. 62.
38. Published in 1946 by Faber.
39. Brown, in an interview with me in Pietermaritzburg on 30 January 1990, said they met in 1951, but I have followed Paton's account here: *Journey Continued*, 62. It was in October 1952 that Paton lived in the hotel at Bulwer.
40. Interview with Peter Brown, Pietermaritzburg, 30 January 1990.
41. 'The Role of White Liberals in South African Politics', review of Paul B. Rich's *White Power and the Liberal Conscience* in *Africa Today*, 1st and 2nd quarter 1985, 115–16.
42. Unpublished letter, Leslie Rubin to the writer, 27 November 1992: PFA.
43. *Journey Continued*, 68.

44. Ibid. 142.
45. 'An Example for Us All', *Contact* 7/5 (10 April 1964), 2.
46. Pamphlet entitled *The Policies of the Liberal Party of South Africa*, publication of the Liberal Party, 1954.
47. Unpublished letter, Paton to Leslie Rubin, 17 February 1954: LR.
48. Paton refers to writing this speech in his letter to Mary Benson, 27 January 1964: MB.
49. At the meeting organized to celebrate Lutuli's Nobel Prize (and which the South African authorities prevented Lutuli from attending) Paton read a praise song: 'You there, Lutuli, they thought your world was small / They thought you lived in Groutville / Now they discover / It is the world you live in'. Quoted in Mary Benson's *A Far Cry*, 134.
50. Unpublished letter, Paton to Mary Benson, 27 January 1964: MB.
51. Unpublished letter, Paton to Rubin, 22 October 1953: LR.
52. Ibid.
53. Though not Edgar Brookes, who at this stage showed no interest in the Liberal Party, nor Rheinallt-Jones, who never joined it.
54. Quoted by Leslie Rubin in 'The Role of White Liberals in South African Politics', review of Paul B. Rich's *White Power and the Liberal Conscience* in *Africa Today* (1st and 2nd quarter 1985), 116.
55. Unpublished letter, Paton to Rubin, 22 October 1953: LR. Paton's letter is somewhat ambiguous on this point; he does not report the 'unconstitutional action' as having been suggested to himself, but to Professor Tom Price of Cape Town, a friend of Leslie Rubin.
56. *Journey Continued*, 68.
57. Interview with Peter Rodda, London, 20 May 1991.
58. *Journey Continued*, 71.
59. Unpublished letter, Paton to Revd A. G. Knight of Toc H, 13 January 1976: APC.
60. Interview with Peter Rodda, London, 20 May 1991.
61. Interview with Pauline Morel, Durban, 29 January 1990.
62. Unpublished letter, Rubin to the writer, 27 November 1992: PFA.
63. Interview with Jonathan Paton, Sydney, 22 June 1993.
64. Nederduitse Gereformeerde Kerk and Nederduitse Hervormde Kerk.
65. 'The Novelist and Christ' was published in *Saturday Review*, xxvii (4 December 1954), 15–16, 56–7.
66. Leopold Marquard (1897–1974), son of a Moderator of the Dutch Reformed Church, was descended on his mother's side from the missionary Andrew Murray. Apprenticed as a carpenter as a youth, he supported himself, went on to become a Rhodes Scholar (1920–23), and founded NUSAS in 1924. During the war he worked in military education. He represented South Africa at the inaugural meeting of UNESCO in 1945, and after the war became editorial manager of Oxford University Press in Cape Town. He was a life member of the South African Institute of Race Relations, and its President in 1957, 1958, and 1969.

67. Ken Hill is the original of Mr Thomson in 'The Hero of Currie Road', in *Knocking on the Door*, 167–74.
68. Interview with Pauline Morel, Durban, 29 January 1990.
69. Unpublished letter, Paton to Dorrie: APC.
70. Both he and Peter Brown thought the universal suffrage decision a mistake at the time, but soon accepted it as logical and right: unpublished letter, Peter Brown to the writer, 4 June 1993: PFA.
71. Interview with Guinevere Ventress, Linton, Cambridgeshire, 2 February 1991.
72. Professor Douglas Irvine was among those who have testified to me of Paton's powers as a speaker.
73. Ian Wyllie tells this story in *Natal Witness*, 27 August 1955, 'Paton: misunderstood liberal moderate'.
74. Interview with David Welsh, Cape Town, 14 January 1992.
75. Interview with Tony Morphet, Cape Town, 12 June 1991.
76. *The Land and People of South Africa* (Philadelphia: Lippincott, 1955, revised, 1964, and Toronto: Longman, 1955), republ. as *South Africa and her People* (London: Lutterworth Press, 1957).
77. Quoted from *South Africa and her People* (London: Lutterworth Press, 1957), 136. The italics are Paton's.
78. Unpublished letter, Paton to Mrs Roosevelt, 3 August 1955: APC. Mrs Roosevelt replied that there was nothing she could do.
79. Interview with Bishop Huddleston, London, 11 May 1991. The most effective of the boycotts seems to have been the refusal of foreign creditor banks to roll over South Africa's loans.
80. T. R. H. Davenport, *South Africa: A Modern History* (4th edn.), 350.
81. Huddleston gives this account of the arrival of the letter in his letter to Paton of 26 June 1967: APC.
82. Quoted by Paton in his unpublished letter to Dorrie of 22 January 1956: APC.
83. He suffered profound depressions in 1960 and 1981, for instance, as his letters to Paton attest.
84. *Journey Continued*, 141.
85. 'Too long a sacrifice / Can make a stone of the heart'—'Easter 1916'.
86. Unpublished letter, Betty Patterson to the writer, 16 March 1993: PFA.
87. Unpublished letter, Edmund Fuller to the writer, 9 June 1992: PFA.
88. Unpublished letter, Paton to Dorrie, 9 November 1955: APC.
89. John Oliver Patterson (1908–1988) trained as an architect before being ordained an Episcopal priest in 1934. After parish work he became Rector and Headmaster of Kent School from 1949 to 1962. He founded St Stephen's School, Rome, in 1962, and was head until his retirement in 1968.
90. Unpublished letter, Paton to John Patterson, undated [January 1958]: BP.
91. Unpublished letter, Dorrie Paton to Jonathan Cape, 1 November 1955: APC.

92. Unpublished letter, Paton to Ursula Niebuhr, 3 June 1985: APC.
93. Unpublished letter, Paton to Dorrie, 21 November 1955: APC.
94. Unpublished letter, Dorrie Paton to Jonathan Cape, 1 November 1955: APC.
95. Unpublished letter, Mary K. Frank to Paton, 21 November 1956: APC. The play opened at the Belasco Theatre on 11 October 1956. Liston Pope, in a letter to Paton of 29 November 1956, reported that the production had technical faults, with long dark pauses for scenery alterations between the too numerous scenes, and that the effect was slow and episodic: APC.
96. Unpublished letter, Paton to Dorrie, 7 January 1956: APC.
97. Unpublished letter, Paton to Dorrie, 1 January 1956.
98. Unpublished letter, David Paton to the writer, 3 August 1993: PFA.
99. Unpublished letter, Paton to Dorrie, 16 January 1956: APC.
100. Edward Callan, *Alan Paton* (revised edition), 76.
101. Unpublished letter, Paton to Dorrie, 29 February 1956: APC.
102. Unpublished letter, Paton to Dorrie, 13 February 1956: APC.
103. Ibid.
104. Unpublished letter, Paton to Dorrie, 12 November 1955: APC.
105. *Journey Continued*, 154.

18. STALKED BY THE STATE

1. Unpublished letter, Paton to the Pattersons, 6 May [1956]: BP.
2. *Journey Continued*, 177. The CIA funded a number of intellectual journals in various countries during the sixties, including *Encounter* in Britain and a major journal in Australia. Edward Callan gives a summary of the harassment of *Contact* staff in *Alan Paton* (revised edition), 87–8, and in his edition of Paton's *Contact* articles, *The Long View*, 50–2.
3. Many of them have been reprinted in *The Long View*, ed. Edward Callan.
4. Unpublished letter, Paton to Rubin, 19 May 1956: LR.
5. Paton refers to this occasion in an unpublished letter to Rubin, 13 November 1968: LR.
6. The unique copy of this ballad, a real literary curiosity, is in the possession of Leslie Rubin.
7. They also arrested members of the Coloured People's Congress and the South African Congress of Trade Unions.
8. T. R. H. Davenport, *South Africa: A Modern History* (4th edn.), 351–2.
9. Unpublished letter, Paton to Irita Van Doren, 24 January 1957: LC.
10. Examples include his correspondence with Gunnar Helander of Gothenborg, Sweden, in April 1957: APC.
11. Unpublished letter, Huddleston to Paton, 16 December 1956: APC.
12. Unpublished letter, Paton to Huddleston, 29 December 1956: TH.
13. Unpublished letters, Stone to Paton, May 1957: APC.
14. Unpublished letter, Paton to Stone, 2 May 1957: Kent.

15. John Collins, *Faith Under Fire*, 207.
16. Unpublished letter, Dorrie Paton to Annie Laurie Williams, 4 April 1957: Kent.
17. The documents generated by this crackpate expedition are preserved in APC, together with Paton's TS entitled 'Lost City of the Kalahari'.
18. 'Towards a Non-Racial Democracy', *Contact* 1/2 (February 1958), 11.
19. *Journey Continued*, 172.
20. Quoted in Jonathan Paton, 'Journeys with my father': JP.
21. Ibid.
22. Interview with Mrs Nancy Fraser, Pinetown, 21 June 1991.
23. Jonathan Paton, 'Journeys with my father': JP.
24. *Journey Continued*, 180.
25. Unpublished letter, Peter Brown to J. H. MacCallum Scott of Pall Mall Press: APC.
26. *Africa South*, August–September 1958, 18.
27. The words are those of a witness to this shameful scene, an Englishman named Waller, whom Paton quotes in an unpublished letter to Bishop Olwin of Northern Rhodesia, 2 December 1958: APC.
28. Ibid.
29. A copy of *David Livingstone* is in APC.
30. The untitled, undated MS is now in APC.
31. Unpublished letter, Paton to Dorrie, 28 June 1959: APC. In *Home and Exile*, 58, Lewis Nkosi claims Paton's story was based on the experience of a *Drum* reporter. Paton's correspondence does not substantiate this.
32. The background to the writing of Mkhumbane is set out in a letter from Malcolm Woolfson to Edward Callan, 15 January 1969: EC.
33. Unpublished letter, Paton to John Stix, 16 July 1957: APC.
34. Interview with Neil Herman, London, 7 May 1991.
35. 'Nationalism and the Theatre', *Contact* 8/3 (March 1965), 2–3. Reprinted in Edward Callan, *The Long View*, 257.
36. *Journey Continued*, 195.
37. Interview with Neil Herman, London, 7 May 1991.
38. I lived in Vanderbijlpark, near Sharpeville, at this time, and vividly remember a double queue, two city blocks long, of whites anxious to obtain firearms, on the day following the Sharpeville killings.
39. *Journey Continued*, 175.
40. *Journey Continued*, 196. Non-Natal members of the party to be arrested included Eric Atwell (Port Elizabeth Committee), John Lang (East Rand Chairman), and Dr Colin Lang (Liberal Party candidate, Transvaal provincial elections, 1959).
41. Gertrude Cohn, 'The Beginnings of the Police State in South Africa: A Personal Experience': GC.
42. *Journey Continued*, 201.
43. Mary Benson, *The African Patriots* (NY: Encyclopedia Britannica Press, 1964), 274.

44. Paton's address to the Liberal Party convention of 1960 is preserved in APC. Though it is dated by Paton, years after, '1961? 1960?' there is no doubt that it was delivered in 1960.
45. *Contact*, 3/8 (16 April 1960), 5.
46. Edward Callan (ed.), Alan Paton, *The Long View*, 35.
47. Unpublished letter, Paton to André Brink, 20 March 1970: APC.
48. Interview with Mrs Eunice ('Dorrie') Arbuthnot, Pietermaritzburg, 31 January 1990.
49. Interview with Mrs Mabel Dent, Pietermaritzburg, 21 June 1991.
50. Interview with Neil Herman, London, 7 May 1991.
51. Unpublished letter, Geoffrey Cantuar to Paton, 13 February 1961: APC.
52. Unpublished letter, Paton to Rubin, 20 August 1959: LR.
53. Unpublished letter, Paton to Rubin, 24 September 1959: LR.
54. I should record that Peter Brown, who knew Paton very well, was quite sure that this sexual licence went no further than jokes: unpublished letter to the writer, 4 June 1993: PFA.
55. Unpublished letter, Paton to Rubin, 4 November 1959: LR.
56. Unpublished letter, Leslie Rubin to the writer, 17 November 1992: PFA.
57. On aerogram to Leslie Rubin, 1 May 1960: LR.
58. Unpublished letter, Paton to Leslie Rubin, 1 May 1960: LR.
59. Guinevere Ventress stood for Ixopo in the Provincial elections at the end of 1959: interview with her, Linton, Cambridgeshire, 2 February 1991.
60. Unpublished letter, Paton to John Collins, 2 May 1960: APC.
61. Unpublished letter, Dorrie Paton to Betty Patterson, 5 May 1960: BP.
62. Unpublished letter, Paton to Marta Hackell [the German translator of *Cry, the Beloved Country*], 11 August 1960: APC.
63. *Journey Continued*, 208.
64. This is suggested by his letter to Mary Benson, 10 September 1960: MB.
65. Unpublished letter, Paton to Mary Benson, 10 September 1960: MB.
66. Interview with Jonathan Paton, Johannesburg, 17 January 1990.
67. Edward Callan (ed.), *The Long View*, 36. Callan also prints Paton's address on this occasion, 167–72.
68. This directive was circular letter A68 of 1960; it was cancelled by another letter, marked 'Uiters Geheim' ('Top Secret'), of 4 January 1962 from the Secretary of Foreign Affairs to all heads of mission: State Archives, Pretoria, SAB, BVV, 98m Ref 79.
69. Unpublished letter, Paton to Dorrie, 23 October 1960: APC.
70. Unpublished letter, Paton to Irita Van Doren, 8 November 1960: LC.
71. Unpublished letter, Paton to Dorrie, 1 November 1960: APC.
72. Unpublished letter, Paton to Dorrie, 13 November 1960: APC.
73. Dorrie Paton's diary, 'France and Italy, Aug 28–Sept 15' [1954]: APC.
74. Matshikiza was now living in London, terribly depressed and lethargic: unpublished letters, Paton to Matshikiza, 16 January 1961: APC, and to Professor Edward Callan, 9 August 1966: EC.
75. This letter is preserved in APC.

76. He quotes this comment in his letter to Canon John Collins, 31 July 1961: APC. He originally made the comment in a letter to the Minister who seized his passport, and subsequently published it in *Contact*, 3/17 (17 December 1960), 2.
77. Paton had to content himself with writing a congratulatory article about Huddleston's new appointment: 'Our New Bishop: Trevor Huddleston', *Contact*, 3/17 (27 August 1960), 5.

19. HARD TIMES

1. *Journey Continued*, 215.
2. Unpublished letter, Dorrie Paton to Elizabeth Patterson, 19 February 1961: BP.
3. Ibid.
4. The first of them, 'The Worst Thing of His Life', had been published in 1951. Edward Callan, *Alan Paton* (rev. edn.), 59.
5. *Journey Continued*, 175–6.
6. *Daily Herald*, 1 June 1961.
7. *Sunday Telegraph*, 28 May 1961.
8. *Times Literary Supplement*, 16 June 1961.
9. *New York Herald Tribune*, 10 April 1961.
10. *New York Times*, 11 April 1961.
11. *Chicago Sun-Times*, 30 April 1961.
12. *New York Herald Tribune*, 10 April 1961.
13. Unpublished letter, Dorrie Paton to Betty Patterson, 16 May 1961: BP.
14. Another dramatic version of *Cry, the Beloved Country* was to be written in 1962 by Felicia Komai, and dedicated to Trevor Huddleston: see the letter from him to Paton of 23 April 1962: APC.
15. 'A Milestone in S. A. Indigenous Theatre' was the headline of the *Daily News*, 13 December 1962, which described *Sponono* as a triumphant success.
16. The letters that passed between him, Krishna Shah, and Mary Frank during January 1964 about this 'investment' are in APC.
17. His foreword to the re-publication of *Sponono* by David Philip, Cape Town, makes this clear: 6.
18. Unpublished letter, Paton to Mary K. Frank, 20 June 1963: APC.
19. These details are given in an unpublished letter, Dorrie Paton to Betty Patterson: BP.
20. *For You Departed* (also published as *Kontakion For You Departed*), 138.
21. Unpublished letter, Paton to Bishop Huddleston, 28 September 1961: TH.
22. T. R. H. Davenport, *South Africa: A Modern History* (4th edn.), 365–6.
23. Unpublished letter, Paton to John Collins, 3 October 1962: APC.
24. Liberals banned after Peter Brown included Harold Head, Walter Hain, Dempsey Noel, Ann Tobias, Eric Harber, Joe Tsele, David Rathswaffo, Saul Bastomsky, Max Thomas, Selby Msimang, Barney Zackon, Alban Thumbran, Fred Prager, David Craighead, Eddie Roux, Samuel Dick, John

Aitchison, Chris Shabalala, Sam Polotho, Michael Francis, Enoch Mnguni, Michael Ndlovu, Jean Hill, Ken Hill, Heather Morkill, and Ruth Hayman. This list is given in *Liberal Opinion*, 4, 6, September 1966, 2. A slightly different list is given in Edward Callan (ed.), *The Long View*, 277–8.

25. Interview published in *Perspective*, 12, 5, May 1963, 29.
26. Correspondence between Paton, his solicitor Abe Goldberg of Durban, and the *Cape Times*, March–June 1961, preserved in APC.
27. Quoted in Edward Callan (ed.), *The Long View*, 146.
28. Unpublished letter, Patrick Duncan to Paton, 10 March 1963: APC. Duncan went on to join the PAC and work for it in Algeria. Dismissed by the PAC, he tried to work for an American church group before dying of anaemia in London in 1967.
29. *Journey Continued*, 222.
30. Unpublished letter, Paton to Duncan, 27 March 1963: APC.
31. *For You Departed* (also published as *Kontakion For You Departed*), 146.
32. Interview with Adrian Leftwich, York, 25 February 1991.
33. *Journey Continued*, 227. This account was read by Leftwich before its publication.
34. Interview with Adrian Leftwich, York, 25 February 1991.
35. Ibid.
36. *Sunday Times*, 24 November 1964: 'Leftwich to Wed State Witness'.
37. *For You Departed* (also published as *Kontakion For You Departed*), 149. In this volume Paton calls Leftwich 'Lester'.
38. Leftwich to Paton, 1 January 1965: APC.
39. Unpublished letter, Paton to Tony Morphet, 7 June 1983: TM.
40. *Journey Continued*, 237. The memorial article he published about Harris, 'John Harris', *Contact* viii. 4 (April 1965), 2–3, accuses him of doing the Liberal Party 'incomputable harm'.
41. Unpublished letter, Paton to Marion Friedmann, 22 February 1965: APC.
42. It was Vigne who organized the memorial service for Paton in St Paul's, London, on 28 June 1988.
43. 'Alan Paton Reports on South Africa', *Commonweal* lxxxii. 10 (28 May 1965).
44. Interview with Tony Morphet, Cape Town, 12 June 1991.
45. Paton published a protest against Brown's banning, an article which also contained a tribute to Brown and to Edgar Brookes, in August 1964: 'Peter Brown', *Contact* vii. 10, 2.
46. *Journey Continued*, 239.
47. Transcript of evidence in the Supreme Court of South Africa (Transvaal Provincial Division, State v Nelson Mandela and Others, before Mr Justice De Wet, 12 June 1964): Wits.
48. An amusing account of this meeting has been published by Jo Thorpe of Durban: 'Annie to the rescue during a storm in a teacup', *Daily News*, 5 September 1964. I am indebted to Ms Thorpe for supplying me with a copy of this article.
49. Statement to Lieut. J. Louw of the SA Police, 6 December 1975: this

statement was made by Paton at the resumption of the case in 1975, and gives most of the details: APC.

50. Donald J. Woods is a sixth-generation South African, born in the Transkei. He was appointed Editor of the *Daily Dispatch* in February 1965. In later years, persecuted by the South African government, he escaped to Britain, but returned to South Africa in the early 1990s.
51. *Journey Continued*, 243.
52. Unpublished letter, Paton to Rubin, 21 January 1966: LR.
53. Ibid.
54. Unpublished letter, Paton to Mary Benson, 7 April 1964: APC.
55. Unpublished letter, Paton to Collins, 23 January 1965: APC.
56. Interview with David Craighead, London, 2 May 1991.
57. Paton quotes the official transcript of Collins's UN speech in a letter to Prof. Z. K. Matthews, 28 October 1965: APC.
58. Unpublished letter, Paton to Laurens van der Post, 25 June 1965: APC.
59. Unpublished letter, Paton to Collins, 3 August 1965: APC.
60. She arrived at the Patons' home on 19 October 1965: *Journey Continued*, 260.
61. Unpublished letter, Paton to Collins, 3 August 1965: APC.
62. Van der Post's letters of 15 December 1969, 16 February 1976, and 21 February 1977 refer to these gifts as coming from his own royalties, but this appears to have been a cover story in case the mail was intercepted. D&A contributions were routinely 'covered' by such subterfuges: interview with Bishop Trevor Huddleston, London, 11 May 1991.
63. Her letter to Paton of 26 January 1968 is the first to raise the issue: APC.
64. His letter of support, half-hearted compared to his usual style, is addressed to the Chairman of the Nobel Peace Prize Committee, 27 May 1970: APC.
65. Unpublished letter, Paton to Diana Collins, 22 February 1968: APC. Paton also wrote to Huddleston, 'I quite frankly cannot recommend a man for a peace prize who encourages guerilla fighting'. 27 February 1968: TH.
66. Unpublished letter, Paton to Diana Collins, 25 June 1969: APC.
67. Unpublished letter, Paton to Lewis Nkosi, 3 August 1965: APC.
68. Unpublished letter, Paton to Lewis Nkosi, 22 March 1969: APC.
69. *Journey Continued*, 259–60.
70. Interview with Eunice ('Dorrie') Arbuthnot, Pietermaritzburg, 31 January 1990.
71. Unpublished letter, Paton to Betty Patterson, 16 March 1966: BP.
72. Dorrie Paton's Diary, 28 February 1966: APC.
73. Ibid. 20 September 1966: APC.
74. Ibid. 25 February 1965: APC.
75. *Journey Continued*, 262.
76. *Star*, 21 March 1966.
77. Unpublished letter, Paton to Diana Collins, 27 April 1966: APC.
78. Unpublished letter, Paton to Matthews, 30 March 1966: APC.

79. Paton actually had the operation on 22 September 1966: Dorrie Paton's Diary of that date, APC.
80. Published in *Liberal Opinion* 4, 7, December 1966, 5.
81. The Phoenix Trust had been set up to help fund the Phoenix Settlement outside Durban, founded by the Mahatma Gandhi and run by his son Manilal. Paton served on it selflessly for many years, despite being constantly at loggerheads with his fellow trustee Mewa Ramgobin, who was married to Gandhi's granddaughter Ela.
82. Interview with Anthony and Maggie Barker, London, 1 May 1991.
83. Unpublished letter, Paton to the British politician Jo Grimond, 13 September 1966: APC. Paton was trying to arrange a visit by Grimond to South Africa.
84. Unpublished letter, Paton to David Craighead, 22 June 1966: APC.
85. Unpublished letter, Paton to Senator Kennedy, 3 November 1966: APC.
86. He told this story to the Liberal Conference in July 1966: it is quoted in *Liberal Opinion*, 4, 6, September 1966, 2. Paton attributed the parable to the Quaker Reginald Reynolds.
87. It was published in New York by Twayne in their World Authors series in 1968, and a revised version appeared in 1982. This remains the best book on Paton's writings.
88. Interview with Edward Callan, Kalamazoo, 5 April 1992.
89. Interview with Anthony and Maggie Barker, London, 1 May 1991.
90. Ibid. Barker wrote a book about his experiences in Zululand: *Giving and Receiving: an Adventure in African Medical Practise* (London: Faith Press, 1959).
91. *The Times*, 5 May 1967.
92. 'In Memoriam: Albert Luthuli', *The Long View* (ed. Edward Callan), 265–7. Details of the Lutuli funeral are given in 'Death of a Chief', by William Redmond Duggan, *The Crisis*, November 1977, 423–5. I am grateful to Mrs Bunny Duggan for drawing my attention to this article.
93. Unpublished letter, Paton to Betty Patterson, 9 February 1967: BP.
94. Unpublished letter, 29 December 1966: APC.
95. Dorrie Paton's Diary, 28 February 1967: APC.
96. *Journey Continued*, 271.
97. Burns would return in 1970 for another brief visit.
98. Interview with Jonathan Paton, Johannesburg, 17 January 1990.
99. *Journey Continued*, 271.
100. Unpublished letter, Paton to Dorrie, 18 December 1955: APC.
101. John Vaughn, O.F.M., Preface to *Francis and Clare: The Complete Works* (New York: Paulist Press), p. xiii. I am grateful to Mrs Frankie Braden of Sydney and Father Geoffrey Plant O.F.M. for drawing my attention to this volume. In England the prayer was widely spread by a Miss M. Berkeley and Lady Brassey, who had been sent a copy from Florence in 1930: Letter from Miss Berkeley to Malcolm Muggeridge, 25 June 1972: APC. The prayer was quoted by Margaret Thatcher on the occasion of her first election victory as Prime Minister and leader of the Conservative Party on 4 May 1979.

102. Unpublished letter, Paton to Malcolm Muggeridge, 3 August 1972: APC.
103. Paton's note in Dorrie's 1967 Diary: APC.
104. Unpublished letter, Paton to Roy Votaw, 28 October 1971: RV.
105. 'Mazeppa'.
106. *For You Departed* (also published as *Kontakion For You Departed*), 155.
107. Interview with Margaret Paton, Johannesburg, 2 February 1992.
108. *Journey Continued*, 273.
109. Paton's entry in Dorrie Paton's Diary, 3 November 1967: APC.
110. Paton's entry in Dorrie Paton's Diary, 2 December 1967: APC.

20. FALL AND RISE

1. Unpublished letter, Pondi Morel to Leslie Rubin, 14 November 1967: LR.
2. Interview with Bob and Guinevere Ventress, Linton, Cambridgeshire, 2 February 1991.
3. *Journey Continued*, 281–2.
4. Unpublished letter, Paton to Betty Patterson, 4 December 1967: BP.
5. Unpublished letter, Paton to Callan, 31 January 1968: EC. Paton sent the introduction on the same day as this letter.
6. Unpublished letter, Paton to Edward Callan, 5 February 1968: EC.
7. Unpublished letter, Paton to Father John Pesce, 20 January 1986: FJP.
8. Note in Dorrie's diary, 1 December 1967: APC.
9. Edward Callan, *Alan Paton* (rev. edn.), 111. The Kiev Kontakion is commonly used in Anglican memorial or burial services.
10. Unpublished letter, Paton to Rubin, 16 February 1968: LR.
11. *Journey Continued*, 279.
12. Unpublished letter, Paton to Mary Benson, 23 March 1968: MB.
13. *Journey Continued*, 279.
14. Durban *Daily News*, 24 July 1968. I am grateful to Mr Roy Rudden for supplying copies of these and other news-clippings of the events surrounding Paton's assault.
15. 'Paton Gives Evidence', Durban *Daily News*, 24 July 1968.
16. *Towards the Mountain*, 16.
17. Interview with Anne Paton, Gillitts, Natal, 29 January 1990.
18. Interview with Jonathan Paton, Sydney, 22 June 1993. The lawyer gave Jonathan Paton this account immediately after the trial concluded, on 25 or 26 July 1968.
19. Interview with a lawyer who cannot be named, Durban, 19 June 1991.
20. Paton in evidence at the trial said he had deliberately pressed the horn: 'Paton Gives Evidence', Durban *Daily News*, 24 July 1968.
21. Ibid.
22. Ibid.
23. Ibid. 25 July 1968.
24. Interviews with David Paton, Johannesburg, 3 February 1990, and Jonathan

Paton, Johannesburg, 14 February 1990.

25. Interview with Jonathan Paton, Johannesburg, 14 February 1990.
26. Ibid.
27. Unpublished letter, Paton to Rubin, 4 November 1959: APC, and Rubin to the writer, 17 November 1992: PFA.
28. Interview with Anne Paton, Gillitts, Natal, 29 January 1990, and 'Paton to Help Son of his Rescuer', Natal *Sunday Tribune* 28 July 1968.
29. Prestwich's letter, 11 March 1968, and Paton's reply, 14 March 1968, are in APC. Other letters of sympathy, and Paton's dismissive replies, include a letter from Leo Marquard (Paton's response 8 April 1968) and letter from Edward Callan, 29 March 1968: APC.
30. Unpublished letter, Paton to his solicitor Brian Hardman of Lester E. Hall & Co, 19 March 1968: APC. He gave Jonathan details of this will in his letter of 18 May 1968: JP.
31. Anne Paton, *Some Sort of a Job: My Life with Alan Paton*, 6.
32. By a curious coincidence, Paddy Hopkins was the first cousin of Paton's daughter-in-law, Nancy, so that Paton and Anne had a link from the first.
33. Anne Paton, *Some Sort of a Job: My Life with Alan Paton*, 7.
34. Interview with Anne Paton, Gillitts, 29 January 1990.
35. Anne Paton, *Some Sort of a Job: My Life with Alan Paton*, 7.
36. Ibid. 8.
37. Quoted in letter from Margaret Snell to the Archivist, Alan Paton Centre, Pietermaritzburg, 10 March 1991. Paton sent Snell the poem from Anerley on 13 February 1953: APC.
38. Anne Paton, *Some Sort of a Job: My Life with Alan Paton*, 9.
39. Ibid.
40. *For You Departed* (also published as *Kontakion For You Departed*), 83–4.
41. Unpublished letter, Anne Paton to the writer, 16 March 1993: PFA.
42. Interview with Anne Paton, Gillitts, Natal, 29 January 1990.
43. Ibid.
44. Anne Paton, *Some Sort of a Job: My Life with Alan Paton*, 14.
45. Unpublished letter, Rubin to the writer, 17 November 1992: PFA.
46. Unpublished letter, Mary Benson to the writer, 28 March 1993: PFA.
47. Interview with Mary Benson, London, 29 April 1991.
48. *Journey Continued*, 282.
49. Interview with Anne Paton, Gillitts, Natal, 29 January 1990, and Anne Paton, *Some Sort of a Job: My Life with Alan Paton*, 13.
50. Anne Paton, *Some Sort of a Job: My Life with Alan Paton*, 14.
51. Interview with Jonathan Paton, Johannesburg, 14 February 1990.
52. *Journey Continued*, 284.
53. Interview with Anne Paton, Gillitts, Natal, 29 January 1990.
54. Interview with Athene Hall, Cape Town, 11 February 1990.
55. Interview with Anne Paton, Gillitts, Natal, 29 January 1990.
56. Unpublished letter, Paton to bishop Philip Russell, 4 September 1975: APC.

Russell's reply is 29 September 1975: APC.

57. *Journey Continued*, 283.
58. Anne Paton, *Some Sort of a Job: My Life with Alan Paton*, 16.
59. Unpublished letter, Paton to Nicholas Paton, 23 May 1968: JP.
60. Unpublished letter, Paton to Pamela Paton, 3 December 1968: JP.
61. Unpublished letter, Paton to Margaret Paton, 5 August 1968: JP.
62. Interview with Colin Gardner, Pietermaritzburg, 21 June 1991. One of those who refused to phone Paton in this way was John Carlyle Mitchell, head of Nottingham Road school and an old friend.
63. Paton gives this figure in an unpublished letter to Jonathan and Margaret Paton, 3 December 1968: JP.
64. Anne Paton doubts that these letters were destroyed at this time; in her view the destruction would have taken place earlier. Unpublished letter, Anne Paton to the writer, 16 March 1993: PFA.
65. Unpublished letter, Paton to Leslie Rubin, 26 August 1969: LR.

21. BETTER THAN RUBIES

1. Unpublished letter, Anne Paton to the writer, 16 March 1993: PFA.
2. Paton's side of the extensive correspondence with his younger son, on which I have drawn for these details, is in Jonathan Paton's possession, while Jonathan's is in APC.
3. Unpublished letter, Paton to Edward Callan, 10 November 1970: EC.
4. Unpublished letter, Paton to Jonathan and Margaret, 24 September 1969: JP.
5. Anne Paton gives details of the difficult emotional triangle which I have sketched here in Anne Paton, *Some Sort of a Job: My Life with Alan Paton*, chapter 3.
6. Interview with Guinevere Ventress, Linton, Cambridgeshire, 2 February 1991.
7. Interview with Anne Paton, Gillitts, Natal, 29 January 1990.
8. *Reality* i. 1, March 1969, 4.
9. Interview with Colin Gardner, Pietermaritzburg, 21 June 1991.
10. Van der Post offered to play this role in a letter to Paton, 26 May 1966, and further correspondence giving evidence of the source of the money includes van der Post's letters of 15 December 1969, 16 February 1976, 2 Nov 1978, and 27 May 1984: all APC. The letter making the link with Defence and Aid plain is Paton to Diana Collins, 27 May 1984: APC.
11. Interview with Colin Gardner, Pietermaritzburg, 21 June 1991.
12. Unpublished letter, Paton to Richard Hirsch, 4 April 1969: APC.
13. He describes some of these activities in a letter to Jonathan Paton, 1 July 1970: JP.
14. Unpublished, untitled poem, dated 26 January 1970: APC.
15. *Knocking on the Door*, 227: the MS is in APC, dated 14 March 1970.
16. Unpublished letter, Paton to Helen Suzman, 2 May 1970: APC.

17. Unpublished letter, Paton to Jonathan Paton, 8 May 1970: JP.
18. As Professor Edward Callan has pointed out, Paton quotes the Whitman poem in the lines on Jeffrey Miller: 'We mourn for you, and yet shall mourn / With Ever-returning spring.'
19. There is a copy of Jahn's score in APC.
20. Unpublished letter, Ogden R. Reid to Paton, 7 May 1971: APC.
21. Bibliography by Edward Callan, introduced by Rolf Italiaander, *Alan Paton* (Hamburg: Hans Christians Verlag, 1970).
22. Interview with Edward Callan, Kalamazoo, 5 April 1992, and Callan's Masterwork Reader's Companion to *Cry, the Beloved Country*, 82–3.
23. Unpublished letter, Paton to Mary Benson, 19 January 1984: MB.
24. Guy Butler, *From a Local Habitation: An Autobiography*, 237.
25. *Burger's Daughter*, 95. The savagery of the simile is unusual for Gordimer, and it illustrates the depth of her feeling about rich whites' passion for radical chic.
26. Unpublished letter, Nadine Gordimer to Paton, 15 November 1979: APC.
27. Unpublished letter, Paton to Edward Callan, 29 July 1980: EC.
28. Unpublished letter, Paton to Edward Callan, 27 January 1981: EC.
29. He told Guy Butler *July's People* was her best book: unpublished letter, Paton to Butler, 17 February 1981: APC.
30. Unpublished letter, Nadine Gordimer to Paton, 21 November 1981: APC.
31. APC.
32. Unpublished letter, Paton to Colonel Van der Post, 7 February 1970: APC.
33. Unpublished letter, Paton to Jonathan Paton, 1 July 1970: JP.
34. Unpublished letter, Anne Paton to the writer, 16 March 1992: PFA.
35. Interview with David Welsh, Cape Town, 14 January 1992.
36. Ibid.
37. Unpublished letter, Anne Paton to Margaret Paton, 3 December 1970: APC.
38. Interview with Sir Laurens van der Post, London, 4 December 1977.
39. Unpublished letter, Paton to Edward Callan, 1 November 1973: EC.
40. Unpublished letter, Paton to Edward Callan, 13 August 1974: EC.
41. Interview with Tony Morphet, Cape Town, 12 June 1991.
42. Robinow's chief publications are *Looking East: Germany Beyond the Vistula* (Terra Mare Office, Hamburg, 1933), *Peter in the Post Office* (John Lane The Bodley Head, London, 1934), *The South African Saturday Book* [co-edited with Eric Rosenthal] (Hutchinson, Johannesburg, 1948) and *Rien Sans Peine: A Look at Five Generations of Robinows* (Privately Printed, Toronto, 1983).
43. Copy of Paton's sermon delivered 30 May 1971: RR.
44. Harvard University News Office Release, 17 June 1971: Kent.
45. *Journey Continued*, 288.
46. Unpublished letter, Anne Paton to the writer, 16 March 1993: PFA.
47. *Journey Continued*, 288.
48. Unpublished letter, Huddleston to Paton, 12 March 1971: APC.
49. Roy Campbell, *Broken Record*, 11.

50. Unpublished letter, Paton to the writer, 26 October 1976: PFA.
51. *Journey Continued*, 289.
52. Interview with Tony Morphet, Cape Town, 12 June 1991.
53. Unpublished Christmas card, Paton to Leslie Rubin, December 1972: LR.
54. *Journey Continued*, 290.
55. Ibid.
56. APC.
57. 'Our Multi-Racial Nation', address delivered by Paton at Rhodes on 7 April 1972: commemorative booklet published by Rhodes University, May 1972: LR.
58. Unpublished letter, Paton to Leo Marquard, 15 April 1972: UCT.
59. 'Our Multi-Racial Nation', address delivered by Paton at Rhodes on 7 April 1972: commemorative booklet published by Rhodes University, May 1972, p. 10: LR.
60. Interview with Professor Colin Gardner, Pietermaritzburg, 21 June 1991.
61. Anne Paton, *Some Sort of a Job: My Life with Alan Paton*, 34.
62. Unpublished poem, undated [11 January 1979]: APC.
63. Anne Paton, *Some Sort of a Job: My Life with Alan Paton*, 35.

22. STORMY TWILIGHT

1. Quoted in *Journey Continued*, 294.
2. Unpublished letter, Paton to Liston Pope, son of his old friend, 18 February 1980: APC.
3. Interview with Anne Paton, Gillitts, Natal, 29 January 1990.
4. Unpublished letter, Paton to Gatsha Buthelezi, 13 June 1973: APC.
5. Unpublished letter, Paton to Mrs Mewa Ramgobin, 14 January 1972: APC.
6. Unpublished letters, Paton to Charles Nupen, 21 May 1974 and 1 July 1974: APC.
7. Unpublished letter, Charles Nupen to Paton, 9 August 1974: APC.
8. Unpublished letter, Paton to Nupen, 15 October 1974: APC.
9. Unpublished letter, Paton to Leslie Rubin, 30 September 1958: LR.
10. Paton to the editor of *The Times*, 6 December 1961: APC.
11. Unpublished letter, Paton and Peter Brown to Mr Walsh, 16 April 1964: APC.
12. Unpublished letter, Paton to Sir Laurens van der Post about the actions of PEN, 25 June 1965: APC.
13. Unpublished letter, Paton to Miss Robinson, 12 November 1965: APC.
14. Unpublished letter, Paton to Trevor Huddleston, 16 November 1969: TH.
15. Transcript of evidence, State v. Nelson Mandela and Others, 12 June 1964: Wits.
16. Harvard University Press release, 17 June 1971: Kent.
17. Anne Paton, *Some Sort of a Job: My Life with Alan Paton*, 40.
18. Unpublished letter, Paton to Leo Marquard, 21 August 1970: UCT.

19. Unpublished letter, Paton to Buthelezi, 20 November 1973: APC.
20. *Reality*, August 1977.
21. On 2 June 1977.
22. Unpublished letter, Edgar Brookes to Paton, 8 September 1977: APC.
23. Unpublished letter, Paton to Brookes, 20 September 1977: APC.
24. 'Hail to the Chief', unpublished doggerel verse, 11 January 1970: APC.
25. Unpublished letter, Mangosuthu Buthelezi to Paton, 7 December 1963: APC.
26. Unpublished letter, Paton to Buthelezi, 18 October 1972: APC.
27. Unpublished letter, Paton to Red and Bunny Duggan, 4 November 1972: BD.
28. 'Gatsha Buthelezi Hoes a Hard Row,' in *Race Relations News*, 36/6, 1.
29. Unpublished letter, Mangosuthu Buthelezi to Paton, 5 April 1977: APC.
30. Paton quotes these words in an unpublished letter to Phebe Leeman, 5 August 1980: APC.
31. Unpublished letter, Paton to Tom Karis, 2 December 1980: APC.
32. Unpublished letter, Buthelezi to Paton, 20 July 1982: APC.
33. Unpublished letter, Paton to Buthelezi, 15 January 1985: APC.
34. Unpublished letter, Buthelezi to Paton, 28 January 1985: APC.
35. Unpublished letter, Paton to Buthelezi: 7 February 1985: APC.
36. *Journey Continued*, 294.
37. 'Paton Attacks Gordimer': *Sunday Times*, 7 December 1974.
38. Unpublished letter, Paton to Mary Benson, 26 March 1975: MB.
39. Unpublished letter, Gordimer to Paton, 1 February 1975: APC.
40. A. J. Lutuli, *Let My People Go*, 139.
41. This cartoon, dated 25 August 1976, is now in APC.
42. Interview with Colin Gardner, Pietermaritzburg, 21 June 1991.
43. 'Novelist Alan Paton Suggests Change (for the Better)', interview with Barbara Hutmacher, *Daily Dispatch*, East London, 21 April 1977.
44. Unpublished letter, Paton to Suzman, 9 April 1974: APC.
45. Unpublished letter, Helen Suzman to Paton, 16 April 1974: APC.

23. THE PUBLIC MAN

1. TS copy of address, 23 November 1974: JP.
2. Unpublished letter, Paton to Butler 9 April 1974: APC.
3. Unpublished letter, Paton to Edward Callan, 13 August 1974: EC.
4. Isaiah 11: 9.
5. Sermon delivered in Great St Mary's, Cambridge, 27 April 1975: PFA.
6. Interview with Tony Morphet, Cape Town, 12 June 1991.
7. Interview with Colin Gardner, Pietermaritzburg, 21 June 1991.
8. Unpublished letter, Paton to Laurens van der Post, 6 February 1975: APC.
9. Unpublished letter, Paton to Trevor Huddleston, 18 March 1976: TH.
10. Unpublished letter, Paton to Rubin, 29 March 1975: LR.

11. This painful process is described in an unpublished letter, Paton to Edward Callan, 11 February 1978: EC. Anne Paton denies that Paton protested or sent a cable; she was the one who wrote to Scribner, as detailed in Anne Paton, *Some Sort of a Job: My Life with Alan Paton*, 153.
12. Unpublished letter, Paton to Aubrey Burns, 23 January 1979: APC.
13. Interview with Peter Rodda, London, 20 May 1991.
14. Interview with Tony Morphet, Cape Town, 12 June 1991.
15. Interview with Adrian Leftwich, York, 25 February 1991.
16. Unpublished letter, Paton to Leftwich, 19 April 1975: AL.
17. Interview with Violet Swann, Cambridge, 23 January 1991.
18. Ibid.
19. Interview with Dr Audrey Glauert, Cambridge, 10 January 1991.
20. Unpublished letter, Prof. J. W. Linnett (Vice-Chancellor of the University of Cambridge) to Trevor Huddleston, 7 August 1974: TH.
21. Unpublished letter, Paton to Adrian Leftwich, 21 June 1975: AL.
22. Unpublished letter, Paton to the editors of *A Tapestry of Fifty Years*, 12 August 1975: APC.
23. Letter from Burns to Paton, 10 April 1976: APC. Paton wrote a foreword to the resulting volume, *Out of a Moving Mist*, which was published in 1977.
24. Interview with Ian Wyllie, Gillitts, 19 June 1991. According to Anne Paton, Paton was on the *Tribune* payroll as an official contributor: unpublished letter to the writer, 16 March 1993.
25. Unpublished letter, Paton to Red and Bunny Duggan, 17 September 1975: BD.
26. Unpublished letter, Paton to Buthelezi, 17 January 1976: APC.
27. Examples are his articles, 'First thing to do: Repent,' Johannesburg *Star*, 22 June 1976, and 'Hate in the Beloved Country', *Observer Magazine*, 31 October 1976.
28. Kruger's reply was printed in the Toronto *Globe & Mail* (which had reprinted the *Star* article) on 3 July 1976.
29. Unpublished letter, Paton to the editor, Toronto *Globe & Mail*, 17 July 1976: APC.
30. It was promptly published in English as *Looking on Darkness* (1975) outside South Africa, and Brink's subsequent novels have been published in English.
31. Unpublished letter, Brink to Paton, 28 April 1980, quotes these comments: APC.
32. Unpublished letter, Paton to Guy Butler, 13 July 1982: APC.
33. Unpublished letter, Brink to Paton, quoting this judgement of Paton's, 28 April 1980: APC.
34. Unpublished letter, Karel Schoeman to Paton, 27 December 1977: APC.
35. Unpublished letter, Karel Schoeman to Paton, 4 December 1976: APC.
36. 'My African Nightmare', *New York Times*, August 1976: EC.
37. 'South African praises U.S. "balance"', *Flint Journal*, 11 May 1977: K.
38. Paton's letter appealing to Cronkite to make the arrangements is dated 15

April 1977: APC.

39. Unpublished letter, Paton to Bunny Duggan, 24 November 1980: BD.
40. Unpublished letter, Paton to Cronkite, 15 April 1977: APC.
41. These details of what he planned to say to Carter are set out in Paton's unpublished letter to Cronkite, 15 April 1977: APC.
42. Paton quotes the Carter statement in an unpublished letter to Gatsha Buthelezi, 26 July 1977: APC.
43. Unpublished letter, Paton to Walter Cronkite, 5 July 1977: APC.
44. Unpublished letter, Paton to Cyrus Vance, 5 July 1977: APC.
45. Red Duggan had died some months before.
46. Biko had been arrested on 18 August 1977. After severe physical violence in the security police cells in Port Elizabeth, he was transported naked and critically ill to Pretoria, where he died of his injuries on 12 September 1977.
47. 'S. African author Paton defends his views', in *Christian Science Monitor*, 25 November 1977.
48. Unpublished letter, Paton to the Revd David Russell, 21 December 1977: APC.
49. Ibid.
50. Paton, 'We shared some thoughts . . . and feelings of guilt', in *Sunday Tribune*, 6 July 1986.
51. Apart from Huber, they were Christopher Probst, Alexander Schmorell, Hans and Sophie Scholl, and Wilhelm Graf.
52. Unpublished letter, Paton to Mary Benson, 9 December 1977: MB.
53. Unpublished letter, Paton to the Revd David Russell, 27 October 1978: APC.
54. Interview with Colin Gardner, Pietermaritzburg, 21 June 1991.

24. UPHILL ALL THE WAY

1. He gives the date of beginning *Ah, But Your Land is Beautiful* in an unpublished letter to Aubrey Burns, 20 January 1981: APC.
2. The *Raj Quartet* consists of *The Jewel in the Crown* (1966), *The Day of the Scorpion* (1968), *The Towers of Silence* (1971), and *A Division of the Spoils* (1975). Paton told Anne that the Scott novels were his inspiration: interview with her, Gillitts, 29 January 1990.
3. Unpublished letter, Paton to Edward Callan, 8 April 1980: EC.
4. Unpublished letter, Paton to Edward Callan, 29 July 1980: EC.
5. Paton's speech at the Cape Town launch of *Ah, But Your Land is Beautiful*, September 1981: APC.
6. He gives the date of completion in an unpublished letter, Paton to Rt Revd Michael Nuttall, 20 January 1981: APC.
7. Unpublished letter, Paton to Edward Callan, 27 January 1981: EC.
8. Interview with Joseph Lelyveld: 'South African Paton writes again, with hope for the beloved country', *Detroit Free Press*, 17 July 1981.

9. This is the line taken by G. D. Killam in *African Book Publishing Record*, vol. viii. 3 (1982).
10. Martin Rubin, 'Alan Paton's South Africa', *The Washington Quarterly*, (Summer 1982), 152–6.
11. Unpublished letter, Paton to Edward Callan, 11 August 1981: EC.
12. Unpublished letter, Paton to the writer, 30 September 1981: PFA.
13. Unpublished letter, Paton to Mary Benson, 24 November 1981: MB.
14. Draft of unpublished letter, Paton to Nadine Gordimer, 9 January 1982: APC. It is not certain that he ever sent this letter.
15. *Journey Continued*, 298.
16. He says as much to Mary Benson, in an unpublished letter, 8 April 1982: MB.
17. Interview with Anne Paton, Gillitts, 29 January 1990.
18. 'Ode to Alan Paton on his 80th birthday, January 1983': APC.
19. Anne Paton, *Some Sort of a Job: My Life with Alan Paton*, 185.
20. This is Wyllie's account. Anne Paton considers it an exaggeration: letter from her to the writer, 16 March 1992: PFA.
21. Interview with Ian Wyllie, Gillitts, 19 June 1991.
22. Anne Paton, *Some Sort of a Job: My Life with Alan Paton*, 185–6.
23. Jonathan might get R12.50 as a birthday present, Margaret R10, Nicholas R7.50 and Pamela R3.50. Jonathan, who spent about R100 on Paton each birthday, found the sums of 50c 'ridiculous and insulting': letter to the writer, 7 July 1993: PFA. Even Anne thought Paton mean in this regard.
24. Interview with Anne Paton, Gillitts, Natal, 29 January 1990. One of these bookcases is now in APC.
25. Interview with Anthony Paton, Johannesburg, 2 February 1992.
26. Ibid.
27. Interviews with Anne Paton, Gillitts, Natal, 29 January 1990, with Anthony Paton, Johannesburg, 2 February 1992, and with Jonathan and Margaret Paton, Johannesburg, 2 February 1992. In later years Jonathan was to come to believe that he had been cut out of his father's final will as a result of this episode: interview with him, Sydney, 6 June 1993.
28. They married in 1993, five years after Paton's death.
29. Unpublished letter, Anne Paton to the writer, 16 March 1993: PFA.
30. Interview with Eunice ('Dorrie') Arbuthnot, Pietermaritzburg, 31 January 1990.
31. Interview with Ian Wyllie, Gillitts, 19 June 1991.
32. Interview with David Paton and Maureen, Johannesburg, 3 February 1990.
33. Interview with Anthony Paton, Johannesburg, 2 February 1992.
34. Anne Paton, *Some Sort of a Job: My Life with Alan Paton* 75.
35. Ibid. 42.
36. Paton tells this story in an unpublished letter to his son Jonathan, 19 March 1983: JP.
37. Unpublished letter, Paton to Pat Poovalingam, 28 February 1983: APC.
38. Unpublished letter, Douglas Livingston to Paton, 15 April 1983.

39. Unpublished letter, Simon Roberts to Paton, 26 April 1982: APC.
40. 'The National Interest: Is there such a thing?' *Leadership SA*, (May 1983), 13–19.
41. P. W. Botha had succeeded Vorster as Prime Minister in September 1976.
42. The Johannesburg *Star*, 25 July 1983.
43. Among these was *Beeld*, which had attacked Paton repeatedly, particularly in 1960 when Bishop Reeves fled the country.
44. He used this phrase in his article explaining his position, and published in the *Financial Mail*, 2 August 1983.
45. Johannesburg *Star*, 1 August 1983: 'I would not vote for SA's constitution, says Paton'.
46. *Tribune*, 29 July 1983: 'I hardly said a word'.
47. Unpublished letter, Paton to Winnie Mandela, 9 August 1983: APC.
48. Interview with Anne Paton, 29 January 1990, and Anne Paton, *Some Sort of a Job: My Life with Alan Paton*, 191. Winnie Mandela was perhaps ashamed of her treatment of Paton, for on p. 81 of her book *Part of My Soul Went With Him* (1984) she paid tribute to Paton for having pleaded in mitigation of her husband's sentence for treason, saying she would never forget him for that.
49. Since remarried, she is now Rebecca Brumley.
50. Jonathan Paton, on a trip to Botswana with Holly Rood and his father in 1980, was struck by the fact that his father flirted with Holly Rood all the way. Interview with Jonathan Paton, Johannesburg, 17 January 1990.
51. Unpublished letter, Paton to Bill Worthington, 23 April 1984: MW.
52. Unpublished letter, Paton to Diana Collins, 27 May 1984: APC.
53. Unpublished letter, Paton to Gonville ffrench-Beytagh, 1 December 1981: APC.
54. Gertrude Cohn was among them: interview with her, London, 7 May 1991.
55. Interview with David Welsh, Cape Town, 14 January 1992.
56. Interview with Peter Brown, Pietermaritzburg, 30 January 1990. Mitchell's version of this remark is 'Paton, that's great, man, you've come back!'
57. Interview with David Welsh, Cape Town, 14 January 1992.

25. JOURNEY COMPLETED

1. Unpublished letter, Paton to Mary Benson, 15 April 1984: MB.
2. Interview with Harry Oppenheimer, Johannesburg, 30 January 1992.
3. Unpublished letter, Paton to P. W. Botha, 10 May 1984: APC.
4. P. W. Botha to Paton, 19 September 1984: APC.
5. Published by Ravan Press, Cape Town, 1984.
6. 'Author: Rich, BOOK: POOR', *Sunday Times*, 24 June 1984.
7. Interview with David Welsh, Cape Town, 14 January 1992.
8. Ibid.
9. Unpublished letter, Paton to Mary Benson 28 October 1984: MB.
10. Unpublished poem, entitled '1950–1984': APC.
11. He admits as much in an interview with the *Financial Mail*, 15 February 1985,

'Alan Paton: Life and Changing Times': 'It's not the ability to write the words [that is lacking]; it's the will to do so'.

12. Unpublished letter, Paton to Ursula Niebuhr, 21 August 1984: LC.
13. The United Democratic Front was an umbrella organization of groups opposed to apartheid, which spoke for the ANC before that organization's unbanning in 1990. Paton thought the UDF 'a faceless organization' run by 'amiable nonentities'. Unpublished letter, Paton to Prof Ben Marais, 12 August 1985: APC.
14. Unpublished letter, Paton to Edward Callan, 28 October 1985: EC.
15. 'Composed a few miles above Tintern Abbey, on revisiting the banks of the Wye'.
16. Unpublished letter, Arthur Mamvura to Paton, no date [1985]: APC.
17. Unpublished letter, Edith Obose to Paton, 12 May 1986: APC.
18. '1985: Doesn't it qualify as the unhappiest year of our lives?' *Sunday Times*, 28 December 1985.
19. Interview with Jonathan Paton, Johannesburg, 17 January 1990.
20. Ibid.
21. Alan Paton, 'We shared some thoughts . . . and feelings of guilt', in *Sunday Tribune*, 6 July 1986.
22. Anne Paton, *Some Sort of a Job: My Life with Alan Paton*, 234.
23. Unpublished letter, Anne Paton to the writer, 16 March 1993: PFA.
24. A common chant during 1986 was reported to be 'Kentucky Fried, Kentucky Fried'.
25. Previously unpublished poem. There exist at least three versions of it, all in APC.
26. Mary Battiata, 'Alan Paton and his cry of alarm', in the *Washington Post*, 25 November 1986, p. E1.
27. It was Conor Cruise O'Brien who had called him 'a wise old baboon', in a private letter to Trevor Huddleston in 1984.
28. Interview with Constance Stuart Larrabee, Chestertown, 8 April 1992.
29. Unpublished letter, Paton to Bill and Margaret Worthington, 25 February 1986: MW.
30. Unpublished letter, Paton to Randolph Vigne, no date [May 1986]: APC.
31. Undated Christmas card, Paton to Mary Benson, [December 1986]: MB.
32. Unpublished letter, Paton to Father John Pesce, 1 September 1987: APC.
33. Unpublished letter, Paton to Mary Benson, 7 December 1987: MB.
34. The word here is, of course, 'judicious'.
35. Interview with Athene Hall, Cape Town, 11 February 1990.
36. Interview with Ian Wyllie, Gillitts, 19 June 1991.
37. Unpublished letter, Paton to Diana Collins, 25 January 1988: APC.
38. Anne Paton, *Some Sort of a Job: My Life with Alan Paton*, 255–6.
39. It was published on the cover of *Leadership*, 7/2 (1988). The last photograph was that taken by Chick Flack, and published in Anne Paton, *Some Sort of a Job: My Life with Alan Paton*, facing p. 89.

40. Interview with Constance Stuart Larrabee, Chestertown, Maryland, 8 April 1992.
41. Unpublished letter, Paton to Ursula Niebuhr, 7 February 1988: LC.
42. *Journey Continued*, 301.
43. Anne Paton, *Some Sort of a Job: My Life with Alan Paton*, 263.
44. Ibid. 265.
45. Her maiden name was Catherine Shallis. A close friend of Jonathan's at university, she had been a leading figure among Liberal students, and in the Natal Liberal Party.
46. Interview with Eunice ('Dorrie') Arbuthnot, Pietermaritzburg, 31 January 1990.
47. 'A tribute to Alan Paton from his son, Jonathan', *The Sunday Star*, 17 April 1988.
48. Anne Paton, *Some Sort of a Job: My Life with Alan Paton*, 266.

CONCLUSION

1. 'PW leads tributes to Alan Paton', *The Citizen*, 13 April 1988.
2. Interview with Eunice ('Dorrie') Arbuthnot, Pietermaritzburg, 31 January 1990.
3. Interview with Anne Paton, Gillitts, 29 January 1990.
4. 'Lintrose', Botha's Hill, is now (1992) owned by Owen and Sheila Clarkson, to whom I am grateful for their kind reception and their generosity in letting me see the house and garden once more.
5. In the Brenthurst Library are the manuscripts of *Cry, the Beloved Country*, *Too Late the Phalarope*, *For You Departed* (also published as *Kontakion For You Departed*), *Hofmeyr*, and *Instrument of Thy Peace*. The Alan Paton Centre is the repository of all Paton's personal papers, letters and memorabilia. A South African film director, Anant Singh, bought the manuscripts of *Ah, But Your Land is Beautiful*, *Towards the Mountain*, and *Journey Continued*.
6. 'He wasn't a saint, says his beloved and "terribly bossy" Anne', *The Sunday Star*, 17 April 1988.
7. He called for such talks many times, notably in an article, 'Solving Our Problems', published in *Contact* vii. 6, 2, on 8 May 1964.
8. The words are those on a tablet to Sir Robert Shirley, Bart., in an old Yorkshire church, and Paton first heard of them from J. H. Hofmeyr. He often quoted them himself, notably in his pamphlet *Federation or Desolation* (SAIRR, 1985), 12.

Select Bibliography

Works By Alan Paton

NOVELS

Cry, the Beloved Country: A Story of Comfort in Desolation (New York: Scribner, 1948; London: Jonathan Cape, 1948; Toronto: Saunders, 1948; London: Penguin, 1958; New York: Collier Macmillan, 1987); *Too Late the Phalarope* (New York: Scribner, 1953; London: Jonathan Cape, 1953; Cape Town: Frederick Cannon, 1953); *Ah, But Your Land Is Beautiful* (Cape Town: David Philip, 1981; New York: Scribner, 1981; London: Jonathan Cape, 1981).

SHORT STORIES

Tales from a Troubled Land (New York: Scribner, 1961); republished as *Debbie Go Home* (London: Jonathan Cape, 1961).

PLAYS

Lost in the Stars, with Maxwell Anderson (New York: Sloane Associates, 1950); *Sponono*, with Krishna Shah (New York: Scribner, 1965; repr. Cape Town: David Philip, 1983).

BIOGRAPHIES

Hofmeyr (London: Oxford University Press, 1964); republished as *South African Tragedy: The Life and Times of Jan Hofmeyr*, abridged by Dudley C. Lunt (New York: Scribner, 1965); this abridgement republished in paperback as *Hofmeyr* (Cape Town: Oxford University Press, 1971); *Apartheid and the Archbishop: The Life and Times of Geoffrey Clayton, Archbishop of Cape Town* (Cape Town: David Philip, 1973; New York: Scribner, 1973; London: Jonathan Cape, 1973).

AUTOBIOGRAPHIES

Kontakion For You Departed (London: Jonathan Cape, 1969), republished as *For You Departed* (New York: Scribner, 1969); *Towards the Mountain* (New York: Scribner, 1980; London: Oxford University Press, 1980; Cape Town: David Philip, 1980); *Journey Continued* (London: Oxford University Press, 1988; New York: Scribner, 1988; Cape Town: David Philip, 1988).

OTHER VOLUMES

The Land and People of South Africa (Philadelphia: Lippincott, 1955, rev. 1964; Toronto: Longman, 1955), republished as *South Africa and her People* (London: Lutterworth Press, 1957, rev. 1970); *South Africa in Transition*, photographs by Dan Weiner (New York: Scribner, 1956); *Hope for South Africa* (London: Pall Mall Press, 1958; New York: Praeger, 1959); *Instrument of Thy Peace: Meditations Prompted by the Prayer of St Francis* (New York: Seabury Press, 1968; London: Collins, 1970); *The Long View*, ed. Edward Callan (New York: Praeger, 1968; London: Pall Mall Press, 1969); *Knocking on the Door: Shorter Writings*, ed. Colin Gardner (Cape Town: David Philip, 1975; New York: Scribner, 1975; London: Rex Collins, 1975); *Diepkloof: Reflections of Diepkloof Reformatory*, compiled and ed. Clyde Broster (Cape Town: David Philip, 1986); *Save the Beloved Country* (Johannesburg: Hans Strydom, 1987; New York: Scribner, 1989).

PAMPHLETS, CONTRIBUTIONS TO BOOKS, AND MISCELLANEOUS PUBLICATIONS

The Non-European Offender, Penal Reform Series, 2 (Johannesburg: South African Institute of Race Relations, 1945); *Freedom as a Reformatory Instrument*, Penal Reform Pamphlet 2 (Pretoria: Penal Reform League of South Africa, 1948); *South Africa Today*, Public Affairs Pamphlet 175 (New York: Public Affairs Committee, 1951; London: Lutterworth Press, 1953); 'Religious Faith and Human Brotherhood', in *Religious Faith and World Culture*, ed. A. W. Loos (New York: Prentice-Hall, 1951); *Salute to My Great Grandchildren* (Johannesburg: St Benedict's Press, 1954); *The Negro in America Today: A Firsthand Report* (Washington: Civil Rights Committee and Education and Research Department, Congress of Industrial Organizations, 1954); 'The Person in Community', in *The Christian Idea of Education: Papers and Discussion*, ed. Edmund Fuller (New Haven, Conn.: Yale University Press, 1957); *The People Wept* (Kloof, Natal: Alan Paton, 1959); *The Charlestown Story* (Johannesburg: The Liberal Party, 1959); 'Trevor Huddleston', in *Thirteen for Christ*, ed. Melville Harcourt (New York: Sheed and Ward, 1963); 'Ein großer Mann mit einem großen Herzen und einem mutigen Geist', in Rolf Italiaander, *Die Friedensmacher* (Kassel: J. G. Oncken Verlag, 1965); *Civil Rights and Present Wrongs: Address to the Civil Rights League, Cape Town, October 1968*, Topical Talks, 20 (Johannesburg: South African Institute of Race Relations, 1968); 'The Challenge of Fear', in *What I Have Learned* (New York: Simon and Schuster, 1968), 19–21, 46; *Towards Racial Justice: Will There Be a Change of Heart?* Alfred and Winifred Hoernlé Memorial Lecture (Johannesburg: South African Institute of Race Relations, 1979); 'The Afrikaner as I Know Him', in *The Afrikaners*, ed. Edwin S. Munger (Cape Town: Tafelberg Publishers, 1979), 24–5; *Case History of a Pinkie*, Topical Talks, 28 (Johannesburg: South African Institute of Race Relations, 1979); *Federation or Desolation*, Alfred and Winifred Hoernlé Memorial Lecture (Johannesburg: South African Institute of Race Relations, 1985).

BOOKS INTRODUCED BY PATON

GRANT, G. C., *The Liquidation of Adams College* ('For Private Circulation', no publisher or year given, but 1957); Paton's introduction, 3–6.

'The South African Novel in English', in *Report of a Conference of Writers, Publishers, Editors and University Teachers of English held at the University of the Witwatersrand, Johannesburg*, 10–12 July 1956 (Johannesburg: Witwatersrand University Press, 1957), 145–57.

KUPER, L., WATTS, H., and DAVIES, R., *Durban: A Study in Racial Ecology* (London: Jonathan Cape, 1958; New York: Columbia University Press, 1958), 13–16.

RUBIN, L., *This Is Apartheid* (London: Christian Action Pamphlets, 1959).

'In a World Ruled by Fear', in R. M. Bartlett (ed.) *They Stand Invincible: Men Who are Reshaping Our World* (New York: Thomas X. Crowell, 1959), 153–70.

HOOPER C., *Brief Authority* (London: Collins, 1960), 15–19.

Non-Racial Democracy: The Policies of the Liberal Party of South Africa (Pietermaritzburg: The Liberal Party, 1962).

BROOKES, E. H., *Three Letters from Africa* (Lebanon, Penn.: Pendle Hill Pamphlet 139, 1965), 3.

PERIODICAL PUBLICATIONS

'Jan Hendrik Hofmeyr: An Appreciation', *South African Opinion* (Sept. 1936), 7, 12; 'New Schools for South Africans', *Mentor* (June 1942), 5–7; 'Punishment and Crime: False Reasoning in Society's Attitude Towards the Offender', *Forum*, 6/25 (Sept. 1943), 5–6; 'Society Aims to Protect Itself: How Effective Is Severity of Punishment as a Deterrent against Crime?', *Forum*, 6/29 (Oct. 1943), 25–7; 'Significance of Social Disapproval', *Forum*, 6/36 (Dec. 1943), 25–7; 'Real Way to Cure Crime: Our Society Must Reform Itself', *Forum*, 6/44 (Jan. 1944), 24–6; 'A Plan for Model Prisons', *Forum*, 6/52 (Mar. 1944), 24–6; 'Education Needs of the Adolescent', 2 parts, *Transvaal Education News*, 40 (Sept. 1944), 13–15; 40 (Oct. 1944), 6–9; 'Who Is Really to Blame for the Crime Wave in South Africa?', *Forum*, 8/37 (Dec. 1945), 7–8; 'The Prevention of Crime', *Race Relations*, 12/3, 4 (1945), 69–77; 'Behandeling van die Oortreder' [Treatment of Offenders], *Nongqai*, 37 (July 1946), 34–5; 'Great Conference of Christians and Jews at Oxford', *Outspan*, 40 (Sept. 1946), 34–5, 103, 105; 'Penal Practice of Some Other Countries', *Nongqai*, 38 (July 1947), 876–8; 38 (Aug. 1947), 1013–17; 'Child in Trouble', *Mentor*, 30/7 (Oct. 1948), 2–9; 'Juvenile Delinquency and Its Treatment', *Community and Crime*, 2 (1948), 52–62; 'From Van Riebeeck to Hofmeyr', *Forum* 13 (Nov. 1948), 12–13; 'Africa Reporting', *New York Times Book Review*, 20 (Feb. 1949), 5, 26; 'Why Did Jan Hofmeyr Not Found a Liberal Party?', *Forum*, 26 (Feb. 1949); 'Towards a Spiritual Community', *Christian Century*, 67 (Mar. 1950), 298–300; 'The Story of *Cry, the Beloved Country*', *SA Publisher and Bookseller*, (1950), 2–3; 'Racial Situation in South Africa: A Struggle Between Justice and Survival', *African World* (Apr. 1950), 17–18; 'Christian Unity: A South African

View', *South African Outlook*, 71 (Oct. 1951), 148–52; 'A Step Forward', *Theoria*, 4 (1952), 6–9; 'The Unrecognized Power', *Saturday Review*, 34 (Nov. 1951), 10–11, 38–40; 'Force Won't Solve South Africa's Problems', *Forum*, 1/1 (Apr. 1952), 4–6; 'South Africa Today', *Southern Review*, 1/1 (Apr. 1952), 1–5; 1/2 (May 1952), 1–4; 1/3 (June 1952), 1–4; 'Africa, Awakening, Challenges the World', *New York Times Magazine*, 6 (July 1952), 6, 25–6; 'The White Man's Dilemma', *Saturday Review*, 26 (May 1953), 12–13, 53–4; 'America and the Challenge of Africa', *Saturday Review*, 26 (May 1953), 9; 'Impending Tragedy', *Life*, 24 (May 1953), 163–70; 'Crisis of White Supremacy', *Forum* (July 1953), 2–3; 'Liberal Approach', *New York Times Magazine* (Aug. 1953), 20–3; 'The English Speaking Churches and the Colour Bar', *Forum*, 2, NS 7 (Oct. 1953), 17–18; 'The Church Amid Racial Tensions', *Christian Century*, 71 (Mar. 1954), 393–4; 'Negro in America Today', *Colliers*, 134 (Oct. 1954), 20, 50–6; 'Negro in the North', *Colliers*, 134 (Oct. 1954), 70–80; 'The Novelist and Christ' (written with Liston Pope), *Saturday Review*, 27 (Dec. 1954), 15–16, 56–7; 'Time, Gentlemen, Time', *Forum*, 4 (Feb. 1955); 'Grim Drama in Johannesburg', *New York Times Magazine* (Feb. 1955), 15, 36, 38, 40; 'Olive Schreiner: The Forerunner', *Forum*, 4/1 (Feb. 1955), 25–9; 'African Advancement: A Problem of Both Copperbelt and Federation', *Optima*, 5 (Dec. 1955), 105–9; 'School in Danger', *Christian Century*, 72 (Dec. 1955), 1524–5; 'Tragedy of the Beloved Country', *Coronet* (May 1956), 40, 64–9; 'The Tragic and Lovely Land of South Africa', *Holiday* (Feb. 1957), 34–42, 119–21; 'Association by Permission', *Africa South* (July–Sept. 1957), 11–20; 'On Trial for Treason', *New Republic*, 137/11 (Nov. 1957), 9–12; 'The Narrowing Gap', *Africa Today* (Nov.–Dec. 1957), 10–12; 'Roy Campbell: Poet and Man', *Theoria*, 60 (1957), 19–31; 'Church, State and Race', *Christian Century*, 75 (Feb. 1958), 248–9; (Mar. 1958), 278–80; 'The Attitude of the Church and the Christian towards the State', *Background Information to Church and Society*, 19 (Mar. 1958); 'The Crusader on a Polo Pony', *Contact*, 1/5 (Apr. 1958); 'Towards a Nonracial Democracy', *Contact*, 1/2 (Feb. 1958), 11; 'Nigeria', *Contact*, 1/2 (Feb. 1958), 11; 'Liberals and the United Party: Sir De Villiers Graaf', *Contact*, 1/4 (Mar. 1958), 11; 'A Calm View of Change', *Contact*, 1/5 (Apr. 1958), 11; 'Liberals and the Nationalist Party', *Contact*, 1/6 (Apr. 1958), 11; 'Nonracialism in a Racial Society', *Contact*, 1/7 (May 1958), 11; 'SABRA Talks with Nonwhites', *Contact*, 1/8 (May 1958), 11; 'Bantu Education: The State Must Not Run the Universities', *Contact*, 1/9 (May 1958), 11; 'Tribute to the Bravest Liberal of Them All', *Contact*, 1/10 (June 1958), 11; 'Racial Juggernaut Moves on Indians', *Contact*, 1/11 (June 1958), 9; 'Group Areas Act Cruelty to the Indians Is Un-Christian', *Contact*, 1/14 (Aug. 1958), 9; 'Hooliganism Reveals the True Nature of Apartheid', *Contact*, 1/16 (Sept. 1958), 9; 'An Open Letter to Dr Verwoerd', *Contact*, 1/17 (Sept. 1958), 9; 'Raise Production by Treating African Labour Humanely', *Contact*, 1/18 (Oct. 1958), 9; 'Verwoerd's Claim to Divine Guidance', *Contact*, 1/19 (Oct. 1958), 9; 'A Foolish Man Imagines an "Indian Menace"', *Contact*, 1/20 (Nov. 1958), 9; 'Alan Paton Invites Mr Ekkis Blindendoof to Take the Short View', *Contact*, 1/21 (Nov. 1958), 9; '*The Cape Argus* and a Planned Utopia: Totalitarianism or Liberalism?', *Contact*, 1/22 (Nov.

1958), 9; 'Michael Scott: What Kind of Man Was He?', *Contact*, 1/23 (Dec. 1958), 9; 'The Accra Conference', *Contact*, 1/24 (Dec. 1958), 9; 'Precepts of a Cabinet Minister', *Contact*, 2/1 (Jan. 1959), 9; 'The Days of White Supremacy are Over', *Contact*, 2/2 (Jan. 1959), 9; 'South Africa, 1959', *Christianity and Crisis*, 19 (May 1959), 64–7; 'Some Thoughts on the Contemporary Novel in Afrikaans', *English Studies in Africa*, 2 (Sept. 1959), 159–66; 'South African Treason Trial', *Atlantic*, 205 (Jan. 1960), 78–81; 'Letter from Alan Paton', *English Studies in Africa*, 3 (Mar. 1960), 102; 'As Blind as Samson Was', *New York Times Magazine*, 10 (Apr. 1960), 9–11, 104–9; 'White Dilemma in Black Africa', *New York Times Magazine* (Sept. 1960), 8, 30–1; 'The Road', *New Yorker*, 26 (Dec. 1960), 32–3; 'Africa, Christianity and the West', *Christianity and Crisis*, 20 (Dec. 1960), 195–8; 'A Man Called Brown', *Contact*, 3/8 (Apr. 1960), 5; 'Our Rulers' Latest Blunder', *Contact*, 3/9 (May 1960), 5; 'Margaret Ballinger', *Contact*, 3/10 (May 1960), 5; 'The United Nations in Africa', *Contact*, 3/11 (June 1960), 5; 'A Nonviolent Third Force', *Contact*, 3/12 (June 1960), 5; 'Keep the Party Clean', *Contact*, 3/13 (July 1960), 5; 'The Congo', *Contact*, 3/15 (July 1960), 5; 'End of an Age', *Contact*, 3/16 (Aug. 1960), 5; 'Our New Bishop: Trevor Huddleston', *Contact*, 3/17 (Aug. 1960), 5; 'Cottlesloe Consultation', *The Compass* (Feb. 1961); 'Republic for which South Africa Stands', *New York Times Magazine* (May 1961), 9, 42–3; 'A Deep Experience', *Contrast*, 1/4 (Dec. 1961), 20–4; 'The Perils of Race Problems', *Egg: A New Magazine*, 1 (Easter 1961), 4; 'South Africa's Black Stallion', *Manchester Guardian Weekly* (Jan. 1962), 11; 'Four Interviews in South Africa: Alan Paton', with Studs Terkel, *Perspective on Ideas and the Arts* (Radio Station WFMT, Chicago, 12/5 (May 1963), 24–9; 'The Abuse of Power', *Liberal Opinion*, 2/4 (Sept. 1963), 5–9; 'Our New Africa', *Maryknoll*, 57 (Nov. 1963), 9–13; 'My Days at Maritzburg College as Schoolboy and Teacher', *Maritzburg College Centenary Supplement*, supplement to *The Natal Mercury*, (Feb. 1963), 16–17, 38; 'A Personal View', *New York Times* (29 Mar. 1964), s. 2, p. 1; 'Liberals Reject Violence', *Liberal Opinion*, 3/4 (Oct. 1964), 1–4; 'The Hofmeyr Biography', *Contrast*, 3 (Oct. 1964), 32–6; 'A Toast to the End of the Colour Bar', *Contact*, 7/2 (Jan. 1964), 2; 'A Dubious Virgin', *Contact*, 7/4 (Mar. 1964), 2; 'An Example for Us All: Elliott Mngadi', *Contact*, 7/5 (Apr. 1964), 2; 'Solving Our Problems', *Contact*, 7/6 (May 1964), 2; 'His Crime is Loyalty (Chief Albert Luthuli)', *Contact*, 7/7 (June 1964), 2; 'Liberalism and Communism', *Contact*, 7/8 (July 1964), 2; 'Intimidation', *Contact*, 7/9 (July 1964), 2; 'Peter Brown', *Contact*, 7/10 (Aug. 1964), 2; 'No Genade at Genadendal', *Contact*, 7/11 (Sept. 1964), 2; 'Something to Be Proud Of', *Contact*, 7/12 (Oct. 1964), 2; 'A Plea For the Freedom of Monkeys', *Contact*, 7/13 (Nov. 1964), 2; 'The Ninety Days', *Contact*, 7/14 (Dec. 1964), 4; 'Ideas Never Die', *Contact*, 8/1 (Jan. 1965), 2; 'St George Deserts to the Dragon', *Contact*, 8/2 (Feb. 1965), 2; 'Nationalism and the Theatre', *Contact*, 8/3 (Mar. 1965), 2; 'John Harris', *Contact*, 8/4 (Apr. 1965), 2; 'Ham-Handed Hildegard', *Contact*, 8/5 (May 1965), 2; 'Defence and Aid', *Contact*, 8/6 (June 1965), 2; 'Alan Paton Reports on South Africa', *Commonweal*, 82/10 (May 1965), repr. as 'A Special Report on the Republic of South Africa, 1965', in *Presbyterian Life*, 18/11 (June 1965), 6–11; 'Beware of Melancholy', *Contact*, 8/7

(July 1965), 2; 'Marquard on Liberalism', *Contact*, 8/8 (Sept. 1965), 5–7; 'Rhodesia', *Contact*, 8/9 (Oct. 1965), 7; 'In the Blossom Land', *Contact*, 8/10 (Dec. 1965), 7; 'Beware of Melancholy', *Christianity and Crisis*, 25 (Nov. 1965), 223–4, reprint of Paton's column 'The Long View', *Contact*, 8/7 (July 1965), 2–3; 'This Strange Phenomenon', *Liberal Opinion*, 4/3 (Aug. 1965), 1–3; 'The Ndamse Affair', *Contact*, 9/1 (Jan. 1966), 7; 'The Long Arm of Persecution', *Contact*, 9/1 (Mar. 1966), 7; 'The Trial', *Contact*, 9/3 (Apr. 1966), 7; 'The Second Quinquennium', *Liberal Opinion*, 4/5 (June 1966), 1–2; 'Waiting for Robert', *Contact*, 9/4 (July 1966), 7; 'Mr. S. E. D. Brown', *Contact*, 9/5 (Sept. 1966), 7; 'On Poverty', *Contact*, 11/6 (Oct. 1966), 5; 'The Challenge of Fear' ('What I Have Learned' series, 16), *Saturday Review*, 50 (Sept. 1967), 19–21, 46; 'In Memoriam: Albert Luthuli', *Christianity and Crisis*, 27 (Sept. 1967), 206–7; 'Dr. Hendrik Verwoerd: A Liberal Assessment', *Natal Daily News* (Sept. 1967); 'A Growth of the Soil', *Contrast*, 5/1 (Nov. 1967), 89–90; 'Authority and Freedom', *Liberal Opinion*, 6/1 (Jan. 1968), 4–6; 'Address to a Meeting of Protest at Johannesburg, 14th April, 1968', *The Black Sash*, 12 (May 1968), 2–5; 'Party Banned in South Africa: A Letter from Alan Paton, National President of the Liberal Party of South Africa', *The Times* (London) (Apr. 1968), 11; 'Smuts', *South African Outlook*, 98 (Oct. 1968), 164–5; 'The Yoke of Racial Inequality', *New York Times*, (Nov. 1968), 38; 'Creative Suffering', *National Catholic Reporter: Lenten Report Book Supplement*, 5/17 (Feb. 1969), 1, 16–17; 'A Message to the People of South Africa', *Christianity and Crisis*, 29/5 (Mar. 1969), 64–8; Editorial, *Reality: A Journal of Liberal Opinion*, 1/1 (Mar. 1969), 2–4; 'Katrina', *Reality: A Journal of Liberal Opinion*, 1/4 (1969), 16–17; 'Black Consciousness', *Reality: A Journal of Liberal Opinion*, 4/1 (1972), 9–10; 'Peter Brown: Rebanned', *Reality: A Journal of Liberal and Radical Opinion*, 4/4 (1972), 5–6; 'Lutuli Memorial Service', *Reality: A Journal of Liberal and Radical Opinion*, 4/4 (1972), 3–4; 'Ode to the New Reality', *Reality: A Journal of Liberal and Radical Opinion*, 4/6 (1973), 3; 'The Threat to the Alice Seminary', *Reality: A Journal of Liberal and Radical Opinion*, 5/3 (1973), 8–10; 'Gatsha Buthelezi Hoes a Hard Row', *Race Relations News*, 36/6 (June 1974), 1; 'The British Heritage in South Africa: A Look at its History and the Question of its Future', *Commonwealth* (Aug./Sept. 1975); 'The Beloved Country Today', *South Africa International*, 7/1 (July 1976), 17; 'An Enduring Scandal', *Reality: A Journal of Liberal and Radical Opinion*, 8/3 (1976), 4; 'William Plomer, Soul of Reticence', *Theoria*, 46 (1976), 1–15; 'Turfloop Testimonies', *Reality: A Journal of Liberal and Radical Opinion*, 8/5 (1977), 9–10; 'The Americans and Us', *Reality: A Journal of Liberal and Radical Opinion*, 9/4 (1977), 4–6; 'The Great Betrayal', *Reality: A Journal of Liberal and Radical Opinion*, 10/4 (1978), 5–8; 'Alan Paton Replies', *Reality: A Journal of Liberal and Radical Opinion*, 10/5 (1978), 7; 'A Patriot's Dilemma: Why I Stay in South Africa', *Commonweal*, 105/22 (Nov. 1978), 714–15; 'The Passing of Page View', *Reality: A Journal of Liberal and Radical Opinion*, 11/3 (1979), 8; 'Review of a Different Gospel', *Reality: A Journal of Liberal and Radical Opinion*, 12/4 (1980), 4–5; 'Marion Friedman', *Reality: A Journal of Liberal and Radical Opinion*, 13/1 (1981), 4; 'Reality', *Reality: A Journal of Liberal and Radical Opinion*, 13/6 (1981), 14; 'The Grand

Old Man Has Gone', *Reality: A Journal of Liberal and Radical Opinion*, 14/3 (1982), 3; 'Z. K. Matthews: A Review of Freedom', *Reality: A Journal of Liberal and Radical Opinion*, 14/3 (1982), 18; 'Apartheid and the Reformed Churches', *Reality: A Journal of Liberal and Radical Opinion*, 14/6 (1982), 10; 'The National Interest: Is There Such a Thing?', *Leadership, South Africa* 2/1 (1983), 12–19; 'Ernie Wentzel', *Reality: A Journal of Liberal and Radical Opinion*, 18/4 (1986), 6; 'The Sol Plaatje Memorial Lecture for 1987 by Alan Paton', *Reality: A Journal of Liberal and Radical Opinion*, 19/6 (1987), 7–10; 'A Literary Remembrance', *Time*, (25 Apr. 1988), 105–6.

BIBLIOGRAPHIES OF ALAN PATON

BENTEL, L., 'Alan Paton: A Bibliography' (Johannesburg: Witwatersrand University School of Librarianship, MA thesis, 1969).

CALLAN, E., *Alan Paton*, introd. Rolf Italiaander, (Hamburg: Hans Christians, 1970).

Other Works Cited

ALEXANDER, P. F., *William Plomer: A Biograhy* (London: Oxford University Press, 1989).

BENSON, M., *The African Patriots* (London: Faber & Faber, 1963), republished as *South Africa: The Struggle for a Birthright* (London: Penguin Books, 1966).

—— *A Far Cry* (London: Viking Penguin, 1989).

BERG, A. S., *Max Perkins, Editor of Genius* (New York: Pocket Books, 1978).

BLACK, M., 'Alan Paton and the Rule of Law', *African Affairs*, 91 (Jan. 1992), 53–72.

BROWN, R. M., 'Alan Paton, Warrior and Man of Grace', *Christianity and Crisis*, 48 (June 1988), 204–6.

BURT, C., *The Young Delinquent* (London: London University Press, 1925).

BUTLER, G., *From a Local Habitation: An Autobiography* (Cape Town: David Philip, 1991).

CALLAN, E., *Alan Paton* (New York: Twayne Publishers, 1968, rev. 1982).

—— 'Alan Paton and the Liberal Party', in A. S. Paton, *The Long View*, ed. E. Callan (New York: Praeger, 1968), 3–44.

—— *Cry, the Beloved Country: A Novel of South Africa* (Boston, Mass.: Twayne, 1991).

CHISHOLM, L., 'Education, Punishment and the Contradictions of Penal Reform: Alan Paton and the Diepkloof Reformatory, 1934–1948', *Journal of Southern African Studies*, 17/1 (Mar. 1991), 23–42.

DANIELS, E., 'Salute to the Memory', *Reality: A Journey of Liberal Opinion*, 20 (1988), 6.

DAVENPORT, T. R. H., *South Africa: A Modern History*, 4th edn. (London: Macmillan, 1977).

DAVIES, H., *A Mirror for the Ministry in Modern Novels* (New York: Oxford University Press, 1959).

DUNCAN, R. M., 'The Suffering Servant in Novels by Paton, Bernanos, and Schwarz-Bart', *Christian Scholar's Review*, 16/2 (Jan. 1987), 122–43.

EGELAND, L., *Bridges of Understanding* (Cape Town: Human & Rousseau, 1977).

FULLER, E., *Books with the Men Behind Them* (New York: Random House, 1962), 83–101.

GARDNER, C., 'Alan Paton: Often Admired, Sometimes Criticized, Usually Misunderstood', *Natalia*, 18 (Dec. 1988), 19–28.

—— 'Paton's Literary Achievement', *Reality*, 20/4 (1988), 8–11.

HAW, S., and FRAME R., *For Hearth and Home: The Story of Maritzburg College 1863–1988* (Pietermaritzburg: MC Publications, 1988).

HOFMEYR, J. H., *South Africa* (London: Ernest Benn, 1931).

HOOPER, M., 'Paton and the Silence of Stephanie', *English Studies in Africa*, 32 (1989), 53–62.

ITALIAANDER, R., 'Das Engagement des Schriftstellers Alan Paton in Südafrika', *Profile und Perspectiven* (Erlangen: Verlag der Ev. luth. Mission, 1970).

LINNEMANN, R. J., 'Alan Paton: Anachronism or Visionary', *Commonwealth Novel in English*, 3/1 (spring–summer 1984), 88–100.

LUTHULI, A., *Let My People Go* (London: Collins, 1962).

MONYE, A. A., '*Cry, the Beloved Country*: Should We Merely Cry?', *Nigeria Magazine*, 144 (1983), 74–83.

MORPHET, T., 'Alan Paton: The Honour of Meditation', *English in Africa*, 10/2 (Oct. 1983), 1–10.

MOSS, R., 'Alan Paton: Bringing a Sense of the Sacred', *World Literature Today*, 57/2 (spring 1983), 233–7.

MUNGER, N., *Touched by Africa* (Pasadena, Calif.: Castle Press, 1983), 29–43.

NASH, A.: 'The Way to the Beloved Country: History and the Individual in Alan Paton's *Towards the Mountain*', *English in Africa*, 10/2 (Oct. 1983), 11–27.

ODUMUH, E., 'The Theme of Love in Alan Paton's *Cry, the Beloved Country*', *Kuka: Journal of Creative and Critical Writing* (1980–1), 41–50.

PATON, A., *Some Sort of a Job: My Life with Alan Paton* (London: Viking Penguin, 1992).

RIVE, R., 'The Liberal Tradition in South African Literature', *Contrast*, 14/3 (July 1983), 19–31.

RUTHERFORD, A., 'Stone People in a Stone Country: Alan Paton's *Too Late the Phalarope*', in R. Welsh (ed.), *Literature and the Art of Creation* (Totowa, NJ: Barnes & Noble, 1988), 140–52.

ST JOHN, R., *Through Malan's Africa* (New York: Doubleday, 1954).

STEVENS, I. N., 'Paton's Narrator Sophie: Justice and Mercy in *Too Late the Phalarope*', *International Fiction Review*, 8/1 (winter 1981), 68–70.

THOMPSON, J. B., 'Poetic Truth in *Too Late the Phalarope*', *English Studies in Africa*, 24/1 (Mar. 1981), 37–44.

WATSON, S., '*Cry, the Beloved Country* and the Failure of Liberal Vision', *English in Africa*, 9/1 (May 1982), 29–44.

WATTS, N. H. Z., 'A Study of Alan Paton's *Too Late the Phalarope*', *Durham University Journal*, 76 (June 1984), 249–54.

Index

Works of literature listed are by Paton unless otherwise identified.